Aquaculture: Enhancing Food Security and Nutrition

Imtiaz Ahmed • Ishtiyaq Ahmad
Editors

Aquaculture: Enhancing Food Security and Nutrition

Springer

Editors
Imtiaz Ahmed
Department of Zoology
University of Kashmir
Srinagar, Jammu and Kashmir, India

Ishtiyaq Ahmad
Department of Zoology
University of Kashmir
Srinagar, Jammu and Kashmir, India

ISBN 978-3-031-92857-4   ISBN 978-3-031-92858-1 (eBook)
https://doi.org/10.1007/978-3-031-92858-1

This Springer imprint is published by the registered company Springer Nature Switzerland AG
The registered company address is: Gewerbestrasse 11, 6330 Cham, Switzerland

Foreword by Arun S. Ninawe

In an era where global food systems face unprecedented challenges, aquaculture emerges as a beacon of hope. It represents the convergence of innovation, sustainability, and resilience, offering a viable solution to feeding a growing global population while ensuring nutritional adequacy. *Aquaculture: Enhancing Food Security and Nutrition* serves as a comprehensive guide to understanding and unlocking the transformative potential of aquaculture. This book presents a detailed and forward-looking examination of aquaculture's role in addressing critical issues of nutrition and food security. Through a collection of expertly written chapters, it provides a holistic view of how nutritional strategies, scientific advancements, and emerging technologies are reshaping the landscape of aquaculture.

Beginning with an introduction to nutritional research in aquaculture, the book sets the stage by emphasizing the importance of tailored diets for aquatic organisms. It highlights the intricate relationship between diet, growth, and health, paving the way for discussions on advanced topics such as gut microbiota, disease prevention, and reproductive performance. These foundational insights are critical for researchers and practitioners aiming to optimize aquaculture systems for both productivity and sustainability. One of the book's key strengths lies in its exploration of cutting-edge topics such as the role of functional and nutritional genomics, epigenetics, and nutrigenomics in aquaculture. These chapters underscore the importance of integrating molecular biology with nutritional science to achieve precision and efficiency in fish health management. Similarly, the discussion on CRISPR-Cas and nanotechnology demonstrates how genetic advancements are paving the way for revolutionary changes in the aquaculture industry.

The book does not shy away from addressing the complex challenges posed by climate change. Chapters on the impacts of climate change on aquaculture nutrition, shellfish, and global food security offer a sobering yet proactive analysis of the risks and adaptive strategies needed to safeguard this vital sector. The emphasis on sustainability is further highlighted in chapters on probiotics, prebiotics, and omega-3 fatty acids, which provide innovative approaches to enhancing both fish and human health. Of particular interest is the discussion on aquaculture's role in improving productivity, income, and nutrition. By linking aquaculture practices to economic development and community well-being, the book paints a comprehensive picture of how this industry can drive positive change at local and global levels.

This volume is not just a repository of knowledge but a call to action. It challenges readers to think critically and innovatively about the future of aquaculture. By addressing questions such as "Does aquaculture have the potential to feed the world's hungry?" and exploring scenarios for global aquaculture, the book provides a framework for tackling some of the most pressing issues of our time.

As we navigate the complexities of global food systems, this book serves as an invaluable resource for scientists, educators, policymakers, and industry stakeholders. It inspires confidence that, through collaboration and innovation, aquaculture can truly enhance food security and nutrition for generations to come.

Dr. Arun S. Ninawe
Former Advisor,
DBT.

Former DBT Advisor, Department of Biotechnology, Arun S. Ninawe
Ministry of Science & Technology
Government of India
New Delhi, Delhi, India

Foreword by Mukhtar Ahmad Khan

Aquaculture is poised to be one of the most significant contributors to addressing the pressing challenges of global food security and malnutrition in the twenty-first century. As the global population continues to grow, the demand for sustainable, nutritious, and accessible food sources intensifies. Aquaculture, the cultivation of aquatic organisms such as fish, shellfish, and seaweed, offers unparalleled opportunities to meet these demands while simultaneously supporting livelihoods, economic growth, and environmental sustainability.

This book, *Aquaculture: Enhancing Food Security and Nutrition,* comes at a critical time. It presents a comprehensive exploration of the multifaceted roles aquaculture can play in improving food systems. It delves into the cutting-edge advancements in nutritional strategies, emerging technologies, and the integration of genomic and epigenetic insights. These chapters collectively provide a roadmap for researchers, policymakers, and practitioners to harness the full potential of aquaculture in enhancing global nutrition.

The book begins by laying the foundation with an introduction to nutritional research in aquaculture. This opening chapter emphasizes the importance of scientific inquiry in understanding the dietary needs of aquatic organisms and the cascading effects on their health, growth, and productivity. From there, the text explores topics ranging from the role of gut microbiota in fish nutrition and health to the impacts of climate change on aquaculture nutrition and food security. These discussions underscore the interconnected nature of aquaculture with broader ecological, economic, and social systems.

Particularly noteworthy are the chapters addressing the role of omega-3 fatty acids and fat-soluble vitamins in aquaculture feeds. These nutrients not only enhance the growth and reproductive performance of aquatic organisms but also offer significant health benefits to humans who consume them. Similarly, the exploration of probiotics, prebiotics, and emerging technologies highlights innovative approaches to improving aquaculture efficiency and sustainability.

In a world increasingly threatened by climate change and resource scarcity, this book's focus on topics such as the impacts of climate change on shellfish and global food security is both timely and essential. The integration of CRISPR-Cas and nanotechnology in aquaculture further exemplifies how technological advancements can revolutionize the sector, making it more resilient and productive.

The chapters on functional genomics, epigenetics, and nutrigenomics represent the cutting edge of scientific research. By leveraging these disciplines, the aquaculture industry can achieve unprecedented levels of precision in nutrition and disease management, ensuring optimal fish health and productivity.

This book is a testament to the collaborative efforts of scientists, industry leaders, and policymakers who are committed to advancing aquaculture as a cornerstone of global food security. It provides a wealth of knowledge that will undoubtedly inspire further research, innovation, and action.

As you journey through the pages of this book, you will discover not only the challenges but also the immense possibilities that aquaculture holds. Together, we can harness these insights to build a more sustainable, nutritious, and equitable food future for all.

Prof. Mukhtar Ahmad Khan
Retired Professor,
Department of Zoology,
Aligarh Muslim university, U.P. India

Retired Professor, Department of Zoology
Aligarh Muslim University
Aligarh, Uttar Pradesh, India

Mukhtar Ahmad Khan

Preface

The significance of aquaculture as a cornerstone of global food security and nutrition cannot be overstated. As the world's population grows and the demand for sustainable food sources intensifies, aquaculture offers a promising solution to alleviate hunger, improve health, and support economic development. Recognizing this potential, *Aquaculture: Enhancing Food Security and Nutrition* explores the multifaceted role of aquaculture in addressing some of the most pressing challenges of our time.

This book is a testament to the incredible advancements in aquaculture nutrition, science, and technology, bringing together the expertise of leading researchers and practitioners in the field. It offers an in-depth analysis of critical topics, ranging from foundational nutritional research to cutting-edge innovations that are shaping the future of aquaculture. Each chapter provides valuable insights into the strategies and tools required to optimize aquaculture systems for both productivity and sustainability.

The book opens with an exploration of nutritional research in aquaculture, setting the stage for discussions on how targeted dietary strategies can enhance growth, health, and reproductive performance in aquatic organisms. Special attention is given to the role of gut microbiota, probiotics, and prebiotics in improving fish nutrition and health, as well as the significance of omega-3 fatty acids and fat-soluble vitamins in aquaculture feeds. These topics highlight the intricate relationship between fish nutrition and human health, emphasizing the broader implications of aquaculture for global well-being.

Emerging technologies and innovations form the backbone of the discussions in this book. Chapters on functional genomics, nutrigenomics, and epigenetics provide a glimpse into the scientific breakthroughs driving precision nutrition and disease management in aquaculture. The inclusion of CRISPR-Cas and nanotechnology further underscores the transformative impact of modern science on the sector. Meanwhile, the book does not shy away from addressing the challenges posed by climate change, examining its implications for aquaculture nutrition, shellfish sustainability, and global food security.

Equally important are the chapters that contextualize aquaculture within the broader framework of global food systems, exploring its role in improving productivity, income, and nutrition. By addressing questions like "Does aquaculture have the potential to feed the world's hungry?" and analyzing scenarios for global aquaculture, the book invites readers to consider the sector's long-term impact.

This book aims to serve as a comprehensive resource for scientists, policymakers, students, and industry professionals. It is our hope that the insights and strategies presented within these pages will inspire collaborative efforts to harness the full potential of aquaculture, ensuring a sustainable, nutritious, and equitable future for all.

This work compiles the impact of aquaculture in addressing the pressing challenges of global food security and nutritional deficits. The book provides a comprehensive understanding of how aquaculture is playing an instrumental role in ensuring a sustainable and nourished food source. In the face of a growing global population and increasing demands for protein-rich diets, traditional means of fish and its products from wild fisheries alone are inadequate which warrants an alternate source that can fulfill the demand for fish for human consumption. Aquaculture emerges as a sustainable solution, promising to bridge the widening gap between the supply and demand for fish and its products. Through meticulously researched insights and data, the book showcases how aquaculture systems are meeting the protein and nutritional needs of people worldwide.

The book, entitled *Aquaculture: Enhancing Food Security and Nutrition*, sheds light on the various forms of aquaculture like freshwater, marine, and brackish water cultivation highlighting their benefits and potential drawbacks. The focus is on in what way aquaculture practices can be environmentally responsible, economically viable, and socially inclusive, paving the itinerary for a balanced and sustainable food ecosystem. By exploring the advancements in aquaculture technologies such as recirculating aquaculture systems (RAS) and integrated multi-trophic aquaculture (IMTA), the book also advocates the environmentally friendly and resource-efficient practices that enhance both food security and nutrition. Moreover, the book underscores the nutritional value of fish and its products, highlighting their rich nutrient composition including proteins, amino acids, fatty acids, vitamins, and minerals, which are fundamental for human health. It also offers insights about the incorporation of fish and its products in the food that can positively impact nutrition and combat malnutrition, especially in vulnerable human populations.

Srinagar, Jammu and Kashmir, India — Imtiaz Ahmed
Srinagar, Jammu and Kashmir, India — Ishtiyaq Ahmad

Acknowledgments

The completion of *Aquaculture: Enhancing Food Security and Nutrition* is the result of dedicated efforts, collaborative spirit, and unwavering support from numerous individuals and institutions. We are deeply grateful to everyone who contributed to the development of this comprehensive resource, which aims to address the challenges and opportunities in aquaculture for a sustainable and nutritious future.

First and foremost, we extend our heartfelt thanks to the esteemed authors and researchers who contributed their knowledge and expertise to the chapters in this book. Their innovative ideas and thorough analyses have enriched the content, making it a valuable resource for academicians, policymakers, and industry professionals alike.

We express our gratitude to the reviewers and editors whose critical insights and constructive feedback were instrumental in refining the chapters. Their meticulous attention to detail ensured that each chapter meets the highest standards of scientific rigor and relevance.

We are particularly thankful to the institutions and organizations that supported this endeavor, either through funding, resources, or encouragement. Their commitment to advancing aquaculture research and education underscores the importance of interdisciplinary collaboration in tackling global food security challenges.

We also acknowledge the inspiration drawn from the tireless efforts of the global aquaculture community. The dedication of scientists, farmers, and practitioners working to improve aquaculture systems and their contributions to food security and nutrition has been a guiding light for this book.

Special thanks go to our families and friends, who provided unwavering support and patience throughout the demanding process of writing and compiling this book. Their encouragement has been a source of strength and motivation.

We are deeply appreciative of the publishing team for their professionalism and efficiency in bringing this book to life. Their efforts in production, design, and distribution will help this work reach a broad audience and maximize its impact.

Finally, we wish to thank the readers of this book, who share our vision of leveraging aquaculture as a sustainable solution for global food security. Your interest and engagement with this work are what make our efforts meaningful.

We hope that this book serves as a catalyst for further innovation and collaboration in the field of aquaculture, inspiring solutions that will contribute to a more sustainable and equitable future.

Contents

Editors and Contributors

Editors

Imtiaz Ahmed PhD, is a Professor at the Department of Zoology at the University of Kashmir, India. His pioneering contributions in the field of research include the determination of complete essential amino acid requirements of Indian major carp, *Cirrhinus mrigala*, at the fingerling stage as a part of his doctoral thesis and the 10 essential amino acids requirement of *Heteropneustes fossilis* and rainbow trout, *Oncorhynchus mykiss*, at fingerling stage as a part of research projects. The key findings of his research projects have been submitted to the funding agency for proper implementation. He has determined and recommended the feeding regime/appropriate ration size of some important cultured fish species for the proper utilization of feed. The protein requirement of two cultured fish species inhabiting Kashmir water bodies, i.e., Common Carp, *Cyprinus carpio* communis and *C. carpio* specularis, was first reported by him. Apart from his studies in protein and amino acid requirements, he has also established quantitative information on the vitamin requirements of different cultured fish species. He is the first to determine the feeding regime and protein and energy maintenance requirements of rainbow trout, *O. mykiss*, from Kashmir waters in lab conditions. He has also established nutritional values of indigenous snow trout *Schizothorax* spp., and seasonal variation in hemato-biochemical composition of fish inhabiting the cold waters of Kashmir valley. Besides, he has also generated data on hemato-biochemical reference intervals of three fish species of the region. Dr. Ahmed has published more than 100 papers in well-reputed international journals and has several book chapters to his credit. Dr. Ahmed has supervised several MPhil and doctoral students in the area of fish nutrition, fish physiology, reproductive biology, and limnology.

Ishtiyaq Ahmad Earned his doctoral degree from the Department of Zoology at the University of Kashmir, Srinagar. He currently holds the position of Assistant Professor (Contractual) at the same Department. His areas of expertise encompass Fish Nutrition, Fish Physiology and Biochemistry, Nutrigenomics, Fish Parasitology, and Molecular Biology. Dr. Ahmad's notable research contribution lies in defining the optimal branched-chain amino acid requirements for rainbow trout. He is the author of over 40 research articles and several book chapters published by esteemed international publishers, including *Scientific Reports*, *Aquaculture Nutrition*, *Aquaculture Research*, *Aquaculture Reports*, *Journal of Applied Ichthyology*, *Frontiers in Marine Science*, and the *International Journal of Biological Macromolecules*. An active participant in the scientific community, Dr. Ahmad has presented his work at various state, national, and international conferences, seminars, workshops, and symposia. He has been the recipient of numerous accolades, including the Young Scientist Award from a national organization and several Best Paper Presentation awards. Additionally, he serves as an editorial member and review editor for prominent international journals. Dr. Ahmad has also contributed as a Senior Research Fellow in a DBT-sponsored project focused on the nutrient requirements of rainbow trout fingerlings. Moreover, he is a recipient of the prestigious CSIR NET award, further solidifying his standing in the field of aquatic sciences.

Contributors

Ishtiyaq Ahmad Department of Zoology, University of Kashmir, Srinagar, Jammu and Kashmir, India

Peerzada Zubair Ahmad Division of Fish Genetics and Biotechnology, Faculty of Fisheries, SKUAST-K, Srinagar, Jammu and Kashmir, India

Imtiaz Ahmed Fish Nutrition Research Laboratory, Department of Zoology, University of Kashmir, Srinagar, Jammu and Kashmir, India

Mukilan Akash Faculty of Fisheries, SKUAST-K, Ganderbal, Jammu and Kashmir, India

Majed Alghamdi Kaust Beacon Development, King Abdullah University of Science and Technology, Jeddah, Saudi Arabia

Y. S. Alhafedh Ministry of Environment, Water and Agriculture, Riyadh, Saudi Arabia

A. Al-Shaikhi Ministry of Environment, Water and Agriculture, Riyadh, Saudi Arabia

Azra Ansai Government High School, Advoi, Goa, India

Z. A. Ansari Retired from CSIR-National Institute of Oceanography, Panaji, Goa, India

Prachi Asgolkar ICAR - Central Institute of Fisheries Education, Mumbai, Maharashtra, India

Oyas Ahmad Asimi Faculty of Fisheries, SKUAST-K, Ganderbal, Jammu and Kashmir, India

Pujadebi Bera Department of Aquaculture, Faculty of Fishery Sciences, West Bengal University of Animal and Fishery Sciences, Kolkata, West Bengal, India

Bharathkumar ICAR-Central Institute of Fisheries Education, Mumbai, Maharashtra, India

Farooz Ahmad Bhat Division of Fisheries Resource Management, FoFy, SKUAST-K, Srinagar, Jammu and Kashmir, India

Sayan Biswas Department of Aquaculture, College of Fisheries Science, Veraval, Kamdhenu University, Gandhinagar, Gujarat, India

Dhitri Borah Department of Zoology, Biswanath College, Chariali, Assam, India

Srinibas Das Department of Fish Nutrition, Faculty of Fishery Sciences, West Bengal University of Animal and Fishery Sciences, Kolkata, West Bengal, India

Hiranmoy Dhara Department of Aquaculture, Faculty of Fishery Sciences, West Bengal University of Animal and Fishery Sciences, Kolkata, West Bengal, India

Menakshi Dhar Department of Zoology, University of Jammu, Jammu, Jammu & Kashmir, India

Oshin Dhillon Division of Fish Genetics and Biotechnology, Faculty of Fisheries, SKUAST-K, Srinagar, Jammu and Kashmir, India

Javeria Fatima Laboratory of Alternative Animal Models, Section of Genetics, Department of Zoology, Faculty of Life Sciences, Aligarh Muslim University, Aligarh, Uttar Pradesh, India

Francesco Fazio Department of Veterinary Sciences, University of Messina, Messina, Italy

João Batista Kochenborger Fernandes Aquaculture Center of UNESP, São Paulo State University (UNESP), Jaboticabal, SP, Brazil

Suvadip Ghara Division of Fish Genetics and Biotechnology, Faculty of Fisheries, SKUAST-K, Srinagar, Jammu and Kashmir, India

Ravi Kumar Goswami Department of Zoology, Hindu College, University of Delhi, New Delhi, Delhi, India

Robert Mervyn Gous School of Agricultural, Earth and Environmental Sciences, University of KwaZulu-Natal, Pietermaritzburg, South Africa

Aamina Hassan Fish Nutrition Research Laboratory, Department of Zoology, University of Kashmir, Srinagar, Jammu and Kashmir, India

Parveena Hassan Cytogenetics and Molecular Biology Lab, Department of Botany, University of Kashmir, Srinagar, Jammu and Kashmir, India

D. Sabina Iqbal Division of Fisheries Resource Management, FoFy, SKUAST-K, Srinagar, Jammu and Kashmir, India

Shagufta Ishtiyaq Fish Nutrition Research Laboratory, Department of Zoology, University of Kashmir, Srinagar, Jammu and Kashmir, India

Yusuf Ja'afar Department of Water Resources, Aquaculture and Fisheries Technology, Federal University of Technology Minna, Minna, Nigeria

Division of Fish Genetics and Biotechnology, Faculty of Fisheries, SKUAST-K, Srinagar, Jammu and Kashmir, India

Kousar Jan Fish Nutrition Research Laboratory, Department of Zoology, University of Kashmir, Srinagar, Jammu and Kashmir, India

Raman Jasrotia Department of Zoology, University of Jammu, Jammu, Jammu & Kashmir, India

Osman Sabri Kesbic Faculty of Veterinary Medicine, Department of Animal Nutrition and Nutritional Diseases, Kastamonu University, Kastamonu, Turkey

Bronson Kumar Khangembam Department of Zoology, Tripura University (A Central University), Agartala, Tripura, India

Kifayat Ullah Khan Department of Zoology, Government College University, Lahore, Pakistan

Aquaculture Center of UNESP, São Paulo State University (UNESP), Jaboticabal, SP, Brazil

Mukhtar A. Khan Fish Nutrition Research Laboratory, Department of Zoology, Aligarh Muslim University, Aligarh, Uttar Pradesh, India

Ganesh Kumar Rani Lakshmi Bai Central Agricultural University, Jhansi, Uttar Pradesh, India

Seema Langer Department of Zoology, University of Jammu, Jammu, Jammu & Kashmir, India

Sandeep K. Malhotra Department of Zoology, University of Allahabad, Uttar Pradesh, India

Sudeshna Mandal Department of Zoology, Visva-Bharati (Central University), Santiniketan, West Bengal, India

Sukalpa Mandal Department of Fish Nutrition, Faculty of Fishery Sciences, West Bengal University of Animal and Fishery Sciences, Kolkata, West Bengal, India

Asra Imtiyaz Mattoo Division of Fisheries Resource Management, FoFy, SKUAST-K, Srinagar, Jammu and Kashmir, India

Asaad H. Mohamed Kaust Beacon Development, King Abdullah University of Science and Technology, Jeddah, Saudi Arabia

Ishrat Mohd Division of Fisheries Resource Management, FoFy, SKUAST-K, Srinagar, Jammu and Kashmir, India

Syed Talia Mushtaq Division of Fisheries Resource Management, FoFy, SKUAST-K, Srinagar, Jammu and Kashmir, India

Naveed Nabi Fish Nutrition Research Laboratory, Department of Zoology, University of Kashmir, Srinagar, Jammu and Kashmir, India

University Centre for Research & Development (UCRD), Chandigarh University, Mohali, Punjab, India

Sana Naz Fish Nutrition Research Laboratory, Department of Zoology, Aligarh Muslim University, Aligarh, Uttar Pradesh, India

Asif Pagarkar Fisheries College Ratnagiri, Shirgaon, Maharashtra, India

Ranjit Singh Patel Division of Fish Genetics and Biotechnology, Faculty of Fisheries, SKUAST-K, Srinagar, Jammu and Kashmir, India

Samaya Prakash Faculty of Fisheries, SKUAST-K, Ganderbal, Jammu and Kashmir, India

Mahpara Qadiri Department of Chemistry, University of Kashmir, Srinagar, Jammu and Kashmir, India

Mohd Ashraf Rather Division of Fish Genetics and Biotechnology, Faculty of Fisheries, SKUAST-K, Srinagar, Jammu and Kashmir, India

Quseen Mushtaq Reshi Fish Nutrition Research Laboratory, Department of Zoology, University of Kashmir, Srinagar, Jammu and Kashmir, India

Surya Kanta Sau Department of Aquaculture, Faculty of Fishery Sciences, West Bengal University of Animal and Fishery Sciences, Kolkata, West Bengal, India

Ubaid Ullah Shah Department of Library Sciences, University of Kashmir, Srinagar, Jammu and Kashmir, India

Dheeraj Sharma Department of Zoology, University of Jammu, Jammu, Jammu & Kashmir, India

Sourabh Sharma Department of Zoology, University of Jammu, Jammu, Jammu & Kashmir, India

Sristi Sharma Department of Zoology, University of Delhi, New Delhi, Delhi, India

Yasir Hasan Siddique Laboratory of Alternative Animal Models, Section of Genetics, Department of Zoology, Faculty of Life Sciences, Aligarh Muslim University, Aligarh, Uttar Pradesh, India

Muzamil Sidiq Fish Nutrition Research Laboratory, Department of Zoology, University of Kashmir, Srinagar, Jammu and Kashmir, India

Samar Pal Singh Department of Zoology, Ramjas College, University of Delhi, New Delhi, Delhi, India

Jaberson Solomon ICAR-Central Institute of Fisheries Education, Mumbai, Maharashtra, India

Iyyappan Somasundaram Faculty of Fisheries, SKUAST-K, Ganderbal, Jammu and Kashmir, India

Ajith Stephen Faculty of Fisheries, SKUAST-K, Ganderbal, Jammu and Kashmir, India

Tincy Varghese ICAR - Central Institute of Fisheries Education, Mumbai, Maharashtra, India

Baskar Veeramaruthu Faculty of Fisheries, SKUAST-K, Ganderbal, Jammu and Kashmir, India

Malik Wasim Division of Fish Genetics and Biotechnology, Faculty of Fisheries, SKUAST-K, Srinagar, Jammu and Kashmir, India

Anita Yadav Department of Zoology, CMP PG College (a Constituent College of the University of Allahabad), Prayagraj, Uttar Pradesh, India

Noorin Zafar Fish Nutrition Research Laboratory, Department of Zoology, Aligarh Muslim University, Aligarh, Uttar Pradesh, India

Seemab Zehra Kaust Beacon Development, King Abdullah University of Science and Technology, Jeddah, Saudi Arabia

Chapter 1
Nutritional Strategies for Improving Growth and Health in Aquaculture

Mukhtar A. Khan, Noorin Zafar, and Sana Naz

1 Introduction

Aquaculture is the fastest-growing food-producing sector globally that meets the demand of the growing population for sustainable food. One of the primary concerns in modern aquaculture is optimizing nutritional strategies to enhance growth rates, improve overall health, and reduce environmental impact. The success of aquaculture operations heavily relies on the provision of appropriate nutrition to farmed fish and shellfish (Lall & Dumas, 2022).

Nutrition plays a pivotal role in determining the productivity and sustainability of aquaculture systems. Well-formulated feeds not only promote optimal growth but also contribute to improved health, stress tolerance, and environmental resilience (Little et al., 2018). The quest for more efficient and sustainable aquaculture practices has driven extensive research into nutritional strategies that maximize growth potential while minimizing environmental footprint.

Recent advances in nutritional science have revealed the complex interplay between dietary components, metabolic processes, and physiological responses in aquatic animals. This understanding has led to the development of novel feed formulations that address specific nutritional needs throughout the life cycle of farmed species. The importance of meeting nutrient requirements in fish feed cannot be overstated, as it directly influences the health, growth, and productivity of aquaculture species. Nutrient imbalances or deficiencies can lead to poor growth, weakened immune systems, and increased susceptibility to diseases, ultimately affecting the overall sustainability and profitability of aquaculture operations. Additionally, a well-balanced feed enhances feed conversion efficiency, reduces waste production, and supports optimal reproduction and survival rates. By tailoring feed formulations

M. A. Khan (✉) · N. Zafar · S. Naz
Fish Nutrition Research Laboratory, Department of Zoology, Aligarh Muslim University, Aligarh, Uttar Pradesh, India

I. Ahmed, I. Ahmad (eds.), *Aquaculture: Enhancing Food Security and Nutrition*, https://doi.org/10.1007/978-3-031-92858-1_1

to the precise nutrient requirements at each developmental stage, aquaculturists can improve the efficiency of production while minimizing environmental impacts, ensuring the long-term viability of the industry.

2 Data on Nutritional Requirements

2.1 Protein and Amino Acid Requirements

Protein and amino acids are essential macronutrients for the growth and health of aquatic organisms. Fish and other aquatic species require proteins for muscle growth, immune response, and various physiological processes. However, not all proteins are equal, and optimizing the composition of proteins and amino acids in aquafeeds is essential for maximizing growth, feed efficiency, and sustainability in aquaculture.

2.1.1 Dietary Protein Requirements

Dietary protein is a crucial and most expensive component in fish feeds that directly affects feed intake, fish growth, and cost of feed. Dietary protein requirement is basically the protein required for maintenance; relative protein concentration required for maximum growth; the efficiency of protein retention in growth; and requirements for maturation and reproduction. Fish protein requirements vary significantly based on species type, fish size, and life stage. In the earlier stage, protein is mainly used for growth and maintenance, and after attaining sexual maturity, it is majorly utilized for reproduction. This distribution will not remain constant throughout the fish's life span (Hardy & Kaushik, 2021) (Table 1.1).

The quality and complexity of dietary proteins directly influence the growth and development of fish, particularly during their early life stages, when their metabolic systems are developing rapidly. The quality of protein affects the efficiency with which fish utilize the amino acids available in the feed and influences overall growth performance.

The complexity of dietary proteins refers to both the amino acid profile and the digestibility of the protein sources. Protein sources like fishmeal, which is rich in high-quality, highly digestible proteins, are often preferred in aquaculture because they provide a balanced amino acid profile. Conversely, plant-based proteins, though abundant, often lack certain essential amino acids such as lysine and methionine and require supplementation to ensure that the fish receive the full spectrum of EAAs for optimal growth.

An interesting aspect of protein quality is its potential influence on epigenetics. Research has shown that feeds with low protein can upregulate de novo DNA methyltransferases, enzymes responsible for adding methyl groups to DNA. This

Table 1.1 Variations in protein requirement of different fish species at different life stages

Fish species	Life stage	Protein requirement (CP) (%)	References
Cyprinus carpio haematopterus	Adult	25–28	Wang et al. (2023)
Cyprinus carpio	Fingerlings	38	Ali and Rawal (2022)
Cyprinus carpio Songpu	Adult	26	Fan et al. (2020)
Carassius auratus gibelio var. CAS III	Juvenile	40.2–42.7	Ye et al. (2015)
	On-growing	33.7–41.8	
Cirrhinus reba	Fingerlings	28.15–28.25	Yadav et al. (2019)
Cyprinus carpio var. Jian	Juvenile	34.10	Liu et al. (2009)
Catla catla	Fry	40	Khan and Jafri (1991)
	Fingerlings	35	
Channa punctatus	Fingerlings	45	Zehra and Khan (2012)
Labeo rohita	Fingerlings	35.37	Debnath et al. (2012)
Cirrhinus mrigala	Fry	36	Singh et al. (2008)
Oncorhynchus mykiss	Fingerlings	45–47	Ahmed and Ahmad (2020)
Cyprinus carpio var. *specularis*	Fingerlings	41.5	Ahmed and Maqbool (2017)

methylation process can modify the expression of genes without changing the underlying DNA sequence. These changes can have long-term effects on the fish's growth, metabolism, and health, such as stunted growth, decreased nutrient metabolism, low immunity, and high disease susceptibility.

High-protein quality feeds are associated with better growth efficiency, meaning that fish can achieve larger sizes with less feed input. This is not only more cost-effective but also reduces the environmental impact of aquaculture. On the other hand, poor-quality protein feeds may result in inefficient growth, requiring more feed and leading to increased waste production. The epigenetic changes induced by low-protein feeds may also have long-term health implications. These effects could reduce their overall productivity in aquaculture systems, making early nutrition critical to maximizing fish health and welfare throughout their lifespan.

In aquaculture, the sustainability of protein sources is an increasing concern. Traditional aquafeeds rely heavily on fishmeal, which is usually derived from wild-caught fish. However, there is a growing push for alternative protein sources, such as plant proteins, algae, and insect meals, to reduce dependence on fishmeal and minimize the ecological impact of aquaculture. Optimizing protein and amino acid compositions of feeds while using sustainable sources of protein is essential for the future of the industry (Fantatto et al., 2024).

The bioavailability of protein and amino acids refers to how well they are absorbed and utilized by the fish. Factors such as feed processing methods, ingredient quality, and the fish's digestive capacity can influence bioavailability. For instance, overcooked proteins may be less digestible, while some amino acids may be degraded during feed storage. Therefore, optimizing protein quality and processing methods can ensure that fish are able to fully utilize the nutrients in the feed.

A balanced and well-formulated amino acid profile does not just promote growth but also improves the fish's overall health and resistance to disease. Amino acids play a role in immune function and tissue repair, and they are essential for synthesizing enzymes and hormones involved in metabolic processes. Ensuring that aquafeeds are nutritionally balanced helps prevent metabolic disorders, growth stunting, and immune deficiencies that can lead to disease outbreaks.

While optimizing protein and amino acid compositions is crucial, it must also be done in a cost-effective manner. High-quality protein sources can be expensive, so fishmeal is often replaced with lower-cost plant-based or alternative protein ingredients. However, these ingredients may require additional nutrient supplementation to achieve the desired amino acid profile, which can increase the overall cost of the feed.

2.1.2 Dietary Amino Acid Requirements

Amino acids are the building units of protein that are fundamental to the health and development of fish, playing a crucial role in growth, metabolism, immune function, and stress responses. The importance of amino acids is particularly highlighted by their classification into essential amino acids (EAAs) and nonessential amino acids (NEAAs), which have distinct roles in supporting the growth and metabolic functions of fish.

EAAs are those that fish cannot synthesize on their own and must be supplied through their diet. These amino acids are pivotal for protein synthesis, cellular functions, and overall growth. The most commonly known EAAs in fish nutrition include arginine, histidine, isoleucine, leucine, lysine, methionine, phenylalanine, threonine, tryptophan, and valine. These amino acids support various physiological processes. EAAs are necessary for the synthesis of proteins required for muscle growth, tissue repair, enzyme function, and structural integrity. A balanced supply of EAAs allows fish to grow at their maximum potential. If any of these essential amino acids are deficient, the overall protein synthesis rate will decrease, leading to impaired growth and a weakened immune system.

NEAAs can be synthesized by fish from precursors available in the diet or from other amino acids. These amino acids include alanine, asparagine, aspartic acid, cysteine, glutamic acid, glutamine, glycine, proline, serine, and tyrosine. While they are not essential in the sense that they do not have to come directly from the feed, they still contribute to the overall metabolic health of the fish.

The amino acid profile of aquafeeds is critical for optimal growth and development. Fish require a range of essential amino acids (EAAs) that cannot be synthesized by the organism and must be obtained from the diet. The ideal amino acid profile in the feed should match the species-specific needs at different growth stages. A well-balanced amino acid profile helps in maximizing protein synthesis, muscle growth, and overall efficiency in converting feed into body mass (Table 1.2).

Table 1.2 Variations in EAA requirement of different fish species at different life stages

Essential amino acids	Fish species	Dietary (CP%)	Requirement		References
			% Dry diet	% CP	
Arginine	*Catla catla*	33	1.67	–	Zehra and Khan (2013a)
	Labeo rohita	40	1.22–1.39	3.05–3.47	Abidi and Khan (2009)
	Cirrhinus mrigala	40	1.8	4.6	Ahmed and Khan (2004a)
	Heteropneustes fossilis	40	2.0–2.1	5.1–5.6	Khan and Abidi (2011a)
		38	1.5–1.6	3.9–4.3	Farhat and Khan (2012)
		40	1.63	4.08	Ahmed (2013a)
	Oncorhynchus mykiss	45	1.76	3.91	Ahmed et al. (2022)
Methionine	*Catla catla*	33	1.28 TSAA	3.8	Zehra and Khan (2016a)
	Labeo rohita	38	2.52–3.13 TSAA	2.52–3.13	Abidi and Khan (2011)
	Cirrhinus mrigala	40	1–1.2 with 1% cys	2–3	Ahmed et al. (2003)
	Heteropneustes fossilis	38	1.47 TSAA	2.7	Farhat and Khan (2014)
		40	1.09	2.73	Ahmed (2014)
Lysine	*Labeo rohita*	40	1.88–1.9	4.5–4.8	Abidi and Khan (2010)
	Cirrhinus mrigala	40	2.3	5.8	Ahmed and Khan (2004b)
	Heteropneustes fossilis	38	2.3	6.1	Farhat and Khan (2013a, b)
		45	1.4	3.1	Khan and Abidi (2011b)
	Channa punctatus	45	2.22–2.40	4.94–5.34	Sharf and Khan (2021)
	Catla catla	33	1.7–1.8	5.15–5.45	Zehra and Khan (2013b)
	Oncorhynchus mykiss	45	2.16	4.80	Ahmed and Ahmad (2021)

(continued)

Table 1.2 (continued)

Essential amino acids	Fish species	Dietary (CP%)	Requirement		References
			% Dry diet	% CP	
Threonine	*Labeo rohita*	40	1.5–1.7	3.8–4.2	Abidi and Khan (2008)
	Cirrhinus mrigala	40	1.8	4.5	Ahmed et al. (2004)
	Heteropneustes fossilis	40	1.27	3.17	Ahmed (2007)
		38	1.62	4.26	Farhat and Khan (2017)
	Channa punctatus	45	2.22–2.30	4.93–5.11	Sharf and Khan (2022a)
	Catla catla	33	1.35–1.48	4.09–4.48	Zehra and Khan (2016b)
Phenylalanine	*Labeo rohita*	40	1.22% + 1.00% tyr	3.05% + 2.50% tyr	Abidi and Khan (2007)
	Catla catla	33	1.01% + 0.68% tyr	3.06% + 2.06% tyr	Zehra and Khan (2014a)
	Channa punctatus	45	0.41% + 1.0% tyr	0.91% + 2.22% tyr	Sharf and Khan (2023a)
	Heteropneustes fossilis	40	1.92% + 4.37% tyr	4.8% + 3.9% tyr	Ahmed (2022)
		40	6.49% + 3.5% tyr	1.58% + 0.8% tyr	Sayed and Ahmed (2022)
Tryptophan	*Catla catla*	33	2.1–2.5	6.4–7.6	Zehra and Khan (2015a, b)
	Cirrhinus mrigala	40	0.4	0.1	Ahmed and Khan (2005a)
	Heteropneustes fossils	38	0.24	0.63	Abidi and Khan (2014b)
		40	0.32	0.80	Ahmed (2012)
	Channa punctatus	45	4.81–5.5	1.06–1.28	Sharf and Khan (2023b)
	Labeo rohita	40	0.36–0.38	0.9–0.95	Abidi and Khan (2010)
	Oncorhynchus mykiss	45	0.4	0.889	Ahmed et al. (2024)
Leucine	*Catla catla*	33	1.57	4.7	Zehra and Khan (2015b)
	Cirrhinus mrigala	40	1.5	3.9	Ahmed and Khan (2006)
	Heteropneustes fossilis	38	1.6	4.3	Farhat and Khan (2014)
	Channa punctatus	35	2	4.62	Sharf and Khan (2023c)
	Labeo rohita	40	1.5–1.6	3.8–4.0	Abidi and Khan (2007)
	Oncorhynchus mykiss	45	1.74	3.87	Ahmad et al. (2021a)

(continued)

Table 1.2 (continued)

Essential amino acids	Fish species	Dietary (CP%)	Requirement % Dry diet	% CP	References
Isoleucine	*Cirrhinus mrigala*	40	1.3	3.2	Ahmed and Khan (2006)
	Channa punctatus	45	1.79–1.83	3.98–4.08	Sharf and Khan (2020)
	Labeo rohita	40	1.5–1.6	3.8–4.0	Khan and Abidi (2007a, b)
	Catla catla	33	1.13–1.18	3.42–3.58	Zehra and Khan (2013c)
	Oncorhynchus mykiss	45	1.34	2.99	Ahmad et al. (2022)
Valine	*Cirrhinus mrigala*	40	1.52	3.8	Ahmed and Khan (2006)
	Labeo rohita	40	1.5	3.75	Abidi and Khan (2004a)
	Catla catla	33	1.02	3.09	Zehra and Khan (2014b)
	Channa punctatus	45	1.59–1.72	3.53–3.82	Sharf and Khan (2023d)
	Oncorhynchus mykiss	45	1.81	4.04	Ahmad et al. (2021b)
Histidine	*Labeo rohita*	40	0.82	2.05	Abidi and Khan (2004b)
	Catla catla	33	0.63–0.68	1.91–2.06	Zehra and Khan (2016c)
	Cirrhinus mrigala	40	0.9	2.1	Ahmed and Khan (2005b)
	Channa punctatus	45	1.25	2.77	Sharf and Khan (2022a, b)
	Heteropneustes fossilis	38	0.94	2.48	Farhat and Khan (2013a, b)
		43	1.5	3.5	Khan and Abidi (2014)
		40	0.54	1.35	Ahmed (2013b)

The concept of limiting amino acids is critical in feed formulation. Limiting amino acids, namely, lysine, methionine, threonine, and tryptophan, are ones that are present in the lowest proportion relative to the required balance, thereby limiting the rate of protein synthesis. For example, if methionine is deficient in the feed, fish cannot fully utilize the other amino acids present. Therefore, optimizing the balance of amino acids to minimize deficiencies is key to achieving the best growth performance and reducing waste, which has environmental benefits as well. High-quality proteins contain a better balance of essential amino acids, while lower-quality proteins may require supplementation with individual amino acids. Plant-based proteins, which are often used in aquafeeds, may not provide the optimal amino acid profile required for fish, so they are often supplemented with synthetic crystalline amino acids or other protein sources, like fishmeal, insect-based proteins, and slaughterhouse waste, to improve feed quality (Nunes et al., 2014; Fantatto et al., 2024).

Fish have a capacity to spare dispensable amino acids when EAAs are scarce. In cases of nutrient stress, the metabolism of fish will prioritize the synthesis and preservation of EAAs at the expense of NEAAs. This means that if EAAs are in short supply, the body may use NEAAs to balance the nutritional deficit, although this can lead to metabolic inefficiencies and impaired growth.

2.2 *Dietary Lipid Requirements*

Lipids play a crucial role in the nutrition and physiology of fish, serving as a primary energy source, structural components of cell membranes, and carriers of fat-soluble vitamins. Lipids are a significant energy source for fish, especially during periods of high-energy demand, such as migration or food scarcity. For instance, salmon utilize stored lipids during their extensive upstream migration to spawn, allowing them to survive for weeks without feeding. Fish metabolize lipids efficiently, which supports their ability to endure prolonged periods of food deprivation.

Fish require specific fatty acids that they cannot synthesize on their own, known as essential fatty acids (EFAs). The requirements differ between freshwater and marine species. Freshwater fish typically require linoleic acid (18:2, n-6) and linolenic acid (18:3, n-3), while marine fish depend on long-chain polyunsaturated fatty acids (PUFAs) such as eicosapentaenoic acid (EPA, 20:5 n-3) and docosahexaenoic acid (DHA, 22:6 n-3). These EFAs are critical for maintaining cellular membrane integrity, growth, reproduction, and overall health.

The optimal dietary lipid requirement varies depending on the species, life stage, and environmental conditions. For instance, increasing dietary lipid levels from 9% to 17% improved growth and feed utilization in juvenile Far Eastern catfish (Kim et al., 2012). In genetically improved farmed tilapia (GIFT), dietary lipid levels influence growth performance, body composition, and metabolic responses (Paul et al., 2022). For sub-adult triploid rainbow trout, lipid levels in the feed affect growth, health status, digestive ability, and expression of growth-related genes. Excessive lipid intake, however, can lead to fat deposition and reduced feed efficiency, while insufficient lipid levels may impair growth and physiological functions (Meng et al., 2019).

Different fish species have varying lipid requirements based on their natural habitats and feeding habits. Salmonids and other cold-water species often require higher levels of long-chain PUFAs like EPA and DHA due to their role in maintaining membrane fluidity at low temperatures. Warmwater species may have different fatty acid profiles and lipid needs (Table 1.3).

Table 1.3 Variations in lipid requirement of different fish species at different life stages

Fish species	Life stafe	Lipid requirement (%)	References
Labeo rohita	Fingerlings	9.5–9.8	Siddiqua and Khan (2022)
Ctenopharyngodon idella	Juvenile	6.5	Jin et al. (2013)
Acipenser schrenckii	Juvenile	20.31	Li et al. (2023)
Clarias magur	Larvae	8	Mir et al. (2020)
Cirrhinus mrigala	Fingerlings	5–7	Jafri et al. (1995)
Heteropneustes fossilis	Fingerlings	7	Anwar and Jafri (1995)

Lipids are essential for the absorption of fat-soluble vitamins (A, D, E, and K) and are integral components of cellular and subcellular membranes. Phospholipids, which contain polyunsaturated fatty acids, are particularly important for maintaining membrane fluidity and function.

Since dietary protein in feeds is the most expensive, it is vital to incorporate it in an amount adequate for normal growth and maintenance. Lipids can effectively conserve protein as they have better bioavailability than any other nutrient in fish. This is referred to as the protein-sparing effect of lipids. In this, the major part of the energy would be obtained from dietary lipids, enabling the protein to be utilized exclusively for growth with minimal inevitable catabolism, thereby resulting in improved growth and nutrient utilization (Thirunavukkarasar et al., 2022).

Lipids, such as vegetable oils or animal fats, are often added to feed formulations to act as binders. They help in forming stable pellets or extruded feeds by improving the binding properties of the ingredients during the pelleting process. This is essential for maintaining the structural integrity of the feed during handling and feeding. They can also provide lubrication during the pelleting or extrusion process, which reduces friction in the machinery, improves the throughput, and prevents wear and tear on equipment. The addition of lipids to feed can also help enhance the texture and sensory characteristics of the feed, thus improving the palatability of fish feed, making it more attractive to the fish. This is particularly important in the aquaculture industry, as it ensures that fish are consuming sufficient amounts of feed for optimal growth and health.

2.3 *Role of Carbohydrates in Fish Feeds*

Carbohydrates are one of the macronutrients in fish feeds, primarily serving as an energy source. However, unlike lipids and proteins, carbohydrates are not considered essential for fish, as they can survive and grow without them.

Fish do not have an absolute dietary requirement for carbohydrates. This is because they can efficiently synthesize glucose through gluconeogenesis from non-carbohydrate sources like proteins and lipids. This contrasts with terrestrial animals, where carbohydrates are a principal energy source. Carbohydrates act as a cost-effective and readily available energy source in fish feeds. They help spare proteins

Table 1.4 Carbohydrate-to-lipid ratio in feeds of different fish species

Fish species	Life stage	Carbohydrate-to-lipid ratio	References
Oreochromis niloticus	Juvenile	1.95	Xie et al. (2017)
Clarias batrachus	Fingerlings	0.02–3.38	Jafri (1998)
Larmichthys crocea	Juvenile	1.34	Zhou et al. (2016)
Pelteobagrus fulvidraco	Juvenile	2.45–5.58	Wang et al. (2014)
Leuciscus idus	Juvenile	2.23–2.96	Ren et al. (2021)

and lipids from being used for energy, allowing these nutrients to be utilized for growth and other physiological functions. However, the ability of fish to utilize carbohydrates varies significantly among species.

The efficiency of carbohydrate utilization depends on the species, their natural diet, and environmental conditions. Warmwater fish such as tilapia and catfish can utilize higher levels of dietary carbohydrates compared to cold-water and marine fish. This is because warmwater species are often omnivorous or herbivorous and have evolved to digest plant-based carbohydrates more effectively. Carnivorous fish like salmon and trout are less equipped to metabolize carbohydrates due to their natural diet being rich in proteins and lipids. Excessive carbohydrate intake in these species can lead to fat deposition and metabolic issues (Li et al., 2022).

While there is no strict requirement, an appropriate amount of digestible carbohydrates are often included in fish feeds in tandem with lipids to spare the use of protein as an energy source and reduce feed costs. The optimal carbohydrate level varies by species; the optimal dietary carbohydrate for hybrid grouper was estimated to be 7.22% (Li et al., 2019). Wuchang bream shows optimal growth performance with dietary carbohydrate levels between 25% and 30% (Zhou et al., 2013) (Table 1.4).

The digestibility of carbohydrates in fish depends on the source and complexity of the carbohydrate. Simple carbohydrates such as glucose and starch are more easily digested and utilized by fish. Complex carbohydrates such as cellulose and hemicellulose are poorly digested, especially by carnivorous fish, as they lack the necessary enzymes to break down plant fibers.

The protein-to-carbohydrate ratio is a key consideration in animal and fish feed formulation, as it helps to balance the energy provided by carbohydrates with the protein needs of the animal. This balance is crucial for optimizing growth, health, feed efficiency, and overall performance.

Feeding fish diets with excessive carbohydrates can lead to fat deposition in the liver and other tissues, reduced feed efficiency and growth in species with low carbohydrate tolerance, and potential metabolic disorders in carnivorous species.

2.4 *Dietary Mineral Requirements*

All animals including fish require minerals for their vital physiological and biochemical functions and to maintain their normal life processes (NRC, 2011; Prabhu et al., 2016). Fish live in a wide range of salinity levels: in freshwater (FW), seawater (SW), and brackish water (BW) environments, and they have the ability to absorb minerals from the surrounding water in addition to their diet.

Minerals are known to interact with other nutrients due to their ability and tendency to form chemical bonds. Interrelationships among mineral elements as revealed by physiological or biochemical consequences are quite evident. Such interactions are broadly identified as positive or synergistic and negative or antagonistic. Direct positive interactions between elements in structural processes, such as the requirement of copper (Cu) and iron (Fe) for hemoglobin formation, calcium (Ca), phosphorus (P), and magnesium (Mg) for the formation of bone hydroxyapatite, and an interaction of manganese (Mn) with zinc (Zn) for the proper conformational shape of RNA molecules in the liver, have been reported. Antagonistic relationships are considered to occur when trace elements with a similar electronic configuration and ionic radius compete for binding sites, such as zinc (Zn) and cadmium (Cd) in metallothionein and Mg/manganese (Mn) substitutions at active sites of enzymes. In the gastrointestinal tract, antagonistic relationships may occur by a simple mechanism, which involves a chemical reaction forming an insoluble complex between minerals such as copper (Cu) and sulfur (S) to form copper sulfide or a mineral and another dietary component such as zinc (Zn) combined with phytic acid to form phytate (Lall & Kaushik, 2021).

There is a need to provide the optimum amount of these minerals in the diet so that their deficiency as well as toxicity can be prevented. Thus, it is necessary to determine the optimum requirement of minerals for different fish species. The exchange of ions across the gills and skin of fish in surrounding water complicates the determination of the quantitative dietary requirement (Lall, 2002; Lall & Kaushik, 2021). Therefore, both the absorption of dietary minerals from the gastrointestinal tract and waterborne mineral absorption are important to consider in order to determine the optimum requirement.

2.4.1 Macrominerals

Phosphorus is an essential mineral for all forms of life, including fish. Fish can obtain a substantial number of required minerals directly from their rearing water, but there is one essential mineral that must be supplied by the diet (Lall, 2002; Zafar & Khan, 2018). Phosphorus is a structural component of hard tissues such as bone, teeth, and scales, as well as a constituent of various coenzymes, phospholipids, and nucleic acids. It is also involved in energy metabolism by virtue of its role in high-energy bonds of ATP (Zafar & Khan, 2018; Musharraf & Khan, 2019b). Dietary requirements for phosphorus have been reported to range rather widely between 0.3% and 0.9% of the diet for various fish species (Prabhu et al., 2016). The first step in using nutritional strategies to lower phosphorus excretion by fish is to obtain accurate data on the bioavailability of phosphorus in common feed ingredients. This step is important because the easiest step in reducing phosphorus excretion is to reduce the proportion lost in feces when indigestible phosphorus levels in the diet are too high. Phosphorus deficiency signs in fish include increased liver or body fat, reduced blood phosphate levels, and poor mineralization of scales (Zafar & Khan, 2018; Cong-mei et al., 2021).

Calcium plays a major role in the development and maintenance of the skeletal system and performs many other physiological functions including the maintenance of acid-base equilibrium (Zimmer et al., 2019). In skeletal tissue, Ca and P are deposited as tricalcium phosphate $Ca_3(PO4)_2$, which then undergoes further crystalline changes to form hydroxyapatite, $Ca_{10}(PO_4)_6(OH)_2$, which is deposited in the organic matrix during mineralization. The ratio of Ca/P in bone may show some changes during development; however, their ratio reported in several fish species ranges from 1.6:1 to 2:1. Fish scale is also a calcified tissue and serves as an internal Ca reservoir during periods of increased Ca demand, such as sexual maturation and starvation (Lall & Kaushik, 2021). The dietary requirement of calcium reported for different fish species is a 1.3–10.2 g/kg diet. The calcium requirement of fish is affected by dietary factors, uptake from water, and species differences (Hossain & Yoshimatsu, 2014).

Magnesium is a critical intracellular divalent cation that plays an essential physiological role in many functions in the body. It forms a key complex with ATP and plays a key role in many important biological processes such as protein synthesis, cell replication, nerve impulse conduction, and energy metabolism (NRC, 2011; Zafar & Khan, 2021). The major proportion (50–70%) of Mg in the body of fish is located in skeletal tissues and scales (Musharraf & Khan, 2018), and the remaining is found within the cells of soft tissues. Magnesium is mobilized from bones and scales when dietary Mg intake is low (Cowey et al., 1977). The optimum dietary magnesium requirements reported for several fish species range from 0.24 to 1.35 g/kg diet (Prabhu et al., 2016). Anorexia, reduced growth, sluggishness, high mortality, reduced magnesium content, calcinosis of the kidney, vertebral deformity, and degeneration of muscle fibers and epithelial cells of the pyloric cecum and gill filaments are the commonly reported magnesium deficiency signs (NRC, 2011).

Potassium exists predominantly in the intracellular fluid and plays a critical role in growth and other physiological activities such as acid-base balance, ion balance, and osmoregulation, as well as nerve and muscle excitability (Lall, 2002; NRC, 2011; Musharraf & Khan, 2019b; Zafar & Khan, 2019a, b). It is required by a variety of enzymes to act in protein synthesis, carbohydrate metabolism, phosphorus metabolic reactions, and nutrient transport (Chen et al., 2016). Deficiency or excess of this mineral may lead to anorexia, stunted growth and protein utilization, convulsions, seizures, and death (Shiau & Hsieh, 2001). Therefore, the diets must contain the optimal level of the aforementioned macrominerals to maintain the health and growth of fish (Table 1.5).

2.4.2 Microminerals

In addition to macro minerals, optimum levels of essential micro minerals are also required for the growth and maintenance of normal health of farmed fish. Four broad biochemical functions of microtrace elements are widely recognized: (a)

Table 1.5 Variations in macromineral requirement of different fish species at different life stages

Fish species	Life stage	Requirement	References
Phosphorus requirement (%)			
Cyprinus carpio var. Jian	Juvenile	0.52	Xie et al. (2017)
Ctenopharyngodon idella	Juvenile	0.85	Liang et al. (2012)
Labeo rohita	Fingerling	0.65	Musharraf and Khan (2019a)
Oreochromis niloticus	Fingerling	0.86	Yao et al. (2014)
Heteropneustes fossilis	Fingerling	0.9–1.1	Zafar and Khan (2018)
Calcium requirement (%)			
Labeo rohita	Fingerling	0.51–0.54	Musharraf and Khan (2020)
Ctenopharyngodon idella	Juvenile	1.0	Liang et al. (2012)
Heterobranchus bidorsalis	Fingerling	0.5	Adebayo and Omitoyin (2013)
Heteropneustes fossilis	Fingerling	0.57–0.68	Zafar and Khan (2019a)
Lateolabrax japonicus		0.42–1.02	Song et al. (2017)
Magnesium requirement (%)			
Labeo rohita	Fingerling	0.045	Musharraf and Khan (2018)
Ctenopharyngodon idella	Juvenile	0.05–0.07	Wang et al. (2011)
Carassius auratus gibelio	Juvenile	0.075	Han et al. (2012)
Oreochromis niloticus × *O. aureus*	Juvenile	0.02	Lin et al. (2013)
Heteropneustes fossilis	Fingerling	0.06–0.07	Zafar and Khan (2021)
Potassium requirement (%)			
Oreochromis niloticus × *O. aureus*	Juvenile	0.2–0.3	Shiau and Hsieh (2001)
Ctenopharyngodon idellus	Juvenile	0.47–0.73	Zhu et al. (2014)
Labeo rohita	Fingerling	0.35	Musharraf and Khan (2019b)
Heteropneustes fossilis	Fingerling	0.54	Zafar and Khan (2019b)
Cirrhinus mrigala	Juvenile	0.71–0.78	Kainat et al. (2023)

catalytic, (b) structural, (c) physiological, and (d) regulatory (Lall & Kaushik, 2021). Copper is widely accepted as an essential trace element required for the cellular functioning of all living organisms. Copper metalloenzymes are involved in Fe metabolism, cellular energy production, protection of cells from free radical damage, collagen synthesis, and melanin production. Fish absorb copper via the gills and digestive tract; however, the diet is considered a major source of copper for growth, development, and essential physiological functions (Bury et al., 2003). Reduced growth, whole-body copper content, liver copper content, and decreased activities of several enzymes, including ceruloplasmin Cu-Zn SOD and cytochrome C oxidase, have been reported in various fish species as copper deficiency signs (Zafar & Khan, 2020a, b; Musharraf and Khan, 2022; Wang et al., 2024). Low concentrations of waterborne copper also affect the hatching of fish eggs by inactivating chorionase activity, which causes osmotic disturbances and also affects the muscular movements necessary to break the eggshell (Johnson et al., 2007; Jezierska et al., 2009).

The inclusion of zinc in optimum amounts is a must, as it is an essential micromineral and has many diverse biochemical functions. The catalytic role of Zn is essential for the biological function of all six classes of more than 300 enzymes (McCall et al., 2000; Zafar & Khan, 2023). Some examples are RNA nucleotide transferases (RNA polymerase I, II, and III), alkaline phosphatase, and carbonic anhydrases. In the structural role of metalloenzyme, Zn ion stabilizes the tertiary structure of enzymes (e.g., Cu-Zn superoxide dismutase). A dietary requirement for Zn has been reported for several fish species. The optimum zinc requirement varies with age, sexual maturity, the composition of diet, water temperature, and water quality. The dietary zinc requirement reported for different fish species ranges from 3.5 to 64.6 mg/kg. Lens cataracts, erosion of fins and skin, growth depression, short body dwarfism, low serum; liver, scale, body, and vertebral Zn concentrations; and high mortality have been reported as zinc deficiency signs in several fish species. On the other hand, excessive dietary zinc levels may become toxic to fish and compete for similar binding sites with other bivalent minerals such as Cu, Fe, Ca, and Cd in the digestive tract during absorption (Luo et al., 2011).

Iron is one of the most investigated essential trace elements and is present in all body cells of vertebrates (NRC, 2011; Zafar & Khan, 2020a, b). It is essential for the functioning of several biochemical processes, which include the electron transfer reaction, gene regulation, binding and transport of oxygen, and regulation of cell growth and differentiation. The most important Fe-containing compounds are the heme proteins, hemoglobin, myoglobin, and cytochromes (Lall & Kaushik, 2021). The Fe requirement reported for certain fish species ranges from 30 to 170 mg/kg diet. Generally, Fe deficiency causes anemia and tissue Fe depletion in fish and other vertebrates (Zafar & Khan, 2020a, b). A characteristic microcytic anemia has been detected in different fish species (NRC, 2011) when fed low-iron diets. Excess of iron can also adversely affect the health of fish. The major effects of Fe toxicity were reduced growth, poor feed utilization, feed refusal, increased mortality, diarrhea, and histopathological damage to liver cells. Iron toxicity occurs from excessive Fe exposure of fish in water, which interferes with Fe homeostatic regulation by causing Fe overload in tissues (Bury et al., 2012).

Manganese (Mn) plays an important role in protein and energy metabolism, bone mineralization, glycosaminoglycan synthesis, cellular defense against free radicals, and metabolic regulation (Aschner & Aschner, 2005). It also acts as an activator of several enzymes, including oxidoreductases, lyases, ligases, hydrolases, kinases, and decarboxylases, and is a constituent of several metalloenzymes (Li & Yang, 2018). Mn and Fe compete for absorption sites. A small fraction of biliary Mn excreted into the intestine may be reabsorbed. Manganese is efficiently absorbed from the diet, but the absorption may be reduced by high levels of Ca and P, fiber, and phytate (Lall, 2002). Manganese uptake takes place through the gills and gastrointestinal tract. High concentrations of Mn caused Na imbalance, reduced the absorption of Ca and P, affected carbohydrate metabolism, and impaired the immune functions of fish (Ye et al., 2009). Manganese requirements of fish range from 2.5 to 25 mg/kg diet. Differences noted in the requirements can be due to differences in manganese uptake in water and bioavailability of manganese from the diets and chemical form of dietary manganese source. Reduced growth, mineralization, skeletal abnormalities, and decreases in the activity of liver Mn-SOD activity have been reported as signs of manganese deficiency in fishes (NRC, 2011; Zafar & Khan, 2019b). High levels of dietary manganese showed changes in feeding behavior, reduced body Fe concentration, and elevated Zn concentration. Mn toxicity also resulted in altering the brain and olfactory nerve histology (Lall & Kaushik, 2021).

Selenium has been documented as an essential micronutrient as well as a toxicant for fish (Uddin et al., 2024). Selenium is one of the major constituents of selenoproteins. The three best-characterized groups of selenoproteins in fish include glutathione peroxidases, thioredoxin reductases, and iodothyronine deiodinases. Fish absorb limited amounts of Se from the environment via the gills and skin under certain conditions; however, the gastrointestinal tract is the primary site for Se absorption (Janz, 2012). After absorption from the diet, Se is transported to the liver and metabolized to selenide and incorporated into selenocysteine for selenoprotein synthesis or converted to selenosugars or methylated metabolites for excretion. Selenium requirements of several fish species have been determined and range between 0.11 and 1.85 mg/kg. Deficiency signs shown by fishes as a result of lower dietary selenium availability may be reduced growth, feed intake, and low enzyme activities (NRC, 2011). Waterborne inorganic forms of Se (selenate and selenite) are not absorbed appreciably through gill membranes, and ingestion of natural food organisms that accumulate Se from the aquatic environment appears to be more toxic (Janz, 2012) (Table 1.6).

These minerals, macro and micro, are supplied in the form of mineral premixes in formulated fish feeds to ensure that fish receive all the necessary minerals for healthy growth and development. These premixes provide the right balance of macro- and trace minerals, supporting critical functions like bone formation, immune response, and metabolic health. By addressing species-specific and life-stage mineral requirements, they help prevent deficiencies that could lead to poor growth or health issues.

2.5 Dietary Vitamin Requirements

Vitamins are essential organic dietary compounds that support the normal growth and health of any organism. They are often not synthesized by fish and must be provided in the diet. Vitamins have been classified into two groups, i.e., water-soluble and fat-soluble. Vitamins A, D, E, and K are the fat-soluble vitamins, while water-soluble vitamins include B complex and vitamin C. Vitamin nutrition plays a key role in optimizing animal performance and supporting the sustainable growth of the industry by improving animal robustness in terms of health and welfare with a concomitant decrease in feed waste. Vitamins act as coenzymes and are also essential for the regulation of metabolic reactions (Berntssen et al., 2016; Khan & Khan, 2019; Soto-Davila et al., 2020). They also have several crucial roles to play in the process of detoxification, bone development, immunity development, flesh quality improvement, and prevention of nutritional disorders (Berntssen et al., 2016; Soto-Davila et al., 2020; Krogdahl et al., 2020). Vitamins are highly sensitive to light,

Table 1.6 Variations in micromineral requirement of different fish species at different life stages

Fish species	Life stage	Requirement	References
Copper requirement (mg/kg)			
Labeo rohita	Fingerling	4.52–4.75	Musharraf and Khan (2022)
Epinephelus malabaricus	Juvenile	4–6	Lin et al. (2008)
Channa punctatus	Fingerling	6.66–6.78	Hameid et al. (2017)
Ctenopharyngodon idella	Fingerling	4.70–4.95	Tang et al. (2013)
Heteropneustes fossilis	Fingerling	5.24–5.68	Zafar and Khan (2020a)
Zinc requirement (mg/kg)			
Labeo rohita	Fingerling	47.85–52.93	Musharraf and Khan (2019c)
Pelteobagrus fulvidraco	Juvenile	17.12–20.86	Luo et al. (2011)
Rachycentron canadum	Fingerling	15	Xu et al. (2007)
Heteropneustes fossilis	Fingerling	26.66–30.64	Zafar and Khan (2023)
Iron requirement (mg/kg)			
Labeo rohita	Fingerling	211–265	Musharraf and Khan (2019d)
Rachycentron canadum	Juvenile	80.5–94.7	Qiao et al. (2013)
Ctenopharyngodon idella	Young	300	Su et al. (2007)
Heteropneustes fossilis	Fingerling	82.51–86.05	Zafar and Khan (2020b)
Manganese requirement (mg/kg)			
Rachycentron canadum	Juvenile	22–25	Liu et al. (2013)
Ctenopharyngodon Idella	Juvenile	20.6	Liang et al. (2015)
Labeo rohita	Fingerling	9.42–10.96	Musharraf and Khan (2021)
Heteropneustes fossilis	Fingerling	7.81–8.14	Zafar and Khan (2019c)
Pelteobagrus fulvidraco	Juvenile	5.5–6.4	Tan et al. (2012)
Selenium requirement (mg/kg)			
Carassius auratus gibelio	Juvenile	0.73–1.19	Zhu et al. (2017)
Rachycentron canadum	Juvenile	0.8	Liu et al. (2010)
Larimichthys croceus	Juvenile	0.38–0.57	Juan et al. (2015)
Oreochromis niloticus	Juvenile	1.06	Lee et al. (2016)
Oncorhynchus mykiss	Fry	0.15–0.38	Hilton et al. (1980)

heat, humidity, oxidizing and reducing agents, acids, and alkalis. Chemical and physical processing technologies have been developed to protect vitamins from degradation during feed manufacturing and storage. The most common technologies involve esterification of vitamins followed by (1) adsorption—typically on silicic acid—(vitamin E), (2) beadlet encapsulation—typically consisting of an emulsion made of gelatine or glycerine coated with starch—(vitamins A, D3), and (3) blending with carriers—typically lignosulfonate and maltodextrin—before spray drying (B-complex vitamins).

Vitamin A plays an important role in the immunological responses of fish, as its deficiency is frequently immunosuppressive (Yang et al., 2008; Zhang et al., 2017). Vitamin D is an important vitamin and plays several important roles, such as antibacterial and antiviral immune responses, muscle contraction, fatty acid oxidation, cell proliferation and differentiation, and secretion of insulin and prolactin (Hideya et al., 2008; Peng et al., 2017; Nunez-Acuna et al., 2018; Soto-Davila et al., 2020). In order to avoid the deficiency signs, an optimum amount of vitamin D should be incorporated in the diets of fish. The best-known action of the vitamin D endocrine system is its role in calcium and phosphate homeostasis. Vitamin D is a key regulator of transcellular calcium uptake (via a luminal calcium channel) and induces cytosolic calcium transport (Bronner, 2009).

Vitamin E (α-tocopherol) plays a critical role in iron metabolism, fish reproductive physiology, transport of amino acids, lipids, and steroidogenesis (El- Sayed & Izquierdo, 2022). It also acts as an antioxidant and plays a notable role in preventing the oxidation of animal tissue (El- Sayed & Izquierdo, 2022). Vitamin E cannot be synthesized endogenously in aquatic animals (NRC, 2011; Mai et al., 2022), so dietary supplementation is necessary. Deficiency of dietary vitamin E in fishes can lead to slower growth, anemia, lower disease resistance, and histopathological alterations (Mai et al., 2022). The dietary vitamin E requirement for different fish species ranges from 50 to 120 mg/kg diets. In addition, lipid oxidation in fish fillets can be decreased by using high doses of dietary vitamin E to increase shelf life and, ultimately, contribute to the reduction of food waste (Faizan et al., 2013). Such an approach illustrates that the concept of optimal vitamin nutrition can further support the sustainable growth of the aquaculture industry by improving flesh quality and nutritional functionality and concomitantly helping to reduce food loss and waste. Vitamin K is best known for its essential role in blood coagulation (Olson, 1999), being responsible for the posttranslational modification and activation of the vitamin K-dependent proteins (Lee et al., 2007; Krossøy et al., 2011). Recommendations for vitamin K inclusion levels in the fish nutrition literature ranged from 0.1 to 10 mg/kg diet for ensuring optimum growth, improved immunity, lack of deficiencies, and decreased feed losses (Krossøy et al., 2011). The instability of vitamin K during feed production and storage might have contributed to the requirement variations (Krossøy et al., 2011; NRC, 2011; Mai et al., 2022). Vitamin K deficiency signs in fish include increased blood coagulation time, reduced growth, anemia, hemorrhages, fin erosion, weak bones, the occurrence of spinal curvature, short tails, and increased mortality (Abdelhamid et al., 2024)

The B-complex vitamins are essential in the metabolism of carbohydrates, amino acids, lipids, and nucleic acids. Thiamine, riboflavin, niacin, and biotin are necessary for optimal energy metabolism and lipid synthesis. Without adequate dietary pyridoxine, folic acid, and vitamin B12, the synthesis and interconversion of amino acids will be inadequate for the optimum synthesis of proteins, and growth rates will be reduced. Insufficient dietary folic acid and vitamin B12 impair nucleic acid synthesis and slow down cell division, consequently limiting growth and delaying the cellular immune response to stop pathogens from infecting the animal. Thiamin deficiency may lead to anorexia, changes in skin coloration, mortality, reproductive disorders, and behavioral abnormalities related to neurological disorders (Harder et al., 2018; Zehra & Khan, 2019; Khan & Khan, 2022). Riboflavin (vitamin B2), niacin (nicotinic acid or vitamin B3), and pantothenic acid (vitamin B5) have also been reported as essential vitamins for fish. Riboflavin is a component of two coenzymes, flavin mononucleotide (FMN) and flavin adenine dinucleotide (FAD), which serve as prosthetic groups of oxidation-reduction enzymes involved in the metabolism of keto acids, fatty acids, and amino acids in the mitochondrial electron transport system (Deng & Wilson, 2003; Li et al., 2010; Huang et al., 2010). Riboflavin acts along with pyridoxine and facilitates the conversion of tryptophan to nicotinic acid. In addition to this, riboflavin plays an essential role in the respiration of poorly vascularized tissues such as the cornea of the eye (Halver, 2002; Zehra & Khan, 2019). Vitamin B6 (pyridoxine) functions as coenzymes, i.e., pyridoxal phosphate (PLP) and pyridoxamine phosphate (PMP), that participate in the protein, lipid, and carbohydrate metabolism (Khan & Khan, 2020; Mai et al., 2022).

Deficiency of vitamin B6 may result in low feed intake, high mortality, and behavioral changes including abnormal swimming, hyperirritability, and ataxia due to the disruption of neurotransmitter synthesis. Organ damage, such as kidney and intestinal lesions, has also been reported as a vitamin B6 deficiency sign (Liu et al., 2022). Biotin is another essential nutrient for fish that serves as a coenzyme for carboxylases critical to the glucose, amino acid, and fatty acid metabolism in all animals (Combs & McClung, 2017; Mai et al., 2022). Histopathological changes and lesions in organs such as gills, liver, and kidney have been reported as the deficiency signs of biotin (Liu et al., 2022). Biotin is also produced by intestinal bacteria in some animals, but no report is yet available on fish. Hence, dietary supplementation is necessary (Yossa et al., 2011; Liu et al. 2022). B12 coenzymes (methylcobalamin and adenosylcobalamin) are important participants in the metabolism in animals, including fish (Hansen et al., 2015; Espe et al., 2020; Khan & Khan, 2021; Mai et al., 2022). Vitamin B12 is required for red blood cell and myelin synthesis (Smith et al., 2018; Calderón-Ospina & Nava-Mesa, 2020; Mai et al., 2022). Reduced feed intake and megaloblastic anemia are reported as B12 deficiency signs in fish (Liu et al., 2022). Vitamin C is an extensively studied water-soluble vitamin in fish nutritional studies. It acts as an antioxidant and helps in reproduction, wound healing, and disease resistance (NRC, 2011; Mai et al., 2022) (Table 1.7).

3 Role of Feed Additives in Fish Feeds

Feed additives are added in minute quantities during feed preparation to improve the feed's digestibility, palatability, and overall quality (Janocha et al. 2022). These additives are crucial to maximize nutrient utilization and growth performance and health status of the fish and enhance productivity. Generally, these are nonnutritious and include phytobiotic compounds, antimicrobial compounds, hormones, organic acids, enzymes, prebiotics, probiotic compounds, antioxidants, and immunostimulants. To ensure the effectiveness of additives and safety of fish, governmental organizations have control over the use of feed additives (Bai et al., 2022).

Table 1.7 Variations in vitamin requirement of different fish species at different life stages

Fish species	Life stage	Requirement	References
Thiamin requirement (mg/kg)			
Cirrhinus mrigala	Fingerling	1.79–3.34	Zehra and Khan (2017a)
Channa punctatus	Fingerling	2.34–2.59	Zehra and Khan (2017b)
Calta calta	Fingerling	0.74–1.46	Khan and Khan (2019)
Cyprinus carpio	Fingerling	0.5	Aoe et al. (1969)
Riboflavin requirement (mg/kg)			
Channa punctatus	Fingerling	5.7–7.7	Zehra and Khan (2017a, b)
Calta catla	Fingerling	7.33–9.67	Khan and Khan (2019)
Ctenopharyngodon idella	Adult	6.65	Chen et al. (2016)
Ictalurus punctatus	Fingerling	3–9	Murai and Andrews (1978)
Cyprinus. Carpio var. Jian	Juvenile	5	Li et al. (2010)
Niacin requirement (mg/kg)			
Heteropneustes fossilis	Fingerling	25	Shaik Mohamed and Ibrahim (2001)
Cyprinus carpio	Fingerling	28–50	Aoe et al. (1967)
Ictalurus punctatus	Fingerling	14	Andrews and Murai (1978)
Cirrhinus mrigala	Fingerling	30	Ahmed (2011)
Ctenopharyngodon idella	Young	34.01–42.08	Li et al. (2015)
Oreochromis niloticus	Fingerling	20.4	Jiang et al. (2014)
Labeo rohita	Fingerling	33	Ahmed (2011)
Catla catla	Fingerling	27.61–40.64	Khan and Khan (2021)
Pyridoxine requirement (mg/kg)			
Cyprinus carpio	Fingerling	5.4	Ogino (1965)
Cyprinus. Carpio var. Jian	Juvenile	6.07	He et al. (2017)
Ictalurus punctatus	Fingerling	3	Andrews and Murai (1979)
Channa punctatus	Fingerling	3.20–6.87	Zehra and Khan (2018)
Heteropneustes fossils	Fingerling	3.21	Shaik Mohammed (2001a)
Cirrhinus mrigala	Fingerling	5.63–8.61	Zehra and Khan (2017a, b)
Catla catla	Fingerling	3.20–6.86	Khan and Khan (2019)
Labeo rohita	Fingerling	100	Akhtar et al. (2012)
Biotin requirement (mg/kg)			
Clarias batrachus	Fingerling	2.49	Shaik Mohammed et al. (2000)
Heteropneustes fossils	Fingerling	0.25	Shaik Mohammed (2001b)
Cyprinus carpio	Fingerling	2.0–2.5	Gunther and Meyer-Burgdorff (1990)
Oreochromis niloticus × O. aureus	Juvenile	0.06	Shiau and Chin (1999)
Lateolabrax japonicus	Fry	0.05	Li et al. (2010)
Catla catla	Fingerling	0.41–0.87	Khan and Khan (2019)
Vitamin C requirement (mg/kg)			

(continued)

Table 1.7 (continued)

Fish species	Life stage	Requirement	References
Cirrhinus mrigala	Fingerling	35.65–41.99	Zehra and Khan (2012)
Clarias batrachus	Fry	69	Mishra and Mukhopadhyay (1996)
Labeo rohita	Fingerling	200	Misra et al. (2007)
Vitamin A requirement (mg/kg)			
Ictalurus punctatus	Fingerling	0.3–0.6	Dupree (1970)
Cyprinus carpio	Juvenile	1.2–6	Aoe et al. (1968)
Oreochromis niloticus × *O. aureus*	Juvenile	1.76–2.09	Hu et al. (2021)
Epinephelus spp.	Juvenile	0.93	Shaik Mohammed et al. (2003)
Cyprinus carpio var. communis	Fingerling	0.1–0.12	Hassan et al. (2022)
Vitamin D requirement (μg/kg)			
Oncorhynchus mykiss	Fingerling	40–60	Barnett et al. (1982a, b)
Ictalurus punctatus	Fingerling	12.5	Lovell and Li (1978)
Oreochromis niloticus × *O. aureus*	Juvenile	9.35	Shiau and Hwang (1993)
Vitamin E requirement (mg/kg)			
Cirrhinus mrigala	Fry	99	Paul et al. (2022)
Catla catla	Fingerling	150	Sinha and Sinha (1994)
Labeo rohita	Fry	131.91	Sau et al. (2004)
Channa punctatus	Fingerling	140–169	Hameid et al. (2012)
Epinephelus malabaricus	Juvenile	61–68	Lin and Shiau (2005a, b)
Cyprinus carpio	Fry	100	Watanabe et al. (1970)
Oreochromis niloticus	Fingerling	80	Quiang et al. (2019)
Ictalurus punctatus		50–100	He et al. (2017)
Ctenopharyngodon idella		100.36	Li et al. (2014)
Vitamin K requirement (mg/kg)			
Ctenopharyngodon idella	Fingerling	1.9	Jiang et al. (2007)
Oreochromis niloticus	Juvenile	2–4	Abdelhamid et al. (2024)

3.1 Phytobiotic Compounds

Phytobiotic or phytogenic compounds are plant derivatives added in the feed to improve the growth and health performance of the fishes. These plant products have different properties, like antioxidant, antimicrobial, anticarcinogenic, analgesic, insecticidal, antiparasitic, anti-coccidial, growth promoters, appetite enhancement, stimulant of secretion of bile, and digestive enzyme activity, etc. (Asimi & Sahu, 2013). These phytobiotic components are heterogeneous feed additives originating from different parts of the plants, viz., leaves, roots, tubers, fruits, and spices. These components are used in different forms (oil, powder, or extract). Different types of phytobiotic compounds have been used in the feeds of various species and resulted in improved growth performance and feeding efficiency (Omoregie et al., 2009), hematological parameters (Thanikachalam et al., 2010), and increased immunostimulation against bacterial infection (Nya et al., 2010; Zilberg et al., 2010).

3.2 Antimicrobial Agents

Aquaculture uses both synthetic and natural antimicrobial medicines, which are rarely used as growth promoters; instead, they are mostly employed as preventative and therapeutic agents (Shao, 2001). These substances exhibit precise cellular targeting and can function as selective ligands for targets associated with disease, which can influence disease-related pathways and maintain a healthy biological network (Chin et al., 2006; Lagunin et al., 2010). Different substances can be used in mitigating antimicrobial activities, such as formalin, which can be used to limit the propagation of monogenetic trematodes, protozoa, and fungi; florfenicol, which has been reported as an effective antimicrobial agent in decreasing the intestinal septicemia-related catfish mortality; calcium chloride, which aids the proper egg hardening; and Fuller's Earth, which reduces the stickiness of fish eggs to improve hatchability (US FDA, 2017).

3.3 Hormones

In aquaculture, hormones are utilized for sex reversal (Taranger et al., 2010) and growth promotion. In aquaculture feeds, hormones can be added for the purpose of sex reversal to create a monosex population for accelerating growth. Hormones like 17β-estradiol, estradiol valerate, 17α-methyl testosterone, or 17α-methyl dihydrotestosterone can be used to feminize a population. Conversely, testosterone is utilized by masculine females (Piferrer, 2001). In species where males grow faster than females and attain larger sizes, tamoxifen, 17α-methyl testosterone, 17α-ethynyl testosterone, and 17α-ethynyl estradiol have all been used to produce an all-male population (Liao et al., 2014).

3.4 Organic Acids

Organic acids are weak carboxylic acids. They partially dissociate in water to form a hydrogen ion and a carboxylate ion. Formic acid, citric acid, benzoic acid, lactic acid, acetic acid, propionic acid, malic acid, and sorbic acid are some common examples of organic acids (NRC, 2011). Organic acids play crucial roles in growth enhancement, nutrient utilization, and disease resistance (Ng et al., 2009). They function by decreasing the pH in the stomach and intestine and increasing the digestive enzyme activity. The organic acids are also known to penetrate into the cell wall of bacteria, resulting in growth inhibition (Nates, 2016). Various factors like fish species, size, age, types and level of organic acids, feed management, and water quality affect the results shown by their dietary inclusion (Fefana, 2014). Citric acid is known to improve the retention of nitrogen and phosphorus in sea bream (Hossain et al., 2007). The addition of sodium diformate in the diets resulted in improved growth performance and feed utilization in hybrid tilapia (Khaled, 2015) and increased the digestibility of protein, lipid, and amino acids in rainbow trout (Morken et al., 2011).

3.5 Enzymes

There are some endogenous factors that affect the utilization and efficiency of feed. The negative effects of anti-nutritional factors present in feed ingredients can affect the digestion of dietary components and growth performance of the fish. These problems can be overcome by exogenous enzymes (Lin et al., 2007; Ebru & Cengiz, 2016). Phytase, carbohydrase, protease, lipase, alpha-amylase, papain, and pepsin are the enzymes that are commonly used in aquaculture feeds. The major portion of the phosphorus in the plant seeds is present in the form of phytate, and it has low bioavailability and digestibility for fish. Hence, the addition of phytase in fish feeds increases the phytate phosphorus digestion and reduces phosphorus excretion (Cao et al., 2007). The enzyme phytase has been effectively incorporated into poultry and swine (Humer et al., 2015) and aquaculture feeds (Lemos & Tacon, 2017) and has notably improved the utilization of phosphorus found in plant-derived meals that are used as sources of dietary protein. The digestible efficiency of the non-starch polysaccharides is increased by the addition of non-starch polysaccharide-degrading enzymes in the feeds (Sinha et al., 2011).

3.6 Probiotics

The use of probiotics as a feed additive in aquaculture feeds has become the subject of investigations during the last decade of the twentieth century, while the use of probiotics and prebiotics in the poultry and swine production industries originated in the 1970s, and success is based on the establishment of a proportional balance of microbial species within the intestine. Probiotics are live microbes supplemented to the fish gut through feeds (Encarnação, 2016) and can also be described as healthy bacteria. Their role in improving the water quality of the aquaculture systems has also been reported (Aly et al., 2016). The use of probiotics in fish feeds can help in maintaining the intestinal microbial balance and improving the utilization of feed. Probiotics are also known for modulating the immune system, regulating the allergic response of the body, and lowering the proliferation of cancer in mammals. Probiotic compounds are known to improve the growth, protein content, and immune status of the rainbow trout, Nile tilapia, and juvenile common dentex (Lara-Flores et al., 2003; Merrifield et al., 2010; Hidalgo et al., 2006).

3.7 *Prebiotics*

Prebiotics are nondigestible feed ingredients that beneficially affect the host by selectively stimulating the growth or activity of one or a limited number of bacterial species already resident in the gut and thus attempt to improve host health. Commonly used prebiotics in aquaculture are inulin, oligofructose, xylooligosaccharide, fructooligosaccharide, mannan oligosaccharide, galactooligosaccharide, and b-glucan. A substance can be used as a prebiotic if it fulfills some criteria, such as being resistant to the upper gut tract, easily fermentable by intestinal microbiota, selectively stimulating the probiotics, and beneficial to host health. The prebiotics can also show antioxidant properties. By strengthening the intestinal epithelial barrier, prebiotics can lessen the amount of pathogens and inflammatory stimuli that enter the bloodstream. This lowers the risk of inflammatory illnesses in fish and helps prevent systemic infections (Ashaolu, 2020). They can reduce the oxidative stress in fish tissues by scavenging free radicals (Devi et al., 2019). The addition of inulin in the diets of Nile tilapia resulted in increased RBC, magnesium, calcium, and iron content; increased length of intestinal villi; and lysozyme activity (Tiengtam et al., 2015). Similarly, fructooligosaccharides enhanced the feed intake and digestibility when added to the feeds of Atlantic salmon (Grisdale-Helland et al., 2008).

4 Feeding Strategies

As much as the knowledge of nutritional requirements of fish is necessary for formulating nutritionally balanced feeds, the deliverance of these nutrients in a way that is optimally utilizable by the fish is key in any aquaculture operation. The successful practice of feeding is determined by the optimal amount of feed to be fed to a particular fish, i.e., ration size, at suitable times in a day, i.e., feeding intensity, and effectively broadcasting the predetermined ration to the fish. This ensures the provision of the right balance of nutrients at each stage of fish for maximum growth and feed efficiency and promotes optimal health. Feed is typically one of the largest costs in fish farming, and reducing 15–20% of waste (Munguti et al., 2014) through efficient feeding strategies not only makes the feed cost-effective but also lowers the environmental burden caused by excess nutrients and organic matter released into the aquatic environment.

4.1 *Ration Size*

The ration size, or feeding rate, of fish refers to the amount of feed provided to fish in aquaculture systems. An optimum ration size gives the best growth and feed conversion ratio. It is a critical factor in ensuring healthy growth, efficient feed utilization, and minimizing waste.

Different species have different feeding habits and requirements. For instance, carnivorous fish might need higher protein content in their feed compared to herbivorous fish. Smaller or younger fish will require less feed than larger, older fish. As fish grow, their nutritional needs increase.

Higher water temperatures generally increase fish metabolism, leading to higher feed consumption. Conversely, colder water temperatures can slow their feeding rate. Water quality, e.g., oxygen levels, pH, and ammonia concentration, also influences feeding behavior and growth. Poor water quality can lead to reduced feeding and hence slower growth. Overfeeding can lead to waste and water quality issues, while underfeeding may hinder growth (Li et al., 2020).

Proper ration sizing is essential to avoid issues like overfeeding, which can lead to excess waste and pollution in the system, or underfeeding, which can result in stunted growth, poor immune function, and higher susceptibility to diseases. Monitoring the fish's response to different rations, adjusting feeding schedules, and regularly checking water quality are essential practices for maintaining an effective and sustainable feeding regimen.

4.2 Feeding Intensity

Feeding frequency for fish in aquaculture is an important aspect of managing their growth, health, and feed efficiency. The optimal feeding frequency depends on several factors, including the species, age, or size of the fish; water temperature; and overall management practices (Gao et al., 2022). Feeding frequency is species-specific. Carnivorous fish like salmon, trout, and tilapia often require more frequent feedings, typically 2–3 times per day, to meet their higher protein needs. Herbivorous fish like carp or tilapia, at certain stages, may require fewer feedings, and once or twice a day is sufficient. Omnivorous fish such as catfish or bass generally have moderate feeding frequency needs, often 2–3 times per day.

The fingerling stage of fish has faster metabolisms and requires more frequent feedings to support their rapid growth, often 4–5 times per day. As they grow, the frequency of feeding can be reduced, whereas adult fish require fewer feedings, generally 2–3 times per day, depending on their size.

Real-time fish feeding intensity assessment (FFIA) is a key technology to effectively improve the feed utilization rate and reduce water pollution. The feeding intensity recognition methods can be of these three types: water-quality sensor-based, machine-vision-based, and acoustic-based (An et al., 2021; Darodes de Tailly et al., 2021).

The water-quality sensor-based method is based on changes in water quality, such as temperature, dissolved oxygen, and pH, that can directly affect the appetite of the fish (Guo et al., 2017). For instance, fish metabolism is influenced by water temperature. In warmer water, fish digest and utilize feed more efficiently, and they might need more frequent feedings, while in colder water, fish metabolism slows down, and they may require fewer feedings, especially in winter or cooler climates (Volkoff & Rønnestad, 2020).

Compared to water-quality parameter changes to assess fish feeding intensity, machine vision techniques are more intuitive, convenient, fast, and highly accurate (Yang et al., 2021). The 3D ResNet-Glore fish feeding intensity quantification network can precisely demarcate four levels of fish feeding behavior through video monitoring (Feng et al., 2022). A multitasking network for continuous monitoring of fish feeding activities and the number of uneaten feed pellets in the fish tank has been developed, and the results showed that it achieved 95.44% accuracy in classifying fish feeding behavior (Wang et al., 2022).

The production of pulsed acoustic signals by fish and shrimp during feeding can be another important indicator of feeding activity (Silva et al., 2019). An audio-spectral model has been developed based on acoustic signals and attentional mechanisms that classify the feeding intensity of fish into four classes (Zeng et al., 2023). The use of hydrophones has no effect on fish feeding and thus provides a basis for developing more effective feeding strategies (Du et al., 2024).

4.3 Feed Dispensing Methods

Delivering the formulated feeds appropriately to the group of fish is an important factor for aquaculture operations. In smaller systems or research setups, broadcast feeding or hand feeding by spreading the feed evenly across the system ensures that each fish has equal access to the feed. The feeding intensity is adjusted manually, often based on visual observation of fish activity and consumption. However, for large setups, the feed must be spread over a larger area, and the manual labor with hand feeding will be physically and monetarily straining. Additionally, fry and small fish require high feeding intensity, making hand-feeding inefficient. In such cases, automatic feeders are used.

Automatic feeders can provide consistent and scheduled feedings over a large area. This ensures that fish get their feed at regular intervals without manual intervention. They can be adjusted based on fish size, species, and environmental conditions (Reis et al., 2021). Feeding can be done by restricted feeding or ad libitum feeding. To feed ad libitum means to supply feed continuously so that all fish have free access to the feed at all times. Restricted feeding can be done by knowing the amount of feed to deliver, i.e., the ration size to the fish group, or there must be some type of feedback to know when most of the fish in the group are fed to satiation at any given point in time.

On-demand feeders are systems where fish are trained to interact with a device, such as a stick hanging in the water, which is connected to a feed hopper. When the fish move or tilt the stick, it triggers a plate below the hopper, releasing food. These feeders are commonly used in low-intensity farms, ponds, and some tanks, as they rely on the fish to activate the feeding process, reducing the need for energy input. Some demand feeders are electronically operated, replacing the mechanical stick with an electronic pressure switch, which starts the motor to release feed when triggered by the fish. These electronic versions still require power but can be programmed to distribute food at specific intervals, helping prevent overfeeding by stopping the fish from "playing" with the system. Demand feeders are also used in research to monitor feeding behavior and appetite. For energy, alternatives like rechargeable batteries, solar cells, and clockworks can be employed, minimizing the need for traditional electric power sources.

Automatic machine feeders can be categorized into three types: feed dispensers, automatic feeders, and complete feeding systems. The key distinction between feed dispensers and automatic feeders is that feed dispensers lack a feed hopper, whereas automatic feeders include a hopper and are more advanced in their operation. A feed dispenser typically dispenses a fixed, weighed amount of food over a specified period, often ranging from 1 to 3 days. It can operate continuously or at fixed intervals, using a simple relay to control the start and stop of the feeding cycle. Mechanisms in feed dispensers usually consist of a rotating belt or scraper, powered by a motor. While such systems are easy to observe and control, they require preweighing of the food, making them more time-consuming and suited for small-scale operations, such as fry and juvenile feeding. On the other hand, automatic feeders contain several main components: a feed hopper, a feed distribution mechanism with a motor, an electrical power supply, and a control unit for operating the feeder. The distribution mechanism, which may include a separate feed spreader motor, is the primary element in automatic feeders, ensuring proper food distribution across larger production areas. These feeders are typically used in more advanced systems, where precise and efficient feeding is crucial for larger or more complex aquaculture operations. Complete feeding systems go beyond automatic feeders by integrating bulk feed transport from silos to the fish production units. These systems are designed for large-scale aquaculture operations, where large quantities of feed need to be delivered efficiently and consistently to multiple feeding units. The system typically includes conveyors, augers, or pneumatic transport to move feed from storage silos directly to the automatic feeders or distribution points in the fish tanks or ponds. This setup helps streamline the feeding process, reduces labor, and ensures a steady, uninterrupted supply of feed to the fish, making it ideal for intensive farming environments.

5 Conclusion

Working out the nutritional requirements of fish is the foundation of effective nutritional strategies in aquaculture. Key nutrients such as protein, lipids, carbohydrates, vitamins, and minerals must be tailored to the specific needs of the cultured species at different life stages in aquafeeds. Non-dietary components such as binders, antioxidants, enzymes, probiotics, and prebiotics in the feeds boost the growth, feed efficiency, and immune system of the fish. Formulating feeds utilizing this knowledge ensures that fish receive a balanced diet, promoting optimal growth, improving feed conversion rates, enhancing disease resistance, and ultimately supporting sustainable aquaculture practices that are both economically viable and environmentally responsible. Appropriate feeding strategies are imperative to increase profit and decrease nutrient loading in the aquaculture operation.

References

Abdelhamid, A. F., Gewida, A. G. A., El-Sayed, A., Badran, M. F., et al. (2024). Impacts of different levels of vitamin K on the growth performance, hematological parameters, and immunological response of juvenile Nile tilapia (*Oreochromis niloticus*). *Aquaculture International, 32*, 477–488.

Abidi, S. F., & Khan, M. A. (2004a). Dietary valine requirement of Indian major carp, *Labeo rohita* (Hamilton) fry. *Journal of Applied Ichthyology, 20*(2), 118–122.

Abidi, S. F., & Khan, M. A. (2004b). Dietary histidine requirement of fingerling Indian major carp, *Labeo rohita* (Hamilton). *Israeli Journal of Aquaculture-Bamidgeh, 56*.

Abidi, S. F., & Khan, M. A. (2007). Dietary leucine requirement of fingerling Indian major carp, *Labeo rohita* (Hamilton). *Aquaculture Research, 38*(5), 478–486.

Abidi, S. F., & Khan, M. A. (2008). Dietary threonine requirement of fingerling Indian major carp, *Labeo rohita* (Hamilton). *Aquaculture Research, 39*, 1498–1505.

Abidi, S. F., & Khan, M. A. (2009). Dietary arginine requirement of fingerling Indian major carp, *Labeo rohita* (Hamilton) based on growth, nutrient retention efficiencies, RNA/DNA ratio and body composition. *Journal of Applied Ichthyology, 25*(6), 707–714.

Abidi, S. F., & Khan, M. A. (2010). Dietary tryptophan requirement of fingerling rohu, *Labeo rohita* (Hamilton), based on growth and body composition. *Journal of World Aquaculture Society, 41*, 700–709.

Abidi, S. F., & Khan, M. A. (2011). Total sulphur amino acid requirement and cystine replacement value for fingerling rohu, *Labeo rohita*: Effects on growth, nutrient retention and body composition. *Aquaculture Nutrition, 17*(2), e583–e594.

Adebayo, A., & Omitoyin, B. O. (2013). Essentiality of calcium supplement in the diets of *Heterobranchus bidorsalis* fingerlings. *International Journal of Fisheries and Aquaculture, 5*, 98–103.

Ahmed, I. (2011). Effect of dietary niacin on growth and body composition of two Indian major carps rohu, *Labeo rohita*, and mrigal, *Cirrhinus mrigala* (Hamilton), fingerlings based on dose-response study. *Aquaculture International, 19*, 567–584.

Ahmad, I., Ahmed, I., & Dar, N. A. (2021a). Effects of dietary leucine levels on growth performance, hematobiochemical parameters, liver profile, intestinal enzyme activities and target of rapamycin signalling pathway related gene expression in rainbow trout, *Oncorhynchus mykiss* fingerlings. *Aquaculture Nutrition, 27*(6), 1837–1852.

Ahmad, I., Ahmed, I., & Dar, N. A. (2021b). Dietary valine improved growth, immunity, enzymatic activities and expression of TOR signaling cascade genes in rainbow trout, *Oncorhynchus mykiss* fingerlings. *Scientific Reports, 11*(1), 22089.

Ahmad, I., Ahmed, I., & Dar, N. A. (2022). Effects of dietary isoleucine on growth performance, enzymatic activities, antioxidant properties and expression of TOR related genes in rainbow trout, *Oncorhynchus mykiss* fingerlings. *Aquaculture Research, 53*(6), 2366–2382.

Ahmed, I. (2007). Dietary amino acid L-threonine requirement of fingerling Indian catfish, *Heteropneustes fossilis* (Bloch) estimated by growth and biochemical parameters. *Aquaculture International, 15*, 337–350.

Ahmed, I. (2012). Dietary amino acid L-tryptophan requirement of fingerling Indian catfish, *Heteropneustes fossilis* (Bloch), estimated by growth and haemato-biochemical parameters. *Fish Physiology and Biochemistry, 38*, 1195–1209.

Ahmed, I. (2013a). Dietary arginine requirement of fingerling Indian catfish (*Heteropneustes fossilis*, Bloch). *Aquaculture International, 21*, 255–271.

Ahmed, I. (2013b). Dietary amino acid l-histidine requirement of fingerling Indian catfish, *Heteropneustes fossilis* (Bloch), estimated by growth and whole body protein and fat composition. *Journal of Applied Ichthyology, 29*(3), 602–609.

Ahmed, I. (2014). Dietary amino acid l-methionine requirement of fingerling Indian catfish, *Heteropneustes fossilis* (Bloch-1974) estimated by growth and haemato-biochemical parameters. *Aquaculture Research, 45*(2), 243–258.

Ahmed, I., & Ahmad, I. (2020). Effect of dietary protein levels on growth performance, hematological profile and biochemical composition of fingerlings rainbow trout, *Oncorhynchus mykiss* reared in Indian himalayan region. *Aquaculture Reports, 16*, 100268.

Ahmed, I., & Ahmad, I. (2021). Dietary lysine modulates growth performance, haemato-biochemical indices, non-specific immune response, intestinal enzymatic activities and antioxidant properties of rainbow trout, *Oncorhynchus mykiss* fingerlings. *Aquaculture Nutrition, 27*, 124–139.

Ahmed, I., & Khan, M. A. (2004a). Dietary arginine requirement of fingerling Indian major carp, *Cirrhinus mrigala* (Hamilton). *Aquaculture Nutrition, 10*(4), 217–225.

Ahmed, I., & Khan, M. A. (2004b). Dietary lysine requirement of fingerling Indian major carp, *Cirrhinus mrigala* (Hamilton). *Aquaculture, 235*(1–4), 499–511.

Ahmed, I., & Khan, M. A. (2005a). Dietary tryptophan requirement of fingerling Indian major carp, *Cirrhinus mrigala* (Hamilton). *Aquaculture Research, 36*(7), 687–695.

Ahmed, I., & Khan, M. A. (2005b). Dietary histidine requirement of fingerling Indian major carp, *Cirrhinus mrigala* (Hamilton). *Aquaculture Nutrition, 11*(5), 359–366.

Ahmed, I., & Khan, M. A. (2006). Dietary branched-chain amino acid valine, isoleucine and leucine requirements of fingerling Indian major carp, *Cirrhinus mrigala* (Hamilton). *British Journal of Nutrition, 96*(3), 450–460.

Ahmed, I., & Maqbool, A. (2017). Effects of dietary protein levels on the growth, feed utilization and haemato-biochemical parameters of freshwater fish, *Cyprinus carpio* var. *specularis*. *Fisheries and Aquaculture Journal, 8*, 1–12.

Ahmed, I., Khan, M. A., & Jafri, A. K. (2003). Dietary methionine requirement of fingerling Indian major carp, *Cirrhinus mrigala* (Hamilton). *Aquaculture International, 11*, 449–462.

Ahmed, I., Khan, M. A., & Jafri, A. K. (2004). Dietary threonine requirement of fingerling Indian major carp, *Cirrhinus mrigala* (Hamilton). *Aquaculture Research, 35*(2), 162–170.

Ahmed, I., Ahmad, I., Malla, B. A., Shah, B. A., Wani, Z. A., & Khan, Y. M. (2022). Dietary arginine modulates growth performance, Hemato-biochemical indices, intestinal enzymes, antioxidant ability and gene expression of TOR and 4E-BP1 in rainbow trout, *Oncorhynchus mykiss* fingerlings. *Frontiers in Marine Science, 9*, 908581.

Ahmed, I., Ahmad, I., & Malla, B. A. (2024). Effects of dietary tryptophan levels on growth performance, plasma profile, intestinal antioxidant capacity and growth related genes in rainbow trout (*Oncorhynchus mykiss*) fingerlings. *Aquaculture, 585*, 740710.

Akhtar, M. S., Pal, A. K., & Sahu, N. P. (2012). Effects of dietary pyridoxine on haemato-immunological responses of *Labeo rohita* fingerlings reared at higher water temperature. *Journal of Animal Physiology and Animal Nutrition, 96*, 581–590.

Ali, B., & Rawal, Y. K. (2022). Evaluation of optimum dietary protein requirement for maximum growth of common carp (*Cyprinus carpio*) fingerlings. *Aquaculture Research, 53*(17), 5915–5924.

Aly, H. A., Abdel-Rahim, M. M., Lotfy, A. M., Abdelaty, B. S., & Sallam, G. M. (2016). The applicability of activated carbon, natural zeolites, and probiotics (EM®) and its effects on ammonia removal efficiency and fry performance of european seabass *Dicentrarchus labrax*. *Journal of Aquactic Research and Development, 7*(459), 2.

An, D., Hao, J., Wei, Y., Wang, Y., & Yu, X. (2021). Application of computer vision in fish intelligent feeding system – A review. *Aquaculture Research, 52*(2), 423–437.

Andrews, J. W., & Murai, T. (1978). Dietary niacin requirements for channel catfish. *Journal of Nutrition, 108*, 1508–1511.

Andrews, J. W., & Murai, T. (1979). Pyridoxine requirements of channel catfish. *Journal of Nutrition, 109*, 533–537.

Anwar, M. F., & Jafri, A. K. (1995). Effect of varying dietary lipid levels on growth, feed conversion, nutrient retention and carcass composition of fingerling catfish *H. fossilis*. *Asian Fisheries Science, 8*, 55–62.

Aoe, H., Masuda, I., & Takada, T. (1967). Water-soluble vitamin requirements of carp. Requirement for niacin. *Bulletin of the Japanese Society of Scientific Fisheries, 33*, 681–685.

Aoe, H., Masuda, I., Mimura, T., Saito, T., Komo, A., & Kitamura, S. (1969). Water-soluble vitamin requirements for carps. 6. Requirement for thiamin and effects of anithiamins. *Bulletin of the Japanese Society of Scientific Fisheries, 35*, 459–465.

Aoe, H., Masuda, I., Saito, T., & Komo, A. (1968). Water soluble vitamin requirements of carp. 1. Requirement for vitamin B2. *Bulletin of the Japanese Society of Scientific Fisheries, 33*, 355–360.

Aschner, J. L., & Aschner, M. (2005). Nutritional aspects of manganese homeostasis. *Molecular Aspect of Medicine, 26*, 353–362.

Ashaolu, T. J. (2020). Immune boosting functional foods and their mechanisms: A critical evaluation of probiotics and prebiotics. *Biomedicine and Pharmacotherapy, 130*, 110625.

Asimi, O. A., & Sahu, N. P. (2013). Herbs/spices as feed additive in aquaculture. *Scientific Journal of Pure and Applied Sciences, 2*(8), 284–292.

Bai, S. C., Hamidoghli, A., & Bae, J. (2022). Feed additives: An overview. In *Feed and feeding practices in aquaculture* (pp. 195–229).

Barnett, B. J., Cho, C. Y., & Slinger, S. J. (1982a). Relative biopotency of ergocalciferol and cholecalciferol and the role of and requirement for vitamin D in rainbow trout (*Salmo gairdneri*). *Journal of Nutrition, 112*, 2011–2019.

Barnett, B. J., Jones, G., Cho, C. Y., & Slinger, S. J. (1982b). The biological activity of 25-hydroxycholecalciferol and 1,25-dehydroxycholecalciferol for rainbow trout (*Salmo gairdneri*). *Journal of Nutrition, 112*, 2020–2026.

Berntssen, M. H. G., Ørnsrud, R., Rasinger, J., Søfteland, L., Lock, E. J., Kolås, K., et al. (2016). Dietary vitamin a supplementationameliorates the effects of ply-aromatic hydrocarbons in Atlantic salmon (*Salmo salar*). *Aquatic Toxicology, 175*, 171–183.

Bronner, F. (2009). Recent developments in intestinal calcium absorption. *Nutrition Reviews, 67*, 109–113.

Bury, N. R., Walker, P. A., & Glover, C. N. (2003). Nutritive metal uptake in teleost fish. *Journal of Experimental Biology, 206*, 11–23.

Bury, N. R., Boyle, D., & Cooper, C. A. (2012). Iron in homeostasis and toxicology of essential metals. In C. M. Wood, A. P. Farrell, & C. J. Brauner (Eds.), *Fish physiology* (Vol. 31A, pp. 201–251). Academic Press.

Calderón-Ospina, C. A., & Nava-Mesa, M. O. (2020). B vitamins in the nervous system: Current knowledge of the biochemical modes of action and synergies of thiamine, pyridoxine, and cobalamin. *CNS Neuroscience & Therapeutics, 26*, 5–13.

Cao, L., Wang, W., Yang, C., Yang, Y., Diana, J., Yakupitiyage, A., et al. (2007). Application of microbial phytase in fish feed. *Enzyme and Microbial Technology, 40*(4), 497–507.

Chen, C., Zhu, W., Wu, F., Liu, M., Tan, Q., Han, D., Xie, S., & Liang, X. (2016). Quantifying the dietary potassium requirement of subadult grass carp (*Ctenopharyngodon idella*). *Aquaculture Nutrition, 22*, 541–549.

Chin, Y. W., Balunas, M. J., Chai, H. B., & Kinghorn, A. D. (2006). Drug discovery from natural sources. *American Association of Pharmaceutical Scientists, 28*, 239–253.

Combs, G. F. J., & McClung, J. P. (2017). *The vitamins: Fundamental aspects in nutrition and health* (5th ed.). Academic Press.

Cong-mei, X., Hai-rui, Y., Zhang, Q., Chen, B.-B., Li, L.-Y., Qiu, X. Y., Qi, T., Liu, J. Q., & Shan, L. L. (2021). Dietary phosphorus requirement of coho salmon (*Oncorhynchus kisutch*) alevins cultured in freshwater. *Aquaculture Nutrition, 27*, 2427–2435.

Cowey, C. B., Knox, D., Adron, J. W., George, S., & Pirie, B. (1977). The production of renal calcinosis by magnesium deficiency in rainbow trout (*Salmo gairdneri*). *British Journal of Nutrition, 38*, 127–135.

Darodes de Tailly, J. B., Keitel, J., Owen, M. A., Alcaraz-Calero, J. M., Alexander, M. E., & Sloman, K. A. (2021). Monitoring methods of feeding behaviour to answer key questions in penaeid shrimp feeding. *Reviews in Aquaculture, 13*(4), 1828–1843.

Debnath, D., Pal, A. K., Sahu, N. P., Yengkokpam, S., & Baruah, K. (2012). Protein requirement of *Labeo rohita* fingerlings based on growth and ammonia excretion. *Journal of the Inland Fisheries Society of India, 44*(2), 12–20.

Deng, D. F., & Wilson, R. P. (2003). Dietary riboflavin requirement of juvenile sunshine bass (*Morone chrysops*♀ × *Morone saxatilis*♂). *Aquaculture, 218*, 695–701.

Devi, G., Harikrishnan, R., Paray, B. A., Al-Sadoon, M. K., Hoseinifar, S. H., & Balasundaram, C. (2019). Effect of symbiotic supplemented diet on innate-adaptive immune response, cytokine gene regulation and antioxidant property in *Labeo rohita* against *Aeromonas hydrophila*. *Fish Shellfish Immunology, 89*, 687–700.

Du, Z., Cui, M., Xu, X., Bai, Z., Han, J., Li, W., Yang, J., Liu, X., Wang, C., & Li, D. (2024). Harnessing multimodal data fusion to advance accurate identification of fish feeding intensity. *Biosystems Engineering, 246*, 135–149.

Dupree, H. K. (1970). *Dietary requirement of vitamin a acetate and beta carotene* (pp. 148–150). U.S. Dept. Resource Publication 88, U.S. Bureau of Sport Fisheries and Wildlife.

Ebru, Y., & Cengiz, K. (2016). Feed additives in aquafeeds, Lucrări Științifice-Universitatea de Științe Agricole și Medicină Veterinară. *Seria Zootehnie, 66*, 155–160.

El-Sayed, A.-F. M., & Izquierdo, M. (2022). The importance of vitamin E for farmed fish - A review. *Reviews in Aquaculture, 14*, 688–703.

Encarnação, P. (2016). Functional feed additives in aquaculture feeds. In *Aquafeed formulation* (pp. 217–237). Academic Press.

Espe, M., Vikeså, V., Helgøy Thomsen, T., Adam, A. C., Saito, T., & Skjærven, K. H. (2020). Atlantic salmon fed a nutrient package of surplus methionine, vitamin B12, folic acid and vitamin B6 improved growth and reduced the relative liver size, but when in excess growth reduced. *Aquaculture Nutrition, 26*, 477–489.

Faizan, M., Stubhaug, I., Menoyo, D., Esatbeyoglu, T., Wagner, A., & Struksnæs, G. (2013). Dietary alpha-tocopherol affects tissue vitamin E and malondialdehyde levels but does not change antioxidant enzymes and fatty acid composition in farmed Atlantic salmon (*Salmo salar* L.). *International Journal for Vitamin and Nutrition Research, 83*, 238–245.

Fan, Z., Wu, D., Li, J., Zhang, Y., Xu, Q., & Wang, L. (2020). Dietary protein requirement for large size Songpu mirror carp (*Cyprinus carpio* Songpu). *Aquaculture Nutrition, 26*(5), 1748–1759.

Fantatto, R. R., Mota, J., Ligeiro, C., Vieira, I., Guilgur, L. G., Santos, M., & Murta, D. (2024). Exploring sustainable alternatives in aquaculture feeding: The role of insects. *Aquaculture Reports, 37*, 102228.

Farhat, & Khan, M. A. (2012). Effects of dietary arginine levels on growth,feed conversion,protein productive value and carcass composition of stinging catfish fingerling *Heteropneustes fossilis* (Bloch). *Aquaculture International, 20*, 935–950.

Farhat, & Khan, M. A. (2013a). Effects of varying levels of dietary L-histidine on growth, feed conversion, protein gain, histidine retention, hematological and body composition in fingerlings stinging catfish, *Heteropneustes fossilis* (Bloch). *Aquaculture*, 130–138.

Farhat, & Khan, M. A. (2013b). Dietary L-lysine requirement of fingerling stinging catfish, *Heteropneustes fossilis* (Bloch) for optimizing growth, feed conversion, protein and lysine deposition. *Aquaculture Research, 44*, 523–533.

Farhat, & Khan, M. A. (2014). Response of fingerling stinging catfish, *Heteropneustes fossilis* (Bloch) to varying levels of dietary l-leucine in relation to growth, feed conversion, protein utilization, leucine retention and blood parameters. *Aquaculture Nutrition, 20*, 291–302.

Farhat, & Khan, M. A. (2017). Growth, feed conversion and body composition of fingerling stinging catfish *Heteropneustes fossilis* (Bloch) fed varying levels of dietary L-threonine. *Aquaculture Research, 48*, 2355–2368.

Fefana. (2014). *Organic acids in animal nutrition* (p. 97). Fefana Publication.

Feng, S., Yang, X., Liu, Y., Zhao, Z., Liu, J., Yan, Y., & Zhou, C. (2022). Fish feeding intensity quantification using machine vision and a lightweight 3D ResNet-GloRe network. *Aquacultural Engineering, 98*, 102244.

Gao, X. Q., Wang, X., Wang, X. Y., Li, H. X., Xu, L., Huang, B., et al. (2022). Effects of different feeding frequencies on the growth, plasma biochemical parameters, stress status, and gastric evacuation of juvenile tiger puffer fish (*Takifugu rubripes*). *Aquaculture, 548*, 737718.

Grisdale-Helland, B., Helland, S. J., & Gatlin, D. M., III. (2008). The effects of dietary supplementation with mannanoligosaccharide, fructooligosaccharide or galactooligosaccharide on the growth and feed utilization of Atlantic salmon *(Salmo salar)*. *Aquaculture, 283*(1–4), 163–167.

Gunther, K. D., & Meyer-Burgdorff, K. H. (1990). Studies on biotin supply to mirror carp (*Cyprinus carpio* L.). *Aquaculture, 84*, 49–60.

Guo, H. Y., Dong, X. Y., Zhang, X. M., Zhang, P. D., & Li, W. T. (2017). Survival, growth and physiological responses of juvenile Japanese flounder (*Paralichthys olivaceus*, Temminck & Schlegel, 1846) exposed to different dissolved oxygen concentrations and stocking densities. *Journal of Applied Ichthyology, 33*(4), 731–739.

Halver, J. E. (2002). The vitamins. In J. E. Halver & R. W. Hardy (Eds.), *Fish nutrition* (3rd ed., pp. 61–141). Academic Press.

Hameid, N. A. H., Abidi, S. F., & Khan, M. A. (2012). Dietary vitamin E requirement for maximizing the growth, conversion efficiency, biochemical composition and haematological status of fingerling *Channa punctatus*. *Aquaculture Research, 43*, 226–238.

Hameid-Abdel, N. A., Zehra, S., & Khan, M. A. (2017). Dietary copper requirement of fingerling *Channa punctatus* (Bloch) based on growth, feed conversion, blood parameters and whole body copper concentration. *Aquaculture Research, 48*, 2787–2797.

Han, D., Liu, H., Liu, M., Xiao, X., Zhu, X., Yang, Y., & Xie, S. (2012). Effect of dietary magnesium supplementation on the growth performance of juvenile gibel carp, *Carassius auratus gibelio*. *Aquaculture Nutrition, 18*, 512–520.

Hansen, A. C., Waagbø, R., & Hemre, G. I. (2015). New B vitamin recommendations in fish when fed plant-based diets. *Aquaculture Nutrition, 21*, 507–527.

Harder, A. M., Ardren, W. R., Evans, A. N., Futia, M. H., Kraft, C. E., Marsden, J. E., Richter, C. A., Rinchard, J., Tillitt, D. E., & Christie, M. R. (2018). Thiamine deficiency in fishes: Causes, consequences, and potential solutions. *Reviews in Fish Biology and Fisheries, 28*, 865–886.

Hardy, R. W., & Kaushik, S. J. (Eds.). (2021). *Fish nutrition*. Academic Press.

Hassan, A., Ahmed, I., & Wani, G. B. (2022). Effect of Supplementation of vitamin a on growth, Haemato-biochemical composition, and antioxidant ability in *Cyprinus carpio* var. communis. *Aquaculture Nutrition*, 8446092.

He, M., Wang, K., Liang, X., Fang, J., Geng, Y., & Chen, Z. (2017). Effects of dietary vitamin E on growth performance as well as intestinal structure and function of channel catfish (*Ictalurus punctatus*, Rafinesque 1818). *Experimental and Therapeutic Medicine, 14*, 5703–5710.

Hidalgo, M. C., Skalli, A., Abellán, E., Arizcun, M., & Cardenete, G. (2006). Dietary intake of probiotics and maslinic acid in juvenile dentex (*Dentex dentex* L.): Effects on growth performance, survival and liver proteolytic activities. *Aquaculture Nutrition, 12*(4), 256–266.

Hideya, T., Nobuo, S., Chiyo, T., Mika, I., Toshio, Y., Akiyoshi, T., & Tatsuya, S. (2008). Prolactin inhibits osteoclastic activity in the goldfish scale: A novel direct action of prolactin in teleosts. *Zoological Science, 25*, 739–745.

Hilton, J. W., Hodson, P. V., & Slinger, S. J. (1980). Some effects of vitamin E and selenium deprivation on tissue enzyme levels and indices of tissue peroxidation in rainbow trout (*Salmo gairdneri*). *Journal of Nutrition, 110*, 2527–2535.

Hossain, M. A., & Yoshimatsu, T. (2014). Dietary calcium requirements of fishes. *Aquaculture Nutrition, 20*, 1–11.

Hossain, M. A., Pandey, A., & Satoh, S. (2007). Effects of organic acids on growth and phosphorus utilization in red sea bream *Pagrus major*. *Fisheries Science, 73*(6), 1309–1317.

Hu, X., Liu, Y., Zhao, Z., Liu, J., Yang, X., Sun, C., et al. (2021). Real-time detection of uneaten feed pellets in underwater images for aquaculture using an improved YOLO-V4 network. *Computers and Electronics in Agriculture, 185*, 106135.

Huang, J. L., Tian, X. W., Yang, H., & Liu, Y. (2010). Effects of dietary riboflavin levels on antioxidant defense of the juvenile grouper (*Epinephelus coioides*). *Fish Physiology and Biochemistry, 36*, 55–62.

Humer, E., Schwarz, C., & Schedle, K. (2015). Phytate in pig and poultry nutrition. *Journal of Animal Physiology and Animal Nutrition, 99*, 605–625.

Jafri, A. K. (1998). Effect of dietary carbohydrate-to-lipid ratio on growth and body composition of walking catfish (*Clarias batrachus*). *Aquaculture, 161*(1–4), 159–168.

Jafri, A. K., Anwar, M. F., Usmani, N., & Alvi, A. S. (1995). Influence of dietary lipid levels on the growth and body composition of fingerlings of an Indian major carp, *C. mrigala* (Hamilton). *Journal of Aquaculture of Tropics, 10*, 151–157.

Janocha, A., Milczarek, A., Kosmalski, M., Gajownik-Mućka, P., & Radzikowski, D. (2022). Effect of feed additives supplementation on the growth performance, gastrointestinal tract characteristics, and carcass composition in Turkey hens. *Animals (Basel), 12*(24), 3464.

Janz, D. M. (2012). Selenium. In C. M. Wood, A. M. Farrell, & C. J. Brauner (Eds.), *Fish physiology: Homeostasis and toxicology of essential metals* (pp. 329–374). Elsevier/Academic Press.

Jezierska, B., Ługowska, K., & Witeska, M. (2009). The effects of heavy metals on embryonic development of fish: A review. *Fish Physiology and Biochemistry, 35*, 625–640.

Jiang, M., Wang, W., Wen, H., Wu, F., Zhao, Z., Liu, A., & Liu, W. (2007). Effects of dietary vitamin K3 on growth, carcass composition and blood coagulation time for grass carp fingerling (*Ctenopharyngodon idellus*). *Freshwater Fish, 37*, 61–64.

Jiang, M., Huang, F., Wen, H., Yang, C., Wu, F., Liu, W., & Tian, J. (2014). Dietary niacin requirement of GIFT tilapia, Oreochromis niloticus, reared in freshwater. *Journal of World Aquaculture Society, 45*, 333–341.

Jin, Y., Tian, L., Zeng, S., Xie, S., Yang, H., Liang, G., & Liu, Y. (2013). Dietary lipid requirement on non-specific immune responses in juvenile grass carp (*Ctenopharyngodon idella*). *Fish & Shellfish Immunology, 34*(5), 1202–1208.

Johnson, A., Carew, E., & Sloman, K. A. (2007). The effects of copper on the morphological and functional development of zebrafish embryos. *Aquatic Toxicology, 84*, 431–438.

Juan, C. J., Zhang WenBing, Z. W., Xu Wei, X. W., Mai KangSen, M. K., & Sun RuiJian, S. R. (2015). Dietary selenium requirement of juvenile large yellow croaker *Larimichthys croceus*. *Acta Hydrobiologica Sinica, 39*, 241–249.

Kainat, S., Shah, S. Z. H., Hussain, M., Nasar, M. F., Fatima, M., Ashraf, S., Aftab, K., Latheef, A., & Hassan, I. (2023). Dietary potassium requirement of *Cirrhinus mrigala* juveniles based on growth performance, gill Na + /K + ATPase activity and antioxidant enzymes. *Research Square*, 1–18.

Khaled, M. (2015). Effect of organic acid salt supplementation on growth performance and feed utilization in practical diets of hybrid tilapia (♀ *O. niloticus x*♂ *O. aureus*) fingerlings. *Egyptian Journal of Animal Production, 52*, 81–88.

Khan, M. A., & Abidi, S. F. (2007a). Total aromatic amino acid requirement of Indian major carp *Labeo rohita* (Hamilton). *Aquaculture, 267*, 111–118.

Khan, M. A., & Abidi, S. F. (2007b). Dietary isoleucine requirement of fingerling Indian major carp, *Labeo rohita* (Hamilton). *Aquaculture Nutrition, 13*(6), 424–430.

Khan, M. A., & Abidi, S. F. (2011a). Dietary arginine requirement of *Heteropneustes fossilis* fry (Bloch) based on growth, nutrient retention and haematological parameters. *Aquaculture Nutrition, 17*, 418–428.
Khan, M. A., & Abidi, S. F. (2011b). Effect of dietary L-lysine levels on growth, feed conversion, lysine retention efficiency and haematological indices of *Heteropneustes fossilis* (Bloch) fry. *Aquaculture Nutrition, 17*, e657–e667.
Khan, M. A., & Abidi, S. F. (2014a). Dietary histidine requirement of Singhi, *Heteropneustes fossilis* fry (Bloch). *Aquaculture Research, 45*, 1341–1345.
Khan, M. A., & Abidi, S. F. (2014b). Dietary L-tryptophan requirement of stinging catfish fingerling *Heteropneustes fossilis* (Bloch). *Aquaculture Research, 45*, 1225–1235.
Khan, M. A., & Jafri, A. K. (1991). Dietary protein requirement of two size classes of the Indian major carp, *Catla catla* Hamilton.
Khan, Y. M., & Khan, M. A. (2019). Dietary biotin requirement of fingerling *Catla catla* (Hamilton) based on growth, feed conversion efficiency, and liver biotin concentration. *Journal of the World Aquaculture Society, 50*, 674–683.
Khan, Y. M., & Khan, M. A. (2021). Dietary niacin requirement of fingerling Indian major carp *Catla catla* (Hamilton). *Aquaculture Nutrition, 27*, 1482–1493.
Khan, Y. M., & Khan, M. A. (2022). Dietary thiamin requirement of fingerling major carp *Catla catla* (Hamilton). *Journal of Animal Physiology and Animal Nurtition, 106*, 939–946.
Kim, K. D., Lim, S. G., Kang, Y. J., Kim, K. W., & Son, M. H. (2012). Effects of dietary protein and lipid levels on growth and body composition of juvenile far eastern catfish *Silurus asotus*. *Asian-Australasian Journal of Animal Sciences, 25*(3), 369.
Krossoy, C., Waagbo, R., & Ornsrud, R. (2011). Vitamin K in fish nutrition. *Aquaculture Nutrition, 11*, 585–595.
Lagunin, A., Filimonov, D., & Poroikov, V. (2010). Multi-targeted natural products evaluation based on biological activity prediction with PASS. *Current Pharmaceutical Design, 16*, 1703–1717.
Lall, S. P. (2002). The minerals. In J. E. Halver & R. W. Hardy (Eds.), *Fish nutrition* (3rd ed., pp. 259–308). Elsevier/Academic Press.
Lall, S. P., & Dumas, A. (2022). Nutritional requirements of cultured fish: Formulating nutritionally adequate feeds. In *Feed and feeding practices in aquaculture* (pp. 65–132). Woodhead Publishing.
Lall, S. P., & Kaushik, S. J. (2021). Nutrition and metabolism of minerals in fish. *Animals, 11*, 2711.
Lara-Flores, M., Olvera-Novoa, M. A., Guzmán-Méndez, B. E., & López-Madrid, W. (2003). Use of the bacteria *Streptococcus faecium* and *lactobacillus acidophilus*, and the yeast *Saccharomyces cerevisiae* as growth promoters in Nile tilapia (*Oreochromis niloticus*). *Aquaculture, 216*(1–4), 193–201.
Lee, N., Sowa, H., Hinoi, E., et al. (2007). Endocrine regulation of energy metabolism by the skeleton. *Cell, 130*, 456–469.
Lee, S., Nambi, R., Won, S., Katya, K., & Bai, S. C. (2016). Dietary selenium requirement and toxicity levels in juvenile Nile tilapia, *Oreochromis niloticus*. *Aquaculture, 464*, 153–158.
Lemos, D., & Tacon, A. G. J. (2017). Use of phytases in fish and shrimp feeds: A review. *Reviews in Aquacuture*, 266–282.
Li, L., & Yang, X. (2018). The essential element manganese, oxidative stress, and metabolic diseases: Links and interactions. *Oxidative Medicine and Cellular Longevity, 5*, 7580707.
Li, W., Zhou, X. Q., Feng, L., Liu, Y., & Jiang, J. (2010). Effect of dietary riboflavin on growth, feed utilization, body composition and intestinal enzyme activities of juvenile Jian carp (*Cyprinus carpio*). *Aquaculture Nutrition, 16*, 137–143.
Li, J., Liang, X. F., Tan, Q., et al. (2014). Effects of vitamin E on growth performance and antioxidant status in juvenile grass carp *Ctenopharyngodon idellus*. *Aquaculture, 430*, 21–27.
Li, S. Q., Feng, L., Jiang, W. D., Liu, Y., Wu, P., Zhao, J., Kuang, S. Y., Jiang, J., Tang, L., Tang, W. N., Zhang, Y. A., & Zhou, X. Q. (2015). Deficiency of dietary niacin decreases digestion and absorption capacities via declining the digestive and brush border enzyme activities and

downregulating those enzyme gene transcriptions related to TOR pathway of the hepatopancreas and intestine in young grass carp (*Ctenopharyngodon idella*). *Aquaculture Nutrition, 22*, 1267–1282.

Li, S., Li, Z., Zhang, J., Sang, C., & Chen, N. (2019). The impacts of dietary carbohydrate levels on growth performance, feed utilization, glycogen accumulation and hepatic glucose metabolism in hybrid grouper (*Epinephelus fuscoguttatus*♀× *E. lanceolatus*♂). *Aquaculture, 512*, 734351.

Li, D., Wang, Z., Wu, S., Miao, Z., Du, L., & Duan, Y. (2020). Automatic recognition methods of fish feeding behavior in aquaculture: A review. *Aquaculture, 528*, 735508.

Li, M. H., Robinson, E. H., & Davis, D. A. (2022). Complete feeds – Intensive systems. In *Feed and feeding practices in aquaculture* (pp. 133–149). Woodhead Publishing.

Li, Y., Li, W., Luo, L., Ren, Y., Xing, W., Xu, G., et al. (2023). Dietary lipid levels affect growth performance, lipid metabolism, antioxidant and immune status of Amur sturgeon, *Acipenser schrenckii*. *Aquaculture Reports, 33*, 101796.

Liang, J. J., Tian, L. X., Liu, Y. J., Yang, H. J., & Liang, G. Y. (2012). Dietary magnesium requirement and effects on growth and tissue magnesium content of juvenile grass carp (*Ctenopharyngodon idella*). *Aquaculture Nutrition, 18*, 56–64.

Liang, J. J., Wang, S., Han, B., Tian, L. X., Yang, H. J., & Liu, Y. J. (2015). Dietary manganese requirement of juvenile grass carp (*Ctenopharyngodon idella* Val.) based on growth and tissue manganese concentration. *Aquaculture Research, 46*, 2991–2998.

Liao, P. H., Chu, S. H., Tu, T. Y., Wang, X. H., Lin, A. Y. C., & Chen, P. J. (2014). Persistent endocrine disruption effects in medaka fish with early life-stage exposure to a triazole-containing aromatase inhibitor (letrozole). *Journal of Hazardous Materials, 277*, 141–149.

Lin, Y. H., & Shiau, S. Y. (2005a). Dietary vitamin E requirement of grouper, *Epinephelus malabaricus*, at two lipid levels, and their effects on immune responses. *Aquaculture, 248*(1–4), 235–244.

Lin, Y. H., & Shiau, S. Y. (2005b). Dietary lipid requirement of grouper, *Epinephelus malabaricus*, and effects on immune response. *Aquaculture, 225*, 243–250.

Lin, S., Mai, K., & Tan, B. (2007). Effects of exogenous enzyme supplementation in diets on growth and feed utilization in tilapia, *Oreochromis niloticus x O. aureus*. *Aquaculture Research, 38*(15), 1645–1653.

Lin, Y. H., Shie, Y. Y., & Shiau, S. Y. (2008). Dietary copper requirements of juvenile grouper, Epinephelus malabaricus. *Aquaculture, 274*, 161–165.

Lin, Y. H., Ku, C. Y., & Shiau, S. Y. (2013). Estimation of dietary magnesium requirements of juvenile tilapia, *Oreochromis niloticus*×*O. aureus*, reared in freshwater and seawater. *Aquaculture, 380-383*, 47–51.

Little, D. C., Young, J. A., Zhang, W., Newton, R. W., Al Mamun, A., & Murray, F. J. (2018). Sustainable intensification of aquaculture value chains between Asia and Europe: A framework for understanding impacts and challenges. *Aquaculture, 493*, 338–354.

Liu, Y., Feng, L., Jiang, J., Liu, Y., & Zhou, X. Q. (2009). Effects of dietary protein levels on the growth performance, digestive capacity and amino acid metabolism of juvenile Jian carp (*Cyprinus carpio* var. Jian). *Aquaculture Research, 40*(9), 1073–1082.

Liu, K., Wang, X., Ai, Q., Mai, K., & Zhang, W. (2010). Dietary selenium requirement for juvenile cobia, *Rachycentron canadum* L. *Aquaculture Research, 41*, e594–e601.

Liu, K., Ai, Q. H., Mai, K. S., Zhang, W. B., Zhang, L., & Zheng, S. X. (2013). Dietary manganese requirement for juvenile cobia, *Rachycentron canadum* L. *Aquaculture Nutrition, 19*, 461–467.

Liu, A., et al. (2022). Vitamin nutrition in salmonid aquaculture: From avoiding deficiencies to enhancing functionalities. *Aquaculture, 561*(2022), 738654.

Lovell, R. T., & Li, Y. P. (1978). Essentiality of vitamin D in diets of channel catfish (*lctalurus punctatus*). *Transactions of the American Fisheries Society, 107*, 809–811.

Luo, Z., Tan, X. Y., Zheng, J. L., Chen, Q. L., & Liu, C. X. (2011). Quantitative dietary zinc requirement of juvenile yellow catfish *Pelteobagrus fulvidraco*, and effects on hepatic intermediary metabolism and antioxidant responses. *Aquaculture, 319*, 150–155.

Mai, K., Waagbø, R., Zhou, X. Q., Ai, Q. H., & Feng, L. (2022). Chapter 3 – Vitamins. In R. W. Hardy & S. J. Kaushik (Eds.), *Fish nutrition* (pp. 57–179). Academic Press.

McCall, K. A., Huang, C.-C., & Fierke, C. A. (2000). Function and mechanism of zinc metalloenzymes. *The Journal of Nutrition, 2000*(130), 1437S–1446S.

Meng, Y., Qian, K., Ma, R., Liu, X., Han, B., Wu, J., et al. (2019). Effects of dietary lipid levels on sub-adult triploid rainbow trout (*Oncorhynchus mykiss*): 1. Growth performance, digestive ability, health status and expression of growth-related genes. *Aquaculture, 513*, 734394.

Merrifield, D. L., Bradley, G., Baker, R. T. M., & Davies, S. J. (2010). Probiotic applications for rainbow trout (*Oncorhynchus mykiss Walbaum*) II. Effects on growth performance, feed utilization, intestinal microbiota and related health criteria postantibiotic treatment. *Aquaculture Nutrition, 16*(5), 496–503.

Mir, I. N., Srivastava, P. P., Bhat, I. A., Jaffar, Y. D., Sushila, N., Sardar, P., et al. (2020). Optimal dietary lipid and protein level for growth and survival of catfish *Clarias magur* larvae. *Aquaculture, 520*, 734678.

Mishra, S., & Mukhopadhyay, P. K. (1996). Ascorbic acid requirement of catfish fry *Clarias batrachus* (Linn.). *Indian Journal Of Fisheries, 43*(2), 157–162.

Misra, C. K., Das, B. K., Mukherjee, S. C., & Pradhan, J. (2007). Effects of dietary vitamin C on immunity, growth and survival of Indian major carp *Labeo rohita* fingerlings. *Aquaculture Nutrition, 13*, 35–44.

Mohamed, J., & Ibrahim, A. (2001). Quantifying the dietary niacin requirement of the Indian catfish, *Heteropneustes fossilis* (Bloch) fingerlings. *Aquaculture Research, 32*, 157–162.

Morken, T., Kraugerud, O. F., Barrows, F. T., Sørensen, M., Storebakken, T., & Øverland, M. (2011). Sodium diformate and extrusion temperature affect nutrient digestibility and physical quality of diets with fish meal and barley protein concentrate for rainbow trout (*Oncorhynchus mykiss*). *Aquaculture, 317*, 138–145.

Munguti, J. M., Musa, S., Orina, P. S., Kyule, D. N., Opiyo, M. A., Charo-Karisa, H., & Ogello, E. O. (2014). An overview of current status of Kenyan fish feed industry and feed management practices, challenges and opportunities.

Murai, T., & Andrews, J. W. (1978). Riboflavin requirement of channel catfish fingerlings. *Journal of Nutrition, 108*, 176–180.

Musharraf, M., & Khan, M. A. (2018). Dietary magnesium requirement for fingerlings of Rohu (*Labeo rohita*). *Aquaculture, 496*, 96–104.

Musharraf, M., & Khan, M. A. (2019a). Dietary phosphorus requirement of fingerling Indian major carp, *Labeo rohita* (Hamilton). *Journal of the World Aquaculture Society, 50*, 469–484.

Musharraf, M., & Khan, M. A. (2019b). Dietary potassium requirement of fingerling Indian major carp, *Labeo rohita* (Hamilton), based on growth parameters, gill Na+-K+ ATPase activity, tissue mineralization and antioxidant activities. *Aquaculture Nutrition, 25*, 271–280.

Musharraf, M., & Khan, M. A. (2019c). Dietary zinc requirement of fingerling Indian major carp, *Labeo rohita* (Hamilton). *Aquaculture, 503*, 489–498.

Musharraf, M., & Khan, M. A. (2019d). Requirement of fingerling Indian major carp, Labeo rohita (Hamilton) for dietary iron based on growth, whole body composition, haematological parameters, tissue iron concentration and serum antioxidant status. *Aquaculture, 504*, 148–157.

Musharraf, M., & Khan, M. A. (2020). Dietary calcium requirement of fingerling Indian major carp, *Labeo rohita* (Hamilton), based on growth performance, tissue mineralization, whole body, and serum biochemical composition. *Aquaculture International, 28*, 1125–1139.

Musharraf, M., & Khan, M. A. (2021). Dietary manganese requirement of fingerling Indian major carp, *Labeo rohita* (Hamilton) estimated by growth, tissue manganese concentration and hepatic manganese-superoxide dismutase activity. *Aquaculture, 540*, 736734.

Musharraf, M., & Khan, M. A. (2022). Estimation of dietary copper requirement of fingerling Indian major carp, *Labeo rohita* (Hamilton). *Aquaculture, 549*, 737742.

Nates, S. F. (2016). In S. F. M. Nates (Ed.), *Feed additives. Aquafeed formulation*. Academic Press.

National Research Council. (2011). *Nutrient requirements of fish and shrimp*. The National Academies Press.

Ng, W. K., Koh, C. B., Sudesh, K., & Siti-Zahrah, A. (2009). Effects of dietary organic acids on growth, nutrient digestibility and gut microflora of red hybrid tilapia, *Oreochromis sp.*, and subsequent survival during a challenge test with *Streptococcus agalactiae*. *Aquaculture Research, 40*(13).

Nunes, A. J., Sá, M. V., Browdy, C. L., & Vazquez-Anon, M. (2014). Practical supplementation of shrimp and fish feeds with crystalline amino acids. *Aquaculture, 431*, 20–27.

Núñez-Acuña, G., et al. (2018). The Atlantic salmon (*Salmo salar*) antimicrobial peptide cathelicidin-2 is a molecular host-associated cue for the salmon louse (*Lepeophtheirus salmonis*). *Scientific Reports, 8*(13), 738.

Nya, E. J., Dawood, Z., & Austin, B. (2010). The garlic component, allicin, prevents disease caused by *Aeromonas hydrophila* in rainbow trout, *Oncorhynchus mykiss* (Walbaum). *Journal of Fish Diseases, 33*, 293–300.

Ogino, C. (1965). B vitamin requirements of carp, Cyprinus carpio: I. Deficiency symptoms and requirements of vitamin B6. *Bulletin of the Japanese Society of Scientific Fisheries, 31*, 546–551.

Olson, R. (1999). Vitamin K. In M. Shils, J. Olson, M. Shike, & A. Ross (Eds.), *Modern nutrition in health and disease* (9th ed., pp. 363–380). Lippincott Williams & Wilkins.

Omoregie, E., Igoche, L., Ojobe, T. O., Absalom, K. V., & Onusiriuka, B. C. (2009). Effect of varying levels of sweet potato (*Ipomea Batatas*) peels on growth, feed utilization and some biochemical responses of the cichlid (*Oreochromis niloticus*). *African Journal of Food, Agriculture, Nutrition and Development, 9*(2), 700–712.

Paul, M., Sardar, P., Sahu, N. P., Jana, P., Deo, A. D., Harikrishna, V., et al. (2022). Effect of dietary lipid level on growth performance, body composition, and physiometabolic responses of genetically improved farmed tilapia (GIFT) juveniles reared in inland ground saline water. *Aquaculture Nutrition, 2022*(1), 5345479.

Peng, X., Shang, G., Wang, W., Chen, X., Lou, Q., Zhai, G., & Yin, Z. (2017). Fatty acid oxidation in zebrafish adipose tissue is promoted by 1α,25(OH)2D3. *Cell Reports, 19*, 1444–1455.

Piferrer, F. (2001). Endocrine sex control strategies for the feminization of teleost fish. *Aquaculture, 197*(1–4), 229–281.

Prabhu, P. A. J., Schrama, J. W., & Kaushik, S. J. (2016). Mineral requirements of fish: A systematic review. *Reviews in Aquaculture, 8*, 172–219.

Qiang, J., Wasipe, A., He, J., et al. (2019). Dietary vitamin E deficiency inhibits fat metabolism, antioxidant capacity, and immune regulation of inflammatory response in genetically improved farmed tilapia (GIFT, *Oreochromis niloticus*) fingerlings following *Streptococcus iniae* infection. *Fish & Shellfish Immunology, 92*, 395–404.

Qiao, Y. G., Tan, B. P., Mai, K. S., Ai, Q. H., Zhang, W. B., & Xu, W. (2013). Evaluation of iron methionine and iron sulphate as dietary iron sources for juvenile cobia (*Rachycentron canadum*). *Aquaculture Nutrition, 19*, 721–730.

Reis, J., Weldon, A., Ito, P., Stites, W., Rhodes, M., & Davis, D. A. (2021). Automated feeding systems for shrimp: Effects of feeding schedules and passive feedback feeding systems. *Aquaculture, 541*, 736800.

Ren, M., Liang, H., Ge, X., Chen, X., Mi, H., & Ji, K. (2021). Effects of dietary carbohydrate-to-lipid ratio on growth performance, plasma parameters and hepatic antioxidant status in ide (Leuciscus idus, Linnaeus, 1758). *Aquaculture Reports, 19*, 100618.

Sau, S. K., Paul, B. N., Mohanta, K. N., & Mohanty, S. N. (2004). Dietary vitamin E requirement, fish performance and carcass composition of rohu (*Labeo rohita*) fry. *Aquaculture, 240*, 359–368.

Sayed, S. F., & Ahmed, I. (2022). Effects of dietary phenylalanine: Tyrosine ratio on growth, DNA/RNA, serum biochemistry, digestive enzyme activities and physiological responses of Heteropneustes fossilis. *Aquaculture Studies, 23*(1).

Shaik Mohamed, J. (2001a). Dietary pyridoxine requirement of the Indian catfish, Heteropneustes fossilis. *Aquaculture, 194*, 327–335.

Shaik Mohamed, J. (2001b). Dietary biotin requirement determined for Indian catfish, *Heteropneustes fossilis* (Bloch) fingerlings. *Aquaculture Research, 32*, 709–716.

Shaik Mohamed, J., Ravisankar, B., & Ibrahim, A. (2000). Quantifying the dietary biotin requirement of the catfish, *Clarias batrachus*. *Aquaculture International, 8*, 9–18.

Shaik Mohamed, J., Sivaram, V., Christopher Roy, T. S., Peter Marian, M., Murugadass, S., & Raffiq Hussain, M. (2003). Dietary vitamin a requirement of juvenile greasy grouper (*Epinephelus tauvina*). *Aquaculture, 219*, 693–701.

Shao, Z. J. (2001). Aquaculture pharmaceuticals and biologicals: Current perspectives and future possibilities. *Advanced Drug Delivery Reviews, 50*(3), 229–243.

Sharf, Y., & Khan, M. A. (2020). Effect of dietary isoleucine level on growth, protein retention efficiency, haematological parameter, lysozyme activity and serum antioxidant status of fingerling *Channa punctatus* (Bloch). *Aquaculture Nutrition, 26*(3), 908–920.

Sharf, Y., & Khan, M. A. (2021). Requirement of fingerling *Channa punctatus* (Bloch) for dietary lysine based on growth, feed conversion and lysine retention efficiency, RNA/DNA ratio, haematological parameters and serum antioxidant activity. *Aquaculture Nutrition, 27*, 140–150.

Sharf, Y., & Khan, M. A. (2022a). Dietary threonine requirement of fingerling *Channa punctatus* (Bloch) based on growth, feed conversion, protein retention efficiency, hematological parameters, and biochemical composition. *Aquaculture, 560*, 738504.

Sharf, Y., & Khan, M. A. (2022b). Dietary histidine requirement for fingerlings of *Channa punctatus* (Bloch). *Aquaculture, 561*, 738732.

Sharf, Y., & Khan, M. A. (2023a). Dietary phenylalanine requirement and tyrosine replacement value for phenylalanine for fingerling *Channa punctatus* based on growth parameters, hematology and antioxidant status. *Aquaculture, 572*, 739503.

Sharf, Y., & Khan, M. A. (2023b). Dietary tryptophan requirement of fingerling *Channa punctatus* (Bloch) based on growth, hematological parameters, intestinal enzymes, non-specific immune response, and antioxidant capacity. *Aquaculture, 562*, 738745.

Sharf, Y., & Khan, M. A. (2023c). Dietary leucine requirement of fingerling *Channa punctatus* (Bloch) based on growth, feed conversion and leucine retention efficiency, hematological parameters, antioxidant and intestinal enzyme activities. *Amino Acids, 55*(4), 451–468.

Sharf, Y., & Khan, M. A. (2023d). Quantification of dietary valine requirement of fingerling *Channa punctatus* (Bloch) based on growth, feed conversion efficiency, carcass composition, hematology, and antioxidant status. *Animal Feed Science and Technology, 306*, 115815.

Shiau, S. Y., & Chin, Y. H. (1999). Estimation of the dietary biotin requirement of juvenile hybrid tilapia, *Oreochromis niloticus* × *O. Aureus*. *Aquaculture, 170*, 71–78.

Shiau, S., & Hsieh, J. (2001). Quantifying the dietary potassium requirement of juvenile hybrid tilapia (*Oreochromis niloticus* × *O. aureus*). *British Journal of Nutrition, 85*, 213–218.

Shiau, S. Y., & Hwang, J. Y. (1993). Vitamin D requirements of juvenile hybrid tilapia, *Oreochromis niloticus* X *O. Aureus*. *Nippon Suisan Gakkaishi, 59*, 553–558.

Siddiqua, K. S., & Khan, M. A. (2022). Effects of dietary lipid levels on growth, feed utilization, RNA/DNA ratio, digestive tract enzyme activity, non-specific immune response and optimum inclusion in feeds for fingerlings of rohu, *Labeo rohita* (Hamilton). *Aquaculture, 554*, 738114.

Silva, J. F., Hamilton, S., Rocha, J. V., Borie, A., Travassos, P., Soares, R., & Peixoto, S. (2019). Acoustic characterization of feeding activity of *Litopenaeus vannamei* in captivity. *Aquaculture, 501*, 76–81.

Singh, R. K., Chavan, S. L., Desai, A. S., & Khandagale, P. A. (2008). Influence of dietary protein levels and water temperature on growth, body composition and nutrient utilization of *Cirrhinus mrigala* (Hamilton, 1822) fry. *Journal of Thermal Biology, 33*(1), 20–26.

Sinha, A., & Sinha, Y. K. P. (1994). Role of vitamin E in growth of an Indian major carp, catla (*Catla catla*). *Journal of the Indian Fisheries Association, 24*, 91–96.

Sinha, A. K., Kumar, V., Makkar, H. P., De Boeck, G., & Becker, K. (2011). Non-starch polysaccharides and their role in fish nutrition-a review. *Food Chemistry, 127*(4), 1409–1426.

Smith, A. D., Warren, M. J., & Refsum, H. (2018). Vitamin B12. *Advances in Food and Nutrition Research, 83*, 215–279.

Song, J. Y., Zhang, C. X., Wang, L., Song, K., Hu, S. C., & Zhang, L. (2017). Effects of dietary calcium levels on growth and tissue mineralization in Japanese seabass, *Lateolabrax japonicus*. *Aquaculture Nutrition, 23*, 637–648.

Soto-Dávila, M., et al. (2020). Effects of vitamin D2 (ergocalciferol) and D3 (cholecalciferol) on Atlantic salmon (*Salmo salar*) primary macrophage immune response to *Aeromonas salmonicida* subsp. salmonicida infection. *Frontiers in Immunology, 10*.

Su, C., Luo, L., Wen, H., Sheng, X., & Li, S. (2007). Effects of dietary iron on growth performance, nutritional composition and some blood indices of grass carp (*Ctenopharyngodon idella*). *Freshwater Fishes, 37*, 48–52.

Tan, X. Y., Xie, P., Luo, Z., Lin, H. Z., Zhao, Y. H., & Xi, W. Q. (2012). Dietary manganese requirement of juvenile yellow catfish *Pelteobagrus fulvidraco*, and effects on whole body mineral composition and hepatic intermediary metabolism. *Aquaculture, 326-329*, 68–73.

Tang, Q. Q., Feng, L., Jiang, W. D., Liu, Y., Jiang, J., Li, S. H., Kuang, S. Y., Tang, L., & Zhou, X. Q. (2013). Effects of dietary copper on growth, digestive, and brush border enzyme activities and antioxidant defense of hepatopancreas and intestine for young grass carp (*Ctenopharyngodon idella*). *Biological Trace Elemental Research, 155*, 370–380.

Taranger, G. L., Carrillo, M., Schulz, R. W., Fontaine, P., Zanuy, S., Felip, A., & Hansen, T. (2010). Control of puberty in farmed fish. *General and Comparative Endocrinology, 165*, 483–515.

Thanikachalam, K., Kasi, M., & Rathinam, X. (2010). Effect of garlic peel on growth, hematological parameters and disease resistance against *Aeromonas hydrophila* in African catfish *Clarias gariepinus* (Bloch) fingerlings. *Asian Pacific Journal of Tropical Medicine, 3*(8), 614–618.

Thirunavukkarasar, R., Kumar, P., Sardar, P., Sahu, N. P., Harikrishna, V., Singha, K. P., Shamna, N., Jacob, J., & Krishna, G. (2022). Protein-sparing effect of dietary lipid: Changes in growth, nutrient utilization, digestion and IGF-I and IGFBP-I expression of genetically improved farmed tilapia (GIFT), reared in inland ground saline water. *Animal Feed Science and Technology, 284*, 115150.

Tiengtam, N., Khempaka, S., Paengkoum, P., & Boonanuntanasarn, S. (2015). Effects of inulin and Jerusalem artichoke (*Helianthus tuberosus*) as prebiotic ingredients in the diet of juvenile Nile tilapia (*Oreochromis niloticus*). *Animal Feed Science and Technology, 207*, 120–129.

Uddin, M. H., Ritu, J. R., Putnala, S. K., Rachamalla, M., Chivers, D. P., & Niyogi, S. (2024). Selenium toxicity in fishes: A current perspective. *Chemosphere, 364*, 143214.

US Food and Drug Administration. (2017). *Approved aquaculture drugs*. US Food and Drug Administration.

Volkoff, H., & Rønnestad, I. (2020). Effects of temperature on feeding and digestive processes in fish. *Temperature, 7*(4), 307–320.

Wang, F. B., Luo, L., Lin, S. M., Li, Y., Chen, S., Wang, Y. G., Wen, H., & Hu, C. J. (2011). Dietary magnesium requirements of juvenile grass carp, *Ctenopharyngodon idella*. *Aquaculture Nutrition, 17*, 691–700.

Wang, L. N., Liu, W. B., Lu, K. L., Xu, W. N., Cai, D. S., Zhang, C. N., & Qian, Y. (2014). Effects of dietary carbohydrate/lipid ratios on non-specific immune responses, oxidative status and liver histology of juvenile yellow catfish *Pelteobagrus fulvidraco*. *Aquaculture, 426*, 41–48.

Wang, Y., Yu, X., Liu, J., An, D., & Wei, Y. (2022). Dynamic feeding method for aquaculture fish using multi-task neural network. *Aquaculture, 551*, 737913.

Wang, S., Tian, J., Jiang, X., Li, C., Ge, Y., Hu, X., et al. (2023). Effects of different dietary protein levels on the growth performance, physicochemical indexes, quality, and molecular expression of yellow river carp (*Cyprinus carpio haematopterus*). *Animals, 13*(7), 1237.

Wang, L., Wang, H., Gao, C., Wang, C., Yan, Y., & Zhou, F. (2024). Dietary copper for fish: Homeostasis, nutritional functions, toxicity, and affecting factors. *Aquaculture, 587*, 740875.

Watanabe, T., Takashima, F., Ogino, C., & Hibiya, T. (1970). Requirement of young carp for α-tocopherol. *Nippon Suisan Gakkaishi, 36*, 972–976.

Xie, D., Yang, L., Yu, R., Chen, F., Lu, R., Qin, C., & Nie, G. (2017). Effects of dietary carbohydrate and lipid levels on growth and hepatic lipid deposition of juvenile tilapia, *Oreochromis niloticus*. *Aquaculture, 479*, 696–703.

Xu, Z., Dong, X., & Liu, C. (2007). Dietary zinc requirement of juvenile cobia (*Rachycentron canadum*). *Fisheries Science, 26*, 138–141.

Yadav, A. K., Mandal, S. C., Patel, A. B., & Maurya, P. K. (2019). Evaluation of dietary protein requirement for the growth performance of minor carp, *Cirrhinus reba* (Hamilton, 1822) fingerlings. *Aquaculture Research, 50*(11), 3343–3349.

Yang, Q. H., Zhou, X. Q., Jiang, J., & Liu, Y. (2008). Effect of dietary vitamin a deficiency on growth performance, feed utilization and immune responses of juvenile Jian carp (*Cyprinus carpio* var. Jian). *Aquaculture Research, 39*, 902–906.

Yang, L., Yu, H., Cheng, Y., Mei, S., Duan, Y., Li, D., & Chen, Y. (2021). A dual attention network based on efficientNet-B2 for short-term fish school feeding behavior analysis in aquaculture. *Computers and Electronics in Agriculture, 187*, 106316.

Yao, Y. F., Jiang, M., Wen, H., Wu, F., Liu, W., Tian, J., & Yang, C. G. (2014). Dietary phosphorus requirement of GIFT strain of Nile tilapia *Oreochromis niloticus* reared in fresh water. *Aquaculture Nutrition, 20*, 273–280.

Ye, C. X., Tian, L. X., Yang, H. J., Liang, J. J., Niu, J., & Liu, Y. J. (2009). Growth performance and tissue mineral content of juvenile grouper (*Epinephelus coioides*) fed diets supplemented with various levels of manganese. *Aquaculture Nutrition, 15*, 608–614.

Ye, W., Han, D., Zhu, X., Yang, Y., Jin, J., & Xie, S. (2015). Comparative studies on dietary protein requirements of juvenile and on-growing gibel carp (*Carassius auratus gibelio*) based on fishmeal-free diets. *Aquaculture Nutrition, 21*(3), 286–299.

Yossa, R., Sarker, P. K., & Vandenberg, G. W. (2011). Preliminary evidence of the contribution of the intestinal microflora to biotin supply in zebrafish *Danio rerio* (Hamilton-Buchanan). *Zebrafish, 8*, 221–227.

Zafar, N., & Khan, M. A. (2018). Determination of dietary phosphorus requirement of stinging catfish *Heteropneustes fossilis* based on feed conversion, growth, vertebrae phosphorus, whole body phosphorus, haematology and antioxidant status. *Aquaculture Nutrition, 24*, 1577–1586.

Zafar, N., & Khan, M. A. (2019a). Quantification of calcium requirement of fingerling *Heteropneustes fossilis* based on feed conversion efficiency, mineralization and serum alkaline phosphatase activity. *Animal Physiology and Animal Nutrition, 103*, 1959–1968.

Zafar, N., & Khan, M. A. (2019b). Growth, body composition, mineralization and Na+/K+-ATPase activity of *Heteropneustes fossilis* fed diets with different levels of potassium. *Archives of Animal Nutrition, 73*, 505–516.

Zafar, N., & Khan, M. A. (2019c). Growth, feed utilization, mineralization and antioxidant response of stinging catfish *Heteropneustes fossilis* fed diets with different levels of manganese. *Aquaculture, 509*, 120–128.

Zafar, N., & Khan, M. A. (2020a). Dietary copper requirement of fingerling *Heteropneustes fossilis* for formulating copper-balanced commercial feeds. *Aquaculture Nutrition, 26*, 248–260.

Zafar, N., & Khan, M. A. (2020b). Effects of dietary iron on growth, haematology, oxidative stress and hepatic ascorbic acid concentration of stinging catfish *Heteropneustes fossilis*. *Aquaculture, 516*, 734642.

Zafar, N., & Khan, M. A. (2021). Effects of dietary magnesium supplementation on growth, feed utilization, nucleic acid ratio and antioxidant status of fingerling *Heteropneustes fossilis*. *Animal Feed Science and Technology, 273*, 114819.

Zafar, N., & Khan, M. A. (2023). Effects of dietary zinc on growth, haematological indices, digestive enzyme activity, tissue mineralization, antioxidant and immune status of fingerling Heteropneustes fossilis. *Biological Trace Element Research, 202*, 1249–1263.

Zehra, S., & Khan, M. A. (2012). Dietary protein requirement for fingerling *Channa punctatus* (Bloch), based on growth, feed conversion, protein retention and biochemical composition. *Aquaculture International, 20*, 383–395.

Zehra, S., & Khan, M. A. (2013a). Dietary arginine requirement of fingerling Indian major carp, *Catla catla* (Hamilton). *Journal of the World Aquaculture Society, 44*(3), 363–373.

Zehra, S., & Khan, M. A. (2013b). Dietary lysine requirement of fingerling *Catla catla* (Hamilton) based on growth, protein deposition, lysine retention efficiency, RNA/DNA ratio and carcass composition. *Fish Physiology and Biochemistry, 39*, 503–512.

Zehra, S., & Khan, M. A. (2013c). Dietary isoleucine requirement of fingerling catla, *Catla catla* (Hamilton), based on growth, protein productive value, isoleucine retention efficiency and carcass composition. *Aquaculture International, 21*, 1243–1259.

Zehra, S., & Khan, M. A. (2014a). Dietary phenylalanine requirement and tyrosine replacement value for phenylalanine for fingerling *Catla catla* (Hamilton). *Aquaculture, 433*, 256–265.

Zehra, S., & Khan, M. A. (2014b). Dietary valine requirement of fingerling *Catla catla*. *Journal of Applied Aquaculture, 26*(3), 232–251.

Zehra, S., & Khan, M. A. (2015a). Dietary leucine requirement of fingerling *Catla catla* (Hamilton) based on growth, feed conversion ratio, RNA/DNA ratio, leucine gain, blood indices and carcass composition. *Aquaculture International, 23*, 577–595.

Zehra, S., & Khan, M. A. (2015b). Dietary tryptophan requirement of fingerling *Catla catla* (Hamilton) based on growth, protein gain, RNA/DNA ratio, haematological parameters and carcass composition. *Aquaculture Nutrition, 21*(5), 690–701.

Zehra, S., & Khan, M. A. (2016a). Total sulphur amino acid requirement and maximum cysteine replacement value for methionine for fingerling *Catla catla* (Hamilton). *Aquaculture Research, 47*(1), 304–317.

Zehra, S., & Khan, M. A. (2016b). Dietary threonine requirement of fingerling I ndian major carp, *Catla catla* (Hamilton) estimated by growth, protein retention efficiency, threonine deposition, haematological parameters and carcass composition. *Aquaculture Research, 47*(1), 253–265.

Zehra, S., & Khan, M. A. (2016c). Dietary histidine requirement of fingerling *Catla catla* (Hamilton) based on growth, protein gain, histidine gain, RNA/DNA ratio, haematological indices and carcass composition. *Aquaculture Research, 47*(4), 1028–1039.

Zehra, S., & Khan, M. A. (2017a). Dietary thiamin and pyridoxine requirements of fingerling Indian major carp, *Cirrhinus mrigala* (Hamilton). *Aquaculture Research, 48*, 4945–4957.

Zehra, S., & Khan, M. A. (2017b). Dietary thiamin requirement of fingerling *Channa punctatus* (Bloch) based on growth, protein gain, liver thiamin storage, RNA/DNA ratio and biochemical composition. *Aquaculture Nutrition, 24*, 1015–1023.

Zehra, S., & Khan, M. A. (2018). Dietary pyridoxine requirement of fingerling *Channa punctatus* (Bloch) based on growth performance, liver pyridoxine concentration, and carcass composition. *Journal of Applied Aquaculture, 30*, 238–255.

Zehra, S., & Khan, M. A. (2019). Quantification of dietary inositol requirement for fingerling *Channa punctatus* (Bloch) based on growth, antioxidant status, hematological tools and liver inositol concentration. *Aquaculture, 512*, 734280.

Zeng, Y., Yang, X., Pan, L., Zhu, W., Wang, D., Zhao, Z., et al. (2023). Fish school feeding behavior quantification using acoustic signal and improved Swin transformer. *Computers and Electronics in Agriculture, 204*, 107580.

Zhang, L., Feng, L., Jiang, W. D., Lui, Y., Wu, P., Kuang, S. Y., et al. (2017). Vitamin A deficiency suppresses fish immune function with differences in different intestinal segments: The role of transcriptional factor *NF-κB* and *p38 mitogen-activated protein kinase* signalling pathways. *British Journal of Nutrition, 117*, 67–82.

Zhou, C., Liu, B., Ge, X., Xie, J., & Xu, P. (2013). Effect of dietary carbohydrate on the growth performance, immune response, hepatic antioxidant abilities and heat shock protein 70 expression of Wuchang bream, *Megalobrama amblycephala*. *Journal of Applied Ichthyology, 29*(6), 1348–1356.

Zhou, P., Wang, M., Xie, F., Deng, D. F., & Zhou, Q. (2016). Effects of dietary carbohydrate to lipid ratios on growth performance, digestive enzyme and hepatic carbohydrate metabolic enzyme activities of large yellow croaker (*Larmichthys crocea*). *Aquaculture, 452*, 45–51.

Zhu, W., Liu, M., Chen, C., Wu, F., Yang, J., Tan, Q., Xie, S., & Liang, X. (2014). Quantifying the dietary potassium requirement of juvenile grass carp (*Ctenopharyngodon idellus*). *Aquaculture, 430*, 218–223.

Zhu, L., Han, F., Zhu, X., Yang, Y., Jin, J., Liu, H., & Xie, S. (2017). Dietary selenium requirement for on-growing gibel carp (*Carassius auratus gibelio* var. CAS III). *Aquaculture Research, 48*, 2841–2851.

Zilberg, D., Tal, A., Froyman, N., Abutbul, S., Dudai, N., & Golan-Goldhirsh, A. (2010). Dried leaves of *Rosmarinus officinalis* as a treatment for *streptococcosis* in tilapia. *Journal of Fish Diseases, 33*(4), 361–369.

Zimmer, A. M., Brix, K. V., & Wood, C. M. (2019). Mechanisms of Ca2+ uptake in freshwater and seawater-acclimated killifish, *Fundulus heteroclitus*, and their response to acute salinity transfer. *Journal of Comparative Physiology B, 189*, 47–60.

Chapter 2
Does Aquaculture Have the Potential to Feed the World's Hunger

Z. A. Ansari, Asif Pagarkar, and Azra Ansari

1 Introduction

Human food requirements in the past have been met from agricultural sources. Aquatic animals such as fish, crustacea, and mollusks have traditionally been caught from inland water and ocean as a source of cheap protein food for the hungry world population particularly in the developing countries (Arino et al., 2013; Tran et al., 2022). According to the Food and Agriculture Organization (FAO, 2020), fish production from aquaculture and captured fisheries accounted for 17% of the animal source protein for human consumption. In India only after the independence, has fisheries together with agriculture been recognized as an important sector. It is now considered an important economic activity with varied resources and their potentials (Gopakumar et al., 1999). In the recent past, Indian fisheries have witnessed a paradigm shift from marine-dominated fisheries to inland fisheries, with the latter emerging as a major contributor to fish production. Fisheries and aquaculture industries are regarded as sunrise sectors and are anticipated to have a big impact on the Indian economy (FAO, 2023). The vibrancy of this sector can be visualized by the 11-fold increase that India achieved in fish production in just six decades, i.e. from 0.75 million tons in 1950–51 to 9.6 million tons during 2012–13. This is prominent because it resulted in an unparalleled average annual growth rate of over 4.5% over the years, which has placed the country at the forefront of global fish production, after China.

Z. A. Ansari (✉)
Retired from CSIR-National Institute of Oceanography, Panaji, Goa, India

A. Pagarkar
College of Fisheries, Ratnagiri, India
e-mail: aupcof@gmail.com

A. Ansari
Government High School, Advoi, Goa, India

I. Ahmed, I. Ahmad (eds.), *Aquaculture: Enhancing Food Security and Nutrition*, https://doi.org/10.1007/978-3-031-92858-1_2

Aquaculture in India has a long history. The traditional practice of fish culture in small ponds in eastern India is known to have existed for hundreds of years; significant advances were made in the State of West Bengal in the early nineteenth century with the controlled breeding of carp in bundhs (tanks or impoundments where riverine conditions are simulated). Fish culture received notable attention in the state of Tamil Nadu as early as 1911, and subsequently, states such as Bengal, Punjab, Uttar Pradesh, Baroda, Mysore, and Hyderabad initiated fish culture through the establishment of Fisheries Departments and support to fishers and farmers for expansion of the sector. The fish culture has evolved as a viable commercial farming practice from the level of traditionally backyard activity over the last three decades with considerable diversification in terms of species and systems. India as the second largest country in aquaculture production, the share of inland fisheries and aquaculture has gone up from 46% in the 1980s to over 85% in recent years in total fish production. Freshwater aquaculture showed an overwhelming tenfold growth from 0.37 million tons in 1980 to 4.03 million tons in 2010; with a mean annual growth rate of over 6%. In 2019, the estimated per capita fish consumption in India was about 7.8 kg.

2 Aquaculture Production Trends in India

Aquaculture is the fastest-growing food production sector with an annual average growth rate of 7.4%, accounting for about 73.6% of the total Indian fish supply (FAO, 2023). Freshwater aquaculture contributes to over 95% of the total aquaculture production in India (Jayasankar, 2018). Freshwater aquaculture comprises the culture of carps, catfishes (air-breathing and non-air breathing), freshwater prawns, culture of pangasius, and tilapia. Thus, the production of carp in freshwater and shrimps in brackish water from the bulk of major areas of aquaculture activity. The three Indian major carps, namely catla (*Catla catla*), rohu (*Labeo rohita*), and mrigal (*Cirrhinus mrigala*) contribute the bulk of production to the extent of 70–75% of the total freshwater fish production, followed by exotic species like silver carp, grass carp, and common carp, which contribute the balance of 25–30% (Gopakumar et al., 1999; Saroj & Khan, 2022). It is estimated that only about 40% of the available area of 2.36 million hectares of ponds and tanks has been put to use and an immense scope for expansion of area exists under freshwater aquaculture (Ayyappan & Jena, 2003). The successful breeding and larval rearing of Indian major carps is a phenomenal achievement. It gave a tremendous boost to the aquaculture industry in India. Today freshwater aquaculture contributes over 95% of the total overall aquaculture production.

The national mean production levels from still-water ponds have gone up from about 600 kg/ha/year in 1974 to over 2900 kg/ha/annum at present and several farmers are even demonstrating higher production levels of 8–12 tons/ha/year (ICAR,

2013). The technologies of induced carp breeding and polyculture in static ponds and tanks virtually revolutionized the freshwater aquaculture sector and turned the sector into a fast-growing commercial sector. The developmental support provided by the Indian Government through a network of Fish Farmers' Development Agencies and Brackish water Fish Farmers' Development Agencies and the research and development programs of the Indian Council of Agricultural Research (ICAR) have been the principal vehicles for this revolutionary development. In addition, support was also provided by various state governments, a host of organizations and agencies like the Marine Products Export Development Authority, financial institutions, etc.

Socioeconomic status depends, to a large extent, on food security and the importance of animal protein in a regular diet. Fish contribute a significant amount of animal protein to the diet of people worldwide (Gareth, 2001). The relative contribution of fish protein as compared to other sources of protein is given in Table 2.5. However, developing countries with less fishery resources and ever-growing population cannot meet their protein requirement. Under these circumstances, aquaculture is the only solution. Fish production by aquaculture through improved technological development and biotechnological interventions can be achieved to feed the world's hungry population (Sreenivasan, 2003).

At the World Food Summit organized by the Food and Agriculture Organisation (FAO) in Rome in 1996, the participating countries committed to reducing the number of malnourished people in the world by half by the year 2015 (Haylor, 2004). It is a well-known fact that fish is rich in protein and essential amino acids. It is also a good source of calcium, vitamins A and B12, and omega-3 fatty acids. Recent research revealed that unsaturated omega-3 fatty acids present in fish oil help to reduce the chance of heart attack, cancer, high blood pressure, and atherosclerosis. These acids have beneficial effects on immune response, membrane structure and function, prostaglandin production, etc. Omega-3 fatty acids are believed to be critical nutrients during pregnancy, especially in the last trimester, when brain and neural development occurs (Basu & Chousey, 2003, Pagarkar et al., 2021). People irrespective of age who do not get sufficient nutrients from cereal-based diets would benefit from the inclusion of fish in the diet. Aquaculture not only supplies dietary essentials for human consumption but provides excellent opportunities for employment and income generation, especially in the more economically backward rural areas. Sixty million people are directly engaged, part-time or full-time, in the primary production of fish, either by fishing or in aquaculture, supporting the livelihoods of 10–12% of the world population (FAO, 2016). Aquaculture currently accounts for over 50% of the global food fish consumption (Subasinghe et al., 2009). Globally India stands second in culture fisheries production. China, with world's one one-fifth of the population, produces one-third of the total fish harvested and two-thirds of the fish cultivated (FAO, 2016). In India, the culture system is based on a three to six species combination, while the Chinese have ten or more species in a single pond, thus maximizing productivity.

3 Indian Fisheries Potential

The marine fishery potential in Indian water has been estimated at 5.31 million tons, constituting about 43.3% demersal, 49.5% pelagic, and 4.3% oceanic group (FAO, 2023). Fisheries contribute to 1.09% of the total gross domestic product (GDP) of India and 6.2% of agricultural GDP. In food security, Indian major carps constitute almost half of the total fish consumption, followed by pelagic (23.7%), freshwater carp 13.2%, shrimp 6.6%, demersal 4.4%, and mollusk 2.7% (FAO, 2023). Indian aquaculture has demonstrated a six- and half-fold growth over the last two decades, with freshwater aquaculture contributing over 95% of the total aquaculture production. India is bestowed with 3.15 million ha of reservoirs, 2.36 million ha of ponds and tanks as well as 0.19 million ha of rivers and canals. Freshwater aquaculture with a share of 34% in inland fisheries in mid 1980s has increased to about 80% in recent years (DADF, 2017). The technologies of induced carp breeding and polyculture in static ponds and tanks have brought about a remarkable upward trend in aquaculture productivity and turned the sector into a fast-growing industry. The research and development programs of the Indian Council of Agricultural Research (ICAR), as well as the development support provided by the India Government through a network of Fish Farmers' Development Agencies (FFDAs) have been the principal vehicles for this development. Additional support has been provided by several other organizations, state departments, and financial institutions. So far, about 0.65 million ha of water area has been brought under fish farming covering 1.1 million beneficiaries. Currently, the average annual yield is around 3 t ha^{-1}. At the same time, training has been imparted to about 0.8 million fishers (DADF, 2017).

Aquaculture production in India has been increasing steadily over the years and during 2015–16 the figure was about 5.77 million t. Besides three Indian major carps, the bulk of the exotic species, namely silver carp (*Hypophthalmichthys molitrix*), grass carp (*Ctenopharyngodon idella*), and common carp (*Cyprinus* carpio) forms the second important group. National freshwater fish productivity has registered a marked rise from 0.6 t ha^{-1} $yr.^{-1}$ (1974) to 3 t ha^{-1} $yr.^{-1}$ at present. Many farmers have demonstrated productivity levels as high as 8–12 t ha^{-1} $yr.^{-1}$ (Jayasankar, 2014). Further, freshwater aquaculture production has begun to diversify, incorporating medium and minor carps, catfishes, and murrels. The only species from the freshwater sector that goes to the export market is the giant freshwater prawn (*Macrobrachium rosenbergii*), while carps and other finfishes are grown for the domestic market. Recently, production of *M. rosenbergii* has plummeted, while that of white-legged shrimp *Litopenaeus vannamei* enhanced considerably. With a national per capita consumption of 11 kg, fish is recognized as one of the chief components of the domestic food security in India (Jayasankar & Barik, 2015). Aquaculture in India brings about socioeconomic development in terms of income and employment through the use of unutilized and underutilized resources in many parts of the country (FAO, 2017).

4 Brackish Water Aquaculture and Its Potential in India

Brackish water aquaculture is also called coastal aquaculture. As far as brackish water aquaculture in India is concerned, it has a long history of traditional practice in the bheries of West Bengal and pokkali fields of Kerala. The scientific farm management and culture of brackish water species in the country was initiated only in the early 1990s, which developed into a major export-oriented sector in subsequent years (Ghoshal et al., 2015). Coastal aquaculture has been emphasized over the past decade as a means to increase the country's fish production. In recent years, brackish water aquaculture has been spreading rapidly and the use of modern technology is increasing production. Most brackish water aquaculture has evolved into a mangrove ecosystem because of congenial environmental parameters (Rao & Ravichandran, 2001). There are many potential brackish water fish species. The most common are sea bass (*Lates calcarifer*), mullets, (*Mugil*.sp.), milk fish (*Chanos chanos*), pearl spot (*Etraplus suratensis*), and catfish. However, commercial farming was confined to a single commodity, shrimp, *Penaeus monodon*, and *L. vannamei* due to their high export potential (Vijayan & Balasubramanian, 2019). It is a multimillion-dollar industry today.

India has an estimated total estuarine area of 3.9 million hectares; of which, 1.2 million hectares of coastal salt-affected lands have been identified to be potentially suitable for brackish water shrimp farming (Vijayan & Balasubramanian, 2019). Of this, about 15–16% of the potential area has been put into aquaculture purposes. Brackish water aquaculture in the country has expanded steadily from a traditional activity to reach the present commercial scale. About nine million hectares of salt-affected land has been estimated in the hot semi-arid and arid eco-region of northern plains and central highlands in the states of Haryana, Rajasthan, Uttar Pradesh, Maharashtra, and Gujarat with surface and sub-soil saline water. Out of the total area, about 10% is being farmed at present. Out of this area, 80% is under a traditional farming system and the remaining is under shrimp farming. The fish production by traditional practice has yielded between 500 and 750 kg/ha/year. Production potential ranging from 0.5 to 3 tons/ha/year has been demonstrated from such waters (ICAR, 2013). The brackish water aquaculture has got a major thrust after the introduction of white-leg shrimp (*L. vannnamei*) in our water. The growth of shrimp aquaculture has been phenomenal, and this resulted in the all-round development of shrimp hatcheries, feed mills, and ancillary industry (Surendran et al., 1991). Under shrimp culture, a semi-intensive or extensive technique is the common practice that gives a good yield. From a production of 90,000 MT in 2010, it rose to 353,413 MT in 2014–15. This has resulted in an increase in export earnings by leap and bound. Total fish and shrimp production from a total brackish water area under cultivation in India is reported at 393,280 MT (Union Govt. India, 2017). This sector has also generated more than 3 lakh job.

5 Total Fish Production Through Aquaculture in India

Aquaculture in India has evolved as a viable commercial farming practice from the level of traditional backyard activity over the last three decades. It is the fastest-growing food production sector in India with an average annual growth rate of 7.4% accounting for about 73.6% of the total Indian fish supply (Swain & Ferosekhan, 2022). The present level of aquaculture production is given in Table 2.1. With considerable diversification in terms of species and systems, it has been showing an impressive annual growth rate of 6–7%. India's aquaculture production basically can be classified into freshwater and brackish water production. There are 429 Fish Farmers Development Agencies (FFDAs) and 39 Brackish Water Fish Farmers Development Agencies (BFDAs) for promoting freshwater and coastal aquaculture. Some of the important species cultured in India are the Indian major carps and shrimp. Besides these, ornamental fish culture and seaweed farming are slowly gaining importance in the aquaculture scenario in the last few years as alternative livelihood-supporting sectors as small-scale activities. As the second largest country in aquaculture production, the share of inland fisheries and aquaculture has gone up from 46% in the 1980s to over 85% in recent years in total fish production. Freshwater aquaculture showed an overwhelming tenfold growth from 0.37 million tons in 1980 to 4.03 million tons in 2010; with a mean annual growth rate of over 6%. Freshwater aquaculture contributes to over 95% of the total aquaculture production in India (Swain & Ferosekhan, 2022).

India recorded the highest-ever export of 1.78 million metric tons worth $7.38 Billion in FY 2023–24 of fisheries and fisheries products. At present, India stands first in inland capture fish production, second in aquaculture fish production, and third in overall fish production. It contributes 8% to the global fish production. Total fish production and its trend in recent years are given in Table 2.2. The fish production in 2021–22 is 16.24 million tons comprising marine fish production of 4.12 million tons and 12.12 million tons from Aquaculture. India is the fourth largest exporter of fish and fisheries products with a growth in exports of 26.73% in terms of quantity between FY 2021–22 and FY 2022–23. This was possible due to

Table 2.1 Aquaculture production in India

Year	Metric tons
1960	44,843
1970	121,671
1980	365,180
1990	1017,136
2000	1908,485
2010	2316,947
2020	3677,584

Source: Handbook of fisheries and aquaculture, ICAR, India 2013)

Table 2.2 Fish production statistic volume in India

Year	Million metric tons
2013	9.04
2014	9.56
2015	10.36
2016	10.76
2017	11.43
2018	12.70
2019	13.57
2020	14.16
2021	14.73
2022	16.25

Source: Keelery (2024). Fish production in India FY 2013–2022

advanced hatcheries in the public and private sectors and the availability of fish seed. In terms of employment, this sector supports the livelihood of over 30 million people. India recorded the highest-ever export of 1.73 MMT worth $8.09 Bn in FY 2022–23 of fisheries and fisheries products. The fisheries and aquaculture sector shall play a vital role in addressing the rising global demand for an affordable quality source of protein.

Since the launch of the Pradhan Mantri Matsya Sampada Yojana (PMMSY) in the year 2020–21, the overall fish production in the country has shown an increasing trend, registering a fish production of 162.48 lakh tons in 2021–22 from 141.64 lakh tons in 2019–20. The Department of Fisheries, Ministry of Fisheries, Animal Husbandry, and Dairying has implemented a flagship scheme "Pradhan Mantri Matsya Sampada Yojana (PMMSY)" to bring about Blue Revolution through sustainable and responsible development of the fisheries sector in India" with the highest ever investment of INR 20,050 Cr for a period of 5 years effective from FY 2020–21 till FY 2024–25 in all states/union territories (UTs). The fisheries sector in India has shown a sustained annual average growth rate of 8.61% for the 8-year period from 2014–15 to 2021–22 (at constant prices). Per capita consumption of fish in India is 6.31 kg per capita. India's top five export destinations for fish and fish products are USA, China, the European Union, Southeast Asia, Japan, and the Middle East. Per capita consumption of fish in India is 6.31 kg per capita. A new sub-scheme of PMMSY named Pradhan Mantri Matsya Kisan Samridhi Sab-Yojana (PM-MKSSY) is to be launched. In the union budget 2023, the target investment of INR 6000 Cr is to enhance the earnings and incomes of fishermen, fish vendors, and micro & small enterprises engaged in the fisheries sector. Considering a phenomenal increase in the Indian population by 2050, India has to produce about 20MT of fish/year, a task possible only through aquaculture.

6 Global Fish Production Through Aquaculture

Among the global food resource, 31.4% of assessed fish stocks are overfished and 51.1% are fully fished (FAO, 2018). Global total fish production stands at 186.6 million metric tons in 2023, of which more than 50% (96 million t) is contributed by the aquaculture sector (FAO, 2023). Global capture fishery is presently at crossroads with over 70% of the resources exploited and therefore aquaculture is the only option to fill up the gap of much of the future demand for fish. Aquaculture is the fastest-growing animal food-producing sector, growing at a rate of more than 7% annually. Many studies have pointed toward a positive scenario for the role of aquaculture in providing the much-needed animal protein to the world population (Tidwell & Allan, 2001; Sugiyama et al., 2004). The importance of freshwater aquaculture in the global aquaculture production is evident as it accounted for 63.8% of the global total of 47,102,391 t (FAO, 2016). Total world fish production from 2002 to 2023 is given in Table 2.3. The fish production through aquaculture has increased from 49.9 million tons in 2007 to 96 million tons in 2023 (FAO, 2023). The same is shown in Table 2.4. The gap between the wild fish catching and production through fish farming has become ever closer. The use of aquaculture in producing fish thus becoming more widespread especially in Asia.

Table 2.3 Global fish production from 2002 to 2023 (in MMT)

Year	MMT
2002	127.8
2003	127.2
2004	134.2
2005	136.4
2006	137.3
2007	140.7
2008	143.1
2009	145.9
2010	148.1
22,011	155.5
2012	157.8
2013	162.9
2014	164.8
2015	169.2
2016	170.9
2017	172.6
2018	177.7
2019	177.8
2020	174.6
2021	178.1
2022	184.6
2023	186.6

Source: FAO book of statics

Table 2.4 World aquaculture production (in MMT)

Year	MMT
2007	49.9
2008	52.9
2009	55.7
2010	59.0
2011	61.8
2012	66.5
2013	70.3
2014	73.4
2015	76.1
2016	80.0
2017	80.1
2018	83.2
2019	85.3
2020	87.5
2021	89.7
2022	92.0
2023	96.0

Source: FAO book of statics

7 Aquaculture Development and Employment Opportunity

The fisheries industry today is a large source of employment for about 200 million people (Gareth, 2001). It plays an important role in the world food economy. Although aquaculture in India has reached the status of an industry, the assessment of human resources required for the fisheries sector has always been a debatable issue, as it is yet to be objectively worked out. A realistic demand for human resources depends on several factors such as policy, future thrust areas of development, recruitment qualifications of major employers, national and international fisheries development scenarios, etc. A database with details of human resources in aquaculture and allied sectors is lacking due to the dispersed nature of aquaculture resources and the nonavailability of a suitable mechanism for data collection. In a study conducted in six major aquaculture-producing Indian States (Andhra Pradesh, Haryana, Karnataka, Orissa, Uttar Pradesh, and West Bengal), Bhatta (2003) reported the age of fish farmers ranged from 38 years in Andhra Pradesh to 58 years in Haryana, with a national average of 47 years. The educational status of these fish farmers varied from 0 to 10 years of schooling, a large percentage of these fish farmers practice aquaculture on a part-time basis with their involvement in the activity ranging from 17 man-days per annum in Karnataka to the highest of 75 man-days in West Bengal. This study also inferred that fish farming, though a part-time activity, contributes a major share of the income of these fish farmers, ranging from 14.98% in Orissa to 95.26% in Andhra Pradesh, with an average of 79.66%.

8 Fisheries Contribution to World Economy

Fisheries play an important role in the world food economy. Fish and seafood play a significant role in supporting the economy, employment, food sources, and livelihood of people across the world. The fisheries sector contributes to the national income in terms of foreign exchange through exports. It provides livelihood, employment, and entrepreneurship to more than 2.8 crores of fishers and fish farmers at the primary level, according to government sources (Union Fishery Ministry, June 2023). As per the estimates of the Central Statistical Organisation (CSO), of the Government of India, the value of GDP from the fisheries sector at current prices during 2011–2012 was Rs 65,541 crores, which is 4.47% of the total GDP of agriculture and allied sectors. Fish contributes substantially to the domestic food security of India which has a per capita consumption of more than 6.00 kg per annum. With freshwater aquaculture being a homestead activity in several parts of the country, besides adding to the nutritional security it also helps in bringing additional income to rural households. The network of 429 FFDAs has brought about 8.05 lakh ha of water under modern fish culture operations benefiting approximately 13,86 lakh beneficiaries. The rapid growth of the sector has generated huge employment opportunities for professional, skilled, and semi-skilled workers for the different support activities such as construction and the management of farms, hatcheries, feed mills, processing units, etc. It has been estimated that over 300,000 jobs have been generated in the brackish water sector alone in the main and supporting areas for shrimp culture, although information on exact numbers involved in aquaculture is not available.

9 Global Demand of Fish as a Source of Nutrition

Fish and fishery products play an important role in food and nutritional security around the world. It is the healthiest food and cheapest source of animal protein in this universe compared to all animal meat. Consumption of fish offers unique nutritional and health benefits. Increased attention is given to fish as a source of essential nutrients in our diet (Khalil & Samples, 2018; Pagarkar et al., 2021). In the past 30 years, the global appetite for fish has doubled. Seafood consumption is generally increasing in many parts of the World, particularly for coastal communities. In many places, fish and shellfish are the only readily available sources of protein that people can self-harvest, often throughout the year. With the rising population, stress on existing resources is increasing tremendously to meet the growing demand for nutritional diets. In such situation, fisheries and aquaculture sector can play a vital role in meeting this growing demand (Alsaleh, 2023). A growing global population, combined with factors such as changing socio-demographics, will place more pressure on the world's resources to provide not only more but also different types of food. Increased demand for animal-based protein in particular is expected to have a negative environmental impact, generating greenhouse gas emissions, requiring more water and more land. It is reported that different factors influence the potential of existing and novel food sources (Hollowed et al., 2013; Delgado et al., 2003). Existing protein sources are primarily hindered by their negative environmental impacts with some concerns around health. However, they offer social and economic benefits, and have a high level of consumer acceptance. Furthermore, recent research emphasizes the role of livestock as part of the solution to greenhouse gas emissions and indicates that animal-based protein has an important role as part of a sustainable diet and as a contributor to food security. Novel proteins require the development of new value chains, and attention to issues such as production costs, food safety, and scalability and consumer acceptance. Furthermore, positive environmental impacts cannot be assumed with novel protein sources and care must be taken to ensure that comparisons between novel and existing protein sources are valid. Greater alignment of political forces, and the involvement of wider stakeholders in a governance role, as well as a development/commercialization role, is required to address both sources of protein and ensure food security.

Fisheries and aquaculture are significant contributors to global animal-source protein production. These sectors are critical contributors to global food security and nutrition in the twenty-first century (FAO, 2022). There are several reports which have shown the relationship between growth and total fish consumption speculated for a period from 1970 to 2010 (FAO, 2000). However, to maintain this position in the future, the production of aquaculture must increase, because global fisheries production is not increasing. It seems likely that the global demand for animal-source proteins will continue to increase, at least in the near future. The Food and Agriculture Organization (FAO) of the United Nations had a statement in The State of World Fisheries and Aquaculture report for 2020 that fisheries and aquaculture produced 17% of global animal-source protein destined for human consumption in 2018. It does reveal that aquaculture produced more live harvest weight of aquatic animals for human consumption than did fisheries, 82,087,000 metric tons versus 74,295,000 metric tons (Boyd et al., 2022.).

10 Role of Aquaculture as a Source of Protein Supply

The contribution of aquatic animal protein to the global, animal-source protein supply and the relative importance of aquaculture to capture fisheries in supplying this protein is relevant in assessments and decisions related to the future of aquatic food production and its security. The meat of terrestrial animals, milk, and eggs resulted in 76,966 Kt crude proteins compared with 13,950 Kt or 15.3% from aquatic animals in 2018. While aquaculture produced a greater tonnage of aquatic animals, capture fisheries resulted in 7135 Kt crude protein while aquaculture yielded 6815 Kt (Table 2.5). Capture fisheries production has not increased in the past two decades, and aquaculture production must increase to assure the growing demand for fisheries products by a larger and more affluent population. We estimated based

Table 2.5 World edible animal source crude protein production (metric tons in 2018)

Source	Metric tons	Percentage
Terrestrial animals	37,391,000	41.1
Milk	30,886,000	34
Eggs	8,686,000	9.6
Fisheries	7,135,000	7.8
Aquaculture	6,815,000	7.5
Total	90,916,000	100

Source: Boyd et al. (2022)

on status quo consumption that aquaculture production would need to increase from 82,087 Kt in 2018 to 129,000 Kt by 2050 to meet the demand of the greater population. About two-thirds of finfish and crustacean production by aquaculture is feed-based and feeds for these species include fishmeal and fish oil as ingredients. Aquaculture feeds require a major portion of the global supply of fishmeal and fish oil. An estimated 71.0% of fishmeal and 73.9% of fish oil are made from the catch with the rest coming from aquatic animal processing waste. The catch of small, pelagic fish from the ocean is not predicted to increase in the future. Aquaculture should reduce its fishmeal and oil use to lessen its dependency on small wild fish important to the integrity of marine food webs and food security for the poor in many coastal areas. Fishmeal and fish oil shortages for use in aquaculture feed will result in a limit on production in the future if goals to lessen their use in feeds are not met.

An estimated 795 million people, mostly in the developing countries, have been suffering from undernourishment in 2014 (FAO, 2015). The protein consumption is expected to nearly double by 2050 and marine-based protein will have a major share. Aquaculture supplies 50% of all fish consumed globally today, and by 2030, it is predicted that aquaculture will be the prime source of fish due firstly to demands from consumers, and secondly due to depletion of wild capture fisheries (FAO, 2016). However, aquaculture needs to become more sustainable. Currently, it is over-dependent on a supply of fishmeal and fish oil, and there are concerns surrounding pollution and water quality, the use of soy and chemicals in aquaculture feed, and habitat destruction. Negative social outcomes are also associated with aquaculture in countries where there are weak regulatory frameworks and there is concern about emerging diseases and disease transmission as a result of increased intensification and globalization (WRAP, 2017). Companies such as Findus are however progressing toward more sustainable aquaculture systems, e.g., they now source shrimp farmed in Sweden that are produced using the "Biofloc" method (Hanson, 2013). This involves feeding shrimp excrement to microorganisms which in turn are consumed by the shrimp as a source of protein. Biofloc can produce up to 40 times more shrimp that conventional shrimp farming (Hanson, 2013). Thus, aquaculture has the potential to be a key element of a circular food system.

Reform of the European Common Fisheries Policy (CFP) in 2015 (Commission Delegated Regulation (EU) No. 1394/2014) further supports the marine sector being a key source of protein. The CFP aims to gradually eliminate the practice of discarding undersized fish as by-products at sea. Landing obligations will be phased in from 2015 to 2019 and will result in greater quantities of catch being landed at port. Undersized fish or under-utilized species such as Blue Whiting are not aesthetically pleasing to consumers and traditionally were thrown back to sea. However, there now exists an exciting opportunity for processors to convert this by-catch or marine "rest raw material" into value-added ingredients due to the high protein and oil content of this by-catch. Several recent publications highlight the potential of this resource as a source of bioactive peptides (Hayes et al., 2016), lipids and biofuel (Adeoti & Hawboldt, 2014), and small molecules/nutrients including vitamins and minerals (Abbey et al., 2016).

11 Future Projections in Fish Production and Consumption

Among the global food resources 31.4% of assessed fish stock are overfished and 51.1% are fully fished (FAO, 2018). The FAO of the United Nation has repeatedly sounded the alarm over a threated stock of wild fish. The consumption of fish as food in developing countries has doubled since 1973 (Delgado et al., 2003). This has burgeoned more fish production. There are many studies that suggested a positive role of aquaculture in providing the much-needed protein food to the world population (Sugiyama et al., 2004; Subasinghe et al., 2009; Jayasankar, 2018). It is believed that about 72% of the world fisheries and aquaculture production available for human consumption in 2030 will be consumed by Asian countries, while the lowest quantity will be consumed in Oceania (FAO, 2022). Regular fish consumption is widely promoted as part of a healthy diet. Fish have a high protein content compared to terrestrial animal meat and have a lower feed conversion rate (FCR) than land animals and more protein can be produced using this lower feed rate from fish (Samples, 2014). Furthermore, fish protein is highly digestible and rich in essential amino acids that are limited in animal-sourced protein including methionine (6.5% compared to 5.7% of total essential amino acids in fish compared to animal meat) and lysine (19.6% compared to 19.0% of total essential amino acids in fish compared to animal meat). Fish and shellfish consumption has several reported health effects including decreased risk of heart diseases, inflammation, and arthritis. The health benefits of fish are mainly linked to the presence of long-chain omega 3 ($n-3$) polyunsaturated fatty acids (PUFAs) (Samples, 2014). However, fish protein is also a rich source of bioactive peptides. These can have a myriad of positive health effects if consumed in appropriate concentrations and bioavailable within the human body. Health effects include control of blood pressure through inhibition of enzymes within the renin-angiotensin-aldosterone system (RAAS), maintenance of bone health, control of inflammation (antioxidant peptides), mental health through the action of opioid peptides and platelet-activating factor acetylhydrolase inhibitory (PAF-AH) peptides, and several other bioactivities (Hayes, 2014). Today, it has been increasingly recognized that the future of food security depends on increased fish production through aquaculture.

12 Role of Aquaculture in World Food Security

Food security is the access to easy food for everyone (FAO, 2000). The problem of food insecurity is multidimensional and needs to be attended by countries facing it. Developing countries with high rates of population growth require an adequate amount of food supply for food security and safety. Aquaculture in India has a role in producing more protein-rich food for its rapidly growing population. Fish appears to occupy the central position in the food insecurity and offers a solution to protein deficiency in developing nations (Tran et al., 2022). However, fish catch from the sea is limited. Therefore, fish production through aquaculture not only guarantees the regular supply of protein-rich food but also a source of employment generation, a regular source of income and foreign exchange (Pradeepkiran, 2019). The FAO report on the state of world fisheries and aquaculture released on 29/6/2022 says fisheries and aquaculture production make a critical contribution to global food security and nutrition in the twenty-first century. According to the state of world Fisheries and Aquaculture, fish production has gone up to all-time high of 214 million tons in 2020. It was 30% higher than average in the 2000s. The aquaculture industry is seen as a more sustainable alternative that can help secure the food supply for the future growing global population. It makes a significant contribution to food security at a global level. It fills the gap between rising global demand for fishery products and the limited increase in capture fisheries production (FAO, 2023). Technological advancement in aquaculture will lead to a significant increase in fish production and supply of protein to the ever-increasing world population. A multistakeholder action and diversification in aquaculture will help a sustainable supply of protein to individual nations and the world in the coming decades (Samples, 2014). Besides, almost 1.4 million marine species contribute to pharmaceuticals, enzymes, cosmetics, and many other value-added products.

The socioeconomic status depends, to a large extent, on the food security and availability of protein in a regular diet. Fish contribute a significant amount of cheap animal protein to the diet of people worldwide (Boyd et al., 2022). However, some developing countries, with limited resources, overfishing, and ever-growing human populations are finding it difficult to meet the required quantity. In such cases, aquaculture can play an important role in the supply of cheap sources of protein. It is expected that in the developing world, a certain increase in the consumption of high- and low-quality fish will take place while fish consumption in the developed world would remain static (Delgado et al., 2003). While fish is seen as a healthy food by consumers, its regular supply is demanding. Aquaculture can play a key role in the regular supply of fish for the world-hungry population through advancement in technology and biotechnological application (Sreenivasan, 2003). With the help of effective fishery sector production, policymakers may ensure long-term food security it is clear that integrated aquaculture management with multiple objectives such as sustainability, economic growth, social development, and biological goals will be vital in future food security and nutrition (Kumar, 2014; Salim et al., 2014). It will play a vital role in food security, particularly in the developing countries.

13 Pollution and Seafood Safety

Marine pollution is a widespread disease. Pollutants such as heavy metals, trace metals, oil hydrocarbons, and microplastics pose a serious threat to the marine ecosystem and human health (Ansari et al., 2012; Pandion et al., 2023; Kibria, 2023; Neethu et al., 2024). These pollutants entering the marine environment from different sources, natural as well as anthropogenic activities, have impaired the coastal water quality and caused heavy loss of marine life in different regions of the world (Lundebye et al., 2021; Vijay Kumar et al., 2022). Heavy metal pollution from industries reaches the water column first and then gets into fish and shellfish (Shenai et al., 2016). In many cases, these metals and other pollutants exceed the standard of food safety and thus pose serious problems and possible long-term risks (Thompson & Darwish, 2019). It is well known, for example, that the Baltic Sea is heavily contaminated with heavy metals due to the presence of heavy industry in the landmass surrounding it that has used the sea as a "sink" for industrial waste for many years. Such chemical contamination, e.g., by heavy metals, such as arsenic, mercury, and cadmium, and polychlorinated biphenyls (PCBs), can result in carcinogenic and toxicological impacts, as well as potentially mitigating the beneficial impact of omega-3 fatty acid (Sidhu, 2003).

Fish and seafood are considered nutritional valuable parts of the human diet, with various health benefits due to their high protein content, nutrients, omega fatty acids, and minimum saturated fats. However, fish and seafood are highly susceptible to environmental contaminants. Plastic comprises the bulk of all marine debris (Ragesh et al., 2024). It has a potential threat to all marine life. Microplastics are gaining access to the human body through seafood (Unuofin & Igwaran, 2023). The burgeoning use of plastic and waste pollution results in microplastic accumulation in fish, adversely affecting fish growth, health (Kibria, 2023), and catch volumes. Consequently, plastic waste pollution poses an increasing threat to the health, food security, food safety, and well-being of fish consumers through the ingestion of contaminated fish and seafood. Moreover, chemical additives such as retardants and plasticizers commonly used in plastics contribute to water, air, and soil contamination with long-term health risks to fish consumers (Tilman & Clark, 2014; Lundebye et al., 2021) and fishing communities. It is clear that fish consumption can meet the nutritional requirement but poses a potential risk to our health due mainly to contaminated seafood brought out from capture fisheries. Fish produced through aquaculture appears safer for human consumption.

14 Effect of Climate Change on Aquaculture

The Paris Climate Agreement of 2015 recognized the threat of climate change and the need for mitigation measures toward food production through fisheries and aquaculture. The short-term climate change includes loss of production and infrastructure due to extreme events such as floods, disease outbreaks, and harmful algal blooms (Cochrane et al., 2009). A significant decrease in fisheries production due to climate change has been reported (FAO, 2019). On the other hand, aquaculture's contribution to global fish production has continued to rise, reaching 82.1 million tons (46%) out of the estimated 176 million tons of global production (FAO, 2020). It is growing sustainably and is expected to reach 53% of the world's fish production. Recently, climate change effects on aquaculture sustainability have gained considerable interest owing to the sector's significant contribution to global food security (Maulu et al., 2021). The aquaculture industry is increasingly being threatened by the predicted effects of climate change. The UN projection on global population growth is a phenomenal increase of 50% since 2000 to 9.5 billion by 2050 (UN, 2015).

It has been reported that climate change will impact water availability, weather patterns, and extreme rain events (De Silva & Soto, 2009). Therefore, the strategies for increasing fish production from aquaculture should be directed toward horizontal and vertical growth of the sector. Under culture expansion, renovation, and management of resources have to be done to improve fish production (FAO, 2019). Apparently, aquaculture in the next few decades will have to be intensified to bridge the gap between demand and supply in a cost-effective manner. A more efficient use of natural resources and more targeted transformative changes are needed to achieve a more sustainable fisheries and aquaculture sector (Tran et al., 2022). The UN 2030 Development Agenda adopted in 2015 outlines 17 Sustainable Development Goals (SDGs) and the organization has continued to put food security at the forefront (Azra et al., 2021). The expected increase in global population and future food security debates has emerged as hot topics. In recent years, the COVID-19 pandemic has demonstrated the fragility of the global food system. We need greater resilience in our food system to support food security against unexpected events. The future demands to provide food security nutrition and an affordable healthy diet by ever-growing human population can safely be met from fisheries and aquaculture.

15 Conclusion

Fisheries play a significant role in supporting the economy, employment, food sources, and livelihood of people across the globe (Alsaleh, 2023). The total world fish production in 2023 has reached 186.6 MMt. India stands third in total fish production in the world. Aquatic foods are contributing more than ever before to food security and nutrition. World capture fisheries do not show the prospect of a high increase in catches. Therefore, world fish production through aquaculture becomes a key factor in global food security with nutritional value. The resource diversification and growth of aquaculture will produce more fish and help us to end global hunger and malnutrition. Responsible aquaculture and prevention and management of aquatic diseases need to be addressed for improving productivity. Achieving a high growth rate on a sustainable basis over a period of 5 years would be possible only with advancement in technology and expansion, intensification, and diversification of culture in the National Aquaculture Development Plan (Gopakumar et al., 1999). The aquatic animals are the highly nutritious and cheapest protein sources, which serve as a valuable supplement in diets by providing essential vitamins, proteins, micronutrients, and minerals, for the poor people. The World Aquaculture sector has to come up with timely strategies to cope with the future challenges of increased fish demand, selective consumers' choices, production of safe and quality fish protein, and tapping the export earnings. It is even more challenging that all these have to be done in the face of increased land and water scarcity, pollution, climate change, competition from other agriculture sectors, shortage of labor, and raw materials besides satisfying the Code of Conduct for Responsible Aquaculture (CCRA). The process of Blue Transformation and revolution in aquatic food production is crucial if we are to achieve the UN Sustainable Development Goals. The aquaculture industry with adequate support and access to finance and technical resources is a widely accepted tool that will help in high return and regular supply of protein-rich food. It is at present considered as a more sustainable alternative that can help secure the food supply for the future growing hungry global population.

References

Abbey, L., Glover-Amengor, M., Atikpo, M. O., Atter, A., & Toppe, J. (2016). Nutrient content of fish powder from low value fish and fish by-products. *Food Science & Nutrition, 5*, 374–379.

Adeoti, I. A., & Hawboldt, K. (2014). A review of lipid extraction from fish processing by-product for use as biofuel. *Biomass and Bioenergy, 63*, 330–340.

Alsaleh, M. (2023). The role of the fisheries industry in the shift towards sustainable food security: A critical study of blue food. *Environmental Science and Pollution Research, 30*, 105575–105595.

Ansari, Z. A., Desilva, C., & Badesab, S. (2012). Total petroleum hydrocarbon in the tissues of some commercially important fishes of the bay of Bengal. *Marine Pollution Bulletin, 64*, 2564–2568.

Arino, J. A., Beltran, A., Herrera, A., & Roncules, R. (2013). Fish and seafood: Nutritional value. In *Encyclopedia of Human Nutrition*. https://doi.org/10.1016/B978-0-12-37083

Ayyappan, S., & Jena, J. K. (2003). Grow-out production of carps in India. *Journal of Applied Aquaculture, 13*(3/4), 251–282.

Azra, M. N., Okomoda, V. T., Tabatabaer, M., Hassan, M., & Ikhwanuddin, M. (2021). The contribution of shellfish aquaculture to global food security: Assessing its characteristic from a future food prospective. *Marine Fisheries, Aquaculture and Living Resources, 8-2021*.

Basu, S. & Chouksey, M. K. (2003) Fish processing and product development, Central Institute of Fisheries Education, Mumbai. 58 pp.

Bhatta, R. (2003). Socio-economic issues in fisheries sector in India. In K. Anjani, K. K. Pradeep, & P. K. Joshi (Eds.), *A profile of people, technologies and policies in fisheries sector in India* (pp. 17–42). National Centre for Agricultural Economics and Policy Research.

Boyd, C. E., McNevin, A. A., & Devis, R. P. (2022). The contribution of fisheries and aquaculture to the total protein supply. *Food Security, 14*, 805–827.

Cochrane, K., De Young, C., Soto, D., & Bahri, T. (2009). Climate change implications for fisheries and aquaculture. *FAO Fisheries and aquaculture technical paper, 530*, 212.

DADF. (2017). *Annual report 2016–17* (162pp.). Department of Animal Husbandry, Dairing and Fisheries. Ministry of Agriculture, Govt. of India.

De Silva, S. S., & Soto, D., 2009. Climate chnge and aquaculture: Potential impact, adaptations and mitigations. In: K. Cochrane, C. De Young, D. Soto, & T. Bahri (Eds.), *Climate change implications for fisheries and aquaculture: Over view of current scientific knowledge* (FAO fisheries and aquaculture technical papers No. 530, 212pp.). Rome: FAO.

Delgado, C. L., Wada, N., Rosegrant, M. W., Meijer, S., & Ahmad, M. (2003). *Outlook for Fish to meeting global demand* (Food policy report). IFPRI.

FAO. (2000). *The state of world fisheries and aquaculture*. Fisheries Department.

FAO. (2015). *The State of Food Insecurity in the World*. http://www.fao.org/3/a4ef2d16-70a7-460a-a9ac-2a65a533269a/i4646e.pdf

FAO. (2016). *The state of world fisheries and aquaculture contributing to food security and nutrition for all* (p. 2016). Rome.

FAO. (2017). *National aquaculture overview: India* (Country profile fact sheets). FAO Fisheries and Aquaculture Department, FAO. http://www.fao.org/fishery/factsheets/en. Accessed 1 Dec 2017.

FAO. (2018). *The state of world fisheries and aquaculture 2018: Contribution to food security and nutrition for all*. FAO.

FAO. (2019). Synthesis of current knowledge, adaptation and mitigation option. In M. Barange (Ed.), *Impact of climate change on fisheries and aquaculture* (FAO fisheries Technical paper 627). FAO.

FAO. (2020). *The state of world fisheries and aquaculture*. FAO. http://www.fao.org/documents/card/en/c/ca9229en

FAO. (2022). *The state of world fisheries and aquaculture*. FAO.

FAO. (2023). *Fisheries and aquaculture country profile-India*. Rome.

Gareth, P. (2001). *Fisheries and the environment. Fisheries subsidies and overfishing: Towards a structured discussion*. UNEP.

Ghoshal, T. K., De, D., Biswas, G., Kumar, P., & Vijayan, K. K. (2015). Brackish water aquaculture: Opportunities and challenges for meeting livelihood demand in Indian Sunderbans. In *The sundarbans: A disaster-prone eco-region*. Springer.

Gopakumar, K., Ayyappan, S., Jena, J. K., Sahoo, S. K., Sarkar, S. K., Satapathy, B. B., & Nayak, P. K. (1999). *National freshwater aquaculture development plan*. Central Institute of Freshwater Aquaculture.

Handbook of Fisheries and Aquaculture. (2013). Indian Council of Agriculture Research publication (ICAR), India.

Hanson, T. (2013). Economic analysis project rising returns for intensive biofloc shrimp system. *Global Aquaculture Advocate, 16*, 24–26.

Hayes, M. (2014). Bioactive peptides and their potential use for the prevention of diseases associated with Alzheimers' disease and mental health disorders: Food for thought? *Annals of Psychiatry and Mental Health, 2*, 1017.

Hayes, M., Mora, L., Hussey, K., & Aluko, R. E. (2016). Boarfish protein recovery using the pH shift process and generation of protein hydrolysates with ACE-I and antihypertensive bioactivities in spontaneously hypertensive rats. *Innovative Food Science & Emerging Technologies, 37*(4). https://doi.org/10.1016/j.ifset.2016.03.014

Haylor, G. (2004). Poverty reduction and aquatic resources. Emerging trends and experiences in Asia-Pacific aquaculture 2003, NACA/FAO. *Proceedings of symposium hosted by the government of Sri Lanka, 21–25 April 2004*. FAO/NACA.

Hollowed, A. B., Barange, M., Beamish, R. J., Brander, K., Cochorane, K., Drinkwater, K., Foremn, M. G. G., Hare, J. A., Jason, H., & Ichi, H. S. (2013). Projected impact of climate change on marine fish and fisheries. *ICES Journal of Marine Science, 70*, 10241037.

Jayasankar, P. (2014). Recent advances in freshwater finfish aquaculture: Prospects and constraints. In V. R. P. Sinha & P. Jayasankar (Eds.), *Aquaculture – New possibilities and constraints* (pp. 1–12). Narendra Publishing House.

Jayasankar, P., & Barik, N. K. (2015). Promotion of PPPs for wider adoption of freshwater aquaculture technologies: Scope and strategies. In: V. R. P. Sinha, K. Gopal, P. Keshavanath & N. R. Kumar (Eds.), *Social entrepreneurship in aquaculture* (pp. 57–67). Narendra Publishing House, New Delhi, India.

Jayasankar, P. (2018). Present status of freshwater aquaculture in India – A review. *Indian Journal of Fisheries, 65*(4), 157–165.

Keelery, S. (2024). *Fish production volume in India FY 2011–2022*. Statista.

Khalil, T. S., & Samples, S. (2018). Nutritional value of fish, lipid, protein, vitamins and minerals.: Review. *Fisheries Science and Aquaculture, 26*, 243–253.

Kibria, G. (2023). Impact of microplastics on fisheries and seafood security. Global analysis nd synthesis. *Science of the Total Environment, 904*, 166652.

Kumar, M. S. (2014). Sustainable management of fisheries and aquaculture for food security and nutrition. Policies requirement and action. *Agricultural Research, 3*, 97–103. https://doi.org/10.1007/s40003-014-0111-0

Lundebye, A. K., Lusher, A. L., & Bank, M. S. (2021). Marine microplastics and seafood: Implication for food security. In *Microplastic in the environment: Pattern & process*. Springer.

Maulu, S., Hasimuna, O. J., Haambiya, L. H., Monde, C., Musuka, C. G., Makorwa, T., Munganga, B. P., Phiri, K. J., & NSekanabo, J. D. (2021). Climate change effect on aquaculture production, sustainability, implications mitigation and adaptation. *Frontiers in Sustainable Food system., 5*, 2021. https://doi.org/10.3389/fsufs.2021.609097

Neethu, K. V., Praved, P. H., Xavier, N., Sankar, N. D., Antony, H., Nandan, S. B., Karthikeyan, P., Marigoudar, S. R., & Sharma, K. V. (2024). Bioaccumulation of heavy metals in seafood resources from the south west coast of India: Human health risk assessment and importance of seafood security. *Journal of Toxicology & Environment Health Science, 16*, 217–231.

Pagarkar, A. U., Satam, S. B., Chogale, N. D., Shinde, K. M., Bhatkar, V. R., Shingare, P. E., Phadake, G. G., Gaikwad, B. V., Metar, S. Y., Sadawarte, V. R., & Bhosale, B. P. (2021). Fish and its nutritional and medicinal importance of fish in human diet. In *Freshwater aquculture, RAFA-2021 on 21–22 Jan. 2021*. Solanke Mahavidyalya, Beed.

Pandion, K., Arunachalam, K. B., Rajagopal, R., Ali, D., Alarifi, S., Chang, S. W., & Ravindran, B. (2023, April). Health risk assessment of heavy metals in the seafood of Kalpakkam coast, South east Bay of Bengal. *Marine Pollution Bulletin, 189*, 114766. https://doi.org/10.1016/j.marpolbul.2023.114766

Pradeepkiran, J. A. (2019). Aquaculture role in global food security with nutritional value: A review. *Translational Animal Science, 3*, 903–910.

Ragesh, S., Abdul Jaleel, K. U., Nikki, R., Razaque, M. A., Ashraf, P. M., Nair, C. K., Abdulaziz, A., & Dinesh Kumar, P. K. (2024). Environmental and ecological risk of microplastic in the surface waters and gastrointestinal track of skipjack tuns (*Katswounus pelamis*) around the Lakshadweep Islands, India. *Environmental Science and Pollution Research*. https://doi.org/10.1007/s11356-024-32564-y

Rao, G. R. M., & Ravichandran, P. (2001). Sustainable brackish water aquaculture. In T. J. Pandian (Ed.), *Sustainable Indian fisheries* (pp. 134–151). New Delhi.

Swain, S., & Ferosekhan, S. (2022). Present status and future scope of freshwater aquaculture sector in India. In *Souvenir of first Indian fisheries outlook* (pp. 84–99). ICAR-CIFRI.

Salim, S. S., Pramod Kiran, R. B., Joshua, N. E., & Kumar, B. (2014). Challenges in food security: The fisheries and aquaculture policy perspectives in India. *Journal of Aquatic Biology & Fisheries, 2*(1), 24–31.

Samples, S. (2014). Towards a more sustainable production of fish as an important source for human nutrition. *Journal of Fisheries & Livestock Production, 2*, 119.

Saroj, S., & Ferose Khan, S. (2022). Present status and future cope of freshwater aquaculture sector in India. In *Souvenir Indian fisheries outlook* (pp. 84–99). ICAR-CMFRI.

Shenai, P. S., Tirodkar, M. U., & Gauns, Z. A. A. (2016). Concentration of heavy metals in commercially important oysters from Goa-central west coast of India. *Bulletin of Environmental Contamination and Toxicology, 97*, 813–819.

Sidhu, K. S. (2003). Health benefits and potential risks related to consumption of fish or fish oil. *Regulatory Toxicology and Pharmacology, 38*, 336–344.

Sreenivasan, P. V. (2003). Food security module and India role of agriculture project international conference. 20–22 October, Rome, Italy.

Subasinghe, R., Soto, D., & Jia, J. (2009). Global aquaculture and its role in sustainable development. *Reviews in Aquaculture, 1*, 2–9. https://doi.org/10.1111/j.1753-5131.2008.01002.x

Sugiyama, S., Staples, D., & Fungi-Smith, S. (2004). *Contribution of fisheries an aquaculture in the Asia Pacific: Status and potential of fisheries and aquaculture in Asia-Pacific* (RAP publication 2004/25, FAO corporate Document). Repository FAO.

Surendran, V., Madhusudhan Reddy, K., & Subba Rao, V. (1991, February). Semi-intensive shrimp farming-TASPARC'S experience at Nellore. *Fishing Chimes, 1991*, 23–29.

Thompson, L. A., & Darwish, W. S. (2019). Environmental chemical contaminants in food. Review of a global problem. *Journal of Toxicology, 2019*, 2345283.

Tidwell, J. H., & Allan, G. L. (2001). Fish as food: Aquaculture's contribution. *EMBO Reports, 2001*(2), 958–963.

Tilman, D., & Clark, M. (2014). Global diets link environmental sustainability and human health. *Nature, 515*, 518–522.

Tran, N., Chu, L., Chan, C. Y., Peart, J., & Nasr-Allah, A. M. (2022). Prospect of fish supply—Demand and its implication for food and nutrition security in Egypt. *Marine Policy, 146*, 105333. https://doi.org/10.1016/j.marpol.2022.105333

United Nations. (2015). *Revision of world population prospects*. United Nations. Available online: https://esa.un.org/unpd/wpp/publications/files/keyfindingswpp2015.pdf

Unuofin, J. O., & Igwaran, A. (2023). Microplastic in sea food: Implication for food security, safety and human health. *Journal of Sea Research, 194*, 2023. https://doi.org/10.1016/j.seares.2023.102410

Vijay Kumar, V., Singh, R., Garagu, S. A., Valappil, N., & Rakhavan, R. (2022, December). Heavy metal contamination, distribution source apportionment in the sediment from Kavvayi estuary, south west cost of India. *Total Environment Research Theme, 3-4*, 100019.

Vijayan, K. K., & Balasubramanian. (2019). *Brackish water aquaculture in India. A driver for blue economy*. https://www.Researchgate.net/publication/34009/750

WRAP (2017). Food futures: From business as usual to business unusual. Available online: http://www.wrap.org.uk/content/food-futures. Accessed on 12 Mar 2017.

Chapter 3
Effective and Responsible Management and Nutritional Strategies and Interventions for Aquaculture Performance Enhancement: Assurance of Food Security and Production Sustainability

Kifayat Ullah Khan, João Batista Kochenborger Fernandes, and Robert Mervyn Gous

3.1 Why Aquaculture?

Food production and consumption have several impacts on the planet earth, particularly in view of the expanding world's population, which could possibly reach up to ten billion by 2050 (FAO, 2017; Tilman et al., 2017; UN DESA, 2015). About 40% of terrestrial land is already cultured or cropped (FAO, 2013), resulting in quick loss of habitat and species genetic diversity (Herrero et al., 2015), inappropriate use of freshwater (Herrero et al., 2015; Mekonnen et al., 2019), enormous pollution in both terrestrial and water ecosystems (Bouwman et al., 2013), and significant release of greenhouse gases into the atmosphere (Tilman & Clark, 2014) over the past century. Moreover, terrestrial animals usually produce enough metabolic heat to regulate their body temperatures and for excretion purposes as well as convert dietary energy to growth less proficiently. On the contrary, aquatic animals do not produce enough metabolic heat, and they use the temperature of their surrounding environment to maintain their body temperatures to optimize their performance and general fitness. Another benefit of rearing aquatic animals is that they usually need

---

K. U. Khan (✉)
Department of Zoology, Government College University, Lahore, Pakistan

Aquaculture Center of UNESP, São Paulo State University (UNESP), Jaboticabal, SP, Brazil
e-mail: kifayatkhan@gcu.edu.pk; kifayat055@gmail.com

J. B. K. Fernandes
Aquaculture Center of UNESP, São Paulo State University (UNESP), Jaboticabal, SP, Brazil

R. M. Gous
School of Agricultural, Earth, and Environmental Sciences, University of KwaZulu-Natal, Pietermaritzburg, South Africa

I. Ahmed, I. Ahmad (eds.), *Aquaculture: Enhancing Food Security and Nutrition*, https://doi.org/10.1007/978-3-031-92858-1_3

a very low amount of energy for the excretion of nitrogenous wastes, which commonly occurs in the form of ammonia release through gills. They also efficiently convert dietary energy into biomass. Some ectotherms are efficient at using certain specific behavioral strategies to regulate their body temperature, although they do not use metabolic heat to do so (Bergman et al., 2019). Overall, aquaculture practices are advantageous over terrestrial animal farming in terms of having minimal negative implications, including land use vicissitudes, greenhouse gas production, and nitrogen effluents for local ecosystems, and producing highly nutritional products for diverse purposes, especially human consumption (FAO, 2017; Tlusty et al., 2018).

3.2 Aquaculture Roles in Global Food Security

Ensuring food security through responsible fisheries and aquaculture practices in response to the growing human population while acquiring both economic and environmental sustainability could be a promising strategy for a healthy and successful existence of life on this planet earth in the long run (FAO, 2017). Several efforts to ensure food security have been studied since 1945 (Shaw, 2007) due to the increasing demand for healthy food in the consumer market. Different food production sectors are working in this regard and are significantly contributing to global food production chains each year. In this scenario, the fisheries and aquaculture sector is playing a crucial role in continuously supplying healthy and sustainable aquatic products to the consumer market, securing an average annual production rate of more than 5.5% (FAO, 2020). For this reason, this sector has the full potential to support the global demand for animal protein in the long run (Engle et al., 2017). Since capture fisheries are in decline due to several challenges such as pollution, overexploitation of wild stocks, and climate change, the aquaculture sector is successfully fulfilling (by about 20%) the demand for aquatic products in the global consumer's market (Stickney, 2005). Aquaculture is expected to continuously grow and ensure the food security and livelihood of the communities (FAO, 2017; Froehlich et al., 2018; HLPE, 2016).

3.3 Nutritional Significance of Aquaculture Products

Farmed aquatic products are very nutrient-rich and efficient compared to food products such as chicken and beef, which are obtained from terrestrial animals. In aquaculture, the feed conversion ratio (FCR), which means how much feed it takes to produce the protein, is a 1:1 ratio. This shows that 1 g of protein could be efficiently obtained by providing a 1-g feed. On the other hand, the FCR in terrestrial animals is considerably variable and remains superior compared to that in aquatic animals.

Consequently, in aquatic animals, protein and energy retention are very high as well (FAO, 2017; NRC, 2011).

As farmed shellfish and fish growth, health, and quality of the rearing water are strictly monitored compared to wild stocks in natural water bodies, aquaculturists have more control over production inputs and outputs. This could have positive implications for both the aquatic and terrestrial environments as well as the culture of aquatic animals. Farmed aquatic animals are usually free of contaminants such as heavy metals, insecticides, and pesticides, as they solely eat artificial feeds that are developed by considering the nutritional requirements of the target species. Artificial feeds, when properly prepared by fulfilling the nutritional requirements of the target species, are processed more healthily compared to natural foods (FAO, 2017; NRC, 2011).

3.4 Systems for Aquaculture Production

Strengthening terrestrial aquaculture by using efficient production systems with proper management and nutritional strategies may lead to optimized and sustainable growth and health. Intensive and semi-intensive aquaculture involves intervention in the rearing process, such as with supplemental feeding and regulating water quality aspects, while extensive aquaculture stock grows on its own without any supplemental feeding or water quality control. Overall, there are very little production costs in extensive aquaculture; however, productivity remains very low in such farming, and achieving harvest size is a lengthy process that might be unable to fulfill the market demand. Contrarily, in intensive and semi-intensive practices, the costs usually remain high, but the productivity is several-fold higher, and harvest size could be achieved in a short time (Costa-Pierce et al., 2010; FAO, 2017).

Intensive and semi-intensive farming is a key strategy to continuously supply diverse aquatic products to the global consumer market. Especially, promoting intensive farming in recirculating aquaculture systems (RASs) and using nutritionally balanced and economically affordable feed could enhance aquaculture production several-fold compared to other types of aquaculture. The latter farming method is very beneficial in regions where there is a shortage of water and land resources. Moreover, in intensive farming, close monitoring of fish behavior enables farmers to quickly detect the symptoms of any disease, stress, or feed deficiency and properly fix any arising issue. Although intensive aquaculture is much more productive, semi-intensive and extensive aquaculture are also significantly contributing to reshaping the degraded aquatic ecosystems, particularly to preserve the natural wetlands in coastal areas (Boyd et al., 2020; Laktuka et al., 2023; Mamun et al., 2021).

Aquaculture sustainability is also linked to exercising all three types of aquaculture instead of depending solely on one type. This could be a valid strategy to cope with the challenges of water and land shortage as well as the degradation of aquatic ecosystems in order to maintain a balance between natural resource stocks and their use (Doublet & Pointereau, 2008; FAO, 2017).

3.5 Integrated Aquaculture: Agriculture or Livestock—Ecosystem-Based Management

Integrating aquaculture with agriculture or livestock farming is a promising sustainable approach, which has already been in practice in several countries, to enhance production assortment, land efficiency, and nutrient recycling on farms. The diversity of production at farms may improve the livelihood of small-scale aquaculturists by providing diverse, quality products to the local communities (Ahmed et al., 2007; FAO, 2017). Conventionally, livestock or agricultural production is estimated in terms of biomass or income per specific area of land. However, Ignowski et al. (2023) have recently introduced nutrition—a sensitive index for farming production, considering production of macro- and micronutrients relative to human needs or other purposes. This index allows exploring the relationship between nutrient production and economic returns across a range of existing aquaculture-integrated systems, highlighting an instinctive estimation of nutrient sensitivity that could be easily understood by investigators and legislators and promoted by farmers, especially among small-scale farm holders. So, production diversity linked to the amalgamation of aquatic and terrestrial foods in integrated aquaculture systems could be advantageous for nutrient utilization and economic returns. This is an efficient way to enhance the role of sustainable aquaculture in ensuring global nutrition security (FAO, 2017; Ignowski et al., 2023).

Integrated aquaculture employment not only enhances production diversity and efficiency of the systems engaged together, but it also has a significant role in lessening aquaculture susceptibility and enhancing its flexibility to climate change. Among integrated aquaculture systems, integrated multi-trophic aquaculture (IMTA) is gaining popularity, which is an innovative technique of rearing finfish together with shellfish (shrimp) or aquatic plants such as algae. Interestingly, we are seeing more applications of this system in many forms throughout the world (Chopin, 2013b; WEF, 2021; Zhang et al., 2011).

The farming of filter feeders, like shellfish, together with finfish can also improve the quality of rearing water. These organisms eat excessive nutrients in the water, which in turn prevents the buildup of effluent by enhancing nutrient recycling. Filter feeders are often integrated into the farming of other species, like finfish, to re-utilize the uneaten feed or fish waste, offsetting the farm's environmental impact. This system is called polyculture or integrated multi-trophic aquaculture (Chopin, 2013a). It has also been suggested that shellfish rebuilding plans and developing small-scale mollusk-based aquaculture setups may alleviate the negative implications of developmental projects or other anthropogenic events occurring near natural aquatic habitats, promoting large-scale eutrophication. Bivalve mollusks have the full potential to perform an estuarine filter role, increasing the clarity of water and helping to remove nitrogen, phosphorus, and other nutrients from the eutrophic zones (FAO, 2017; Rice, 2001).

3.6 Potential Management and Development Plans for Successful Aquaculture

The success of aquaculture practices in a region is absolutely subject to the establishment of potential management and development plans. Otherwise, any aquaculture system could fail. The plans should include founding a responsible and sustainable aquaculture while safeguarding food security, livelihood, and well-being of a community or a nation, particularly the most vulnerable segments of communities, including children and women. The management and development plans must ensure sustainable utilization of water, land, and infrastructure resources in accordance with the sustainable development goals (SDGs) of the United Nations (Costa-Pierce et al., 2010; FAO, 2017).

Appropriate organizational and infrastructure development with skilled professionals is necessary for the overall success of this sector. Proper training modules for the government, concerned institutions, communities, professionals, researchers, aquaculturists, and private sector may be recognized and duly approved by concerned authorities. The plan may promote a complete strategy toward sustainable aquaculture management and development for all factors, from export and import of aquatic organisms or aquatic products and certification of aquatic products to hatchery and farming licensing, animal handling and processing, as well as product marketing and waste disposal. The plan should especially ensure the due rights and safety of the aquaculturists and their families. The aquaculture plan may consist of a sound commitment to benefit the most vulnerable segments of the community, the children and women. Overall, it should be flexible and may consist of international-level modernized and responsible aquaculture legislation, protocols, procedures, plans, guidelines, quality control, biosecurity, action against climate change, ensuring production sustainability, products nutritional value and safety, community well-being and livelihood, ensuring due rights and safety of the fishermen and their families, and developing required permitting and certification systems (FAO, 2017, 2018a, b; Laktuka et al., 2023).

3.7 Critical Factors Affecting the Efficiency and Sustainability of Aquaculture Systems

Water quality, type of rearing system, stocking density, seed quality, feeding practices, including feeding frequency and feeding rate, feed quality, including feed texture, palatability, nutritional composition, digestion and assimilation, feed price (NRC, 2011; Oshira et al (2024)), as well as harvest marketing strategies (Khan et al. 2020a, b, 2022), are among the most important factors affecting the efficiency and sustainability of aquaculture systems (FAO, 2017).

3.7.1 Management of Water Quality in Aquaculture

For optimum performance of aquatic organisms and the overall success of aquaculture activities, water quality improvements are a significant win. Good water quality for one species could be lethal to another; therefore, much research has been done over the years to quantify the optimum range of water quality variables for each culturable species (Stickney, 2005). Potential managemental practices, including waste disposal, are key to maintaining water quality aspects within the optimum range required for a species. This includes regular cleaning and maintenance of tanks, ponds, cages, or pens, with appropriate stocking densities and feeding practices, including feed quality, palatability, as well as feeding frequency and rate. Aquaculturists must be conscious of possible pollution sources and may take steps to lessen nutrient and chemical runoff from surrounding land as well as from supplemental feed. Regular biomonitoring and software use for data collection, tracking, reporting, and prediction systems, along with trained workers, are mandatory to make this activity successful and sustainable (FAO, 2017; Tziortzioti et al., 2019).

Monitoring and assessment of water quality aspects, particularly temperature, pH, dissolved oxygen (DO_2), ammonia, and turbidity, on a regular basis has a crucial role in successful and healthy aquaculture operations. As fish bodies depend on external aquatic conditions to regulate their metabolism and other physiological and biochemical mechanisms accordingly, these aspects have a direct and indirect impact on aquaculture performance. Farmers must also be aware of the appropriate water quality range for the cultured species and take potential steps to maintain it throughout the growing season up to harvest. It is important to note that different species of fish have different water quality requirements, so it is essential to know the ideal requirements of the target species (Lindholm-Lehto, 2023; Mwegoha et al., 2010).

Ammonia, one of the most important water quality aspects, is usually amassed in aquaculture systems followed by microbial decay of the uneaten feed, organic matter, and fertilization practices. Microbial decay may result in low dissolved oxygen levels in tanks or ponds and may transform ionized ammonia (NH_4^+) into un-ionized ammonia (NH_3), which is highly toxic to aquatic organisms and may cause damage to tissues and organs, induce stress, and disturb the general homeostasis (Stickney, 2005). The absence of enough dissolved oxygen increases toxicity to aquaculture species. In an aqueous ammonia solution, the unionized form of ammonia exists in equilibrium with the ionized form of ammonia and hydroxide ions. The unionized ammonia has high lipid solubility, and it is capable of diffusing quite readily across the cell membranes. The nitrifying bacteria (*Nitrosomonas* and *Nitrobacter*) use ammonia as an energy source and oxidize it to nitrite and nitrate (Chumkiew et al., 2019; Timmons et al., 2018).

Temperature is another important water quality parameter that affects the survival, feed intake, metabolism, growth, and general health conditions of aquatic animals. Cold winds can abruptly disbalance the pond water temperature because deeper water comes to the surface and surface water goes down, inducing life-threatening consequences for the aquatic organisms (Stickney, 2005). Every aquatic animal species has a specific optimum range of temperature, and if it exceeds or drops in relation to that optimum range, this may induce undesirable changes in the feeding behavior, appetite, activity, metabolism, growth, and health, as well as reproductive performance (Drake et al., 2010; Olaniyi et al., 2022).

Dissolved oxygen (DO_2) represents the concentration of atmospheric oxygen being saturated in the rearing water. Water receives oxygen directly from the atmosphere through the diffusion process and via photosynthesis from aquatic plants in the surrounding environment. Aquatic animals require optimum oxygen concentration in water for survival and successful growth, so DO_2 plays a significant role in successful aquaculture practices. The adequate amount of DO_2 in rearing water should be at the rate of about 5 mg/L or above for warm-water species and about 7 or above for cold-water species, with an overall 3 mg/L as a threshold but overall not less than 2 mg/L (Stickney, 2005). Significantly, lower levels of DO_2 (such as 1 mg/L or below) induce oxidative stress in aquatic animals; they will begin to move toward the water surface to breathe for air. Feed intake will decline and mortalities could occur if there are high stocking densities, concentrated rearing systems, or increased temperature or ammonia concentration in water (Buentello et al., 2000; Zhang et al., 2011).

The pH values represent the magnitude of water alkalinity or acidity, ranging from 0 to 14. Water with a pH of 7 shows that it is neutral by nature. The pH values less than 7 are considered acidic, while those greater than 7 are alkaline. Inappropriate pH levels may lead to mortalities, and therefore, it is essential to maintain the pH levels within the optimum range required for a species. Lower pH can harm shells, gills, eyes, skin, and other aspects of aquatic organisms and make it difficult for them to perform their body activities efficiently. Higher pH concentrations may enhance the toxicity of total ammonia in the rearing water, which could be lethal for fish and other aquatic animals (Boyd et al., 2016; Olaniyi et al., 2022). The pH of water can be maintained within the optimum range for a species by liming ponds before stocking. Liming slightly increases pH and aids in the elimination of pathogens by destroying their membranes and homeostasis. This adjustment helps keep the pH within the optimal range throughout the growing season. Various forms of agriculture lime are used for this purpose, including crushed mollusk shells (Stickney, 2005).

Turbidity is another important water quality index, which measures the magnitude of cloudiness or clarity of water. Suspended solid particles, decaying organic matter, uneaten feed, poor feed quality, or higher plankton concentration may result in water turbidity. Turbid water stops light entry and may reduce the photosynthesis process. The suspended solid particles can block the gills of the fish and eventually trigger their death. These sediments also engross heat from the sun, which raises the temperature of rearing water and declines photosynthetic activity, resulting in lower DO_2 levels and an increase in the concentrations of CO_2 and other harmful gases beyond the optimum range. This way, the survival and performance of aquatic animals could severely be affected in a negative way (Stickney, 2005; Tziortzioti et al., 2019).

3.7.2 *Impact of Aquaculture Culture Systems on Production Sustainability*

3.7.2.1 Extensive Aquaculture

This is an ancient type of aquaculture, which is still in practice due to its traditional background and wide acceptance among the rural populations settled near aquatic bodies. This is the simplest aquaculture practice, where a confined zone is created by fixing a net, cage, pen, or fence in a large water body where aquatic organisms are cultured. Human intervention is only limited to confining an area, stocking, escape prevention, and hence harvesting and marketing. The output of such systems is usually very low, but this is an ecosystem-friendly practice that could be easily, but in a responsible way, exercised in sensitive areas of the waters (Anras, 2008; Lekang, 2013).

Although this is an environment-friendly practice, environmental stressors and climate change could have negative implications for the growth and health of such systems. Another issue could be "escape," as cultured animals may escape to the wild environment, posing adverse environmental and economic consequences, and therefore it must be properly monitored and managed. Aquatic animals cultured in extensive systems could have higher acceptability among consumers compared to the aquatic organisms reared in semi-intensive or intensive systems. The reason is that many people still think that aquatic animals reared on artificial feed are "not organic," as highlighted by Stickney (2005), so for them, the labeling of the products of extensive aquaculture is one of the better alternatives to aquatic products of intensified systems for sustaining the net profit of this environment-friendly practice. This is the only way forward to cope with the market demand for products from the "wild" in certain regions in order to maintain the overall sustainability of aquaculture (Anras, 2008; FAO, 2018a, b).

3.7.2.2 Semi-intensive Aquaculture Systems

These systems are partially managed through human intervention. Lekang (2013) stated that semi-intensive systems are actually the combination of extensive and intensive systems, such as the production of small fish in intensive rearing systems followed by growing in an extensive culture area. The inputs involved are water source, ponds, stocking, aeration, partially artificial feeding, fertilization, treatments, removal of excessive vegetation from inside ponds and its banks in order to regulate the water quality, waste management, and finally harvesting and marketing (Boyd, 2018). If there is a continuous supply of water (flow-through system), then the rearing facility would need a little management of water and waste, but if there is a semi-static water system, then effective water and waste management is necessary. Feeding is done a minimum of two times a week and fertilization once per week; however, overall, all these inputs vary from region to region and species to species (Oddsson, 2020). The productivity in such systems is higher than in extensive systems but remains lower than in intensified systems. However, such systems could be integrated with agriculture or livestock farming to increase productivity and minimize the negative consequences of both types of farming for aquatic and terrestrial ecosystems (Stickney, 2005).

In modern semi-intensive aquaculture, feeding fish with nutritionally balanced and economically feasible artificial feed is the real key to sustainable and healthy farming. Although, like extensive aquaculture, this type of aquaculture could be negatively affected by the worsening climatic conditions, proper water monitoring and management, controlled feeding using quality feed, effective waste management, and stopping nutrient-rich water from entering into natural water bodies are some of the important factors that could be helpful in maintaining sustainability and efficiency of semi-intensive systems (Anras, 2008; FAO, 2017).

3.7.2.3 Intensive Aquaculture Systems

These systems are absolutely controlled through human intervention. The major factors or inputs involved here are water, rearing facility, artificial feeding, treatments, full control of water quality aspects, waste management, regular monitoring, harvesting, and finally marketing of aquatic products (Oddsson, 2020). If a closed system such as recirculating aquaculture is used, then it is necessary to effectively monitor water quality aspects and properly manage the waste produced in rearing ponds. This could be easily done by connecting the rearing facility with a biological filter. For a constant temperature, using a heat exchange system is a good idea, and oxygen could be supplied by using oxygen pumps. Some cold-water species, such as salmon and trout, are cultured in raceways with a continuous water flow (flow-through systems), having minimum environmental effects because the circulation of water helps maintain the water quality variables within a specific range (Stickney, 2005). Overall, intensive aquaculture could give optimum and healthy production if it is properly operated. Moreover, the production of any desired aquatic species could be obtained in a short time compared to extensive and semi-intensive practices. On the other hand, inappropriate handling and management practices could result in unsustainable and unhealthy production in such systems, with harsh environmental consequences (FAO, 2017; Brune et al., 2003).

It has been recently described that the physicochemical characteristics of any type of aquaculture, if they do not remain within the optimum range, could be a potential source of greenhouse gas emissions as well as unsustainable and unhealthy growth (FAO, 2017; da Silva et al., 2018). One of the major issues in any aquaculture system is inappropriate stocking densities, particularly when the number of individuals per cage or tank exceeds the recommended level. This may lead to overcrowding and hence deteriorate the water quality and negatively affect the performance and health of aquatic animals (Ali et al., 2006; Diao et al., 2023; Liu et al., 2016). Moreover, higher stocking densities in aquaculture systems may lead to increased production of greenhouse gases such as methane gas, as highlighted by da Silva et al. (2018). Sustainable intensification of aquaculture facilities and their appropriate integration with livestock farming and agriculture or employing functional recirculating aquaculture systems, especially for intensive farming, is a key to aquaculture success (Engle et al., 2017). Furthermore, this strategy could also help to reduce the pressure on water resources from the aquaculture sector, as highlighted by Costa-Pierce et al. (2010).

3.7.3 *Quality Seed for Aquaculture*

Any aquaculture system is primarily dependent on seed availability (Stickney, 2005). However, the effectiveness and health of aquaculture operations are subject to the morphological and genetic quality of the seed (Gjedrem, 2012). Seeds of different species for stocking in different types of aquaculture systems are usually obtained from both captivity (hatcheries) and the wild. For aquatic species with undetermined breeding techniques and technologies and perhaps lower seed production capability, quality seed could be a challenging issue because commercial production of that species would largely depend on seed from the wild with natural genetic makeup (Mair, 2002). Seed quality is affected by different factors, including nongenetic and genetic factors, as well as breeding and handling procedures and environmental factors (Little et al., 2002). Quality seeds in terms of fast growth, high yield, optimum disease resistance, and high nutritional value could be produced by employing sustainable and effective breeding techniques at hatcheries involving genetic interventions and selective approaches (Hishamunda et al., 2009). The supply of quality seed to rearing farms could greatly contribute to promoting sustainable and healthy aquaculture practices in different regions (FAO, 2017; Tongdee et al., 2017).

The available literature demonstrates that selective breeding and genetic manipulations have brought key improvements to potential aquaculture. Predictions of trade tendencies are made, and the position of such innovative techniques and procedures has been comprehensively elaborated to trigger further advancements in aquaculture performance (Janssen et al., 2017). Genetically improved seeds could give optimum performance in a short time, which is highly desired in the ongoing issues related to food security and food safety. Moreover, culturing genetically improved species could help us enhance the sustainability of aquaculture production systems (Stickney, 2005).

3.7.4 *Potential Nutritional Strategies for Aquaculture Performance*

Optimum nutrition in terms of feed quality and feeding practices has a direct impact on aquaculture growth and health as well as the rearing environment (FAO, 2017; Hardy & Gatlin, 2002). Nutritionally healthy and economically sustainable aquafeed, along with effective feeding practices including feeding frequencies and feeding rates, are the main factors that greatly contribute to obtaining aquatic products with high nutritional value in a relatively short time (Cho & Bureau, 2001). Inappropriate nutritional practices coupled with poor management strategies would show a suppressive impact on aquaculture performances (Sugiura & Hardy, 2000) and could also result in increased waste production and hence high greenhouse gases (da Silva et al., 2018).

As capture fisheries are declining due to several factors such as overexploitation, aquatic pollution, and climate change, aquaculture is becoming an important alternative to wild aquatic resources. Earlier in this chapter, we discussed that there are different types of aquaculture systems, such as extensive, semi-intensive, and intensive farming. Each type has its own advantages and disadvantages; overall, semi-intensive farming is popular among farmers, particularly in developing countries. Although intensive farming is very productive, it needs well-organized systems and skilled farmers for proper handling, so currently this type of aquaculture is mostly practiced in advanced regions of the world. However, due to the increasing demand for animal meat, intensive farming should be promoted everywhere. Intensive farming connected with recirculating aquaculture systems could be a good fit for regions with lower water resources. Considering the importance of semi-intensive and intensive aquaculture, it is worth describing that the success of both systems depends on effective management and optimum nutritional practices (FAO, 2017; Laktuka et al., 2023; Stickney, 2005).

Previously, the primary objective of aquaculturists was to achieve maximum performance, while the economic and environmental aspects were somewhat neglected. However, with the rising issues of aquatic and terrestrial pollution, the United Nations Organization has set certain guidelines (SDGs) for aquaculture, emphasizing sustainability across environmental, economic, and community dimensions. Current research often follows these guidelines, and the obtained data are shared with farmers to accordingly adjust and optimize their nutritional practices, using the locally available protein and energy sources in a responsible way, in order to achieve sustainable and healthy performance (Boyd et al., 2020; FAO, 2017).

3.7.5 Feeding Strategies for Enhanced Optimization of Aquaculture Growth and Health

3.7.5.1 High-Quality Aquafeed for Maximum Performance

The quality of feed is a critical factor for maximum growth performance and optimum health conditions of aquatic animals. Usually, highly nutritious diets are required to get maximum performance without any negative effects. It is important to always use high-quality aquafeed from trustworthy dealers or markets. Protein quality must be optimum, such as the essential amino acid (EAA) balance in feed with optimum energy, which is a prerequisite to successful and sustainable aquaculture nutrition. On the other hand, working with absolute protein contents in aquafeed without fixing the EAA balance or improper energy levels could greatly decline the feed utilization efficiency and overall performance (NRC, 2011). So, if the objective of an aquaculturist is to get maximum aquaculture growth, this could probably be obtained by using a diet with high nutritional quality. However, high-quality diets could also negatively impact the rearing system efficiency, so, in this case, the intensified and semi-intensified aquaculture systems should be properly monitored and cleaned in order to maintain the quality of rearing water within an optimum range. Working with high-quality diets could give us maximum biological output in terms of higher growth and nutritional value; however, the economic performance could be unsustainable, and there could be high waste production. So, such high-quality diets are only ideal for scientific research or producing quality seeds for commercial aquaculture activities (FAO, 2017; Khan et al. 2019, 2020a, b, 2022).

3.7.5.2 Development of Aquafeed for Sustainable and Healthy Production

In the literature on aquaculture nutrition, it has already been documented that aquafeed costs account for about 60% of total production costs. The protein content of feed alone contributes about 70% to the total feeding costs (NRC, 2011). Interestingly, in some recent research, it has been evidenced and validated that changes in total feed cost and different marketing strategies employed to sell the harvest have a great influence on the harvest net profit. These studies reported that feed components, particularly those rich in proteins, synthetic essential amino acids, or oil components, have a great impact on the total feed cost and, hence, gross margin. These studies have concluded that for sustainable aquaculture, the nutritional requirements, particularly protein concentration and quality (essential amino acid balance) in aquafeed, are subject to the purpose of farming, total feed cost, and harvest marketing strategy (Khan et al. 2020a, b, 2022). Moreover, various biological analyses were performed to determine the effects of the aquafeed developed to obtain maximum gross margin with sustainable growth, and these diets were found to be absolutely healthy and safe for fish (Oshiro et al., 2024). It was suggested that the essential amino acid balance must be provided in feed, mostly through locally available conventional protein-rich ingredients; otherwise, nutritional responses and hence sustainability of aquaculture production will be negatively affected (Khan et al. 2020a, b, 2022).

According to these recent data, the way aquaculture harvest is marketed in the consumer market has a great impact on the efficiency and sustainability of the rearing systems. Supplying fillets to the consumer market gives more net profit compared to whole fish or other processed parts, so at the time of stocking, fish farmers should define the destiny of their final yield, as the farmer must know whether the aquatic organisms will be supplied to the market as a whole body or in processed parts (sliced fish or fillet). Accordingly, the feed should be adjusted. Furthermore, it was observed that large-sized fish gave higher net profit compared to the younger (preadult and early-stage juvenile) fish. This information seems very innovative from a commercial point of view, which in aquaculture was not evidenced in the past. These studies strongly suggest that aquaculturists should adjust aquafeed according to the fluctuations in feed price and aquatic product value in the consumer market. These studies have been designed according to the UN Sustainable Development Goals, with the objective of acquiring enhanced optimization of aquaculture performance. The results obtained clearly show that to avoid unsustainable aquaculture performance, aquaculturists should adjust aquafeed according to fluctuations in feed and meat consumer markets.

Aquaculturists should culture species with high market demand, and at the end of the harvest, they should supply processed parts, particularly fillets, slices, and eviscerated forms compared to whole fish, to the consumer markets. Farmers need to perform harvesting only when the aquatic organisms attain market size; otherwise, unsustainable net profit will be the ultimate outcome. Moreover, aquaculturists need to adjust nutrients, particularly protein concentration in aquafeed, by considering the marketing strategy that will be adopted at the end of harvest. If the strategy of a farmer is to supply processed parts to consumer markets, protein concentration in the aquafeed should be kept slightly higher compared to if the plan is to supply whole fish to consumer markets. In both cases, feeding costs will slightly deviate, but net income will always be high in the previous case, which is selling harvest in the form of processed parts. In this strategy, the core objective is to obtain maximum net income with sustainable growth and health dimensions. The execution of this approach could be innovative toward promoting sustainable and healthy aquaculture practices. The findings of the recent advances in aquaculture nutrition (Khan et al. 2020a, b, 2022) are described in Tables 3.1, 3.2, and 3.3.

Table 3.1 Effect of change in the prices of protein-rich ingredients on total aquafeed cost and hence gross margin

Protein-rich ingredients	Fluctuation in price	Total feeding cost	Revenue	Harvest gross margin
Fish meal	Increase ↑	Increase ↑	Decrease ↓	Decrease ↓
Soy bean meal	Increase ↑	Increase ↑	Decrease ↓	Decrease ↓
Synthetic amino acids	Increase ↑	Increase ↑	Decrease ↓	Decrease ↓
Fish or other oil	Increase ↑	Increase ↑	Decrease ↓	Decrease ↓
Fish meal	Decrease ↓	Decrease ↓	Increase ↑	Increase ↑
Soy bean meal	Decrease ↓	Decrease ↓	Increase ↑	Increase ↑
Synthetic amino acids	Decrease ↓	Decrease ↓	Increase ↑	Increase ↑
Fish or other type oil	Decrease ↓	Decrease ↓	Increase ↑	Increase ↑

Table 3.2 Effect of change in harvest product prices or marketing strategy on gross margin

Harvest products	Fluctuation in price	Harvest revenue	Gross margin
Whole fish	Increase ↑	Increase ↑	Increase ↑
Sliced fish	Increase ↑	Increase ↑	Increase ↑
Fillet	Increase ↑	Increase ↑	Increase ↑
Whole fish	Decrease ↓	Decrease ↓	Decrease ↓
Sliced fish	Decrease ↓	Decrease ↓	Decrease ↓
Fillet	Decrease ↓	Decrease ↓	Decrease ↓

Table 3.3 Effect of animal size (growth phase) and type of harvest product on gross margin

(a) Effect of animal size on gross margin		
Fish size	Commercial/market value	Gross margin
Mature/adult	Highest	Highest
Later stage juvenile/preadult	Moderate	Moderate
(b) Effect of aquaculture product on gross margin		
Early-stage juvenile	Lowest	Lowest
Whole-body fish	Lowest	Lowest
Sliced fish	Moderate	Moderate
Fillet	Highest	Highest

3.7.5.2.1 Sustainable Alternatives to Conventional Protein and Energy Ingredients

Fish meal and fish oil are conventional protein and energy sources used in aquafeed. Aquaculture uses around 60% of the current fish meal production and about 80% of fish oil production (FAO, 2020; Lazzarotto et al., 2018). Most fish used for fish meal production are raised for human consumption, so the use of fish meal or fish oil is no longer sustainable as they cause huge dependencies on imports, increase in feed costs, and hence a great decline in gross margins (Boyd & McNevin, 2015). Since a long time ago, wild stocks have already been used for several purposes, such as human consumption, medicinal use, and recreational activities. The increasing aquaculture practices may result in an increasing demand for protein and energy feedstuffs such as fish meal and fish oil. These feed ingredients are most often obtained from marine resources that are wild stocks of fish (Abdel-Tawwab et al., 2020; Alfiko et al., 2022; Basto et al., 2021). The dependence of aquafeed on marine stocks is highlighted by pooling about 21 million tons of marine fish catches, with 76% fixed for the aquafeed industry, putting a huge burden on the wild stocks for purposes other than human consumption (Iaconisi et al., 2018). The continuous use of marine sources, the fish, to get fish meal and fish oil for expanding aquaculture operations could put an extra burden on aquatic populations of oceans and seas (Boyd et al., 2019). In this regard, it is necessary to employ potential nutritional strategies for the enhanced optimization of inland aquaculture practices in order to enhance the captive-reared stocks. This could be done by using several locally available potential feed additives of animal and plant origin in aquafeed. Since feeding costs have a great role in total production costs, adjusting aquafeed by considering the demand for feed and harvest, and hence their prices in the consumer markets, could greatly help to maintain the sustainability of aquaculture production systems (FAO, 2017; Khan et al., 2020a, b, 2022; Oshiro et al., 2024).

The wide acceptance of fish oil and fish meal in aquafeed is because of their better fatty acid pattern and well-balanced essential amino acid profile, high digestibility, and better palatability, respectively. These qualities are vital for essential nutrient intake, digestion, and hence absorption and assimilation (Fantatto et al., 2024). Although fish meal and fish oil result in promising aquaculture performance, the large-scale use of such ingredients not only greatly declines the sustainability of aquaculture activities but also results in the overexploitation of wild stocks (Alfiko et al., 2022). Given that, searching for alternative protein feedstuffs that have good nutritional quality and are economical is critical for enhancing sustainable and healthy aquaculture operations to obtain aquatic products for diverse purposes (Bories et al., 2007; Edwards et al., 2019). Some plant sources, such as soybeans, are already used in aquafeed as an alternative to fish meal and fish oil; however, soybean contains antinutritional factors, is low in some essential amino acids, and has insufficient availability in temperate regions. So, these limitations make this ingredient unsustainable for aquaculture (NRC, 2011). Now, researchers and aquaculturists are working on identifying sustainable potential protein and energy ingredients with minimum environmental impacts.

Several alternative protein and energy sources of plant and animal origin are investigated in aquaculture, which has shown promising results (NRC, 2011). Recently, researchers have been working on some novel alternatives, such as single-cell protein meal (SCPM). The microbial biomass (SCPM) is produced by using high-yielding methanotroph bacteria (Battistuzzi & Hedges, 2009; Sillman et al., 2019). These microbes metabolize methane as a source of carbon to integrate into biomass (Strong et al., 2015). SCPM has shown promising growth and health results in various aquaculture species, and this novel ingredient has the full potential to replace about 75% of fish meal in a carnivore diet, the barramundi (*Lates calcarifer*). This finding could help us promote successful intensive and semi-intensive farming of aquatic species, particularly carnivores, by using such novel feed ingredients in their feed. Furthermore, this approach is very effective in aquaculture waste management and could contribute to decreasing greenhouse gas emissions from aquaculture systems (Woolley et al., 2023).

In the last two decades, several studies have investigated the impact of replacing fish meal with insect meal on aquaculture growth and health. Compared to plant-based ingredients, which possess antinutritional factors and a comparatively weak amino acid profile (Daniel, 2018), insect meal possesses a good nutritional profile and is sustainable (Alfiko et al., 2022). Like microbes in the case of single-cell protein meal, insects also potentially extract nutrients from aquaculture wastes, contributing to environmental and economic sustainability. Although some insect species, for example, yellow mealworms, possess high-fat content, they could be used in aquafeed in the form of defatted powder after defatting them (Alfiko et al., 2022). The defatting process is usually accomplished by exerting some kind of mechanical pressure by applying aqueous and solvent-oriented procedures such as Soxhlet and supercritical carbon dioxide (Rumpold et al., 2017).

Since insects are naturally available only during specific seasons of the year, insect farming in controlled conditions is a good option to make available their populations throughout the year, as reported by Cadinu et al. (2020), DiGiacomo and Leury (2019), Hawkey et al. (2021), van der Fels-Klerx et al. (2018), and van Huis (2020). Although several European countries, such as Norway, already allow insect meal addition to aquafeed, insects are still not widely used in aquafeed in several other regions due to diverse concerns, particularly ethical and religious debates. An increasing trend of using insect meal in aquafeed as a source of protein or energy could greatly decline its prices in the feed market, which will result in a reduction in total feeding costs and hence total production costs, while net profit (sustainability) will be ultimately enhanced (Alfiko et al., 2022). However, fish consumers' preference and aquaculturists acceptance are the two main points that could have a significant impact on insect meal inclusion in the aquafeed and the overall sustainability of aquaculture (Bazoche & Poret, 2021; Gasco et al., 2019).

Seaweeds, or macroalgae, are another good alternative protein and energy source for aquafeed. They can be used as potential protein or energy ingredients in aquafeed for freshwater species because they are a good source of proteins, minerals, fibers, vitamins, and omega-3 fatty acids (MacArtain et al., 2007; Radulovich et al., 2015). By using seaweeds in the diets of freshwater species, it is possible to improve their nutritional value in order to reduce the increasing pressure on marine stocks, which are enormously utilized for diverse purposes, particularly fish meal and fish oil (Mwendwa et al., 2023). It is important to highlight that alternative protein or energy sources must meet the nutritional requirements of the animals in order to ensure their optimum growth, health, and well-being (Engle & D'Abramo, 2018). Their local availability is also significant because imported ingredients are usually unsustainable. Besides some nutritional concerns, no doubt, soybean is a good alternative to fish meal and fish oil; however, due to its insufficient availability or lower nutritional quality in some regions, aquaculturists usually import it from countries such as the United States and Brazil, which could not be a sustainable way in the long run (NRC, 2011).

Diverse plant-origin alternatives, including soy protein concentrates, canola protein concentrates, canola meal, corn gluten meal, wheat peas and gluten meal, and cottonseed meal, have also been examined and are widely used in aquafeed (Gatlin et al., 2007). Overall, alternative protein and energy sources could be a sustainable approach, but aquaculturists need to be careful about their use in aquafeed. The nutritional quality of some alternative protein and energy sources could be low in terms of essential amino acids and fat profiles, which could adversely affect aquaculture performance. In this regard, some organic nutrients, such as carnitine, which have an important role in energy metabolism (carbohydrates and fats), should be added to aquafeed at an optimum level. In normal conditions, when conventional protein and energy sources are used, such as fish meal and fish oil, carnitine is formed from lysine and methionine with the help of ascorbic acid and some other important compounds. If the alternative sources are deficient in the nutrients required for carnitine synthesis, the dietary supplementation of synthetic carnitine or carnitine synthesis-facilitating ingredients becomes necessary in order to help the metabolism of energy (carbohydrates and fats) that comes from the alternative sources (Harpaz, 2005; Lee et al., 2020).

In addition, feed wastes still represent a sticking point in the aquaculture industry. Bad feed quality with inappropriate feeding management entails that a part of the feed is lost to the environment instead of being directed toward animal growth. The execution of precision fish farming, which applies to control-engineering principles and mathematical models in aquaculture, is promising (FAO, 2017; Føre et al., 2018). This approach may result in much lower feed waste, costs, and environmental footprint with sustainable aquaculture systems efficiency. The nitrogen, phosphorus, and organic solids in ponds or tanks are the major pollutants of concern. To lower their losses to farm wastewater, proper attention and planning are required to be done about improving the overall quality of aquafeed, including the nutritional value and texture of feed, and reducing the amount of uneaten feed by optimizing feeding frequencies and feed utilization efficiencies to improve the retention of these nutrients in the fish body (Boyd et al., 2019). Moreover, aquafeeds should be developed by considering a wide range of factors such as feedstuff prices, market demand of species, and marketing strategies. That is how the end product of harvest is. In order to overcome these issues, further studies that target fish nutrition, feeding strategies, alternative potential raw materials (Bories et al., 2007), and precision (Føre et al., 2018) fish farming are required (FAO, 2017).

3.7.5.2.2 Immunostimulants for Healthy Aquaculture

Intensive and semi-intensive aquaculture systems could suffer from disease outbreaks. As a preventive measure against viral and bacterial diseases, vaccination (injection-based) is highly recommended (Mondal & Thomas, 2022). Antibiotics (González-Gaya et al., 2022) and certain drugs and chemicals (Douet et al., 2009) at recommended concentrations against diverse bacterial, fungal, and parasitic infections are allowed to be used in aquaculture. Overall, vaccination is a safe way, although it needs skilled professionals to perform this procedure with minimum handling stress (Mondal & Thomas, 2022). Likewise, immunostimulant addition at optimum rates to aquafeed is also a routine matter in aquaculture and is widely accepted. On the other hand, although antibiotics and chemicals are usually accepted at recommended concentrations, their use in aquaculture is considered unsustainable and unhealthy. So, to avoid the implications of using such unsustainable and unhealthy substances in aquaculture, vaccination via injections (against both bacterial and viral infections of aquatic organisms) and the incorporation of immunostimulants in aquafeed are preferably recommended by nutritionists. In the case of immunostimulants, several kinds of locally available novel substances, such as potential plants (whole or parts) or their derivatives (functional compounds), as well as vitamins and minerals, could be added at optimum concentrations as functional additives to aquafeed. Aquafeed containing any functional ingredient is termed functional feed, which has beneficial implications for aquaculture performance (Vidanarachchi et al., 2005).

The optimum concentrations of immunostimulants and vaccines with a specific administration duration are required to be accurately determined for different growth phases of commercially important aquatic animals (Vallejos-Vidal et al., 2016; Vidanarachchi et al., 2005; Tacon et al., 2011; Tacon & Metian, 2008). Vaccination has a direct role in stimulating adaptive immunity, while immunostimulants greatly contribute to enhancing nonspecific immunity via enhancing the activities of innate immune cells as well as components of humoral and complement systems. The innate immune system then ultimately provokes the adaptive response. The initiation of these immune systems provides solid protection to aquatic animals against diverse infectious diseases (Vijayaram et al., 2024). Overall, through sustainable intensification along with integrative mechanisms and optimum nutrition, the production efficiency of aquaculture systems could be enhanced by several folds (Boyd et al., 2020; Tacon & Metian, 2008).

3.8 Sustainable Solutions to Enhanced Optimization of Aquaculture Performance and Mitigation of Environmental Concerns

In the long run, food-producing practices, including aquaculture, should be responsible and sustainable to provide quality products with minimum environmental concerns (Costa-Pierce et al., 2010; Porter et al., 2014). Sustainable techniques such as recirculating aquaculture systems could be employed to increase the efficiency and environmental sustainability of the intensified and semi-intensive systems. In the intensive and semi-intensive aquaculture systems (RAS), aquafeed leads to about 75% of the nitrogen, which usually accumulates in the form of total ammonia (56.2%), feces (13.8%), and uneaten feed (5%) (Deng et al., 2020). Overfeeding increases waste and nitrogen load, which not only disrupts overall water quality but also leads to economic losses. Unnecessary nutrient load in aquatic systems has deleterious effects on production efficiency and environmental sustainability (Rojas-Tirado et al., 2018; Stickney, 2005; Zhou et al., 2018). It has been well-established that biofilters associated with recirculating aquaculture systems help in the removal of suspended solid particles produced from waste and uneaten feed while microbes feed on the uneaten food as well as decaying organic matter, maintaining a healthy environment for aquaculture growth (FAO, 2017; Zhao et al., 2020).

Moreover, if the aquaculture facility is provided by a flow-through system, then working with hydroponics or agriculture is a sustainable solution to minimize the adverse impact of the aquaculture wastes. Such wastewater is rich in nutrients, particularly nitrogen and phosphorus, which, when not absorbed by the body, are released into the surrounding water. Uneaten feed and aquaculture wastes are also the major sources of these nutrients. Usually, the apparent availability of nitrogen and phosphorus in the diet declines when certain feed ingredients higher in these nutrients are employed in aquafeed. As a counterstrategy, developing aquafeed by using the digestible values of nutrients at appropriate concentrations is necessary for the optimum performance of aquatic species at a specific growth stage (FAO, 2017; Hardy & Gatlin III, 2002).

Extensive aquaculture setups should be built in open waters, which are less sensitive in terms of aquatic ecosystem quality and where there is a good circulation of water. Such areas should also be free of potential predators that could harm the stocked aquatic organisms. Public access should be minimal in such areas. Extensive aquaculture setups also have a good role in establishing habitats for various aquatic organisms, particularly invertebrates and algal communities. Therefore, such strategies may help to grow aquatic organisms without any significant investment or environmental damage (Hansen et al., 2007; Stickney, 2005).

According to the recommendations of Boyd et al. (2020), Boyd and McNevin (2015), Cashion et al. (2017), Costa-Pierce et al. (2010), da Silva et al. (2018), Doublet and Pointereau (2008), Edwards et al. (2019), Engle et al. (2017), FAO (2017, 2018a, b, 2020); Hardy and Gatlin (2002), Khan et al. (2020a, b, 2022), Laktuka et al. (2023), Mishra et al. (2024), Oshiro et al. (2024), NRC (2011), Rice (2001), Stickney (2005), Tacon and Metian (2015), Tacon et al. (2011), Tilman and Clark (2014), Timmons et al. (2018), Tlusty et al. (2018), and WEF (2018, 2021), the long term enhanced optimization of aquaculture performance in terms of economic, environmental and community sustainability could be successfully achieved through the following potential management and nutritional strategies:

8.1 Strengthening land-based (inland) aquaculture could help reduce the increasing pressure on marine stocks from different sectors. In several regions of the world, land-based aquaculture is still not widely practiced due to limited skills, a lack of advanced technologies and trained professionals, and probably low investment opportunities in this sector. In such regions, land-based aquaculture could be enhanced through support from United Nations Food Security Programs and intervention by developed countries, providing enough investment, skilled professionals, and advanced technologies.

8.2 Seas and oceans make up 70% of the planet earth's surface, but only about 2% is being utilized for food purposes. Mariculture is a somewhat neglected area in several countries that needs to be properly promoted and effectively practiced in open waters (offshore). Most effluent-related issues exist in the coastal water, where the water is shallow and currents are weak. In contrast, offshore waters are deeper, have strong water currents, and usually contain fewer nutrients and lower biodiversity. Therefore, offshore aquaculture makes possible a quicker dispersal of effluents into the deep water, which would have much lower environmental implications. Cage or pen culture should be an ideal option for offshore aquaculture; however, this type of aquaculture should be properly legalized by local governments and international authorities. Responsible offshore aquaculture with appropriate feeding and operational practices could greatly contribute to enhancing marine production by several folds.

8.3 Feed only when aquatic organisms want to eat, and the feed must be nutritionally balanced. Furthermore, feeding must be done considering several important factors such as water quality (temperature, pH, dissolved oxygen, ammonia, total suspended particles, etc.), species, age, and rearing system. Feeding frequency and feeding rate should be kept at optimum. In the case of feeding rate, hand-feeding with an apparent satiation strategy should be preferably practiced. Similarly, the prices of feed ingredients, which have a great impact on total feeding costs, and aquatic products in the market are also important to consider during planning a feeding strategy. The nutritional quality, texture, palatability, digestibility, economic sustainability, and environmental impact of aquafeed, along with feeding rates and frequencies, should be properly investigated for aquaculture species. This strategy is efficient in enhancing aquaculture performance through healthy and sustainable aquafeed with a minimum impact on the surrounding aquatic and terrestrial ecosystems.

8.4 Searching for potential alternative protein and energy sources that are economically and environmentally sustainable and nutritionally safe for aquaculture has been a focus of researchers and aquaculturists in response to the increasing demand for aquatic products and huge pressure on wild stocks. The aquafeed industry still relies on some conventional but expensive protein and energy ingredients such as fish meal and fish oil. These ingredients are mostly obtained from marine fishes, which has resulted in the overexploitation of several important marine fish species. On the other hand, capture fisheries also have enhanced anthropogenic activities in wild environments, which are adversely affecting the aquatic ecosystems. Furthermore, their use in aquafeed enhances feed costs and ultimately declines the sustainability of aquaculture.

Producing aquatic products is not enoughproducing quality aquatic food in a sustainable way is the need of this era and, of course, of the upcoming times. Alternative protein and energy sources include seaweeds, insects, grains (such as corn, wheat, etc.), legumes (such as peas, lupins, beans, etc.), oilseeds (sunflower, rapeseed, soybean, cottonseed, etc.), certain grass species, vegetables, and some other herbs and shrubs. These sources are also good in having antioxidative, anti-inflammatory, and antimicrobial activities, so they are most often utilized to develop functional feed for different aquatic species. However, there are certain issues with using alternative protein and energy sources due to the presence of high fats, antinutritional factors, and a lack of essential amino acid balance, as well as reduced palatability and digestibility. The issues of antinutritional factors and overall feed quality could be improved by using some special kinds of procedures, such as the extrusion of canola and soybean at a high temperature or genetic and selective breeding techniques. Although using alternative protein and energy sources is a sustainable approach, unraveling their impacts on aquaculture growth and health is necessary to check their safety and nutritional quality before being used in commercial-scale aquafeed development. Only optimum levels of alternative protein or energy sources that are nutritionally safe for cultured organisms should be incorporated in aquafeed to avoid any negative consequences in terms of economic loss, lower nutritional quality of the aquatic products, or environmental impact.

8.5 Encouraging the utilization of locally available resources in aquaculture, including feed ingredients and native aquatic species, is essential. Target species for aquaculture should be chosen by considering their demand in the consumer market of a region. Exotic species and hybrids could also be considered for commercial aquaculture only if there is a high demand in the consumer market, although preferably exotic species and hybrids should be cultured in intensive or semi-intensive systems far from open waters in order to avoid their escape.

8.6 Water quality management has a critical role in aquaculture growth, health, and sustainability, as well as the efficiency of production systems. For water quality management, aquaculturists must regularly monitor the physicochemical characteristics of aquaculture facilities and consider sustainable strategies, if necessary. The following approaches are very efficient for managing water quality and getting maximum economic benefits from aquaculture.

8.6.1 Working with recirculating aquaculture systems (RASs) is a potential way to achieve sustainable and healthy aquaculture production. Moreover, in areas with low groundwater resources, RAS could be an ideal option for aquaculture activities. Biological filters connected with recirculating aquaculture systems contribute to enhancing system efficiency by minimizing effluent discharge. This approach is significant regarding minimizing the environmental impact of aquaculture as well, and such systems also help in securing the available water resources.

8.6.2 Employing multi-trophic aquaculture is an efficient approach to reduce the accumulation of nutrients by stocking filter feeders (animal and plant species) along with target species, which could act as artificial filters. The combination of shellfish, grass carp, and seaweed culture with desired fish species could greatly improve the quality of water while also contributing to extra economic benefits. The byproducts produced from desired species culture serve as food for the filter feeders, contributing to effective and sustainable waste management of aquaculture facilities.

8.6.3 Integrated farming is another promising approach to minimize the environmental concerns related to aquaculture. In this practice, aquaculture systems are connected with livestock farming or agriculture (hydroponics or aquaponics), such as livestock or agriculture, and aquaculture systems benefit from each other. Inputs used in these systems are almost completely utilized, helping in minimizing waste production and production costs.

8.7 Certification and labeling/tagging (information about aquatic species, its nutritional quality, safety, manufacturing, and expiry dates) of aquatic products is also important to attract consumers to the aquatic food market. Antibiotic use should be done according to the recommended dosages, and their unnecessary use should be avoided in order to prevent antibiotic resistance in aquaculture. The incorporation of optimum levels of naturally available immunostimulants in aquafeed, followed by vaccination, should be encouraged. Similarly, the hormones, drugs, or chemicals that are used in aquaculture for diverse purposes should be properly investigated to check their safety for aquatic organisms and for humans, who are the ultimate consumers. Potential biomonitoring in combination with sustainable handling practices is a worthy approach to ensure responsible and healthy aquaculture.

8.8 Utilization of renewable energy to operate aquaculture, instead of depending on electricity or natural gas-based operating systems, can be highly effective. By employing renewable energy sources such as solar panels or wind turbines, aquaculturists can efficiently operate heating and cooling systems, as well as water and oxygen pumps. Paddle-like wheels fixed in ponds, used to enhance dissolved oxygen concentrations, could also be efficiently operated by renewable energy sources instead of using tractor engines. This is eco-friendly approach could greatly reduce the total production costs, which will definitely enhance production sustainability.

8.9 Promoting awareness among consumers that aquaculture does not cause any harm to the ecosystem or contribute to overfishing, while motivating the consumption of healthy aquatic products, could help build a soft image of aquaculture in the community. At the same time, ensuring the nutritional quality and safety of aquatic food that comes from farms could encourage supermarkets to supply farm-reared aquatic organisms instead of depending on captured aquatic organisms. The latter approach has severe negative implications for ecosystems and consumers. So, changing the perception of the public about aquatic products could be a game changer to enhance the market demand for aquaculture products. Market demand will ultimately boost the worth of aquaculture products, so large-scale expansion of land-associated aquaculture activities could become a reality with the intervention of investors and the private sector.

8.10 Encouraging small-scale aquaculture while securing the due rights of the fishermen and farmers could be a good idea for promoting sustainable aquaculture in the long run. Local and federal governments should facilitate aquaculturists developing small-scale aquaculture systems and provide them with interest-free loans to buy advanced technologies and build advanced rearing setups. This community-oriented approach could be very successful in uplifting the microeconomic status of the communities.

3.9 Conclusion

Aquaculture has the full potential to fulfill the increasing demand for animal protein in the consumer market. Economic, environmental, and community issues could be mitigated by employing responsible, healthy, and sustainable aquaculture practices such as integrating aquaculture with livestock and agriculture sectors or using recirculating aquaculture systems, particularly for intensified production. Potential management and nutritional strategies could significantly uplift aquaculture performance in terms of enhanced optimization of economic, environmental, and community sustainability. Moreover, aquaculture has several advantages over terrestrial animal farming, particularly in terms of high nutritional value and higher feed utilization efficiency, and greenhouse gas emissions from aquaculture systems could be efficiently managed by adopting effective waste management approaches. It is strongly recommended that aquaculturists execute responsible, healthy, and sustainable management and nutritional strategies such as regular checks of water quality, feeding fish according to species and body size, the harvest marketing strategy, and only using nutritionally balanced feed. This could be the only way forward toward promoting responsible, healthy, and sustainable aquaculture in order to minimize the pressure on capture fisheries as well as to benefit communities, particularly the most vulnerable segments such as women and children, by providing them with quality food and employment opportunities to secure their well-being and livelihood.

References

Abdel-Tawwab, M., Khalil, R. H., Metwally, A. A., Shakweer, M. S., Khallaf, M. A., & Abdel-Latif, H. M. R. (2020). Effects of black soldier fly (*Hermetia illucens* L.) larvae meal on growth performance, organs-somatic indices, body composition, and hemato-biochemical variables of European seabass, *Dicentrarchus labrax*. *Aquaculture, 522*, 735136. https://doi.org/10.1016/J.AQUACULTURE.2020.735136

Ahmed, N., Wahab, M. A., & Thilsted, S. H. (2007). Integrated aquaculture–agriculture systems in Bangladesh: Potential for sustainable livelihoods and nutritional security of the rural poor. *Aquaculture Asia, 12*, 14–22.

Alfiko, Y., Xie, D., Astuti, R. T., Wong, J., & Wang, L. (2022). Insects as a feed ingredient for fish culture: Status and trends. *Aquaculture and Fisheries, 7*(2), 166–178. https://doi.org/10.1016/j.aaf.2021.10.004

Ali, M. H., Mateo, L. G., & Aragon, M. L. (2006). Effect of stocking density on growth and yield of GIFT tilapia under rice-fish production system. *Bangladesh Journal of Fisheries, 10*, 35–39.

Anras, L. (2008). *SEACASE national report of extensive and semi-intensive production practices in France—D19*. Forum des Marais Atlantiques (FMA).

Basto, A., Valente, L. M. P., Conde-Sieira, M., & Soengas, J. L. (2021). Central regulation of food intake is not affected by inclusion of defatted *Tenebrio molitor* larvae meal in diets for European seabass (*Dicentrarchus labrax*). *Aquaculture, 544*, 737088. https://doi.org/10.1016/j.aquaculture.2021.737088

Battistuzzi, F. U., & Hedges, S. B. (2009). A major clade of prokaryotes with ancient adaptations to life on land. *Molecular Biology and Evolution, 26*(2), 335–343. https://doi.org/10.1093/molbev/msn247

Bazoche, P., & Poret, S. (2021). Acceptability of insects in animal feed: A survey of French consumers. *Journal of Consumer Behaviour, 20*(2), 251–270. https://doi.org/10.1002/cb.1845

Bergman, J. N., Bennett, J. R., Binley, A. D., Cooke, S. J., Fyson, V., Hlina, B. L., Reid, C. H., Vala, M. A., & Madliger, C. L. (2019). Scaling from individual physiological measures to population-level demographic change: Case studies and future directions for conservation management. *Biological Conservation, 238*, 108242.

Bories, G., Brantom, P., de Barberà, J. B., Chesson, A., Cocconcelli, P. S., Debski, B., Franklin, A., Gropp, J., Halle, I., Hogstrand, C., de Knecht, J., Leng, L., Haldorsen, A. L., Mantovani, A., Mézes, M., Nebbia, C., Rambeck, W., Rychen, G., von Wright, A., & Wester, P. (2007). Safety and efficacy of Panaferd-AX (red carotenoid rich bacterium *Paracoccus carotinifaciens*) as feed additive for salmon and trout. Scientific opinion of the panel on additives and products or substances used in animal feed. *EFSA Journal, 546*, 1–30.

Bouwman, A. F., Beusen, A. H. W., Overbeek, C. C., Bureau, D. P., Pawlowski, M., & Glibert, P. M. (2013). Hindcasts and future projections of global inland and coastal nitrogen and phosphorus loads due to finfish aquaculture. *Reviews in Fisheries Science, 21*(2), 112–156. https://doi.org/10.1080/10641262.2013.790340

Boyd, C. E. (2018). Aquaculture pond fertilization. *CAB Reviews, 13*(002), 1–12.

Boyd, C. E., & McNevin, A. (2015). *Aquaculture, resource use, and the environment.* Wiley-Blackwell.
Boyd, C. E., Tucker, C. S., & Somridhivej, B. (2016). Alkalinity and hardness: Critical but elusive concepts in aquaculture. *Journal of the World Aquaculture Society, 47*, 6–41. https://doi.org/10.1111/jwas.1224
Boyd, C. E., McNevin, A. A., & Tucker, C. S. (2019). Resource use and the environment. In J. Lucas, P. Southgate, & C. Tucker (Eds.), *Aquaculture: Farming aquatic animals and plants* (3rd ed., pp. 93–112). Wiley.
Boyd, C. E., D'Abramo, L. R., Glencross, B. D., Glencross, B. D., Huyben, D. C., Juarez, L. M., Lockwood, G. S., McNevin, A. A., Tacon, A. G. J., Teletchea, F., Tomasso, J. R., Jr., Tucker, C. S., & Valenti, W. C. (2020). Achieving sustainable aquaculture: historical and current perspectives and future needs and challenges. *Journal of the World Aquaculture Society, 51*(3), 578–633. https://doi.org/10.1111/jwas.12714
Brune, D. E., Schwartz, G., Eversole, A. G., Collier, J. A., & Schwedler, T. E. (2003). Intensification of pond aquaculture and high-rate photosynthetic systems. *Aquacultural Engineering, 28*, 65–86.
Buentello, J. A., Gatlin, D. M., & Neill, W. H. (2000). Effects of water temperature and dissolved oxygen on daily feed consumption, feed utilization and growth of channel catfish (*Ictalurus punctatus*). *Aquaculture, 182*, 339–352.
Cadinu, L. A., Barra, P., Torre, F., Delogu, F., & Madau, F. A. (2020). Insect rearing: Potential, challenges, and circularity. *Sustainability, 12*(11), 4567. https://doi.org/10.3390/su12114567
Cashion, T., Le Manach, F., Zeller, D., & Pauly, D. (2017). Most fish destined for fishmeal production are food-grade fish. *Fish and Fisheries, 18*, 837–844.
Cho, C. Y., & Bureau, D. P. (2001). A review of diet formulation strategies and feeding systems to reduce excretory and feed wastes in aquaculture. *Aquaculture Research, 32*, 349–360. https://doi.org/10.1046/j.1355-557x.2001.00027.x
Chopin, T. (2013a). Integrated multi-trophic aquaculture. Ancient, adaptable concept focuses on ecological integration. *Global Aquaculture Advocate, 16*(2), 16–19.
Chopin, T. (2013b). Aquaculture, integrated multi-trophic (IMTA). In P. Christou, R. Savin, B. A. Costa-Pierce, I. Misztal, & C. B. A. Whitelaw (Eds.), *Sustainable food production* (pp. 184–205). Springer.
Chumkiew, S., Jaroensutasinee, M., Tina, F. W., Jaroensutasinee, K., & Koad, P. (2019). Physical factors affecting oyster diversity and distribution in southern Thailand. *Journal of Environmental Biology, 40*, 0308.
Costa-Pierce, B. A., Bartley, D. M., Hasan, M., Yusoff, F., Kaushik, S. J., Rana, K., Lemos, D., Bueno, P., & Yakupitiyage, A. (2010). In R. P. Subasinghe, J. R. Arthur, D. M. Bartley, S. S. Silva, M. Halwart, N. Hishamunda, C. V. Mohan, & P. Sorgeloos (Eds.), *Responsible use of resources for sustainable aquaculture—Panel review 1.1* (pp. 113–147). FAO, and NACA.
da Silva, M. G., Packer, A. P., Sampaio, F. G., Marani, L., Mariano, E. V. C., Pazianotto, R. A. A., Ferreira, W. J., & Alvalá, P. C. (2018). Impact of intensive fish farming on methane emission in a tropical hydropower reservoir. *Climatic Change, 150*, 195–210. https://doi.org/10.1007/s10584-018-2281-4
Daniel, N. (2018). A review on replacing fish meal in aqua feeds using plant protein sources. *International Journal of Fisheries and Aquatic Studies, 6*, 164–179.
Deng, M., Dai, Z., Senbati, Y., Li, L., Song, K., & He, X. (2020). Aerobic denitrification microbial community and function in zero-discharge recirculating aquaculture system using a single biofloc-based suspended growth reactor: Influence of the carbon-to-nitrogen ratio. *Frontiers in Microbiology, 11*, 1760.
Diao, W., Jia, R., Hou, Y., Dong, Y., Li, B., & Zhu, J. (2023). Effects of stocking density on the growth performance, physiological parameters, antioxidant status and lipid metabolism of *Pelteobagrus fulvidraco* in the integrated rice-fish farming system. *Animals (Basel), 13*(11), 1721. https://doi.org/10.3390/ani13111721

DiGiacomo, K., & Leury, B. J. (2019). Review: Insect meal: A future source of protein feed for pigs? *Animals, 13*(12), 3022–3030. https://doi.org/10.1017/S1751731119001873

Doublet, S., & Pointereau, P. (2008). *SoCo—Sustainable agriculture and soil conservation, relevant practices* (86pages). JRC-IES.

Douet, D.-G., Giraud, E., & Le Bris, H. (2009). Environmental aspects of drug and chemical use in aquaculture: An overview. In C. Rogers & B. Basurco (Eds.), *The use of veterinary drugs and vaccines in Mediterranean aquaculture* (Options Méditerranéennes: Série a. Séminaires Méditerranéens; n. 86) (pp. 105–126). CIHEAM.

Drake, S. L., Drake, M. A., Sanderson, R., Daniels, H. V., & Yates, M. D. (2010). The effect of purging time on the sensory properties of aquacultured southern flounder (*Paralichthys lethostigma*). *Journal of Sensory Studies, 25*, 246–259.

Edwards, P., Zhang, W., Belton, B., & Little, D. C. (2019). Misunderstandings, myths, and mantras in aquaculture: Its contribution to world food supplies has been systematically over reported. *Marine Policy, 106*, 103547.

Engle, C., & D'Abramo, L. (2018). Showcasing research focusing on sustainability of aquaculture enterprises and global food security. *Journal of the World Aquaculture Society, 47*(3), 311–313.

Engle, C. R., McNevin, A., Racine, P., Boyd, C. E., Paungkaew, D., Viriyatum, R., Tinh, H. Q., & Minh, H. N. (2017). Economics of sustainable intensification of aquaculture: Evidence from shrimp farms in Vietnam and Thailand. *Journal of the World Aquaculture Society, 48*, 227–239.

Fantatto, R. R., Mota, J., Ligeiro, C., Vieiria, I., Guilgur, L. G., Santos, M., & Murta, D. (2024). Exploring sustainable alternatives in aquaculture feeding: The role of insects. *Aquaculture Reports, 37*, 102228. https://doi.org/10.1016/j.aqrep.2024.102228

FAO. (2013). *FAOSTAT database collections*. Food and Agriculture Organization of the United Nations. Available: faostat.fao.org

FAO. (2017). *Climate-smart agriculture sourcebook* (2nd ed.). Available at: https://www.fao.org/climate-smart-agriculture-sourcebook/about/significant-developments/en/

FAO. (2018a). *Solomon Islands national aquaculture management and development plan 2018–2023. Policy for 2018–2023*. Available at: https://www.fao.org/faolex/results/details/en/c/LEX-FAOC215102/

FAO. (2018b). *The state of world fisheries and aquaculture: meeting the sustainable development goals* (Doc nr. CC BY-NC-SA 3.0 IGO). FAO.

FAO. (2020). *FAO Aquaculture Newsletter No. 61*. Retrieved from: http://www.fao.org/faoterm/collection/aquaculture/en/

Føre, M., Frank, K., Norton, T., Svendsen, E., Alfredsen, J. A., Dempster, T., Eguiraun, H., Watson, W., Stahl, A., Sunde, L. M., Schellewald, C., Skøien, K. R., Alver, M. O., & Berckmans, D. (2018). Precision fish farming: A new framework to improve production in aquaculture. *Biosystems Engineering, 173*, 176–193. https://doi.org/10.1016/j.biosystemseng.2017.10.014

Froehlich, H. E., Gentry, R. R., & Halpern, B. S. (2018). Global change in marine aquaculture production potential under climate change. *Nature Ecology and Evolution, 2*, 1745–1750. https://doi.org/10.1038/s41559-018-0669-1

Gasco, L., Biasato, I., Dabbou, S., Schiavone, A., & Gai, F. (2019). Quality and consumer acceptance of products from insect-fed animals. In G. Sogari, C. Mora, & D. Menozzi (Eds.), *Edible insects in the food sector*. Springer. https://doi.org/10.1007/978-3-030-22522-3_6

Gatlin, D. M. I. I. I., Barrows, F. T., Brown, P., Dabrowski, K., Gaylord, T. G., Hardy, R. W., Herman, E., Hu, G., Krogdahl, Å., Nelson, R., et al. (2007). Expanding the utilization of sustainable plant products in aquafeeds: A review. *Aquaculture Research, 38*, 551–579. https://doi.org/10.1111/j.1365-2109.2007.01704.x

Gjedrem, T. (2012). Genetic improvement for the development of efficient global aquaculture: A personal opinion review. *Aquaculture, 344–349*, 12–22. https://doi.org/10.1016/j.aquaculture.2012.03.003

González-Gaya, B., García-Bueno, N., Buelow, E., Marin, M., & Rico, A. (2022). Effects of aquaculture waste feeds and antibiotics on marine benthic ecosystems in the Mediterranean

Sea. *Science of the Total Environment, 806*(Part 2), 151190. https://doi.org/10.1016/j.scitotenv.2021.151190

Hansen, J. W., Baethgen, W., Osgood, D., Ceccato, P., & Ngugi, R. K. (2007). Innovations in climate risk management: Protecting and building rural livelihoods in a variable and changing climate. *Journal of Semi-Arid Tropical Agricultural Research, 4*(1), 38.

Hardy, R. W., & Gatlin, D. M. (2002). Nutritional strategies to reduce nutrient losses in intensive aquaculture. In L. E. Cruz-Suárez, D. Ricque-Marie, M. Tapia-Salazar, M. G. Gaxiola-Cortés, & N. Simoes (Eds.), *Avances en Nutrición Acuícola VI. Memorias del VI Simposium Internacional de Nutrición Acuícola (Spanish). 3 al 6 de Septiembre del 2002. Cancún, Quintana Roo, México.*

Harpaz, S. (2005). L-carnitine and its attributed functions in fish culture and nutrition—A review. *Aquaculture, 249*(1–4), 3–21. https://doi.org/10.1016/j.aquaculture.2005.04.007

Hawkey, K. J., Lopez-Viso, C., Brameld, J. M., Parr, T., & Salter, A. M. (2021). Insects: A potential source of protein and other nutrients for feed and food. *Annual Review of Animal Biosciences, 9*(1), 333–354. https://doi.org/10.1146/annurev-animal-021419-083930

Herrero, M., Wirsenius, S., Henderson, B., Rigolot, C., Thornton, P., Havlík, P., de Boer, I., & Gerber, P. J. (2015). Livestock and the environment: What have we learned in the past decade? *Annual Review of Environment and Resources, 40*, 177–202.

High Level Panel of Expers (HLPE). (2016). *Sustainable agricultural development for food security and nutrition: What roles for livestock?* A report by the high level panel of experts on Food Security and Nutrition of the Committee on World Food Security.

Hishamunda, N., Ridler, N. B., Bueno, P., & Yap, W. G. (2009). Commercial aquaculture in Southeast Asia: Some policy lessons. *Food Policy, 34*(1), 102–107. https://doi.org/10.1016/j.foodpol.2008.06.006

Iaconisi, V., Bonelli, A., Pupino, R., Gai, F., & Parisi, G. (2018). Mealworm as dietary protein source for rainbow trout: Body and fillet quality traits. *Aquaculture, 484*, 197–204. https://doi.org/10.1016/j.aquaculture.2017.11.034

Ignowski, L., Belton, B., Ali, H., & Thilsted, S. H. (2023). Integrated aquatic and terrestrial food production enhances micronutrient and economic productivity for nutrition-sensitive food systems. *Nature Food, 4*, 866–873. https://doi.org/10.1038/s43016-023-00840-8

Janssen, K., Chavanne, H., Berentsen, P., & Komen, H. (2017). Impact of selective breeding on European aquaculture. *Aquaculture, 472*, 8–16.

Khan, K. U., Tellechea Rodrigues, A., Menegasso Mansano, C. F., de Almeida Queiroz, D. M., Kazue Sakomura, N., de Souza Romaneli, R., do Nascimento, T. M. T., & Kochenborger Fernandes, J. (2019). Dietary protein quality and proper protein to energy ratios: A bio–economic approach in aquaculture feeding practices. *Latin American Journal of Aquatic Research, 47*(2), 232–239. https://doi.org/10.3856/vol47-issue2-fulltext-3

Khan, K. U., Gous, R. M., Mansano, C. F. M., Nascimento, T. M. T., Romaneli, R. S., Rodrigues, A. T., Sakomura, N. K., & Fernandes, J. B. K. (2020a). Response of juvenile pacu (*Piaractus mesopotamicus* Holmberg, 1887) to balanced digestible protein. *Aquaculture Research, 51*, 3729–3740. https://doi.org/10.1111/are.14722

Khan, K. U., Gous, R. M., Sakomura, N. K., Azevedo, J. M., Nascimento, T. M. T., Mansano, C. F. M., Romaneli, R. S., Oliveira, T. S., Boaratti, A. Z., & Fernandes, J. B. K. (2020b). Modeling optimum levels of balanced digestible protein for adult pacu (*Piaractus mesopotamicus*) and the development of an economic optimization procedure for different marketing strategies of the final yield. *Aquaculture, 526*, 735368. https://doi.org/10.1016/j.aquaculture.2020.735368

Khan, K. U., Gous, R. M., Nascimento, T. M. T., Sakomura, N. K., Mansano, C. F. M., Queiroz, D. M. A., Rodrigues, A. T., Boaratti, A. Z., Romaneli, R. S., & Fernandes, J. B. K. (2022). Modeling of nutritional behavior and harvest profit margin in fattening pacu (*Piaractus mesopotamicus*) fed on balanced digestible protein contents and raised in an intensive recirculating aquaculture system. *Animal Feed Science and Technology, 292*, 115430. https://doi.org/10.1016/j.anifeedsci.2022.115430

Laktuka, K., Kalnbalkite, A., Sniega, L., Logins, K., & Lauka, D. (2023). Towards the sustainable intensification of aquaculture: Exploring possible ways forward. *Sustainability, 15*(24), 16952. https://doi.org/10.3390/su152416952

Lazzarotto, V., Médale, F., Larroquet, L., & Corraze, G. (2018). Long-term dietary replacement of fishmeal and fish oil in diets for rainbow trout (*Oncorhynchus mykiss*): Effects on growth, whole body fatty acids and intestinal and hepatic gene expression. *PLoS One, 13*(1), e0190730. https://doi.org/10.1371/journal.pone.0190730

Lee, S., Kumar, V., Cleveland, B., Romano, N., Vemuri, G. N., Yadav, A. K., & Hardy, R. W. (2020). Fishmeal alternative from renewable CO2 for rainbow trout feed. *Aquaculture Research, 51*, 4065–4074. https://doi.org/10.1111/are.14749

Lekang, O.-I. (2013). *Aquaculture engineering* (2nd ed.). Wiley-Blackwell.

Lindholm-Lehto, P. (2023). Water quality monitoring in recirculating aquaculture systems. *Aquaculture, Fish and Fisheries, 3*(2), 113–131.

Little, D. C., Satapornvanit, A., & Edwards, P. (2002). Freshwater fish seed quality in Asia. In P. Edwards, D. C. Little, & H. Demaine (Eds.), *Rural aquaculture* (pp. 185–195, 385 pages). CABI Publishing.

Liu, B., Jia, R., Zhao, K., Wang, G., Lei, J., & Huang, B. (2016). Stocking density effects on growth and stress response of juvenile turbot (*Scophthalmus maximus*) reared in land-based recirculating aquaculture system. *Acta Oceanologica Sinica, 36*, 31–38. https://doi.org/10.1007/s13131-017-0976-4

MacArtain, P., Gill, C. I., Brooks, M., Campbell, R., & Rowland, I. R. (2007). Nutritional value of edible seaweeds. *Nutrition Reviews, 65*(12), 535–543. https://doi.org/10.1111/j.1753-4887.2007.tb00278.x

Mair, G. (2002). Topical issues in genetic diversity and breeding: Genes and fish: Supply of good quality fish seed for sustainable aquaculture. *Aquaculture Asia, VII*(2), 25–27.

Mamun, A., Murray, F. J., Sprague, M., McAdam, B. J., Roos, N., de Roos, B., Pounds, A., & Little, D. C. (2021). Export-driven, extensive coastal aquaculture can benefit nutritionally vulnerable people. *Frontiers in Sustainable Food Systems, 5*, 713140.

Mekonnen, E., Brehanu, G., & Yitayew, T. (2019). Biological aspects, catch and length distribution of African catfish, *Clarias gariepinus* and common carp, *Cyprinus Carpio* in in Lake Lugo, south Wollo, Ethiopia. *Ethiopian Journal of Science and Technology, 12*(3), 185–202. https://doi.org/10.4314/ejst.v12i3.1

Mishra, S., Devi, S., Nottanalan, H., & Kantal, D. (2024). Chapter 9: Unraveling the complexities of antinutritional factors: Implications for fish nutrition and feeding practices. In *Handbook of sustainable aquaculture and fisheries*. Elite Publishing House.

Mondal, H., & Thomas, J. (2022). A review on the recent advances and application of vaccines against fish pathogens in aquaculture. *Aquaculture International, 30*, 1971–2000. https://doi.org/10.1007/s10499-022-00884-w

Mwegoha, W. J. S., Kaseva, M. E., & Sabai, S. M. M. (2010). Mathematical modeling of dissolved oxygen in fish ponds. *African Journal of Environmental Science and Technology, 4*, 623638.

Mwendwa, R., Wawire, M., & Peter, K. (2023). Potential for use of seaweed as a fish feed ingredient: A review. *The Journal of Agricultural Science, 15*, 96. https://doi.org/10.5539/jas.v15n2p96

NRC. (2011). *Nutrient requirements of fish and shrimp*. National Academies Press. https://doi.org/10.17226/13039

Oddsson, G. V. (2020). A definition of aquaculture intensity based on production functions—The aquaculture production intensity scale (APIS). *Water, 12*(3), 765. https://doi.org/10.3390/w12030765

Olaniyi, A. A., Siriwardena, S. N., & Nasr-Allah, A. (2022). *Water quality management*. WorldFish. Program Brief: 2022-35. Available at: https://digitalarchive.worldfishcenter.org/server/api/core/bitstreams/16b1428d-5d8f-46a9-937f-67bac707a10c/content

Oshiro, A. M., Khan, K. U., Carli, G. C., Pahor-Filho, E., Sakomura, N. K., Fernandes, J. B. K., & Takahashi, L. S. (2024). Effect of digestible protein on intermediate metabolism, hepatic enzyme activities, energy reserves, and growth performance of pacu (*Piaractus mesopotamicus*) in the finishing growth phase. *Fish Physiology and Biochemistry, 50*, 259–271. https://doi.org/10.1007/s10695-023-01243-z

Porter, J. R., Xie, L., Challinor, A. J., Cochrane, K., Howden, S. M., Iqbal, M. M., Lobell, D. B., & Travasso, M. I. (2014). Food security and food production systems. In C. B. Field, V. R. Barros, D. J. Dokken, K. J. Mach, M. D. Mastrandrea, T. E. Bilir, M. Chatterjee, K. L. Ebi, Y. O. Estrada, R. C. Genova, B. Girma, E. S. Kissel, A. N. Levy, S. MacCracken, P. R. Mastrandrea, & L. L. White (Eds.), *Climate change 2014: impacts, adaptation, and vulnerability. Part A: global and sectoral aspects* (Contribution of Working Group II to the Fifth Assessment Report of the Intergovernmental Panel on Climate Change) (pp. 485–533). Cambridge University Press.

Radulovich, R., Umanzor, S., Cabrera, R., & Mata, R. (2015). Tropical seaweeds for human food, their cultivation and its effect on biodiversity enrichment. *Aquaculture, 436*, 40–46. https://doi.org/10.1016/j.aquaculture.2014.10.032

Rice, M. A. (2001). Environmental impacts of shellfish aquaculture: Filter feeding to control eutrophication. In M. Tlusty, D. Bengtson, H. O. Halvorson, S. Oktay, J. Pearce, & R. Rheualt (Eds.), *Marine aquaculture and the marine environment: A meeting for the stakeholders in the northeast. Held Jan. 11-13, 2001 at the Univ. of Massachusetts Boston* (pp. 77–86). Cape Cod Press.

Rojas-Tirado, P., Pedersen, P. B., Vadstein, O., & Pedersen, L. (2018). Changes in microbial water quality in RAS following altered feed loading. *Aquacultural Engineering, 81*, 80–88.

Rumpold, B. A., Klocke, M., & Schlüter, O. (2017). Insect biodiversity: Underutilized bioresource for sustainable applications in life sciences. *Regional Environmental Change, 17*, 1445–1454. https://doi.org/10.1007/s10113-016-0967-6

Shaw, R. (2007). Community-based climate change adaptation in Vietnam: Inter-linkages of environment, disaster, and human security. In S. Sonak (Ed.), *Multiple dimension of global environmental changes*. TERI.

Sillman, J., Nygren, L., Kahiluoto, H., Ruuskanen, V., Tamminen, A., Bajamundi, C., Nappa, M., Wuokko, M., Lindh, T., Vainikka, P., Pitkänen, J.-P., & Ahola, J. (2019). Bacterial protein for food and feed generated via renewable energy and direct air capture of CO2: Can it reduce land and water use? *Global Food Security, 22*, 25–32. https://doi.org/10.1016/j.gfs.2019.09.007

Stickney, R. R. (2005). *Aquaculture: An introductory text* (320 pages). Oxford University Press Paperback. ASIN:B0050X678Q.

Strong, P. J., Xie, S., & Clarke, W. P. (2015). Methane as a resource: Can the methanotrophs add value? *Environmental Science & Technology, 49*, 4001–4018. https://doi.org/10.1021/es504242n

Sugiura, S. H., & Hardy, R. W. (2000). Environmentally-friendly feeds. In R. R. Stickney (Ed.), *Encyclopedia of aquaculture* (pp. 299–310). Wiley.

Tacon, A. G. J., & Metian, M. (2008). Aquaculture feed and food safety: The role of FAO and codex Alimentarius. *The New York Academy of Sciences, 1140*, 50–59.

Tacon, A. G. J., & Metian, M. (2015). Feed matters: Satisfying the feed demand of aquaculture. *Reviews in Fisheries Science & Aquaculture, 23*(1), 1–10.

Tacon, A. G. J., Hasan, M. R., & Metian, M. (2011). *Demand and supply of feed ingredients for farmed fish and crustaceans: Trends and prospects* (FAO fisheries and aquaculture technical paper No. 564). Food and Agriculture Organization.

Tilman, D., & Clark, M. (2014). Global diets link environmental sustainability and human health. *Nature, 515*, 518–522.

Tilman, D., Clark, M., Williams, D. R., Kimmel, K., Polasky, S., & Packer, C. (2017). Future threats to biodiversity and pathways to their prevention. *Nature, 546*, 73–81.

Timmons, M. B., Guerdat, T., & Vinci, B. J. (2018). Water quality. In *Recirculating aquaculture* (4th ed., 2758 pages). Ithaca Publishing Company LLC.

Tlusty, M., Tyedmers, P., Ziegler, F., Jonell, M., Henriksson, P. J. G., Newton, R., Little, D., Fry, J., Love, D., & Cao, L. (2018). Commentary: Comparing efficiency in aquatic and terrestrial animal production systems. *Environmental Research Letters, 13*, 128001. https://doi.org/10.1088/1748-9326/aae945

Tongdee, N., Klinsukhon, S., & Sulit, V. (2017). Status and trends of capture fisheries and aquaculture in Southeast Asia. In *The southeast Asian state of fisheries and aquaculture 2017* (pp. 1–40). Secretariat, Southeast Asian Fisheries Development Center. http://hdl.handle.net/20.500.12066/6178

Tziortzioti, C., Amaxilatis, D., Mavrommati, I., & Chatzigiannakis, I. (2019). IoT sensors in sea water environment: Ahoy! Experiences from a short summer trial. *Electronic Notes in Theoretical Computer Science, 343*, 117–130.

United Nations DESA. (2015). *World population prospects: The 2015 revision, key findings and advance tables* (Report ESA/P/WP 241). United Nations. Available at: https://esa.un.org/unpd/wpp/publications/files/key_findings_wpp_2015.pdf

Vallejos-Vidal, E., Reyes-López, F., Teles, M., & MacKenzie, S. (2016). The response of fish to immunostimulant diets. *Fish & Shellfish Immunology, 56*, 34–69.

van der Fels-Klerx, H. J., Camenzuli, L., Belluco, S., Meijer, N., & Ricci, A. (2018). Food safety issues related to uses of insects for feeds and foods. *Comprehensive Reviews in Food Science and Food Safety, 17*(5), 1172–1183. https://doi.org/10.1111/1541-4337.12385

Van Huis, A. (2020). Insects as food and feed, a new emerging agricultural sector: A review. *Journal of Insects as Food and Feed, 6*(1), 27–44. https://doi.org/10.3920/JIFF2019.0017

Vidanarachchi, J. K., Mikkelsen, L. L., Sims, I., Iji, P. A., & Choct, M. (2005). Phytobiotics: Alternatives to antibiotic growth promoters in monogastric animal feeds. *Recent Advances in Animal Nutrition – Australia, 15*, 131–144.

Vijayaram, S., Ringø, E., Zuorro, A., van Doan, H., & Sun, Y. (2024). Beneficial roles of nutrients as immunostimulants in aquaculture: A review. *Aquaculture and Fisheries, 9*(5), 707–720. https://doi.org/10.1016/j.aaf.2023.02.001

WEF. (2018). *5 ways to net a sustainable future for aquaculture*. Available at: https://www.weforum.org/stories/2018/09/5-ways-to-guarantee-sustainable-aquaculture/

WEF. (2021). *Food and water. What is integrated aquaculture and how can it help feed the world?* Available at: https://www.weforum.org/stories/2021/08/integrated-aquaculture

Woolley, L., Chaklader, M. R., Pilmer, L., Stephens, F., Wingate, C., Salini, M., & Partridge, G. (2023). Gas to protein: Microbial single cell protein is an alternative to fishmeal in aquaculture. *Science of the Total Environment, 859*(Part 1), 160141. https://doi.org/10.1016/j.scitotenv.2022.160141

Zhang, S.-Y., Li, G., Wu, H.-B., Liu, X.-G., Yao, Y.-H., Tao, L., & Liu, H. (2011). An integrated recirculating aquaculture system (RAS) for land-based fish farming: The effects on water quality and fish production. *Aquacultural Engineering, 45*, 93102.

Zhao, M., Yao, D., Li, S., Zhang, Y., & Aweya, J. J. (2020). Effects of ammonia on shrimp physiology and immunity: A review. *Reviews in Aquaculture, 12*(4), 2194–2211.

Zhou, C., Lin, K., Xu, D., Chen, L., Guo, Q., Sun, C., & Yang, X. (2018). Near infrared computer vision and neuro-fuzzy model-based feeding decision system for fish in aquaculture. *Computers and Electronics in Agriculture, 146*, 114–124.

Chapter 4
Nutritional Approaches for Prevention and Treatment of Various Diseases in Aquaculture

Shagufta Ishtiyaq and Imtiaz Ahmed

1 Introduction

Over the past few decades, aquaculture sector has significantly evolved in increasing of fish production along with crustaceans, molluscs, and algae (Rana, 1997). Due to this, aquaculture is now considered as one of the leading sectors for aquatic food production worldwide. As a result, sizeable human population relies on fish as well as seafood items as sources of essential proteins and micronutrients (Reverter et al., 2014). It has been suggested that by 2050, the global human population will exceed 9 billion, necessitating a 50% increase in the demand for fish as a protein source to meet this growing need (Syanya, 2018). Although various factors, such as the widespread adoption of mixed aquafeeds and advancements in culture system fortification, contribute to the speedy extension of contemporary aquaculture (Dawood et al., 2014), this sector still faces significant challenges with the emergence of diseases during the farming activities. The expenses associated with treating infections and the resulting fish mortality contribute to increased production costs, ultimately hindering growth and sustainability of aquaculture operations (Alfred et al., 2020).

Presently, outbreak of diseases is considered as one of the leading causes, in reduction in aquatic yield, underscoring the need for a provision to control the advancements of any disease and its impact on fish production, thereby improving faster growth and healthier aquaculture production (Ahmad & Ahmed, 2022). Moreover, these diseases are impeding social and economic progress in numerous nations (Bondad-Reantaso et al., 2005). Various pathways exist for infectious diseases to enter a country, including the introduction of recognized exotic diseases, unexpected fluctuations in the patterns of prevailing endemic diseases, or the

S. Ishtiyaq · I. Ahmed (✉)
Fish Nutrition Research Laboratory, Department of Zoology, University of Kashmir, Srinagar, Jammu and Kashmir, India

I. Ahmed, I. Ahmad (eds.), *Aquaculture: Enhancing Food Security and Nutrition*, https://doi.org/10.1007/978-3-031-92858-1_4

emergence of formerly unidentified infections. Effectively managing these disease challenges relies heavily on proactive measures such as risk assessment, early detection, and prompt response (WHO, 2006). Furthermore, recent advancements in the use of veterinary drugs and chemical additives have significantly enhanced disease prevention and control efforts (Nomoto, 2005).

In the realm of fish, a disease is characterized as a disturbance in which the regular physiological processes of multiple fish organs are altered, leading to enduring and persistent symptoms (Hernandez et al., 2009). The occurrence and subsequent rise of fish diseases in aquaculture is closely linked to the existence of various harmful bacteria in water and the deficiency of essential nutrients in the fish feed. Besides, stressed fish, such as those experiencing malnutrition, are more susceptible to illness (Alfred et al., 2020). Utilization of balanced feed together with essential nutrients is considered as innovative approach to boost the immune systems of aquatic animals including fish and diminish the chance of disease occurrence in their environment, offering cost-effective solutions without posing any toxicity risks (Vijayaram et al., 2023). Recently, it has been observed that in fish aquaculture, vaccines and immunostimulants have emerged as novel strategies for disease management. Due to this, there has been a shift towards immune-boosting approaches, including immunostimulants, to prevent disease transmission in aquaculture, partly influenced by public concerns regarding the use of antibiotics in farmed animals (Cuesta & Esteban, 2021). In essence, nutrition refers to the biochemical and physiological mechanisms through which organisms obtain the essential energy, nutrients, and other vital components necessary for growth, survival, and reproduction. As outlined by the National Geographic Society in 2017, nutrients are broadly categorized into two groups: macronutrients, comprising carbohydrates, dietary fiber, fats, and proteins; and micronutrients, encompassing vitamins and minerals needed in small amounts, measured in milligrams or micrograms but play a vital role in overall development of fish health and its quality.

2 Risks Associated with the Use of Antibiotics in Aquaculture

In aquaculture practices, antibiotics are employed pre-emptively to avoid diseases by improving either the water or the host before contagious diseases manifest (Christensen et al., 2006). Although the prophylactic practice of antibiotics in disease prevention is financially advantageous due to its ability to avert losses and expedite host growth, it also comes with significant drawbacks. Like Overreliance on antibiotics can lead to the advent of antibiotic-resistant bacteria. These resilient strains arise from the evolutionary pressure to survive, resulting in genetic alterations that enable bacteria to withstand antibiotic treatments (Anderson et al., 2003). Since certain antibiotics administered to fish for treatment are identical to those used in human medicine, bacteria causing diseases in humans may also develop resistance to antibiotics due to the treatment of fish with these medications (Angulo, 2000). By virtue of this concern, excessive antibiotic usage in treating aquatic animals could lead to public health challenges. Consequently, numerous countries have prohibited or are in the process of prohibiting the incorporation of antibiotics in aquafeeds as a standard method for enhancing growth. Moreover, the extensive utilization of various antibiotics together with nonbiodegradable ones used in human medicine results in their persistence in the aquatic environment, employing selective pressure over prolonged durations (Iwashita et al., 2015). Furthermore, effective gene transfer linked to antibiotic resistance among bacteria in aquaculture settings and pathogenic bacteria linked to animals and humans in terrestrial environments, as well as between microorganisms in animals and pathogens affecting humans, has also been observed. In addition, the overuse of antibiotics in marketable aquaculture leads to the presence of residual, unmetabolized compounds in the environment. This could lead to the inadvertent ingestion of therapeutic contaminants by fish, potentially disrupting their normal bacterial flora and impacting their vulnerability to infections, antibiotic-resistant microorganisms, allergies, and toxicity (Cabello, 2006). In recent years, the use of antibiotics and chemotherapeutics to combat several such diseases which led to numerous complications, including the proliferation of drug-resistant pathogens, suppression of the immune system in fish besides environmental risks (Allameh et al., 2016). Consequently, there is growing encouragement for an alternative strategy aimed at bolstering disease resistance, immune responses, and overall health benefits through the utilization of probiotics, prebiotics, and other feed additives with diverse well-being-endorsing capabilities for various aquatic species. So, in order to avoid such disadvantages and risk from aquaculture, some alternative strategies have to be developed by employing the safe and sound procedure to prevent the outbreak of fish disease.

3 Nutrients as Immune Stimulants in Aquaculture

The nonspecific immune system recognizes self-cells from nonself-cells through a range of cellular and humoral elements, enabling it to eradicate undesirable substances like microbes and toxins directly (Magnadottir, 2006). Thrombocytes are a response to phagocytosis and the primary protecting cells of fish (Tavares-Dias et al., 1999). Macrophages and neutrophils as part of their phagocytosis response combat infections by releasing reactive oxygen species. Humoral immune response not only aids in eliminating or shielding against the proliferation and dissemination of harmful microorganisms but also facilitates the recognition of specificities. To mitigate infection risks in fish farming, the use of immunostimulants and vaccine carriers has been considered as vital and safe strategy in the field of aquaculture. Despite this, relatively few studies have explored these conditions. However, a substantial body of research indicates that enhancing the intensity, expediting the process, and prolonging the duration of the specific immunological response are all achievable objectives (Anderson, 1992). Several workers have investigated the utilization of diverse immunostimulants to enhance disease resistance and immune response, along with the colonization of gut microbes in different aquatic animals (Ringø et al., 2010). Immunostimulants are also considered as safe for application in aquatic settings, as they are nontoxic and do not induce any adverse properties on the system (Ali et al., 2021) Conversely, it has also been suggested that the generation of immune-boosting substances, comprising of macronutrients, vitamins, chitosan polysaccharides, nanoelements, and various biological component, can be achieved through advance innovative means as well (Koch et al., 2021).

Enhancing both humoral and cellular immune response in head-kidney, blood and gene expression, as well as improving response of immune system to infections in aquatic animals, are all benefits of immune-stimulating feed additives (Sakai et al., 2020). Among the numerous components and extracts evaluated as supplements for various diets, macroalgae and their extracts stand out for their high polysaccharide content including alginic acid, laminarin, and fucoidan along with various bioactive molecules such as vitamins and polyphenols. These substances are recognized for their ability to promote the growth and overall health of various fish species (Sheikhzadeh et al., 2012). Studies have shown that efficient feed additives prompt the natural defense mechanisms of aquatic animals through two main mechanisms, i.e., by the regulation of commensal microbiome of the gut, which subsequently influences immunity and directly triggering both the innate (nonspecific) and adaptive immune systems (Thepot et al., 2021). Presently, spurred by ongoing discussions among consumers regarding the use of antibiotics in animals including fish raised under captivity, the adoption of immunostimulants derived from nutrients is beneficial in preventing disease outbreaks in aquaculture practices (Montero-Rocha et al., 2006). Various categories of substances, like amino acids, vitamins, and minerals, are acknowledged for their ability to stimulate the immune system along with growth and development and thus enhance the aquaculture output economically (Siwicki et al., 1998) (Table 4.1).

Table 4.1 Commonly used immunostimulants in aquaculture

Glucan	Capable of triggering a nonspecific immune response in carp	Yano et al. (1991)
Levamisole	An efficient remedy for combating camallanus roundworm infestations in freshwater tropical fish. Augments the nonspecific immune response in fish. Employed as a supplement alongside vaccines	Baba et al. (1993) and Jeney and Anderson (1993)
Chitin	Enhanced defense against *Aeromonas salmonicida* in brook trout	Stickney (2000)
Chitosan	Provides safeguarding for salmonids and carps from bacterial illnesses	Siwicki et al. (1994)
Muramyl dipeptide	Amplifies antibody activity, triggers polyclonal activation of lymphocytes, and activates macrophages	Mastan (2015)
Lentinan, oligosaccharide, and schizophyllan	Boosts both cellular and noncellular defense mechanisms, such as lysozyme, phagocyte, and complement activities in fish	Mastan (2015)

4 Feed Additives as Immune Stimulants in Aquaculture

Feed additives are described as nonnutritive components or ingredients incorporated during feed formulations to modify the physical, chemical attributes of fish diet. They are helpful in maintaining the overall metabolism or impact the overall metabolism or quality of aquatic animals and their resulting products (Barrows, 2000). These additives encompass a wide range of chemical compositions and are utilized to varying degrees in commercial aquatic diet formulations, including pellet binders, preservatives such as antimicrobial agents, and antioxidants besides feeding enhancers (NRC, 2011a, b). These additives are frequently incorporated to attain and uphold ideal physical and chemical attributes related to the fish and its products. Additionally, other additives, such as probiotics, prebiotics, acidifiers, and extracts derived from plants or animals, which can directly impact fish performance or product quality, are commercially accessible for aquatic animals. An effective approach to boost both growth performance and immune resilience in fish involves employing functional feed additives as a viable alternative (Dawood et al., 2015).

Numerous studies indicated that varying feed components and alterations in diet makeup can impact the structure of the gut and the balance of microbiota, thus affecting digestive and absorptive capabilities of the fish. Changes in the composition of intestinal microbiota and subsequent decrease in beneficial gut microorganisms could play a role in gut-related disease development. Thus, maintaining a healthy gut flora is crucial for optimizing feed efficiency, promoting fish growth, and ensuring its health (Ringo & Olsen, 1999). In contemporary times, there exist more eco-friendly methods to regulate gut microflora and enhance fish performance. These methods involve incorporating nutraceuticals or functional foods into feeds to adjust the health and performance of farmed fish.

5 Organic Acids as Fish Feed Additives for the Management and Prevention of Diseases

The practice of dietary acidification through the incorporation of organic acids has garnered significant importance in animal nutrition. Organic acids have emerged as a capable feed supplement for enhancing both gut health and its performance. Varieties of acids, including acetic, butyric, citric, formic, lactic, propionic, malic, and sorbic acids, along with their salts, have been utilized as acidifiers in fish feed formulations (NRC, 2011a, b). Organic acids are incorporated in feedstuffs and compound feed in various forms. Liquid acids and blends are typically sprayed onto the feed, while solid acids and acid salts are added directly or through specialized premixtures in the feed. The mechanism by which organic acids operate in the intestinal tract involves two primary actions: First, they lower the pH in the stomach and small intestine, thereby enhancing the activity of digestive enzymes and creating an unfavorable environment for pathogens; second, they inhibit the growth of gram-negative bacteria by dissociating into acids and generating anions within bacterial cells (Luckstadt, 2008). Many organic acid feed supplements are added to diets primarily for their antimicrobial properties. When the pH drops below 5, the growth of various gram-negative bacteria is stunted. Organic acids, being weak acids, retain a portion of their molecules undissociated, influenced by the acid's pKa value and the surrounding pH. Additionally, small acids are lipophilic, allowing them to pass through the cell membrane of gram-negative bacteria. Once inside the relatively alkaline cytoplasm, they dissociate, leading to a decrease in pH. The alteration in pH causes disruption of function, effectively inhibiting the growth of intraluminal microbes, particularly pathogens (FEFANA, 2014). An emerging approach to regulate the gut microflora and also mitigate the pathogens involves combining of organic acids with other antimicrobial compounds like phytochemicals or permeabilizers. This strategy aims to capitalize on potential synergistic effects to more efficiently combat pathogenic bacteria. While investigating the antimicrobial properties of thymol and carvacrol in combination with acetic acid against *Salmonella typhimurium*. It has been observed that when combined with acetic acid, the antibacterial effectiveness of thymol (100 mg/L) plus acetic acid (0.10%) and carvacrol (100 μL/L) plus acetic acid (0.10%) was comparable to that achieved by using higher concentrations of thymol (400 mg/L) and carvacrol (400 μL/L) alone. This illustrates a notable cooperative antimicrobial effect (Zhou et al., 2007). Studies have shown that the addition of a permeabilizer substance to a combination of organic acids (formic, propionic, and acetic) and a phytochemical (cinnamaldehyde) significantly enhanced their inhibitory effects against *Aeromonas veronni* and *Yersinia ruckeri*. When this synergistic mixture was added to tilapia feed at a concentration of 1 g/kg, fish consuming the supplemented feed showed notably higher growth rates and improved feed conversion ratios (FCR) (Riemensperger and Santos, 2011a).

6 Role of Probiotics in the Management of Diseases in Aquaculture

Probiotics are commonly referred to as "friendly," "beneficial," "good," or "helpful" bacteria, due to their role in maintaining gut health. A recent definition describes probiotics as "live microorganisms" that, when consumed in sufficient quantities, provide health benefits to the host (FAO/WHO, 2001). They are live microbial additives given to animals for positive impact on their intestinal microbial equilibrium. The interface between probiotics and the digestive system, including its natural microorganisms, has been thoroughly researched, and their beneficial effects on the health of aquatic animals including fish have been extensively documented and scientifically evaluated. An increasing trend in the aquaculture industry involves employing "probiotics," a microbial strategy aimed at boosting disease resistance, improving survival rates, and enhancing growth through more efficient feed conversion. Various forms of probiotics are utilized in aquaculture, with bacterial cells being among them. Typical probiotic microorganisms include *Bacillus* species, *Lactobacillus* species, *Enterococcus* species, *Carno bacterium* species, and the yeast *Saccharomyces cerevisiae*, among others. Encouraging the use of immune stimulants, non-specific immune enhancers, probiotics and vaccines can help reduce the need for antimicrobial agents in aquaculture practices. These avenues are regarded as significant areas of potential research for controlling diseases in aquaculture (Suguna, 2010). Probiotic organisms, whether alive, deceased, or components of microbial cells, are introduced through feed or rearing water in aquaculture. They enhance the host's health by enhancing their resistance to disease and promoting growth, feed efficiency, stress response, and overall vitality, largely by balancing microbial populations in the host or its surroundings (Merrifield et al., 2010). These probiotics encompass a variety of bacteria, bacteriophages, microalgae, and yeast, commonly employed in aquaculture through water or feed supplementation (Llewellyn et al., 2014). As per existing literature, it has been clearly documented that probiotics are extensively been involved in two major perspectives:

6.1 Enhancement of Disease Resistance

Probiotics like microorganisms, including lactic acid bacteria, *Brevibacillus brevis*, *Vagococcus fluvialis,* and *Vibrio harveyi*, adhere to the mucosal epithelium of the gastrointestinal track and aid in antimicrobial resistance in addition to immunomodulation (Luis-Villasenor et al., 2011).

Another way that probiotics use to improve feed digestibility is by raising the levels of several digestive enzymes, including proteases, amylases, and alginate lyases (Zokaeifar et al., 2012). In addition, they generate lysozyme, organic acids, fatty acids, hydrogen peroxide, bacteriocins, siderophores, biotin, and vitamin B_{12}, all of which are advantageous to the health of the host principally fish (Vine et al., 2006).

6.2 Modulation of Immune Parameters

Studies indicate that probiotics have the potential to influence key components of the nonspecific immune system of fish, including monocytes, macrophages, neutrophils, and natural killer cells. Previous studies have also shown an increase in various types of white blood cells (WBCS), such as monocytes and lymphocytes, as well as other immune cells like macrophages and granulocytes, in different fish species after its treatment to the fish (Kumar et al., 2008). Probiotics, beneficial bacteria, not only combat harmful pathogens, but also help in regulation of the host's immune system. They contain conserved microbe-associated molecular patterns (MAMPs), such as peptidoglycan (PGN), lipoteichoic acids (LTA), S-layer protein A (SlpA), exopolysaccharides (EPS), flagellin, and microbial nucleic acids. These MAMPs are recognized by specific pattern recognition receptors (PRRs), triggering a signaling cascade, that leads to the production of cytokines, chemokines, and other molecules, thereby activating the immune response in the fish (Remus et al., 2012). Over recent years, there has been growing interest in understanding how probiotics interact with the intestinal immune system. Researchers have explored the potential connection between the recognition of probiotics through Toll-like receptor (TLR) signaling and the activation of intestinal immunity. For instance, studies have indicated that the TLR_2 signaling pathway plays a role in recognizing probiotic strains like *Psychrobacter* sp. SE6, leading to subsequent immune responses in grouper (*Epinephelus coioides*) (Sun et al., 2014).

6.3 Competition for Nutrients

One of the mechanisms by which probiotics suppress pathogens is through competing for nutrients (Ringø & Song, 2016). Most bacteria require iron for their growth, yet the availability of iron in animal tissues and body fluids is limited (Verschuere et al., 2000). Siderophores, which are iron-binding molecules, assist bacteria in acquiring the iron they need for growth. There is a clear correlation between the production of siderophores and the virulence of certain pathogens (Gram et al., 1999).

6.4 Production of Antibacterial Substances

In aquaculture, probiotics serve as a substitute for antibiotics and chemicals (Heo et al., 2013). Although the exact mechanism by which probiotics exhibit antibacterial effects is still unclear, numerous studies have shown that probiotics can produce antibiotic compounds (Moriarty, 1998). Additionally, the reduction in pH resulting from the production of organic acids can inhibit the growth of pathogenic bacteria (Ma et al., 2009). Production of short chain fatty acids and bactericidal proteins by commonly used probiotic, i.e., *lactobacillus, i*s known to enhance immune response as well as disease resistance (Faramarzi et al., 2011). As a result, probiotics can safeguard aquatic animals from pathogenic threats by producing antibiotic compounds.

7 Phytocompounds as Better Alternatives for Antibiotics in Disease Management in Aquaculture

Aquatic habitats are open and connected hydrodynamically. Microbes can spread passively over a distance of more than 50 km through the water column and ultimately inhabit a different aquatic habitat (Kragesteen et al., 2018; Cantrell et al., 2020). The concurrent use of antibiotics in aquaculture practices has led to growing problem of antibiotic resistance among different aquaculture species which has ultimately spread to other unaffected microbial communities across interconnected ecosystems (Preena et al., 2020). Phytocompounds provide a way for natural alternative to these antibiotics as they are known to play an important role in the control of various aquaculture diseases. Different plant extracts have been utilized and tested over time to determine their efficacy for various microbial diseases making it a potential candidate for use in disease management in aquaculture. Over time various plant compounds such as alkaloids, flavonoids, phenolic acids, terpenoids, saponins have been tested over time against various microbial agents. The different biological compounds have varied levels of effects including promotion of growth activity, modulation of immune response, release of various antioxidants, antibacterial, antiviral as well as anti-parasitic effects (Zhu, 2020) (Table 4.2).

8 Conclusion

In aquaculture, intervention in nutritional methods provides viable paths for illness prevention and treatment, offering long-term substitutes for chemical and antibiotic therapies. Probiotics, prebiotics, immune stimulants, functional foods, and nutritional supplements can all help aquatic species perform better overall and strengthen their immune systems while lowering their risk of illness. Optimizing these nutritional therapies, effectiveness, dose, and compatibility with current aquafeed formulations and production processes is still a challenge. To further understand the mechanisms of action, assess the long-term impacts, and create financially viable plans for implementing nutritional approaches into aquaculture operations, more research is required. Advanced aquaculture practices can minimize their environmental effect and protect public health while improving the sustainability, resilience, and profitability of their operations by utilizing nutritional interventions.

Table 4.2 Various plant compounds used in the management of various aquaculture problems

Plant part used	Effect of the substitution	Species	References
Diet supplemented with 0.4% and 0.8% of *A.vera* extract	Promotes growth rate in juveniles	*Cyprinus carpio*	Khanal et al. (2021)
Consumption of *A. vera* gel	Increase in lactobacillus population in gut	*Cyprinus carpio*	Parsa et al. (2016)
Vanillin of nettle extract	Increase in growth rate and feed palatability	*Oncorhynchus mykiss*	Zare and Mirzakhani (2021)
Olive leaf extract at 0.1% and 0.25%	Expression of growth related genes (IG & IGF 1), increased levels of digestive enzymes	*Cyprinus carpio*	Zemheri-Navruz et al. (2020)
Diet supplemented with oregano	Improved gut microbiota and provides resistance against *Aeromonas hydrophila*	Koi carp, *Cyprinus carpio*	Zhang et al. (2020)
Diet supplemented with *Aegle marmelos* fruit extract	Increased antioxidants and decreased cholesterol	*Oreochromis niloticus*	Wangkahart et al. (2022)
Dietary supplementation of combination of *Boesenbergia pandurata*, *Solanum ferox* and *Zingiber Zerumbet* plant extracts	Treatment of *A. hydrophila* and *Pseudomonas fluorescens*	Tilapia	Hardi et al. (2019)
Dietary supplementation of combination of thymol and chitosan nanoparticles	Improved antibacterial activity	*O. niloticus*	Abd El-Naby et al. (2020)
Dietary supplementation of guava, *Psidium guajava*	Increased protease and lysozyme activity	Hybrid tilapia *O. niloticus*×*O. mossambicus*	Ceballos-Francisco et al. (2020)
Dietary supplementation of *P. guajava*	Improved gut morphology, enhanced growth rate	*O. niloticus*	Omitoyin et al. (2019)
Dietary inclusion of *Z. officinale*	Increased nonspecific immune response to *Yersinia ruckeri*	*O. mykiss*	Soltanian et al. (2019)
Tannic acid	Causes the disruption of plasma membrane of parasite *Ichthyophthirius multifiliis*	Zebra fish, *Danio rerio*	Alavinia et al. (2018)

References

Abd El-Naby, A. S., Al-Sagheer, A. A., Negm, S. S., & Naiel, M. A. E. (2020). Dietary combination of chitosan nanoparticle and thymol affects feed utilization, digestive enzymes, antioxidant status, and intestinal morphology of *Orechromis niloticus*. *Aquaculture, 515*, 734577.

Ahmad, I., & Ahmed, I. (2022). Ulceration in fish: Causes, diagnosis and prevention. In *Bacterial fish diseases* (pp. 337–350). Academic Press.

Alavinia, S. J., Mirzargar, S. S., Rahmati-Holasoo, H., & Mousavi, H. E. (2018). The in vitro and in vivo effect of tannic acid on *Ichthyophthirius multifiliis* in zebrafish (*Danio rerio*) to treat ichthyophthiriasis. *Journal of Fish Diseases, 41*, 1793–1802.

Alfred, O., Shaahu, A., Orban, D. A., & Egwenomhe, M. (2020). An overview on understanding the basic concept of fish diseases in aquaculture. *IRE Journal, 4*, 83–91.

Ali, M. F., Nakahara, S., Otsu, Y., Ido, A., Miura, C., & Miura, T. (2021). Effects of functional polysaccharide from silkworm as an immunostimulant on transcriptional profiling and disease resistance in fish. *Journal of Insects as Food and Feed, 8*, 1221–1233.

Allameh, S. K., Yusoff, F. M., Ringø, E., Daud, H. M., Saad, C. R., & Ideris, A. (2016). Effects of dietary mono-and multiprobiotic strains on growth performance, gut bacteria and body composition of Javanese carp (*Puntius gonionotus*, Bleeker 1850). *Aquaculture Nutrition, 22*, 367–373.

Anderson, D. P. (1992). Immunostimulants, adjuvants, and vaccine carriers in fish: Applications to aquaculture. *Annual Review of Fish Diseases, 2*, 281–307.

Anderson, A. D., Nelson, J. M., Rossiter, S., & Angulo, F. J. (2003). Public health consequences of use of antimicrobial agents in food animals in the United States. *Microbial Drug Resistance, 9*, 373–379.

Angulo, F. (2000). Antimicrobial agents in aquaculture: Potential impact on public health. *APUA Newsletter, 18*, 4–5.

Baba, T., Watase, Y., & Yoshinaga, Y. (1993). Activation of mononuclear phagocyte function by levamisole immersion in carp. *Nippon Suisan Gakkaishi, 59*, 301–307.

Barrows, F. T. (2000). Feed additives. In R. P. Stickney (Ed.), *Encyclopedia of aquaculture* (pp. 335–340). Wiley.

Bondad-Reantaso, M. G., Subasinghe, R. P., Arthur, J. R., Ogawa, K., Chinabut, S., Adlard, R., Tan, Z., & Shariff, M. (2005). Disease and health management in Asian aquaculture. *Veterinary Parasitology, 132*, 249–272.

Cabello, F. C. (2006). Heavy use of prophylactic antibiotics in aquaculture: A growing problem for human and animal health and for the environment. *Environmental Microbiology, 8*, 1137–1144.

Cantrell, D. L., Groner, M. L., Ben-Horin, T., Grant, J., & Revie, C. W. (2020). Modelling pathogen dispersal in marine fish and shellfish. *Trends in Parasitology, 36*(3), 239–249.

Ceballos-Francisco, D., Castillo, Y., Rosa, F. D. L., Vásquez, W., Reyes-Santiago, R., Cuello, A., Cuesta, A., & Esteban, M. A. (2020). Bactericidal effect on skin mucosa of dietary guava (*Psidium guajava* L.) leaves in hybrid tilapia (*Orechromis niloticus* × *O. mossambicus*). *Journal of Ethnopharmacology, 259*, 112838.

Christensen, A. M., Ingerslev, F., & Baun, A. (2006). Ecotoxicity of mixtures of antibiotics used in aquacultures. *Environmental Toxicology and Chemistry, 25*, 2208–2215.

Cuesta, A., & Esteban, M. Á. (2021). Pathologies in farmed tilapia and the use of immunostimulants and vaccines to prevent or treat diseases: An overview. *Biology and Aquaculture of Tilapia, 10*, 119–136.

Dawood, M. A., El-Dakar, A., Mohsen, M., Abdelraouf, E., Koshio, S., Ishikawa, M., & Yokoyama, S. (2014). Effects of using exogenous digestive enzymes or natural enhancer mixture on growth, feed utilization, and body composition of rabbitfish, *Siganus rivulatus*. *Journal of Agriculture, Science and Technology, 1*, 4.

Dawood, M. A. O., Koshio, S., Ishikawa, M., & Yokoyama, S. (2015). Interaction effects of dietary supplementation of heat-killed lactobacillus plantarum and b-glucan on growth performance, digestibility and immune response of juvenile red sea bream, *Pagrus major*. *Fish & Shellfish Immunology, 45*, 33–42.

FAO/WHO. (2001). *Report of a joint FAO/WHO expert consultation on evaluation of health and nutritional properties of probiotics in food including powder Milk with live lactic acid bacteria*. Cordoba, Argentina.

Faramarzi, M., Kiaalvandi, S., & Iranshahi, F. (2011). The effect of probiotics on growth performance and body composition of common carp, (*Cyprinus carpio*). *Journal of Animal and Veterinary Advances, 10*, 2408–2413.

FEFANA. (2014). *Organic acids in animal nutrition* (97pp.). Fefana Publication.

Gram, L., Melchiorsen, J., Spanggaard, B., Huber, I., & Nielsen, T. F. (1999). Inhibition of *vibrio anguillarum* by *Pseudomonas fluorescens* AH2, a possible probiotic treatment of fish. *Applied and Environmental Microbiology, 65*, 969–973.

Hardi, E. H., Nugroho, R. A., Isnansetyo, A., Agriandini, M., Kusuma, I. W., & Sidik, A. S. (2019). Simultaneous administration of *Boesenbergia pandurata* and vaccination to stimulate immune response in tilapia, *Oreochromis niloticus. Pakistan Journal of Biological Sciences, 22*, 419–426.

Heo, W. S., Kim, Y. R., Kim, E. Y., Bai, S. C., & Kong, I. S. (2013). Effects of dietary probiotic, *Lactococcus lactis* subsp. lactis I2, supplementation on the growth and immune response of olive flounder (*Paralichthys olivaceus*). *Aquaculture, 376*, 20–24.

Hernandez, E., Figueroa, J., & Iregui, C. (2009). Streptococcosis on a red tilapia, *Oreochromis* sp., farm: A case study. *Journal of Fish Diseases, 32*, 247–252.

Iwashita, M. K. P., Addo, S., & Terhune, J. S. (2015). Use of pre-and probiotics in finfish aquaculture. In A. Davis (Ed.), *Feed and feeding practices in aquaculture* (p. 235). Woodhead Publishing.

Jeney, G., & Anderson, D. P. (1993). Enhanced immune response and protection in rainbow trout to *Aeromonas salmonicida* bacterin following prior immersion in immunostimulants. *Fish & Shellfish Immunology, 3*, 51–58.

Khanal, M., Lamichhane, S., Bhattarai, A., Kayastha, B. L., & Labh, S. N. (2021). Extract of aloe vera (*Aloe barbaden*-sis Miller) enhances the growth, protein contents, and gastrosomatic index (GaSI) of common carp *Cyprinus carpio. Journal of Nutrition and Metabolism, 2021*, 8029413.

Koch, J. F., de Oliveira, C. A., & Zanuzzo, F. S. (2021). Dietary β-glucan (MacroGard®) improves innate immune responses and disease resistance in *Nile tilapia* regardless of the administration period. *Fish & Shellfish Immunology, 112*, 56–63.

Kragesteen, T. J., Simonsen, K., Visser, A. W., & Andersen, K. H. (2018). Identifying salmon lice transmission characteristics between Faroese salmon farms. *Aquaculture Environment Interactions, 10*, 49–60.

Kumar, R., Mukherjee, S., Ranjan, R., & Nayak, S. (2008). Enhanced innate immune parameters in *Labeo rohita* (Ham.) following oral administration of *Bacillus subtilis. Fish & Shellfish Immunology, 24*, 168–172.

Llewellyn, M. S., Boutin, S., Hoseinifar, S. H., & Derome, N. (2014). Teleost microbiomes: The state of the art in their characterization, manipulation and importance in aquaculture and fisheries. *Frontiers in Microbiology, 5*, 207.

Luckstadt, C. (2008). The use of acidifiers in fish nutrition. *CABI Reviews, 3*, 1–8.

Luis-Villasenor, I. E., Macias-Rodriguez, M. E., Gomez-Gil, B., Ascencio-Valle, F., & Campa-Cordova, A. I. (2011). Beneficial effects of four bacillus strains on the larval cultivation of *Litopenaeus vannamei. Aquaculture, 321*, 136–144.

Ma, C. W., Cho, Y. S., & Oh, K. H. (2009). Removal of pathogenic bacteria and nitrogens by *Lactobacillus* spp. JK-8 and JK-11. *Aquaculture, 287*, 266–270.

Magnadottir, B. (2006). Innate immunity of fish (overview). *Fish & Shellfish Immunology, 20*, 137–151.

Mastan, S. A. (2015). Use of immunostimulants in aquaculture disease management. *International Journal of Fisheries and Aquatic Studies, 2*, 277–280.

Merrifield, D. L., Dimitroglou, A., Foey, A., Davies, S. J., Baker, R. T., Bøgwald, J., Castex, M., & Ringø, E. (2010). The current status and future focus of probiotic and prebiotic applications for salmonids. *Aquaculture, 302*, 1–8.

Montero-Rocha, A., McIntosh, D., Sanchez-Merino, R., & Flores, I. (2006). Immunostimulation of white shrimp (*Litopenaeus vannamei*) following dietary administration of Ergosan. *Journal of Invertebrate Pathology, 91*, 188–194.

Moriarty, D. (1998). Control of luminous vibrio species in penaeid aquaculture ponds. *Aquaculture, 164*, 351–358.

National Research Council (NRC). (2011a). *Nutrient requirements of fish and shrimp*. National Academies Press.

National Research Council (NRC). (2011b). *Nutrient requirements of fish* (376pp.). National Academy Press.

Nomoto, K. (2005). Prevention of infections by probiotics. *Journal of Bioscience and Bioengineering, 100*, 583–592.

Omitoyin, B. O., Ajani, E. K., Orisasona, O., Bassey, H. E., Kareem, K. O., & Osho, F. E. (2019). Effect of guava *Psidium guajava* (L.) aqueous extract diet on growth performance, intestinal morphology, immune response and survival of *Oreochromis niloticus* challenged with *Aeromonas hydrophila*. *Aquaculture Research, 50*, 1851–1861.

Parsa, A., Bahramian, S., & Sharifpour, I. (2016). Effect of oral consumption of Aloe vera gel on intestinal micro-flora and liver tissue of rainbow trout. *Iranian Journal of Fisheries Sciences, 15*, 590–596.

Preena, P. G., Swaminathan, T. R., Kumar, V. J., & Singh, I. S. (2020). Antimicrobial resistance in aquaculture: a crisis for concern. *Biologia, 75*, 1497–1517.

Rana, K. J. (1997). Status of global production and production trends. *FAO Fisheries Circular, 886*, 3–16.

Remus, D. M., van Kranenburg, R., van Swam, I. I., Taverne, N., Bongers, R. S., Wels, M., Wells, J. M., Bron, P. A., & Kleerebezem, M. (2012). Impact of 4 *Lactobacillus plantarum* capsular polysaccharide clusters on surface glycan composition and host cell signaling. *Microbial Cell Factories, 11*, 149.

Reverter, M., Bontemps, N., Lecchini, D., Banaigs, B., & Sasal, P. (2014). Use of plant extracts in fish aquaculture as an alternative to chemotherapy: Current status and future perspectives. *Aquaculture, 433*, 50–61.

Riemensperger, A., & Santos, G. (2011a). Breakthrough in natural growth promotion. *International Aquafeed*, 18–20.

Riemensperger, A., & Santos, G. (2011b). Maintaining health in shrimp culture. *Aqua Culture Asia Pacific, 28-30*

Ringo, E., & Olsen, R. E. (1999). The effect of diet on aerobic bacterial flora associated with intestine of Arctic charr (*Salvelinus alpinus* L.). *Journal of Applied Microbiology, 86*, 22–28.

Ringø, E., & Song, S. K. (2016). Application of dietary supplements (synbiotics and probiotics in combination with plant products and β-glucans) in aquaculture. *Aquaculture Nutrition, 22*, 4–24.

Ringø, E., Løvmo, L., Kristiansen, M., Bakken, Y., Salinas, I., Myklebust, R., Olsen, R. E., & Mayhew, T. M. (2010). Lactic acid bacteria vs. pathogens in the gastrointestinal tract of fish: A review. *Aquaculture Research, 41*(4), 451–467.

Sakai, M., Hikima, J. I., & Kono, T. (2020). Fish cytokines: Current research and applications. *Fisheries Science, 87*, 1–9.

Sheikhzadeh, N., Heidarieh, M., Pashaki, A. K., Nofouzi, K., Farshbafi, M. A., & Akbari, M. (2012). Hilyses® fermented Saccharomyces cerevisiae, enhances the growth performance and skin non-specific immune parameters in rainbow trout *(Oncorhynchus mykiss). Fish & Shellfish Immunology, 32*, 1083–1087.

Siwicki, A. K., Anderson, D. P., & Rumsey, G. L. (1994). Dietary intake of immunostimulants by rainbow trout affects non-specific immunity and protection against furunculosis. *Veterinary Immunology and Immunopathology, 41*, 125–139.

Siwicki, A. K., Morand, M., Terech-Majewska, E., Niemczuk, W., Kazun, K., & Glabski, E. (1998). Influence of immunostimulants on the effectiveness of vaccines in fish: In vitro and in vivo study. *Journal of Applied Ichthyology, 14*, 225–227.

Soltanian, M., Langrodi, H. F., & Nejad, M. M. (2019). The use of Zingiber officinale extract against *Yersinia ruckeri* and its effects on the antioxidant status and immune response in *Oncorhynchus mykiss*. *International Journal of Aquatic Biology, 795*, 301–304.

Sun, Y. Z., Xia, H. Q., Yang, H. L., Wang, Y. L., & Zou, W. C. (2014). TLR2 signaling may play a key role in the probiotic modulation of intestinal microbiota in grouper *Epinephelus coioides*. *Aquaculture, 430*, 50–56.

Syanya, F. J. (2018). An investigation on causal elements for project sustainability among selected nongovernmental organizations in the education sector: A case of Nairobi County. *IOSR Journal of Humanities and Social Science, 23*, 63–105.

Tavares-Dias, M., Schalch, S. H., Martins, M. L., Silva, E. D., Moraes, F. R., & Perecin, D. (1999). Hematologia de teleosteos brasileiros com infecçao parasitaria. I. Variaveis do Leporinus macrocephalus Garavelo e Britski, 1988 e (Anostomidae) Piaractus mesopotamicus Holmberg, 1887 (Characidae). *Acta Scientiarum, 21*, 337–342.

Thepot, V., Campbell, A. H., Rimmer, M. A., & Paul, N. A. (2021). Meta-analysis of the use of seaweeds and their extracts as immunostimulants for fish: A systematic review. *Reviews in Aquaculture, 13*, 907–933.

Verschuere, L., Rombaut, G., Sorgeloos, P., & Verstraete, W. (2000). Probiotic bacteria as biological control agents in aquaculture. *Microbiology and Molecular Biology Reviews, 64*, 655–671.

Vijayaram, S., Ringø, E., Zuorro, A., van Doan, H., & Sun, Y. (2023). *Beneficial roles of nutrients as immunostimulants in aquaculture: A review* (Vol. 9, p. 707). Aquaculture and Fisheries.

Vine, N. G., Leukes, W. D., & Kaiser, H. (2006). Probiotics in marine larviculture. *FEMS Microbiology Reviews, 30*, 404–427.

Wangkahart, E., Wachiraamonloed, S., Lee, P.-T., Subramani, P. A., Qi, Z., & Wang, B. (2022). Impacts of *Aeglemarmelos* fruit extract as a medicinal herb on growth performance, antioxidant and immune responses, digestive enzymes, and disease resistance against *Streptococcus agalactiae* in Nile tilapia *Oreochromis niloticus*. *Fish & Shellfish Immunology, 120*, 402–410.

World Health Organization. (2006). *Report of a joint FAO/OIE/WHO Expert Consultation on antimicrobial use in aquaculture and antimicrobial resistance*, Seoul, Republic of Korea, 13–16 June 2006.

Yano, T., Matsnyama, H., & Mangindann, R. E. P. (1991). Polysaccharide induced protection of carp, *Cyprinus carpio* against bacterial infection. *Journal of Fish Diseases, 14*, 557–582.

Zare, M. K., & Mirzakhani, S. V. (2021). Growth, body proximate composition and selected blood parameters of rainbow trout *Onchorhynchus mykiss* fingerlings fed on a diet supplemented with nettle *Urtica dioica* and tarragon *Artemisia dracunculus*. *Aquaculture Research, 52*, 5691–5702.

Zemheri-Navruz, F., Acar, U., & Yilmaz, S. (2020). Dietary supplementation of alive leaf extract enhances growth performance, digestive enzyme activity and growth related genes expression in common carp *Cyprinus carpio*. *General and Comparative Endocrinology, 296*, 113541.

Zhang, R., Wang, X. W., Liu, L. L., Cao, Y. C., & Zhu, H. (2020). Dietary oregano essential oil improved the immune response, activity of digestive enzymes, and intestinal microbiota of the koi carp, *Cyprinus carpio*. *Aquaculture, 518*, 734781.

Zhou, F., Ji, B., Zhang, H., Jiang, H., Yang, Z., Li, J., Ren, Y., & Yan, W. (2007). Synergistic effect of thymol and carvacrol combined with chelators and organic acids against *Salmonella typhimurium*. *Journal of Food Protection, 70*, 1704–1709.

Zhu, F. (2020). A review on the application of herbal medicines in the disease control of aquatic animals. *Aquaculture, 526*, 735422.

Zokaeifar, H., Balcazar, J. L., Saad, C. R., Kamarudin, M. S., Sijam, K., Arshad, A., & Nejat, N. (2012). Effects of Bacillus subtilis on the growth performance, digestive enzymes, immune gene expression and disease resistance of white shrimp, *Litopenaeus vannamei*. *Fish & Shellfish Immunology, 33*, 683–689.

Chapter 5
Role of Prebiotics and Probiotics in Aquaculture Nutrition

Baskar Veeramaruthu, Oyas Ahmad Asimi, Samaya Prakash, Jaberson Solomon, Ajith Stephen, Iyyappan Somasundaram, Mukilan Akash, and Bharathkumar

1 Introduction

Aquaculture plays a pivotal role in addressing the increasing global demand for food fish and ensuring food security, especially as the global population is expected to reach 9.7 billion by 2050 (Ahmed et al., 2024). Over recent decades, aquaculture has experienced remarkable growth, emerging as one of the fastest-growing sectors within the global food industry (Ahmed & Ahmad, 2020). By 2022, global aquaculture production had reached around 94 million tons, accounting for 51% of total fish production and surpassing the output of capture fisheries (FAO, 2024). As this sector advances, one of the key challenges for farmers is optimizing fish nutrition to achieve robust growth rates while mitigating disease outbreaks. This has driven a transition from extensive farming systems to semi-intensive and intensive practices. However, intensification often leads to a higher incidence of diseases. Historically, antibiotics and chemotherapeutic agents have been used to manage these issues. Yet, their application has raised significant concerns, including the emergence of drug-resistant pathogens, compromised immune responses in fish, and adverse environmental effects. To address these challenges, alternative nutritional strategies are gaining traction, focusing on enhancing the health and disease resistance of aquatic species. Among these strategies, prebiotics and probiotics have emerged as promising tools for improving fish nutrition and overall health.

Prebiotics are nondigestible compounds that stimulate beneficial alterations in the gut microbiota, contributing to improved nutritional status and overall health in fish (Ringø et al., 2014). The fermentation byproducts produced from prebiotics act

B. Veeramaruthu (✉) · O. A. Asimi · S. Prakash · A. Stephen · I. Somasundaram · M. Akash
Faculty of Fisheries, SKUAST-K, Rangil, Ganderbal, India
e-mail: sbvmarudhu@gmail.com

J. Solomon · Bharathkumar
ICAR-Central Institute of Fisheries Education, Mumbai, India

I. Ahmed, I. Ahmad (eds.), *Aquaculture: Enhancing Food Security and Nutrition*, https://doi.org/10.1007/978-3-031-92858-1_5

as essential energy sources for gut bacteria, thereby enhancing nutrient absorption and metabolic efficiency in fish (Roberfroid, 1993; Song et al., 2014). Studies have consistently demonstrated that prebiotics can enhance growth performance, gut microbiota diversity, immune function, and disease resistance in aquaculture species (Dawood & Koshio, 2016; Ringø et al., 2010, 2014). Similarly, probiotics, described by Fuller (1989) as live microbial feed additives that improve the host's intestinal microbial balance, play a critical role in fish nutrition. The FAO and WHO emphasize that probiotics provide health benefits when administered in sufficient quantities. These beneficial microorganisms serve as a viable alternative to antibiotics in aquaculture, supporting nutrient utilization, gut health, and immune system functionality. However, the success of probiotics depends on several factors, including their source, dosage, and the duration of supplementation (Dawood & Koshio, 2016). It is essential to understand their mechanisms of action, as probiotic effects can be highly strain- and host-specific. While interest in probiotics is growing worldwide, their application across diverse fish species remains relatively underexplored. Although existing research extensively covers the roles of prebiotics and probiotics in aquaculture, this review seeks to consolidate knowledge on their application across various fish species. It will emphasize their potential benefits, identify research gaps, and provide direction for future studies, ultimately advancing sustainable aquaculture practices through improved fish nutrition.

2 Probiotics and Prebiotics

Probiotics are live microorganisms that, when administered in suitable quantities, confer health benefits to the host by promoting a balanced gut microbiota. The term "probiotic," derived from the Greek words "pro" and "bios," translates to "for life" (Gismondo et al., 1999) and was initially defined as organisms or substances that enhance intestinal health (Parker, 1974). In aquaculture, probiotics include a diverse range of beneficial microorganisms such as bacteria, bacteriophages, microalgae, and yeast, delivered via feed or water to improve fish health and minimize reliance on antibiotics.

Prebiotics, on the other hand, are indigestible dietary components that support the proliferation and activity of beneficial gut bacteria, fostering a healthy intestinal environment. As nondigestible carbohydrates, prebiotics encompass a variety of compounds, including fructo-oligosaccharides, mannanoligosaccharides, and inulin, differentiated by their molecular structure or degree of polymerization (Mussatto & Mancilha, 2007). These compounds selectively stimulate beneficial bacterial populations, improving gut health, immune responses, and potentially promoting growth in aquaculture species (Gibson & Roberfroid, 1995). Although the use of prebiotics in aquaculture is relatively new, they have shown promise in interacting with immune cell receptors, such as beta-glucan receptors, triggering pathways that enhance immune defenses (Brown et al., 2002).

3 Probiotics and Prebiotics as Functional Ingredients in Aquaculture

Prebiotics and probiotics are vital functional ingredients in aquaculture, significantly contributing to fish health, growth, and immunity. Probiotics, as live microorganisms, enhance fish health through mechanisms such as the competitive exclusion of pathogens, improved digestion, and immune system stimulation (Edun & Akinrotimi, 2011). They have been shown to promote growth performance, feed efficiency, digestive enzyme activity, gut microbial balance, and disease resistance. Additionally, probiotics can boost antioxidant capacity while reducing stress and deformities in fish. However, challenges such as the sensitivity of probiotics to feed processing and their potential to disrupt native microbial communities may limit their effectiveness (Lauzon et al., 2014).

Prebiotics, first defined by Gibson and Roberfroid in 1995, provide an alternative that complements probiotics. These indigestible compounds bypass digestion in the upper gastrointestinal tract, undergoing fermentation by gut bacteria to produce short-chain fatty acids (SCFAs) such as acetic, propionic, and butyric acids. These SCFAs lower gut pH, suppress harmful bacteria, and serve as energy sources for the host (Huazano-García & Lopez, 2013). Prebiotics such as inulin, fructooligosaccharides (FOS), and galactooligosaccharides (GOS) are known to enhance the growth of beneficial gut bacteria, improve nutrient utilization, and strengthen immune responses. Unlike probiotics, prebiotics do not involve live cultures, making them a stable and reliable addition to aquafeeds.

The combined use of prebiotics and probiotics, as depicted in Fig. 5.1, demonstrates their synergistic effects in aquaculture systems. Together, they enhance fish growth, immunity, and resilience to disease while reducing dependence on antibiotics. This integration supports sustainable aquaculture practices by improving productivity, ensuring fish welfare, and mitigating environmental impacts.

4 Role of Probiotics in Aquaculture Nutrition

4.1 *Growth Promotion*

Probiotics have gained recognition as a valuable tool in aquaculture, offering benefits such as improved health and growth performance across a range of cultured species. According to Irianto and Austin (2002), probiotics contribute to growth enhancement in fish and shrimp, though it remains unclear whether these effects are due to appetite stimulation, increased digestibility, or a combination of both. They

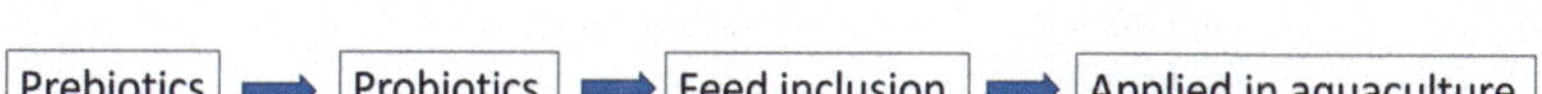

Fig. 5.1 Flow of prebiotic and probiotic use in aquaculture

also highlighted the necessity of assessing the palatability of probiotic formulations for different aquaculture species. In a study by El-Haroun et al. (2006), Nile tilapia (*Oreochromis niloticus*) showed notable growth improvements when fed a diet supplemented with the probiotic *Streptococcus* strain. Over a nine-week trial, fish fed with a 2% probiotic formulation exhibited increases in crude protein and fat content, with body weight rising from 0.154 g to 6.164 g—equivalent to a 115.3% improvement in growth.

Probiotics have also shown promising results in shellfish culture. For example, Macey and Coyne (2005) isolated two yeast strains and a bacterial strain from the digestive tract of abalone (*Haliotis midae*), formulating a probiotic-enriched diet with these microorganisms. After an eight-month feeding period, growth rates improved by 8% in smaller abalone (20 mm) and by 34% in larger ones (67 mm). Additionally, survival rates significantly increased, with 62% of treated abalone surviving exposure to the pathogen *Vibrio anguillarum* compared to just 25% in the untreated group. As illustrated in Table 5.1 these findings emphasize the potential of probiotics to enhance growth, survival, and disease resistance in aquaculture species, highlighting their importance in fostering sustainable and efficient aquaculture practices.

4.2 Digestion and Absorption

Probiotics play a crucial role in improving the digestive processes of aquatic animals by contributing essential extracellular enzymes and enhancing nutrient utilization, as highlighted in Table 5.1. Probiotic strains are known to produce enzymes such as lipases, proteases, and amylases, which facilitate the breakdown of dietary

Table 5.1 Role of specific probiotics in aquaculture nutrition

Role	Probiotic strain	Species	Reference
Growth promotion	*Bacillus* sp.	*Cyprinus carpio*	Lin et al. (2012)
	Lactobacillus lactis	Catfish	
	Streptococcus thermophilus	*Macrobrachium rosenbergii*	
Disease resistance	*Saccharomyces cerevisiae*	*Oncorhynchus mykiss*	Sun et al. (2012)
	Pseudomonas fluorescens	*Anguilla anguilla*	
	Lactococcus lactis	*Epinephelus coioides*	
Digestion enhancement	*Bacillus* sp.	*Macrobrachium rosenbergii*	Tapia-Paniagua et al. (2012)
	Lactobacillus acidophilus	*Clarias gariepinus*	
	Shewanella putrefaciens	*Solea senegalensi*	

components and promote efficient digestion (Balcazar et al., 2006). For example, the inclusion of *Bacillus cereus* in the diet of juvenile common dentex (*Dentex dentex*) has been shown to improve digestive efficiency by increasing the activity of these enzymes. Similarly, supplementation with *Bacillus subtilis* and *B. licheniformis* in the diet of rainbow trout (*Oncorhynchus mykiss*) has optimized digestive processes by enhancing nutrient breakdown (Merrifield et al., 2010).

Additionally, the use of probiotic yeast (*Saccharomyces cerevisiae*) in the diet of European sea bass larvae (*Dicentrarchus labrax*) has demonstrated significant impacts on the activity of key digestive enzymes and related gene expression patterns, highlighting the direct influence of probiotics on digestive physiology (Tovar-Ramirez et al., 2010). These studies underscore the capacity of probiotics to enhance digestive efficiency, ensuring improved nutrient absorption and utilization in aquaculture species, thereby supporting overall growth and health.

4.3 Gut Microbiota

Probiotic applications significantly influence the gut microbiota of fish by introducing beneficial Gram-positive and Gram-negative bacteria. Commonly used Gram-positive bacteria include *Bacillus*, *Micrococcus*, and various lactic acid bacteria such as *Carnobacterium*, *Enterococcus*, *Lactobacillus*, and *Weissella*. Meanwhile, Gram-negative bacteria like *Aeromonas*, *Photorhodobacterium*, *Pseudomonas*, *Phaeobacter*, *Shewanella*, and *Vibrio* are also utilized. In addition to bacterial probiotics, yeast species such as *Saccharomyces cerevisiae*, *Debaryomyces hansenii*, and *Rhodotorula glutinis* have been shown to modulate the indigenous gut microbiota of fish.

Probiotics establish themselves in various regions of the gastrointestinal (GI) tract, including the pyloric caeca, stomach, and intestine, where they interact with the native microbiota. These probiotic populations can reside on the epithelial surface, within intestinal mucus, or within the digesta, influencing the composition and abundance of gut microorganisms. The modulation of gut microbiota by probiotics is effective across all developmental stages of fish, from larvae (via live prey) to fry, fingerlings, juveniles, and adults. This adaptability underscores the potential of probiotics to enhance gut health and microbiota balance throughout the fish's entire life cycle.

4.4 Disease Resistance

Probiotics offer a promising approach to disease management and infection prevention in aquaculture species, as highlighted in Table 5.2. These beneficial microorganisms create a protective barrier against opportunistic pathogens by producing bactericidal or bacteriostatic compounds within the host's intestine. Their direct

Table 5.2 Role of specific prebiotics in growth promotion

Prebiotic	Species	Effects	Reference
MOS (Bio-Mos®)	Gilthead sea bream	FBW, SGR,FCR, PER	Dimitroglou et al. (2010a)
MOS (Bio-Mos®)	European sea bass	FBW, SGR, relative growth, fish length, CF, survival, FCR, FI	Torrecillas et al. (2007)
MOS (Bio-Mos®)	White sea bream larvae	Growth, survival	Dimitroglou et al. (2010a)
Inulin (Methodo chemicals	Sharpsnout sea bream	FBW, WG, SGR, FI, FCR, PER, dry matter, protein	Piccolo et al. (2011)

FBW Final body weight, *WG* Weight gain, *SGR* Specific growth rate, *FI* Feed intake, *FCR* Feed conversion ratio, *PER* Protein efficiency ratio

mechanisms of action include the secretion of bacteriocins, antibiotics, siderophores, hydrogen peroxide, digestive enzymes, and organic acids, which collectively lower intestinal pH and inhibit pathogenic bacterial growth (Nair et al., 2017).

Research by Ghosh et al. (2008) demonstrated that probiotic concentrations between 10,610^6 and 10,810^8 cells/g promote the proliferation of beneficial gut microbiota in ornamental fish species, such as those in the *Xiphophorus* and *Poecilia* genera, while simultaneously reducing harmful heterotrophic microbes. Moreover, probiotics have been investigated for their public health benefits, including reducing enteric infections and minimizing contamination in fish products. These live, non-pathogenic microorganisms not only support a healthy intestinal microbiome but also act as immunomodulators and facilitate the early colonization of beneficial bacteria in aquaculture species (Gul & Alsayeqh, 2022).

5 Role of Prebiotics in Aquaculture Nutrition

5.1 Growth Performance and Feed Utilization

Prebiotics have demonstrated notable impacts on growth performance and feed utilization in various fish species, as depicted in Table 5.2, primarily by enhancing nutrient availability through mechanisms such as improved digestive enzyme activity or changes in gut morphology (Xu et al., 2009; Wu et al., 2013). For instance, studies on allogynogenetic crucian carp, Caspian roach, and blunt snout bream linked enhanced growth to increased digestive enzyme activity. Conversely, Anguiano et al. (2013) reported that in red drum (Sciaenops ocellatus) and hybrid striped bass (Morone chrysops × M. saxatilis), nutrient availability improvements were more closely associated with gut structural modifications than enzyme activity. These variations underscore the influence of several factors, including the source and dosage of prebiotics, fish species, developmental stage, diet composition, and rearing conditions, on their effectiveness (Merrifield et al., 2010; Ringø et al., 2010, 2014; Dimitroglou et al., 2010b).

Mediterranean fish species such as gilthead sea bream and European sea bass have shown diverse responses to prebiotic supplementation. For European sea bass, studies demonstrated improved growth performance and feed conversion efficiency at specific Bio-Mos and XOS levels (Torrecillas et al., 2007, 2011, 2015). However, in gilthead sea bream, outcomes were inconsistent, with some studies reporting reduced growth or no significant effect at varying dosages. Additionally, prebiotic efficacy appeared diet-dependent; for example, XOS enhanced growth in European sea bass fed PF-based diets but not FM-based diets (Guerreiro et al., 2015). Despite these mixed results, prebiotics such as MOS and XOS show considerable potential for improving growth performance and feed efficiency. However, their effects remain highly species-specific and context-dependent, necessitating further research to determine optimal application strategies across diverse aquaculture systems.

5.2 *Digestive Enhancement*

Prebiotics have been shown to influence digestive enzyme activities in fish, as shown in Table 5.3 by modulating gut microbiota, which plays a key role in the digestion of feedstuffs. The increased activity of enzymes such as amylase, lipase, and protease in fish fed prebiotics has been linked to enhanced microbial enzymatic production in the gut (Merrifield & Rodiles, 2015; Ray et al., 2012). Notably, bacteria such as Bacillus sp., Aeromonas, and Pseudomonas have been identified as major contributors to exogenous enzyme activity, facilitating nutrient digestion. For instance, *Hypophthalmichthys molitrix* showed significant lipase activity due to gut bacteria (Bairagi et al., 2002). Similarly, studies on common carp fed short-chain fructooligosaccharides (scFOS) revealed enhanced amylase and lipase activities, attributed to increased exogenous microbial enzyme production (Hoseinifar et al., 2016).

However, the effects of prebiotics on digestive enzymes are not consistent across all species and conditions. For example, gilthead sea bream fed diets supplemented with inulin or scFOS showed no significant changes in amylase, trypsin, or alkaline phosphatase activities, regardless of dietary concentration or rearing temperature

Table 5.3 Role of specific prebiotics in digestion enhancement

Prebiotic	Species	Effects	Reference
Inulin (sigma®)	Gilthead sea bream	Gut expression of amylase, trypsin	Cerezuela et al. (2013)
scFOS (PROFEED®)	White sea bream	Gut trypsin, amylase (15 days), lipase	Guerreiro et al. (2017)
GOS (Quingdao FTZ United International Inc.)	White sea bream	Gut trypsin (15 days), amylase (15 days), lipase	Guerreiro et al. (2017)
XOS (Quingdao FTZ United International Inc.)	White sea bream	Gut TAP and lipase (15 days), trypsin, amylase	Guerreiro et al. (2017)

(Cerezuela et al., 2013). In white sea bream, supplementation with scFOS, xylooligosaccharides (XOS), or galactooligosaccharides (GOS) resulted in only minor effects on digestive enzyme activities. While scFOS and XOS increased total protease activity early in the trial, amylase activity decreased over longer durations (Guerreiro et al., 2017).

5.3 *Gut Microbiota*

Prebiotics are known to improve fish performance, health, and disease resistance by influencing gut bacterial communities, which are vital for digestion, nutrition, and immunity. Unlike warm-blooded animals, fish gut microbiota is affected by environmental factors such as temperature, salinity, and microbial populations (Merrifield & Rodiles, 2015). Studies have shown that gut bacteria play a key role in fish development, with the absence of gut microbes leading to developmental issues (Bates et al., 2006).

Prebiotics exert their effects by promoting beneficial bacteria like Lactobacillus and Bifidobacterium, which ferment the prebiotics and produce short-chain fatty acids (SCFAs), enhancing gut health. Research on Mediterranean fish species such as gilthead sea bream and white sea bream has shown that prebiotics like MOS, inulin, and scFOS can alter gut microbiota composition. For example, MOS increased bacterial diversity in fish fed a fishmeal-based diet, while inulin decreased bacterial richness. However, other studies found no significant effects on gut microbiota with scFOS or XOS in white sea bream (Guerreiro et al., 2017). Some examples have been mentioned in Table 5.4

Temperature also influences prebiotic effects on gut microbiota, with higher temperatures increasing bacterial population similarity (Guerreiro et al., 2016). These findings highlight the complexity of prebiotic effects, which are influenced by the type of prebiotic, fish species, and environmental conditions. In conclusion, prebiotics can modulate fish gut microbiota, potentially improving health and performance. However, species-specific responses and environmental factors make further research necessary to fully understand their effects.

Table 5.4 Role of specific prebiotics in gut microbiota

Prebiotic	Species	Effects	Reference
MOS (Bio-Mos)	Gilthead sea bream	Gut microbiota richness and diversity	Dimitroglou et al. (2010a)
Inulin (Sigma, St. Louis, MO, USA)	Gilthead sea bream	Gut microbiota diversity	Cerezuela et al. (2013)
GOS (Quingdao FTZ United International Inc.)	White sea bream	Gut microbiota, richness and diversity	Guerreiro et al. (2017)
XOS (Quingdao FTZ United International Inc.)	White sea bream	Gut microbiota, richness, diversity	Guerreiro et al. (2017)

5.4 Disease Resistance

As depicted in Table 5.5, prebiotics play a crucial role in enhancing disease resistance in aquaculture species by promoting gut health and supporting the innate immune system. Prebiotics, such as mannan-oligosaccharides (MOS), galacto-oligosaccharides (GOS), and inulin, have demonstrated effectiveness against bacterial infections and environmental stressors. MOS improves survival by preventing pathogenic bacterial attachment to gut epithelium and promoting their excretion. For instance, Senegalese sole and European sea bass fed MOS-enriched diets showed significantly reduced mortality rates during bacterial challenges, such as pasteurellosis and Vibrio infections (Torrecillas et al., 2012). Similarly, inulin at 1% improved the survival of gilthead sea bream during *Photobacterium damselae* challenges (Cerezuela et al., 2012), and XOS at 0.5–1% enhanced the survival of European sea bass against *Aeromonas hydrophila* while improving immune parameters (Abdelmalek et al., 2015). Prebiotics also strengthen gut morphology and osmoregulatory functions, critical in stress conditions, as demonstrated in white sea bream larvae exposed to salinity challenges. Despite promising results, further studies are necessary to confirm these mechanisms and optimize prebiotic formulations for improved disease resistance across diverse aquaculture species.

5.5 Constraints of Prebiotic and Probiotics in Aquaculture Nutrition

The use of probiotics and prebiotics has garnered significant attention as a promising approach in aquaculture; however, their application is sometimes hindered by insufficient understanding of their mechanisms of action (Sutriana, 2017). As noted by Verschuere et al. (2000), employing probiotics as biological control agents can be viewed as a form of risk management that may not yield substantial benefits when aquaculture is conducted under suboptimal conditions or in the absence of

Table 5.5 Role of specific prebiotics in disease resistance

Prebiotic	Species	Effects	Reference
Mannan oligosaccharide (Bio-Mos®)	Gilthead sea bream	Improvement in Total RBC, WBC, and thrombocytes counts	Gültepe et al. (2012)
Mannan oligosaccharide (Bio-Mos®)	European sea bass	Gut mucus lysozyme Skin mucus lysozyme, gut and skin mucus bactericidal activity	Torrecillas et al. (2011)
Mannan oligosaccharide	European sea bass	Head kidney leucocyte viability, phagocytic ability	Torrecillas et al. (2015)
Inulin	Gilthead Sea bream	Head kidney leucocyte viability, phagocytic ability and capacity, peroxidase and cytotoxicity activities, respiratory burst	Cerezuela et al. (2013)

pathogens. During the cultivation of aquatic species, the bacterial populations in the gastrointestinal tract are significantly more abundant than those in the surrounding water. Administering probiotics to aquatic animals has been utilized to enhance their immune response against bacterial infections. Tuan et al. (2013) highlighted the importance of determining the appropriate timing for probiotic administration, particularly in treating large populations of fish. According to Hoseinifar et al. (2016), advancements in probiotic strategies have been proposed and implemented as alternatives to antibiotics in aquaculture. For instance, the application of *Lactobacillus lactis* in *Acipenser persicus* influenced hematological parameters by reducing blood lymphocyte levels and increasing granulocyte levels. Conversely, the effects of prebiotics on modulating immune responses in fish remain inconsistent. For example, dietary oligofructose had no significant impact on serum glucose, total protein, RBC counts, mean cell hemoglobin concentration, mean corpuscular volume, or enzymatic activities such as alkaline phosphatase, alanine aminotransferase, and serum lactate dehydrogenase in beluga. Similarly, the administration of fructooligosaccharides in juvenile stellate sturgeon over an 11-week period significantly enhanced hemato-immunological parameters (e.g., RBC, WBC, hemoglobin, hematocrit, and lymphocyte levels) but did not notably improve leucocyte respiratory burst activity. Despite these promising findings, research highlighting the limitations of probiotics and prebiotics in aquaculture systems remains scarce.

5.6 Future Perspectives of Probiotics and Prebiotics

The future of probiotics in aquaculture involves their potential to reduce dependency on antibiotics while improving fish health and productivity. However, critical areas require attention, including their efficiency under stressful conditions and the possibility of inducing pathogenicity. Research should prioritize developing strain-specific probiotics with enhanced storage stability and cost-effectiveness for large-scale applications (Hoseinifar et al., 2018). Additionally, establishing quality control protocols and increasing awareness among farmers through government-led training programs are crucial for practical implementation. Probiotics have shown promise in enhancing fish reproductive performance, highlighting the need for further studies to explore their impact on fertility and other physiological processes (Banerjee & Ray, 2017).

Prebiotics offer significant potential in aquaculture, but further research is needed to clarify their mechanisms of action and influence on gut microbiota composition and activity (Hoseinifar et al., 2016). Advanced techniques such as next-generation sequencing, real-time PCR, and fluorescent in situ hybridization should be utilized to quantify microbial changes and determine the prebiotic index for better cross-study comparisons. Studies must also investigate the effects of environmental factors, such as temperature, on prebiotic efficacy and identify optimal dosages tailored to species, size, and age (Guerreiro et al., 2016). Furthermore, examining gene expression pathways linked to prebiotic effects will deepen our understanding, while potential negative impacts, like inulin-induced gut histomorphology changes, must be addressed to ensure safe application (Akter et al., 2016).

6 Conclusion

Probiotics and prebiotics have demonstrated significant potential in promoting health, enhancing disease resistance, and supporting sustainable aquaculture practices. Probiotics have been shown to positively impact disease resistance in fish and shellfish, presenting a promising alternative to antibiotics in aquaculture. However, their benefits are species-specific, influenced by the immune systems of aquatic organisms and their unique vulnerabilities to disease outbreaks. The development of effective and environmentally friendly disease biocontrol methods using probiotics remains a critical area of focus. While encouraging results have been reported in controlling bacterial diseases, further research is required to explore their efficacy against viral and parasitic pathogens. Additionally, understanding the molecular mechanisms by which probiotics interact with the host's immune system is vital for optimizing their application and developing targeted interventions. Similarly, prebiotics hold significant promise in modulating gut microbiota, improving gut health, and enhancing immunity in aquatic species. Prebiotics stimulate the growth of beneficial gut bacteria, thereby indirectly supporting disease resistance and nutrient absorption. However, there remains a gap in knowledge regarding their role in parasite and viral biocontrol, as most studies focus on bacterial pathogens. Further research is needed to unravel the molecular interactions between prebiotics, gut microbiota, and host immunity. Studies addressing optimal prebiotic formulations, dosages, and their effects under various environmental conditions will contribute to more effective strategies for enhancing aquaculture resilience. Incorporating both probiotics and prebiotics into aquaculture systems offers a synergistic approach to disease management and health improvement. Future research focusing on the integration of these functional feed additives, along with advancements in their formulations, can lead to innovative solutions for biocontrol, fostering sustainable and productive aquaculture practices.

References

Abdelmalek, B. E., Driss, D., Kallel, F., Guargouri, M., Missaoui, H., Chaabouni, S. E., et al. (2015). Effect of xylan oligosaccharides generated from corncobs on food acceptability, growth performance, haematology and immunological parameters of *Dicentrarchus labrax* fingerlings. *Fish Physiology and Biochemistry, 41*, 1587–1596.

Ahmed, I., & Ahmad, I. (2020). Effect of dietary protein levels on growth performance, hematological profile and biochemical composition of fingerlings rainbow trout, *Oncorhynchus mykiss* reared in Indian himalayan region. *Aquaculture Reports, 16*, 100268.

Ahmed, I., Ahmad, I., & Malla, B. A. (2024). Effects of dietary tryptophan levels on growth performance, plasma profile, intestinal antioxidant capacity and growth related genes in rainbow trout (*Oncorhynchus mykiss*) fingerlings. *Aquaculture*, 740710.

Akter, M. N., Sutriana, A., Talpur, A. D., & Hashim, R. (2016). Dietary supplementation with mannan oligosaccharide influences growth, digestive enzymes, gut morphology, and microbiota in juvenile striped catfish, *Pangasianodon hypophthalmus*. *Aquaculture International, 24*, 127–144.

Anguiano, M., Pohlenz, C., Buentello, A., & Gatlin, D. M. (2013). The effects of prebiotics on the digestive enzymes and gut histomorphology of red drum (*Sciaenops ocellatus*) and hybrid striped bass (*Morone chrysops*× *M. Saxatilis*). *British Journal of Nutrition, 109*(4), 623–629.

Bairagi, A., Ghosh, K. S., Sen, S. K., & Ray, A. K. (2002). Enzyme producing bacterial flora isolated from fish digestive tracts. *Aquaculture International, 10*, 109–121.

Balcázar, J. L., De Blas, I., Ruiz-Zarzuela, I., Cunningham, D., Vendrell, D., & Múzquiz, J. L. (2006). The role of probiotics in aquaculture. *Veterinary Microbiology, 114*(3–4), 173–186.

Banerjee, G., & Ray, A. K. (2017). The advancement of probiotics research and its application in fish farming industries. *Research in Veterinary Science, 115*, 66–77.

Bates, J. M., Mittge, E., Kuhlman, J., Baden, K. N., Cheesman, S. E., & Guillemin, K. (2006). Distinct signals from the microbiota promote different aspects of zebrafish gut differentiation. *Developmental Biology, 297*(2), 374–386.

Brown, G. D., Taylor, P. R., Reid, D. M., Willment, J. A., Williams, D. L., Martinez-Pomares, L., et al. (2002). Dectin-1 is a major β-glucan receptor on macrophages. *The Journal of Experimental Medicine, 196*(3), 407–412.

Cerezuela, R., Guardiola, F. A., Meseguer, J., & Esteban, M. Á. (2012). Increases in immune parameters by inulin and *Bacillus subtilis* dietary administration to gilthead seabream (*Sparus aurata* L.) did not correlate with disease resistance to *Photobacterium damselae*. *Fish & Shellfish Immunology, 32*(6), 1032–1040.

Cerezuela, R., Meseguer, J., & Esteban, M. Á. (2013). Effects of dietary inulin, Bacillus subtilis and microalgae on intestinal gene expression in gilthead seabream (*Sparus aurata* L.). *Fish & Shellfish Immunology, 34*(3), 843–848.

Dawood, M. A., & Koshio, S. (2016). Recent advances in the role of probiotics and prebiotics in carp aquaculture: A review. *Aquaculture, 454*, 243–251.

Dimitroglou, A., Merrifield, D. L., Spring, P., Sweetman, J., Moate, R., & Davies, S. J. (2010a). Effects of mannan oligosaccharide (MOS) supplementation on growth performance, feed utilisation, intestinal histology and gut microbiota of gilthead sea bream (*Sparus aurata*). *Aquaculture, 300*(1–4), 182–188.

Dimitroglou, A., Davies, S. J., Sweetman, J., Divanach, P., & Chatzifotis, S. (2010b). Dietary supplementation of mannan oligosaccharide on white sea bream (*Diplodus sargus* L.) larvae: Effects on development, gut morphology and salinity tolerance. *Aquaculture Research, 41*(9), e245–e251.

Edun, O. M., & Akinrotimi, O. (2011). The use of probiotics in aquaculture. *Nigerian Journal of Biotechnology, 22*, 34–39.

El-Haroun, E. R., Goda, A. S., & Kabir Chowdhury, M. A. (2006). Effect of dietary probiotic Biogen® supplementation as a growth promoter on growth performance and feed utilization of Nile tilapia *Oreochromis niloticus* (L.). *Aquaculture Research, 37*(14), 1473–1480.

FAO. (2024). *FAO Fisheries and Aquaculture* Report No. 1432 NFIFO/R 1432 (En).

Fuller, R. (1989). Probiotics in man and animals. *The Journal of Applied Bacteriology, 66*(5), 365–378.

Ghosh, S., Sinha, A., & Sahu, C. (2008). Dietary probiotic supplementation in growth and health of live-bearing ornamental fishes. *Aquaculture Nutrition, 14*(4), 289–299.

Gibson, G. R., & Roberfroid, M. B. (1995). Dietary modulation of the human colonic microbiota: Introducing the concept of prebiotics. *The Journal of Nutrition, 125*(6), 1401–1412.

Gismondo, M. R., Drago, L., & Lombardi, A. (1999). Review of probiotics available to modify gastrointestinal flora. *International Journal of Antimicrobial Agents, 12*(4), 287–292.

Guerreiro, I., Enes, P., & Oliva-Teles, A. (2015). Effects of short-chain fructooligosaccharides (scFOS) and rearing temperature on growth performance and hepatic intermediary metabolism in gilthead sea bream (*Sparus aurata*) juveniles. *Fish Physiology and Biochemistry, 41*, 1333–1344.

Guerreiro, I., Serra, C. R., Enes, P., Couto, A., Salvador, A., Costas, B., & Oliva-Teles, A. (2016). Effect of short chain fructooligosaccharides (scFOS) on immunological status and gut microbiota of gilthead sea bream (*Sparus aurata*) reared at two temperatures. *Fish & Shellfish Immunology, 49*, 122–131.

Guerreiro, I., Serra, C. R., Pousão-Ferreira, P., Oliva-Teles, A., & Enes, P. (2017). Prebiotics effect on growth performance, hepatic intermediary metabolism, gut microbiota and digestive enzymes of white sea bream (*Diplodus sargus*). *Aquaculture Nutrition, 24*(1), 153–163.

Gul, S. T., & Alsayeqh, A. F. (2022). Probiotics as an alternative approach to antibiotics for safe poultry meat production. *Pakistan Veterinary Journal, 42*(3).

Gültepe, N., Hisar, O., Salnur, S., Hoşsu, B., Tanrikul, T. T., & Aydın, S. (2012). Preliminary assessment of dietary mannanoligosaccharides on growth performance and health status of gilthead seabream *Sparus auratus*. *Journal of Aquatic Animal Health, 24*(1), 37–42.

Hoseinifar, S. H., Eshaghzadeh, H., Vahabzadeh, H., & Peykaran Mana, N. (2016). Modulation of growth performances, survival, digestive enzyme activities and intestinal microbiota in common carp (*Cyprinus carpio*) larvae using short chain fructooligosaccharide. *Aquaculture Research, 47*(10), 3246–3253.

Hoseinifar, S. H., Sun, Y. Z., Wang, A., & Zhou, Z. (2018). Probiotics as means of diseases control in aquaculture, a review of current knowledge and future perspectives. *Frontiers in Microbiology, 9*, 2429.

Huazano-García, A., & López, M. G. (2013). Metabolism of short chain fatty acids in the colon and faeces of mice after a supplementation of diets with agave fructans. *Lipid Metabolism, 8*, 163–182.

Irianto, A., & Austin, B. (2002). Probiotics in aquaculture. *Journal of Fish Diseases, 25*(11), 633–642.

Lauzon, H. L., Dimitroglou, A., Merrifield, D. L., Ringø, E., & Davies, S. J. (2014). Probiotics and prebiotics: Concepts, definitions and history. Aquaculture nutrition: gut health, probiotics and prebiotics, 169–184.

Lin, S., Mao, S., Guan, Y., Luo, L., Luo, L., & Pan, Y. (2012). Effects of dietary chitosan oligosaccharides and *Bacillus coagulans* on the growth, innate immunity and resistance of koi (*Cyprinus carpio koi*). *Aquaculture, 342*, 36–41. https://doi.org/10.1016/j.aquaculture.2012.02.009

Macey, B. M., & Coyne, V. E. (2005). Improved growth rate and disease resistance in farmed *Haliotis midae* through probiotic treatment. *Aquaculture, 245*(1–4), 249–261.

Merrifield, D. L., & Rodiles, A. (2015). The fish microbiome and its interactions with mucosal tissues. In *Mucosal health in aquaculture* (pp. 273–295). Academic.

Merrifield, D. L., Dimitroglou, A., Foey, A., Davies, S. J., Baker, R. T., Bøgwald, J., et al. (2010). The current status and future focus of probiotic and prebiotic applications for salmonids. *Aquaculture, 302*(1–2), 1–18.

Mussatto, S. I., & Mancilha, I. M. (2007). Non-digestible oligosaccharides: A review. *Carbohydrate polymers, 68*(3), 587–597.

Nair, M. S., Amalaradjou, M. A., & Venkitanarayanan, K. (2017). Antivirulence properties of probiotics in combating microbial pathogenesis. *Advances in Applied Microbiology, 98*, 1–29.

Parker, R. B. (1974). Probiotics, the other half of the antibiotics story. *Animal Nutrition and Health, 29*, 4–8.

Piccolo, G., Centoducati, G., Marono, S., Bovera, F., Tudisco, R., & Nizza, A. (2011). Effects of the partial substitution of fish meal by soy bean meal with or without mannanoligosaccharide and fructooligosaccharide on the growth and feed utilization of sharpsnout seabream, *Diplodus puntazzo* (Cetti, 1777): Preliminary results. *Italian Journal of Animal Science, 10*(3), e37.

Ray, A. K., Ghosh, K., & Ringø, E. J. A. N. (2012). Enzyme-producing bacteria isolated from fish gut: A review. *Aquaculture Nutrition, 18*(5), 465–492.

Ringø, E., Olsen, R. E., Gifstad, T. Ø., Dalmo, R. A., Amlund, H., Hemre, G. I., & Bakke, A. M. (2010). Prebiotics in aquaculture: A review. *Aquaculture Nutrition, 16*(2), 117–136.

Ringø, E., Dimitroglou, A., Hoseinifar, S. H., & Davies, S. J. (2014). Prebiotics in finfish: An update. *Aquaculture nutrition: gut health, probiotics and prebiotics*, 360–400.
Roberfroid, M. (1993). Dietary fiber, inulin, and oligofructose: A review comparing their physiological effects. *Critical Reviews in Food Science & Nutrition, 33*(2), 103–148.
Song, S. K., Beck, B. R., Kim, D., Park, J., Kim, J., Kim, H. D., & Ringø, E. (2014). Prebiotics as immunostimulants in aquaculture: A review. *Fish & Shellfish Immunology, 40*(1), 40–48.
Sun, Y. Z., Yang, H. L., Ma, R. L., & Zhai, S. W. (2012). Does dietary administration of *Lactococcus lactis* modulate the gut microbiota of grouper, *Epinephelus coioides*. *Journal of the World Aquaculture Society, 43*(2), 198–207.
Sutriana, A. M. A. L. I. A. (2017). *The use of selected prebiotics and probiotic in feed development for striped catfish (Pangasianodon hypophthalmus, Sauvage, 1878) juveniles: Effects on growth parameters and health status.* Doctoral dissertation, Ph. D. dissertation, Universiti Sains Malaysia, Malaysia
Tapia-Paniagua, S. T., Diaz-Rosales, P., Leon-Rubio, J. M., Garcia de La Banda, I., Lobo, C., Alarcon, F. J., & Morinigo, M. A. (2012). Use of the probiotic Shewanella putrefaciens Pdp11 on the culture of Senegalese sole (*Solea senegalensis*, kaup 1858) and gilthead seabream (*Sparus aurata* L.). *Aquaculture International, 20*, 1025–1039. https://doi.org/10.1007/s10499-012-9509-5
Torrecillas, S., Makol, A., Caballero, M. J., Montero, D., Robaina, L., Real, F., et al. (2007). Immune stimulation and improved infection resistance in European sea bass (*Dicentrarchus labrax*) fed mannan oligosaccharides. *Fish & Shellfish Immunology, 23*(5), 969–981.
Torrecillas, S., Makol, A., Caballero, M. J., Montero, D., Ginés, R., Sweetman, J., & Izquierdo, M. (2011). Improved feed utilization, intestinal mucus production and immune parameters in sea bass (*Dicentrarchus labrax*) fed mannan oligosaccharides (MOS). *Aquaculture Nutrition, 17*(2), 223–233.
Torrecillas, S., Makol, A., Caballero, M. J., Montero, D., Dhanasiri, A. K. S., Sweetman, J., & Izquierdo, M. (2012). Effects on mortality and stress response in European sea bass, *Dicentrarchus labrax* (L.), fed mannan oligosaccharides (MOS) after *vibrio anguillarum* exposure. *Journal of Fish Diseases, 35*(8), 591–602.
Torrecillas, S., Montero, D., Caballero, M. J., Robaina, L., Zamorano, M. J., Sweetman, J., & Izquierdo, M. (2015). Effects of dietary concentrated mannan oligosaccharides supplementation on growth, gut mucosal immune system and liver lipid metabolism of European sea bass (*Dicentrarchus labrax*) juveniles. *Fish & Shellfish Immunology, 42*(2), 508–516.
Tovar-Ramírez, D., Mazurais, D., Gatesoupe, J. F., Quazuguel, P., Cahu, C. L., & Zambonino-Infante, J. L. (2010). Dietary probiotic live yeast modulates antioxidant enzyme activities and gene expression of sea bass (*Dicentrarchus labrax*) larvae. *Aquaculture, 300*(1–4), 142–147.
Tuan, T. N., Duc, P. M., & Hatai, K. (2013). Overview of the use of probiotics in aquaculture. *International Journal of Research in Fisheries and Aquaculture, 3*(3), 89–97.
Verschuere, L., Rombaut, G., Sorgeloos, P., & Verstraete, W. (2000). Probiotic bacteria as biological control agents in aquaculture. *Microbiology and Molecular Biology Reviews, 64*(4), 655–671.
Wu, Y., Liu, W. B., Li, H. Y., Xu, W. N., He, J. X., Li, X. F., & Jiang, G. Z. (2013). Effects of dietary supplementation of fructooligosaccharide on growth performance, body composition, intestinal enzymes activities and histology of blunt snout bream (*Megalobrama amblycephala*) fingerlings. *Aquaculture Nutrition, 19*(6), 886–894.
Xu, B., Wang, Y., Li, J., & Lin, Q. (2009). Effect of prebiotic xylooligosaccharides on growth performances and digestive enzyme activities of allogynogenetic crucian carp (*Carassius auratus gibelio*). *Fish Physiology and Biochemistry, 35*, 351–357.

Chapter 6
Global Aquaculture: Scenarios and Nutritional Implications

Naveed Nabi, Imtiaz Ahmed, Mahpara Qadir, and Quseen Mushtaq Reshi

1 Introduction

Aquaculture holds the potential to improve the health of our planet and the population it thrives on, as long as it is achieved within the boundaries that are socially responsible and environmentally friendly and take into consideration animal welfare and food safety standards. With an augmented pace in global aquaculture, the production from this sector has enhanced nearly five times faster than landings from capture fisheries (Kongkeo, 2001). It is regarded as an essential part of a robust global food system that helps to report challenges related to nutrition, environmental sustainability, and food security. Asia remains the leading producer of aquaculture, accounting nearly 92% of the live-weight volume of seaweeds and animals in the year 2017 (FAO, 2019). Asian aquaculture is interestingly more diversified than other regions in terms of cultivated species and production systems (Bush et al., 2019). Among the top-ranked global nations depicting huge species diversity in aquaculture, China leads by a varied margin. For instance, a wide range of Chinese aquaculture production systems cultured nearly 86 diverse species of aquatic

N. Nabi (✉)
Department of Biological Sciences, Indian Institute of Science Education and Research, Mohali, Punjab, India

University Centre for Research & Development (UCRD), Chandigarh University, Gharaun, Mohali, Punjab, India
e-mail: darnaveed4@gmail.com

I. Ahmed (✉) · Q. M. Reshi
Fish Nutrition Research Laboratory, Department of Zoology, University of Kashmir, Srinagar, Jammu and Kashmir, India

M. Qadir
Department of Chemistry, University of Kashmir, Srinagar, Jammu and Kashmir, India

I. Ahmed, I. Ahmad (eds.), *Aquaculture: Enhancing Food Security and Nutrition*, https://doi.org/10.1007/978-3-031-92858-1_6

organisms during the year 2017, whereas Norwegian aquaculture which heavily relies on marine cage systems could cultivate only 13 different species (Metian et al., 2020). The critical development trends show that this sector continues to strengthen, diversify, and use new species in order to modify the traditional practices. Asian aquaculture has evolved into a substantial food production sector, constituting traditional, modern, and hybrid practice cultures that vibrantly contribute to national economies, especially by upliftment of rural farmer communities. Asia dominates the world in aquaculture production, and as a key private player, it holds a repute as a better-organized economic sector which cuts across the various social barriers including caste and gender (Brown et al., 2021; Brugere et al., 2023).

Aquaculture has emerged as a foundation of food security worldwide, providing a substantial share of the seafood and world's fish. As the demand for aquatic products continues to grow, it is crucial to explore future scenarios of aquaculture for understanding its growing role in human nutrition. Hence this chapter scrutinizes numerous potential situations, including sustainability efforts, advancements in technology and global market dynamics, in order to assess how these advances could outline the future scenario of aquaculture and its impact to meet the nutritional needs.

The worldwide demand for seafood (i.e., freshwater and marine foods) is rising, and as a result, the percentage of farm-based production is enhancing. The diversification of cultured species and a variety of cultivation methods employed influence the socioeconomic status of farming community and promote employment opportunities in aquaculture enterprise, besides helping improve food availability and poverty alleviation (Jessica et al., 2023). Aquaculture holds a remarkable potential to address diet-related diseases and malnutrition by increasing the universal supply of quality food. The availability of nutrients from aquaculture depends upon species selected and feed used. In order to decrease the dependence on wild fisheries, feed fortification, huge progress has been made in the development of sustainable and novel aquaculture feeds to improve growth and nutrient profiles of a wide range of cultured aquatic species (Fiorella et al., 2021). The pathways and possible role of aquaculture in acquiring the targets of global food security and nutrition are enlisted below:

1.1 Increasing Fish Supply and Its Diversification

Fish are considered as a rich source of protein and other essential nutrients; therefore, it forms an inseparable element of the human diet (Troell et al., 2014). Aquaculture significantly increases the global supply of seafood and fish, as wild fish shares face depletion and overfishing problems, it provides ecological alternative, assisting to meet the rising demand for protein. Feeding a predicted global population of 9 billion people in 2050 poses a huge burden that has drawn in many scholars, food authorities, merchants, farmers, specialists, and government bodies around the world. Fish and other cultured sources of meat could help to meet not only the global nutraceutical needs, but also the nutrition demands of the most vulnerable segments of the society (Golden et al., 2017). The development of aquaculture makes it easier for people to broaden their avenues of sustenance, and this is most important in areas where traditional food production is under threat due to climate changes (Cheung et al., 2013), overfishing (Muir & Young, 1998), bad management (Garcia & Rosenberg, 2010), and destruction of the ecosystem (Delgado et al., 2003). As a result, its importance is continuously growing due to the increasing requirement for seafood (Costello et al., 2020). Fulfilling these obligations of overall responsible management of aquaculture production, sustainable practices, and variability will be very important for food supply in the future. In the last 30 years, aquaculture production and, subsequently, diversification patterns have been different across regions due to variations in natural resources, infrastructure, technology, consumer preferences, demand, and government intervention in each country.

1.2 Nutritional Value

Fish and seafood, obtained from aquaculture, acts as a rich source of essential nutrients, including vitamins, omega-3 fatty acids, high-quality minerals, and proteins, which are critical in preventing malnutrition and maintaining health, mostly in areas where other sources of essential nutrients and proteins are limited. With a minimum of 845 million people expected to be nutritionally dependent on seafood, it currently contributes 20% of animal protein essential to the development of human health and cognition (Bene et al., 2015; Golden et al. 2016). The role of aquaculture in providing nutritional values is very varied and is not only affected by the available culturable species and their feed (Fry et al., 2016; Tacon 2020) but also by the available sociocultural, political economic, and ecological systems of production and distribution. On the other hand, the comprehensive understanding of aquaculture related to nutrition should be taken into consideration (Ahmed & Ahmad, 2020; Ahmed et al., 2024). This means examining not only the nutritional quality of the seafood produced but also understanding the entire supply chain and the economic benefits involved. By integrating these aspects, we can better evaluate how aquaculture contributes to nutrition and overall well-being. For instance, research has shown that certain types of production systems yielded aquaculture products of enhanced nutritional value with reduced amounts (<1%) of greenhouse gas emissions, similar to emissions from sheep production (Gephart et al., 2016, 2021a, b; Hallström et al., 2019; MacLeod et al., 2020).

1.3 Achieving Food Security and Reducing Pressure on Capture Fisheries

Cultivating fish not only reduces pressure on wild fish populations but also helps to preserve biodiversity and maintain the ecological balance of freshwater and marine environments (Cottrell et al., 2018). It also improves the global food supply and holds great potential to combat malnutrition and diet-related disorders (Reshi et al., 2025; Khan et al., 2025). The key areas for continuous research and development have led to the advancement in the development of novel feeds for aquaculture purposes. The extent to which aquaculture may contribute to human food requirements depends largely on the quantity and quality of fish produced, both of which rely on the production technology employed for aquaculture practice as well as price and trade dynamics in addition to its environmental sustainability (Fiorella et al., 2021). Aquaculture can improve regional and local food security by providing a constant and reliable food source, particularly in developing countries, where food imports can be unreliable and costly (Brown et al., 2021; Wanjaria et al., 2025).

1.4 Economic Benefits

The aquaculture industry provides economic benefits to the people by providing millions of jobs worldwide, ranging from farming, retail, processing to distribution, which contributes to the economies of many countries, particularly those with major rural and coastal populations. Several kinds of initiatives if taken at the governmental and nongovernmental levels can help in endless growth of aquaculture with huge economic returns. The economic support to small-scale investors can improve the means of livelihood and foster innovations in culture practices (Krause et al., 2015). Moreover, by using this approach, aquaculture growth can be achieved on a large-scale that is going to benefit the research and development supported by private players. At the governmental level, the removal of possible restrictions to growth of aquaculture such as regulatory guidelines and legislation pertaining to environments and/or labor laws can be reviewed (Raworth, 2017a, b).

1.5 Sustainable Practices and Adaptability

The diverse aquaculture sector significantly contributes toward accomplishing the Sustainable Development Goals (SDGs) 2030 and can progressively do so in the future. Its primary function is to provide nutrition, food security and economic opportunities to the underprivileged sections of society (Troell et al., 2023). Advances in practices such as integrated multi-trophic aquaculture (IMTA) and technologies like recirculating aquaculture systems (RAS) have led to establishment of more ecological approaches. These practices help to diminish environmental impacts and thereby help in improving resource efficiency (Henriksson et al., 2021). Aquaculture practices can be adapted to numerous environments, including coastal areas, land-based systems, and offshore locations (Hazell & Wood, 2008). Through technological advancements that merge traditional and modern methods, alongside improved site selection and management practices, the industry continues to evolve. Inland aquaculture practices carried out in ponds are poised to remain the primary source of global aquatic food production. This helps to address the challenges of local food security by allowing production in areas where traditional agriculture practices might not be viable.

2 Current Trends in Global Aquaculture

2.1 Growth, Diversification, and Sustainability

Aquaculture is a burgeoning sector in the global food production landscape, exhibiting remarkable growth trajectories. Predominantly, its influence is pronounced in developing countries, particularly across the Asian continent, which is responsible for over 85% of the worldwide output. In Asia, aquaculture often consists of small-scale farming practices, predominantly managed and operated within family-based structures. This sector has not only generated substantial direct and indirect income streams for millions, but the majority of these beneficiaries reside in rural locales, thereby contributing to the socioeconomic vitality of these areas. Moreover, for certain Asian nations, aquaculture assumes a pivotal role as a principal source of foreign exchange earnings. It currently supplies more than half of the fish consumed globally, underscoring its profound impact on human dietary sustenance. This significant contribution to food security is further accentuated by its role in poverty alleviation. Recognized on a global scale, the aquaculture sector is now widely regarded as a fundamental component of primary food operations (Sevna & Brian, 2009).

Despite a huge growth in aquaculture sector globally, there is need for increase in diversification of species being farmed that includes not only traditional fish like tilapia and salmon but also includes diverse species like seaweed, shellfish, and other-known species. The increase in diversification will help to address the market demands and at the same time regulate the dependence on a single kind of farmed product. As the aquaculture industry grows, emphasis must be laid on healthy practices that are ecologically friendly with a minimum impact on the surrounding environment to ensure sustainability in socioeconomic and ecological perspective (Naylor, 2000). It is increasingly acknowledged that sustainable aquaculture development and responsible production cannot be effectively attained in the absence of comprehensive stakeholder involvement, particularly that of the manufacturers, within the decision-making and regulatory frameworks (Carter, 2018). The active participation of these entities is paramount to facilitating the transition toward enhanced self-regulation, a process which can be significantly bolstered through the vigilant oversight of relevant governmental bodies (Carter, 2018). Such methodologies are essential in fostering the adoption of "better management" practices within the producer community, which in turn can have remarkable impacts on poverty alleviation, food security enhancement, and the realization of sustainable development objectives (Subasinghe et al., 2008).

2.2 *Technological Advancements and Integration with Agriculture*

The conventional aquaculture practices are not sufficient enough to cater global food security challenges. Therefore, the technological advancements in the form of offshore aquaculture, recirculating aquaculture systems (RAS), and precision feeding are trending practices that not only bridge the gap between demand and supply but also help to reduce the environmental footprint. Additionally, innovations like integrated multi-trophic aquaculture (IMTA) systems play an important role internationally in ecological expansions of aquaculture operations within a balanced ecosystem approach in order to respond to a universal increasing seafood demand with a new paradigm in designing the most efficient food production system. IMTA is a technique that combines fish farming with the farming of other species like shellfish and seaweed. IMTA serves several advantages including the reduction of waste generation in addition to nutrient recycling (Qiu et al., 2022). Hence, this recyclable and sustainable option must be further promoted to ensure a stable income especially among the coastal fishers which are adversely affected by the impact of climate change.

2.3 *Increased Consumer Awareness in Emerging Global Market*

Aquaculture is regarded as one of the fastest growing food-producing sectors in the world. It is emerging, increasing, and intensifying in almost all provinces of the world. The present world aquaculture is not confined to only traditional producers, but it is continuously expanding due to emerging markets in Africa, Asia, and Latin America, which are driven by the factors like increasing protein demand and growing populations. There is a correlation between the persistent and fast growth of the world's population and the high fish demand and seafood consumption that is seen and experienced in almost all parts of the world. Despite the fact that the capture fisheries are in a stable state and big, the maximum potential of many primary fishing areas has been reached. Consequently, the supply from capture fisheries is disrupted and inadequate in meeting the demand for aquatic foodstuffs in the world, and aquaculture provides the essential link between them. With the growth in aquaculture industry across the globe, consumers are becoming more conscious about the sustainability and origins of their food. This consciousness is a driving call for practiced sustainable seafood and transparency in aquaculture practices. Hence, consumers prefer to choose products that are environmentally sustainable (i.e., obtained from well managed and assessed stocks), that include, certified aquaculture and eco-labelled products. The study has conveyed that the trade, market, and customer preference segments are going to be the major influencers of the growth and development of this sector, with production of harmless and quality goods as the clear needs. Thus, the focal place of the improvement will be the implementation of regulation and the strengthening of the governance that enable such fundamental concepts as healthier and ecological food choices to be the mainstay along with making the hospitality service providers as the means to aware the population about the seafood sustainability by evolving menus with ecological recipes (Teixeira & Silva, 2023). Besides, the improved communication with consumers would help advance their awareness toward sustainable consumption, providing clear and objective messages. With this approach, it is amicable to address the great challenge and simultaneously ensure that the food production, its processing, and distribution are made in a manner that are sustainable in terms of social, economic, and environmental framework.

3 Challenges Facing Global Aquaculture

3.1 Disease Management and Antibiotic Resistance

Despite developments, growing aquaculture production will come with environmental costs. The impact of aquaculture on environment includes factors associated with water reliance, energy use, feed inputs, genetic risks, disease, and pollutant release. These factors affect the environment in a number of ways depending upon the type of production system, its efficiency in feed utilization, and waste dispersal. In addition, the required amounts of feed among highly diverse culture of algae, shellfish, and finfish raised in water environments of different salinity (freshwater, brackish, and marine), and a wide range of technologies also pose varying effects on the surrounding environments (Troell et al., 2014; Tacon, 2020). Importantly, one of the key challenges that world aquaculture faces is the prevention and control of diseases and health management. The disease outbreak due to aquaculture operations can devastate fish populations that may lead to huge loss of production in aquaculture. Disease outbreaks in aquaculture settings pose serious effects including mass mortality of cultured fish which in turn leads to financial losses, reduced incomes and loss of livelihood to farmers. Studies revealed that among the developing nations, almost 50% of production loss are emancipated. This issue is particularly critical because 90% of the aquaculture operations are located in these regions, exacerbating the challenges faced by the industry. According to an estimate, 6 billion dollars of annual loss of revenues has occurred due to disease outbreak. For example, in Chile, infection due to salmon anemia alone costs 2 billion dollars and 20,000 workers lost their jobs. China is one of the leading countries in aquaculture production led to 15% loss of the total fish production to diseases (Leung & Bates, 2013). As "prevention is better than cure," it is logical to emphasize on preventing the existence of disease rather than treating it (Romero et al., 2012). According to Peeler and Taylor (2011), the challenges facing aquaculture because of dearth of water availability, exponential industrial growth, and massive climate change drive the need for epidemiological methods to ensure safety of aquatic animal health. The uses of better management practices, nonspecific immune-stimulants, genetically resistant stock, vaccine, probiotics, prebiotics, dietary supplements, products obtained from medicinal plants, biological control, antimicrobial agents, and water disinfection are the best approaches to control the infectious diseases of fish (Kumar et al., 2016). Moreover, managing fish diseases while declining the use of antibiotics and looking for alternative sources of feeds that are nonconventional, containing plant-based protein replacements in a possible ratio so that the reliance on fish oil and fish meal derived from wild-caught fish is considerably reduced in addition to paving way for prevention of antibiotic resistance in cultured fish. Though there is ongoing research into alternative feeds, such as plant-based proteins, insect meals, and antibiotic-free feeds, transitioning away from traditional feeds is a feat to be achieved yet.

3.2 *Regulating Aquaculture Policies and Technological Advancements to Mitigate the Effects of Global Climate Change*

Aquaculture faces several other challenges, including the hazardous effects of climate change due to lack of decision making in updating regulatory frameworks and policy issue guidelines amid global effects of climate change. The multi-dimensional impacts of weather change on aquaculture which includes the shifts in water temperature, ocean acidification, sea-level rise, harmful algal blooms, and extreme weather events influence fish health, growth rates, and the distribution of aquaculture operations (Yadav et al., 2024). Moreover, inconsistent guidelines and lack of standardization across nations hinder international trade creating challenges for safeguarding food safety and sustainability. Hence, there is need for ongoing discourse on climate-resilient aquaculture which emphasizes an immediate need for practical measures to secure the upcoming sustainability of food production sector (Yadav et al.2024). Some of the innovative climate-resilient and scientific interventions aquaculture strategies which are meant at strengthening the adaptive capacity of aquaculture practices are species diversification, selective breeding, and implementation of sustainable and innovative aquaculture systems such as recirculating aquaculture systems (RAS and aquaponics). Besides, the lack of species diversification in global aquaculture can make the sector susceptible to disease occurrences or market instabilities, and diversifying the species cultivated can help alleviate these risks. Moreover, the benefits of aquaculture in some sections may not be equally distributed. To ensure equitable development regulatory frameworks and implementing agencies must ensure that small-scale and local fish farmers have equal access to markets, resources and technology, which is important for justifiable development. All these approaches work toward stimulating aquaculture systems against climate-induced disturbances, thereby modifying risks and ensuring sustained production.

3.3 Sustainable Development Goals: Building Public Perception and Trust

Concerns such as food safety, animal welfare concerns, and perceived environmental impacts can affect public opinion of aquaculture. It is necessary to address these concerns proactively and transparently that is needed for maintaining consumer trust (Krause et al., 2015). Standards dealing with above mentioned issues can be in conflict, and looking for ways to reduce these conflicts is vital for international standards that govern animal welfare development. Animal welfare is frequently regarded as an ethical concern, and addressing it poses significant challenges under contemporary trade agreements. The interconnection between animal health and welfare elucidates its critical role in food safety. The practical application of animal welfare standards holds the capacity to mitigate on-farm risks, thereby enhancing food safety. This is primarily achieved through decreased prevalence of infectious diseases within farms, reduced stress-induced immunosuppression in livestock, and diminished shedding of human pathogens by farm animals. Moreover, a reduction in antibiotic use and the emergence of antibiotic resistance can also be attributed to improved animal welfare practices. The persistent health issues affecting farm animals remain substantial threats to their welfare, and the prevalence of these diseases can serve as indicators of animal welfare conditions (de Passillé & Rushen, 2005). Furthermore, the integration of hazard analysis and critical control point (HACCP)-based methods into animal welfare frameworks would facilitate a more harmonious convergence of animal welfare and food safety standards. This integration is essential for identifying and managing risks associated with animal health, thereby ensuring a consistent and reliable approach to safeguarding animal welfare and public health. Additionally, there is a pressing need to incorporate aquaculture more effectively into discussions surrounding the global food system. This integration should involve a nuanced examination of aquaculture's multifaceted roles, which would assist in the establishment of comprehensive standards at various scales. This approach is vital for evaluating the current and prospective contributions of aquaculture to the Sustainable Development Goals (SDGs). Enhanced consideration of aquaculture in these dialogues will not only improve animal welfare practices within the sector but also foster a better understanding of its overall impact on global sustainability and food security objectives. Being a distinctive sector, aquaculture unites all aquatic systems (marine, freshwater, and brackish systems) with firm interrelation with terrestrial ecosystems (Troell et al., 2023). SDGs include identification and construction of already existing systems which are being monitoring so that it can capture SDG-relevant aquaculture statistics at a national level and discussion of how a comprehensive aquaculture policy will be framed to meet the SDGs that will help countries to prioritize activities for successful well-being.

In conclusion, it is worth mentioning that while aquaculture holds a promise for improving food security globally and nutrition, but addressing these trends is crucial for safeguarding its sustainable development and continued contribution to global food systems. Moving toward sustainable food systems is a complex problem that necessitates high level coherence, co-ordination and integration of national food policy. Hence in India, the environmental sustainability was integrated with national food policy and the analysis of food policies was conducted, and its findings led to planning of ministerial responsibility, in order to estimate the budget, its allocation, and relevant sustainable development goals for its successful coordination. As a result, 52 policies were identified, under the charge of 10 ministries with relevance to 6 SDGs.

4 Exploring Future Scenarios for Aquaculture and Its Impact on Human Nutrition

The future trajectories of aquaculture are contingent upon an integration of current trends, technological innovations, and potential hurdles, which together will shape its impact on human nutrition. Presently, almost 20% of the global animal protein intake is sourced from seafood, which is abundant in minerals, vitamins, and omega-3 fatty acids and assumed as critical components of a healthy diet that cater to approximately 900 million individuals worldwide (Bene et al., 2015; Golden et al., 2016). The nutritional contribution of aquaculture is not uniform, as it is influenced by various factors such as the diversity of farmed species, as well as the economic, social, and environmental conditions governing production and distribution systems (Fry et al., 2016; Tacon, 2020). Consequently, a systematic analysis of the nutritional value of seafood and its economic worth throughout the supply chain is imperative to fully comprehend its significance. The concept of nutrition-sensitive aquaculture must be situated within the broader framework of food systems to appreciate the interplay between nutrition and the environment (Halpern et al., 2019). This holistic perspective advocates for a balanced evaluation of both the nutritional and ecological performance of cultured fish species, thereby avoiding a simplistic approach that merely compares individual species (Tlusty et al., 2019). Instead, it underscores the importance of considering the diversity of aquatic products within the context of local consumption patterns and production scales.

Nutrition-sensitive food production, including aquaculture, seeks to optimize nutritional outcomes by emphasizing the multifaceted benefits derived from a variety of food sources and acknowledging their societal implications (Uccello et al., 2017). In essence, nutrition-sensitive aquaculture is characterized by a food system that enhances public health via the cultivation of diverse and nutrient-rich seafood, ensures equitable access to safe and nutritious diets, and is mindful of the broader food system's impact on livelihoods and ecosystems (Gephart et al., 2020). To achieve this, there must be a transition from viewing aquaculture solely as a means to produce seafood to recognizing it as a tool for fostering well-being, necessitating an understanding of cultural, socioeconomic, and environmental dimensions (Krishna Bahadur et al., 2018). Identifying the conditions and policies that could facilitate such a shift requires the exploration of potential development trajectories for aquaculture and a comprehensive analysis of the constraints and opportunities inherent in each scenario. To forecast the future of aquaculture, researchers utilize demand and supply models that project consumption and production levels using historical data on consumption patterns and price elasticity (Tran et al., 2019). These models are particularly useful for short-term projections but can be further refined when integrated with qualitative scenarios that elucidate the underlying factors influencing nutrition-sensitive aquaculture (Reilly and Willenbockel, 2010). These factors include the interplay of private and public investments in shaping trade policy, knowledge dissemination, technological advancements, and consumer responsiveness to information and marketing strategies (Garlock et al., 2020). Scenarios serve as plausible narratives of potential future developments, grounded in coherent and internally consistent analyses rather than predictions or forecasts per se (Nakićenović, 2000). They offer valuable insights into the conditions that may either promote or hinder nutrition-sensitive aquaculture and are essential for informed decision-making in this dynamic and complex sector.

The qualitative setups for future aquaculture are established through a process of expert elicitation which is focused to achieve within time duration of 2030–2050. The specific settings required to underpin nutrition-sensitive aquaculture are highlighted within the scenarios, along with the country at an individual level fact, production here highlights the complexity of things even on a global level, and that even implies that markets and production systems can diverge while bringing the scenarios to toward a global perspective. The variety of paths that can simultaneously exist in the current aquaculture fabrication systems showing key elements of each scenario trajectory are highlighted. The first step is to converge the focal issue for the development of aquaculture and to quantify how different trajectories would affect the impact of aquaculture on human nutrition. Second, we need to understand what drove drivers of change are introduced for the aquaculture sector. Subsequently, a range effects on human nutrition. Here are some possible scenarios and their consequences for human nutrition.

4.1 *Sustainable and Diversified Aquaculture*

The advances in sustainable practices lead to the widespread implementation of alternative feed sources, integrated multi-trophic aquaculture (IMTA), and recirculating aquaculture systems (RAS), with the following implications:

(a) *Increased variety:* A wider range of sea food species including fish and underutilized seaweeds and other species become available and thereby enhance nutrient diversity and dietary intake.
(b) *Improved nutritional quality:* The nutritional quality of farmed fish is improved due to healthier farming practices and sustainable feed sources that are the potential sources of minerals, omega-3 fatty acids, and vitamins.

4.2 Technological Integration and Innovation

The advancement in technologies, such as genetically modified fish, precision, advancements in breeding techniques and aquaculture have impacts following on human nutrition:

(a) *Improved nutrient profiles*: The significant health benefits can be provided by utilizing genetically modified or selectively bred fish with enhanced nutritional profiles (e.g., enhanced vitamin levels and higher omega-3 content).
(b) *Efficient production:* Precision aquaculture can make it more accessible for diverse range of populations by optimizing feed efficiency and growth conditions that could lead to more reliable and reasonable seafood.
(c) *Diminished environmental impact:* Technological innovations may alleviate environmental impacts which will ensure the growth in aquaculture that are not going to compromise the health of ecosystems or human nutrition.

4.3 Global Expansion and Market Integration

Aquaculture participates in global food systems, which leads to its significant expansion in emerging markets and is driven by growing demand and investment; its impact on human nutrition is outlined below:

(a) *Increased protein availability*: Expansion in emerging markets could address protein deficiencies and improve overall dietary quality in regions with high population growth.
(b) *Economic growth:* The expansion of local aquaculture industries could improve economic stability and thereby provide better access to affordable and nutritious food.
(c) *Cultural and dietary shifts*: Better availability of diverse seafood may result in changes in dietary patterns that laid more emphasis on seafood and fish in traditionally non-seafood-consuming regions.

4.4 Climate Change and Resilience

The advanced practices used in aquaculture have remarkable adaption to the challenges of climate change and resistant systems influencing human nutrition in the following ways:

(a) *Stable food supply:* Resilient aquaculture systems can survive in climate variations and environmental changes ensure a steady supply of nutritious seafood.
(b) *Nutrient-rich alternatives:* Development of new aquaculture practices or species that can thrive under changing conditions could offer alternative sources of essential nutrients.

(c) *Enhanced local production:* Regional aquaculture adaptations may reduce dependence on global supply chains, making nutritious seafood more available locally.

4.5 Ethical and Social Considerations

Growing awareness of ethical and social issues has applicability to human nutrition as it leads to greater emphasis on fair labor practices, animal welfare, and equitable distribution of benefits:

(a) *Ethical choices:* Consumers increasingly prioritize seafood from operations with high standards for animal welfare and social responsibility, potentially influencing market dynamics and promoting higher quality and more responsibly produced food.
(b) *Access and equity:* Efforts to address social inequities in the aquaculture sector could improve access to nutritious seafood for marginalized communities.
(c) *Consumer trust:* Enhanced transparency and ethical practices could build consumer trust and promote more widespread consumption of farmed seafood.

4.6 Alternative Proteins and Novel Foods

Growth in alternative protein sources, such as insect-based feeds, lab-grown fish, and plant-based seafood substitutes, complements traditional aquaculture and its impact on human nutrition is given below:

(a) *Broadened nutritional options:* The introduction of novel food products can diversify dietary options, providing additional sources of protein and essential nutrients for human consumption.
(b) *Reduced pressure on resources:* Alternatives to traditional fish feeds and products may alleviate pressure on marine resources, supporting the sustainability of aquaculture and its nutritional contributions.
(c) *Innovative solutions:* Lab-grown fish and plant-based alternatives could offer new ways to meet human nutritional needs while addressing some of the environmental and ethical concerns associated with traditional aquaculture.

5 Conclusion

Given that the world population is expected to reach 9.7 billion by 2050 (United Nations, World Population Prospect 2019), correspondingly global demand for animal-derived foods would rise by as much as 88%, most of which might be consumed at levels beyond guidelines for healthy eating. The greatest challenges faced by humanity today are how to feed a growing population a healthy (nutritious) and sustainable diet. The future of aquaculture holds varied possibilities, each with diverse effects for human nutrition. Advancements in technology, global integration, and sustainability could enhance the availability and quality of seafood while addressing challenges related to social equity, climate change, and environmental impact. Besides, the extent of globalization and the economic growth strategy will significantly influence the shape of aquaculture. As these scenarios unfold, they will shape the role of aquaculture in meeting global nutritional needs and contributing to food security. The macro-policies framed by development organizations, investors, and nations look toward aquaculture in order to meet rising seafood demands. While some of the scenarios presented hold promise for their potential contribution to nutrition-sensitive aquaculture, simply put, in order for any of them to materialize requires a level of public policy commitment. These commitments need to be made and maintained in the future by the nations through devising policies toward sustainability, as long-established fish increases in production which would lead to prices declines improving accessibility for all. However, those policies must adapt haltingly and without delay into a larger world where liberal internationalism trumps nationalist individualism. As with most things aquaculture too is on the cusp of natural opportunities to purposefully steer its development trajectory via consumer education policy development or market-based regulation toward cleaner and more nutrition-sensitive shores in the coming years. Given its importance to global food and nutrition security, current scenarios prompt discussion among researchers, policymakers, and advocacy groups on how to shape a nutrition-sensitive aquaculture path forward.

References

Ahmed, I., & Ahmad, I. (2020). Effect of dietary protein levels on growth performance, hematological profile and biochemical composition of fingerlings rainbow trout, *Oncorhynchus mykiss* reared in Indian himalayan region. *Aquaculture Reports, 16*, 100268.

Ahmed, I., Ahmad, I., & Malla, B. A. (2024). Effects of dietary tryptophan levels on growth performance, plasma profile, intestinal antioxidant capacity and growth related genes in rainbow trout (*Oncorhynchus mykiss*) fingerlings. *Aquaculture*, 740710.

Bene, C., Barange, M., Subasinghe, R., Pinstrup-Andersen, P., Merino, G., Hemre, G.-I., & Williams, M. (2015). Feeding 9 billion by 2050– Putting fish back on the menu. *Food Security, 7*, 261–274. https://doi.org/10.1007/s12571-015-0427-z

Brown, K. A., Venkateshmurthy, N. S., Law, C., Harris, F., Kadiyala, S., Shankar, B., Mohan, S., Prabhakaran, D., & Knai, C. (2021). Moving towards sustainable food systems: A review of Indian food policy budgets. *Global Food Security, 28*, 100462.

Brugere, C., Bansal, T., Kruijssen, F., & Williams, M. (2023). Humanizing aquaculture development: Putting social and human concerns at the center of future aquaculture developmentsese. *J World Aquac Soc, 54*, 482–526. https://doi.org/10.1111/jwas.12959

Bush, S. R., Belton, B., Little, D. C., & Islam, M. S. (2019). Emerging trends in aquaculture value chainresearch. *Aquaculture, 498*, 428–434. https://doi.org/10.1016/j.aquaculture.2018.08.077

Carter, C. (2018). *The politics of aquaculture: Sustainability interdependence, territory and regulation in fish farming*. Routledge.

Cheung, W. W. L., Watson, R., & Pauly, D. (2013). Signature of ocean warming in global fisheries catch. *Nature, 497*(7449), 365–368. https://doi.org/10.1038/nature12156

Costello, C., Cao, L., Gelcich, S., Cisneros-Mata, M., Free, C. M., Froehlich, H. E., Golden, C. D., Ishimura, G., Maier, J., Macadam-Somer, I., Mangin, T., Melnychuk, M. C., Miyahara, M., de Moor, C. L., Naylor, R., Nøstbakken, L., Ojea, E., O'Reilly, E., Parma, A. M., Plantinga, A. J., Thilsted, S. H., & Lubchenco, J. (2020). The future of food from the sea. *Nature, 588*, 95–100. https://doi.org/10.1038/s41586-020-2616-y

Cottrell, R. S., Fleming, A., Fulton, E. A., Nash, K. L., Watson, R. A., & Blanchard, J. L. (2018). Considering land-sea interactions and trade-offs for food and biodiversity. *Global Change Biology, 24*, 580–596. https://doi.org/10.1111/gcb.13873

de Passillé, A. M., & Rushen, J. (2005). Food safety and environmental issues in animal welfare. *Revue Scientifique et Technique, 24*(2), 757–766.

Delgado, C. L., Wada, N., Rosegrant, M. W., Meijer, S., & Ahmed, M. (2003). *Fish to 2020 supply and demand in changing global markets*. World Fish Center Technical Report 62.

FAO. (2019). Fisheries and aquaculture software. *FishStatJ: Software for Fishery and Aquaculture Statistical Time Series*. FAO Fisheries Division. http://www.fao.org/fishery/statistics/software/fishstatj/en

Fiorella, K. J., Okronipa, H., Baker, K., & Heilpern, S. (2021). Contemporary aquaculture: Implications for human nutrition. *Current Opinion Biotechnol, 70*, 83–90.

Fry, J. P., Love, D. C., MacDonald, G. K., West, P. C., Engstrom, P. M., Nachman, K. E., & Lawrence, R. S. (2016). Environmental health impacts of feeding crops to farmed fish. *Environment International, 91*, 201–214.

Garcia, S. M., & Rosenberg, A. A. (2010). Food security and marine capture fisheries: Characteristics, trends, drivers and future perspectives. *Philosophical Transactions of the Royal Society, 365*(1554), 2869–2880. https://doi.org/10.1098/rstb.2010.0171

Garlock, T., Asche, F., Anderson, J., Bjørndal, T., Kumar, G., Lorenzen, K., Ropicki, A., Smith, M. D., & Tveterås, R. (2020). A global blue revolution: Aquaculture growth across regions, species, and countries. *Reviews in Fisheries Science and Aquaculture, 28*(1), 107–116.

Gephart, J. A., Davis, K., Emery, K. A., Leach, A., Galloway, J., & Pace, M. (2016). The environmental cost of subsistence: Optimizing diets to minimize 275 footprints. *Science of the Total Environment, 553*, 120–127.

Gephart, J. A., Golden, C. D., Asche, F., Belton, B., Brugere, C., Froehlich, H. E., ... & Allison, E. H. (2020). Scenarios for global aquaculture and its role in human nutrition. *Reviews in Fisheries Science & Aquaculture, 29(1), 122–138.* https://doi.org/10.1080/23308249.2020.1782342

Gephart, J. A., Golden, C. D., Asche, F., Belton, B., Brugere, C., Froehlich, H. E., Fry, J. P., Halpern, B. S., Hicks, C. C., Jones, R. C., Klinger, D. H., Little, D. C., McCauley, D. J., Thilsted, S. H., Troell, M., & Allison, E. H. (2021a). Scenarios for global aquaculture and its role in human nutrition. *Reviews in Fisheries Science and Aquaculture, 29*(1), 122–138.

Gephart, J. A., Henriksson, P. J. G., Parker, R. W. R., Shepon, A., Gorospe, K. D., Bergman, K., Eshel, G., Golden, C. D., Halpern, B. S., Hornborg, S., Jonell, M., Metian, M., Mifflin, K., Newton, R., Tyedmers, P., Zhang, W., Ziegler, F., & Troell, M. (2021b). Environmental performance of blue foods. *Nature, 597*(7876), 360–365.

Golden, C. D., Allison, E. H., Cheung, W. W. L., Dey, M. M., Halpern, B. S., McCauley, D. J., Smith, M., Vaitla, B., Zeller, D., & Myers, S. S. (2016). Nutrition: Fall in fish catch threatens human health. *Nature, 534*(7607), 317–320.

Golden, C. D., Seto, K. L., Dey, M. M., Chen, O. L., Gephart, J. A., Myers, S. S., Smith, M., Vaitla, B., & Allison, E. H. (2017). Does aquaculture support the needs of nutritionally vulnerable nations? *Frontiers in Marine Science, 4*, 1–7. https://doi.org/10.3389/fmars.2017.00159

Hallström, E., Bergman, K., Mifflin, K., Parker, R., Tyedmers, P., Troell, M., & Ziegler, F. (2019). Combined climate and nutritional performance of seafoods. *Journal of Cleaner Production, 230*, 402–411.

Halpern, B. S., Cottrell, R. S., Blanchard, J. L., Bouwman, L., Froehlich, H. E., Gephart, J. A., Sand Jacobsen, N., Kuempel, C. D., McIntyre, P. B., & Metian, M. (2019). Opinion: Putting all foods on the same table: Achieving sustainable food systems requires full accounting. *Proceedings of the National Academy of Sciences USA, 116*(37), 18152–18156.

Hazell, P., & Wood, S. (2008). Drivers of change in global agriculture. *Philosophical Transactions of the Royal Society of London. Series B, Biological Sciences, 363*(1491), 495–515. https://doi.org/10.1098/rstb.2007.2166

Henriksson, P. J. G., Troell, M., Banks, L. K., Belton, B., Beveridge, M. C. M., Klinger, D. H., Pelletier, N., Phillips, M. J., & Tran, N. (2021). Interventions for improving the productivity and environmental performance of global aquaculture for future food security. *One Earth, 4*(9), 1220–1232.

Jessica, F., Christophe, B., & Stephen, D. (2023). Achieving food security through a food systems lens. In *Resilience and food security in a food systems context (pp. 31–52). Springer International Publishing, Cham.*

Kongkeo, H. (2001). Current status and development trends of aquaculture in the Asian region. In R. P. Subasinghe, P. Bueno, M. J. Phillips, C. Hough, & M. G. S. E. A. JR (Eds.), *(2000) Aquaculture in the third millennium. Technical proceedings of the conference on aquaculture in the third millennium, Bangkok, Thailand* (pp. 267–293). NACA/FAO.

Khan, S. K., Dutta, J., Ahmad, I., Rather, M. A., Badroo, I. A., & Taj, S. (2025). Fish protein: A nutritional solution for global food security. In J. K. Sundaray, M. A. Rather, I. Ahmad, & A. Amin (Eds.), *Food security, nutrition and sustainability through aquaculture technologies*. Springer. https://doi.org/10.1007/978-3-031-75830-0_27

Krause, G., Brugere, C., Diedrich, A., Ebeling, M. W., Ferse, S. C., Mikkelsen, E., Agúndez, J. A. P., Stead, S. M., Stybel, N., & Troell, M. (2015). A revolution without people? Closing the people–policy gap in aquaculture development. *Aquaculture, 447*, 44–55.

Krishna Bahadur, K., Dias. G. M., Veeramani, A., Swanton, C. J., Fraser, D., Steinke, D., Lee, E., Wittman, H., Farber J. M., Dunfield, K., McCann, K., Anand, M., Campbell, M., Rooney, N., Raine, N. E., Acker, R. V., Hanner, R., Pascoal, S., Sharif, S., Benton, T. G., & Fraser, E. D. G. (2018). When too much isn't enough: Does current food production meet global nutritional needs? *PLoS ONE 13(10) e0205683*. https://doi.org/10.1371/journal.pone.0205683

Kumar, V., Roy, S., Meena, D. K., & Sarkar, U. K. (2016). Application of probiotics in shrimp aquaculture: Importance, mechanisms of action, and methods of administration. *Reviews in Fisheries Science & Aquaculture, 24*(4), 342–368.

Leung, T. L. F., & Bates, A. E. (2013). More rapid and severe disease outbreaks for aquaculture at the tropics: Implications for food security. *Journal of Applied Economics, 50*(1), 215–222.

MacLeod, M. J., Hasan, M. R., Robb, D. H. F., & Mamun-Ur-Rashid, M. (2020). Quantifying greenhouse gas emissions from global aquaculture. *Scientific Reports, 10*, 11679. https://doi.org/10.1038/s41598-020-68231-8

Metian, M., Troell, M., Christensen, V., Steenbeek, J., & Pouil, S. (2020). Mapping diversity of species in global aquacult. *Reviews in Aquaculture, 12*, 1090–1100.

Muir, J. F., & Young, J. A. (1998). Aquaculture and marine fisheries: Will capture fisheries remain competitive? *Journal of Northwest Atlantic Fishery Science, 23*, 157–174. https://doi.org/10.2960/J.v23.a10

Naki cenovic N, Intergovernmental Panel on Climate Change. (2000). *Special report on emissions scenarios: A special report of working group III of the intergovernmental panel on climate change. Cambridge*. Cambridge University Press.

Naylor, R. L. (2000). Effect of aquaculture on world fish supplies. *Nature, 405*, 1017–1024.

Peeler, E. J., & Taylor, N. J. (2011). The application of epidemiology in aquatic animal health - opportunities and challenges. *Veterinary Research, 42*(1), 1–15.

Qiu, X., Carter, C. G., Hilder, P. E., & Hadley, S. (2022). A dynamic nutrient mass balance model for optimizing waste treatment in RAS and associated IMTA system. *Aquaculture, 555*, 738216. https://doi.org/10.1016/j.aquaculture.2022.738216
Raworth, K. (2017a). A doughnut for the Anthropocene: Humanity's compass in the 21st century. *Lancet Planet Health, 1*(2), e48–e49. https://doi.org/10.1016/S2542-5196(17)30028-1
Raworth, K. (2017b). *Doughnut economics: Seven ways to think like a 21st century economist*. Chelsea Green Publishing.
Reilly, M., & Willenbockel, D. (2010). Managing uncertainty: A review of food system scenario analysis and modelling. *Philosophical Transactions of the Royal Society of London. Series B, Biological Sciences, 365*(1554), 3049–3063. https://doi.org/10.1098/rstb.2010.0141
Reshi, Q. M., Ahmad, I., Nabi, N., Ahmed, I., & Sidiq, M. (2025). Global food insecurity: Role of fishes to compensate food insecurity and nutritional demand. In J. K. Sundaray, M. A. Rather, I. Ahmad, & A. Amin (Eds.), *Food security, nutrition and sustainability through aquaculture technologies*. Springer. https://doi.org/10.1007/978-3-031-75830-0_5
Rohana, S., Soto, D., & Jia, J. (2008). Global aquaculture and its role in sustainable development. *Reviews in Aquaculture, 1*(1), 1–18.
Romero, J., Feijoo, C. G., & Navarrete, P. (2012). Antibiotics in aquaculture use, abuse and alternatives. *Health and Environment in Aquaculture*, 160–198. https://doi.org/10.5772/28157
Sena, D. S., & Brian, D. (2009). Aquaculture successes in Asia: Contributing to sustained development and poverty alleviation. *Success stories in Asian Aquaculture*. https://doi.org/10.1007/978-90-481-3087-0_1
Tacon, A. G. J. (2020). Trends in global aquaculture and aqua feed production: 2000–2017. *Reviews in Fisheries Science and Aquaculture, 28*(1), 43–56. https://doi.org/10.1080/23308249.2019.1649634
Teixeira, C. M., & Silva, P. M. (2023). The huge dilemma: How to increase seafood consumption for health benefits without impacting fisheries' sustainability? *International Journal of Food Science & Technology, 59*(2), 661–672. https://doi.org/10.1111/ijfs.16841
Tlusty, M. F., Tyedmers, P., Bailey, M., Ziegler, F., Henriksson, P. J. G., Bene, C., Bush, S., Newton, R., Asche, F., & Little, D. C. (2019). Reframing the sustainable seafood narrative. *Glob Environ Change, 59*, 101991. https://doi.org/10.1016/j.gloenvcha.2019.101991
Tran, N., Chu, L., Chan, C. Y., Genschick, S., Phillips, M. J., & Kefi, A. S. (2019). Fish supply and demand for food security in sub-Saharan Africa: An analysis of the Zambian fish sector. *Marine Policy, 99*, 343–350. https://doi.org/10.1016/j.marpol.2018.11.009
Troell, M., Costa-Pierce, B., Stead, S., Cottrell, R. S., Brugere, C., Farmery, A. K., Little, D. C., Strand, Å., Pullin, R., Soto, D., Beveridge, M., Salie, K., Dresdner, J., Moraes-Valenti, P., Blanchard, J., James, P., Yossa, R., Allison, E., Devaney, C., & Barg, U. (2023). Perspectives on aquaculture's contribution to the sustainable development goals for improved human and planetary health. *JWAS, 54*(2), 251–342. https://doi.org/10.1111/jwas.12946
Troell, M., Naylor, R. L., Metian, M., Beveridge, M., Tyedmers, P. H., Folke, C., & De Zeeuw, A. (2014). Does aquaculture add resilience to the global food system? *PNAS, 111*(37), 13257–13263.
Uccello, E., Kauffmann, D., Calo, M., & Streissel, M. (2017). *Nutrition-sensitive agriculture and food systems in practice*. FAO.
United Nations. (2019). *World population prospects*. https://www.un.org/development/desa/pd/news/world-populationprospects-
Yadav, N. K., Patel, A. B., & Singh, S. K. (2024). Climate change effects on aquaculture production and its sustainable management through climate-resilient adaptation strategies: A review. *Environmental Science and Pollution Research, 31*, 31731–31751.
Wanjari, R. N., Hamid, I., Abass, Z., Magloo, A. H., & Ahmad, I. (2025). Ecosystem Services of Aquatic Biodiversity for food production and livelihoods. In J. K. Sundaray, M. A. Rather, I. Ahmad, & A. Amin (Eds.), *Food security, nutrition and sustainability through aquaculture technologies*. Springer. https://doi.org/10.1007/978-3-031-75830-0_11

Chapter 7
Fish Health in Aquaculture

Francesco Fazio and Osman Sabri Kesbiç

1 Introduction

The well-being of fish is a crucial factor for ensuring the long-term viability and financial success of aquaculture. Healthy fish populations in aquaculture systems demonstrate excellent growth performance and efficient conversion of feed, hence increasing productivity and playing a crucial role in satisfying market needs (Stentiford et al., 2020). Ensuring the absence of diseases in fish contributes to economic stability by decreasing expenses related to treatment and lowering death rates. From an ecological sustainability standpoint, fish that are in good condition have a higher capacity to withstand illnesses caused by stress, which reduces the need for antibiotics and other medicines. As a result, this helps to minimize adverse effects on the environment (Assefa & Abunna, 2018). In addition, the presence of healthy fish helps enhance food safety by decreasing the chances of pathogens that might harm human health and by aiding in the prevention of antibiotic resistance (Oliveira et al., 2012). Consumers and regulatory organizations are increasingly expecting fish to meet high standards of health, which is in line with ethical and animal welfare concerns. Investments in fish health management help aquaculture enterprises comply with regulatory norms, which in turn improve their market access and competitiveness (Schneider et al., 2012). Hence, the well-being of fish in aquaculture is crucial for ensuring high production, economic viability, environmental conservation, and adherence to ethical norms. Efficient and sustainable

F. Fazio (✉)
Department of Veterinary Sciences, University of Messina, Viale Giovanni Palatucci, Messina, Italy
e-mail: ffazio@unime.it

O. S. Kesbiç
Faculty of Veterinary Medicine, Department of Animal Nutrition and Nutritional Diseases, Kastamonu University, Kastamonu, Turkey

I. Ahmed, I. Ahmad (eds.), *Aquaculture: Enhancing Food Security and Nutrition*, https://doi.org/10.1007/978-3-031-92858-1_7

aquaculture relies heavily on the implementation of effective fish health management practices. This holistic approach encompasses various strategies such as disease prevention, management of water quality, implementation of appropriate nutrition and feeding practices, optimization of stocking density, adoption of biosecurity measures, reduction of stress, regular health monitoring, and implementation of sustainable practices (Beveridge, 2004; Murray & Peeler, 2005; Barton & Iwama, 1991). These principles safeguard the well-being and survival of fish populations in aquaculture, hence improving efficiency and long-term viability.

The welfare of organisms in aquaculture systems is of utmost importance for aquaculture to move closer to meeting global seafood demand and for the sustainability of the sector. This book chapter focuses on the factors affecting the health of organisms in aquaculture systems and the potential outcomes and relationships that can occur when their health is impaired.

2 The Economic Significance of Fish Health

The first expected focus in the remainder of this section is to mention factors that directly affect fish health, e.g., water quality and/or disease. These are undoubtedly the first topics that come to mind when thinking about fish health, and we will certainly mention them in this chapter, but first, the relationship between fish health and economics. Because aquaculture is an economic enterprise, fish health has a very important impact on the costs of this enterprise.

Healthy fish populations grow faster, have better feed conversion ratios, and reach a size where they can be sold profitably faster. This increases production, reducing costs and increasing revenues for aquaculture operations. Research shows that maintaining optimum health conditions in aquaculture can significantly increase production efficiency and profitability (Bogard et al., 2016). Healthy fish populations also have a positive impact on market prices and profitability. Consumers are willing to pay higher prices for high-quality and healthy fish, which increases the market prices of healthy fish produced. Furthermore, reduced disease outbreaks increase the supply of healthy fish by reducing mortality rates, which improves market stability and profitability. Research indicates that consumers prefer healthy fish produced with welfare considerations, which generally corresponds to better managed and healthier aquaculture practices (Menozzi et al., 2020). Preventive measures such as vaccination, biosecurity protocols, and regular health monitoring reduce the incidence of diseases, reducing treatment costs and losses due to fish mortality. Research shows that effective health management not only reduces disease-related losses but also increases overall operational efficiency and profitability (Thorvaldsen et al., 2020). By minimizing disease outbreaks, aquaculture operations can maintain higher survival rates and consistent production levels. Research shows that implementing comprehensive health management strategies improves economic outcomes and protects investment in fish stocks by reducing the frequency and severity of disease outbreaks (Fiedler et al., 2016). Consequently, maintaining fish health in aquaculture is critical to maximizing revenues through higher yields and market prices and managing costs by preventing disease-related losses. Investments in health management are economically beneficial and lead to more sustainable and profitable aquaculture practices.

3 Key Factor Affecting Fish Health

Protecting the health of fish in the best possible way can be managed efficiently by managing several parameters that interact directly with each other. Maintaining optimum water quality is vital for maintaining the health of fish. For this reason, it is very important to provide qualified data to make models and to produce these models in determining the area where fish production will take place. Because in order to avoid stress and disease in fish, the appropriate temperature, pH, oxygen levels, ammonia, nitrate, nitrite, and chlorine levels should be at appropriate levels for the species, so the systems are constantly monitored and these values are constantly recorded (Svobodová, 1993; Hoque et al., 2002; Boyd, 2017).

Feeding is another parameter that is one of the most important factors for fish to be healthy. An optimal diet is crucial for enhancing fish immunity and promoting overall health (Ahmed & Ahmad, 2020; Ahmed et al., 2024). The efficacy of fish feed not only influences the growth and maturation of fish but also has an impact on the aquatic environment via its effects on nutrient use and waste management. Utilizing high-quality fish meals may effectively decrease nitrogen and phosphorus burdens, which play a vital role in preserving water quality (Lovell, 1991; Oliva-Teles, 2012). As we mentioned at the beginning of this chapter, the key to good health is the management of many interrelated systems. Feeding is determined by the individual needs of each species, which include not only essential nutritional requirements like protein, fat, and ash, but also the ability of the animal to digest the feed and its contents. Therefore, it is equally crucial to develop a fish feed that fulfills the fish's requirements and to manufacture this feed using appropriate technologies. Alternatively, if feed that cannot be digested is used, it will not only result in financial losses due to higher feeding expenses, but it will also contribute to water pollution as it is excreted as feces with high levels of nutrients (Davidson et al., 2013).

Stock density is one of the most challenging situations to manage in intensive aquaculture systems. One of the main objectives of intensive aquaculture is to achieve maximum production per unit water volume. However, this situation can turn into a big loss by developing very quickly with the slightest disruption or unexpected problem. High-density aquaculture systems often experience disease outbreaks, rendering fish susceptible to a range of diseases, including parasites, bacteria, and fungus. For this reason, the main focus in intensive systems is based on the principle of keeping the pathogens away from the system rather than on controlling the disease. Implementing efficient health management techniques, such as biosecurity measures and hygiene practices, is crucial in order to decrease the occurrence of infectious illnesses (Liu et al., 2017; Long et al., 2019).

Genetic control is crucial for improving the health of fish in aquaculture systems. By using selective breeding techniques, it is feasible to cultivate fish strains that possess heightened disease resistance, enhanced adaptability to the particular requirements of an aquaculture setting, and demonstrate superior general health and development rates (Gjedrem et al., 2012). Essential components of genetic management include disease resistance, stress tolerance, as well as development and reproduction. By selectively breeding fish for illness resistance, death rates may be decreased and the reliance on medical treatments like antibiotics can be minimized, thereby lowering the danger of antibiotic resistance. Fish that possess the ability to endure typical stresses seen in aquaculture, such as handling, transportation, and fluctuating water conditions, are expected to have enhanced immune systems and superior general health (You et al., 2020). Breeding programs often prioritize characteristics like growth rate and reproductive success to enhance output. However, it is crucial to also evaluate health and welfare factors in order to avoid any adverse impacts on fish resilience (Olesen et al., 2003; Grimsrud et al., 2013). Furthermore, the conservation of genetic diversity helps to mitigate the adverse consequences of inbreeding and support that a population maintains its ability to adjust to environmental changes and the appearance of novel diseases. Advanced genetic technologies, such as gene editing and genomic selection, enhance the capacity to manage diseases and enhance health outcomes via selective breeding. Responsible management of genetic resources is crucial for maintaining sustainable fish populations that flourish in their rearing circumstances, while also taking into account the ethical implications of genetic modification (Gratacap et al., 2019). By incorporating genetic management into comprehensive health management systems, aquaculture operations can generate more resilient and disease-resistant fish populations, resulting in enhanced animal well-being and economic consequences.

4 Role of Nutrition in Fish Health

The health and development of fish in aquaculture are greatly affected by the quality and balance of their nutrition. It is important that their diet closely resemble their natural food sources in order to promote optimum physiological functioning, illness resistance, and growth performance. Optimal nutrition ensures that fish get the appropriate ratios of proteins, lipids, carbs, vitamins, and micronutrients that are vital for sustaining biological processes. Insufficient or unbalanced meals may undermine health, rendering fish more vulnerable to illnesses, impairing their immune systems, reducing growth rates, and inducing physiological abnormalities. On the other hand, excessive intake of nutrients, especially in large amounts, may result in the buildup of waste, which has negative effects on both the health of fish and the environment. An optimally balanced diet enhances fish immune responses to pathogens and enhances their capacity to adapt to the challenges of an aquaculture environment. Additionally, it facilitates their efficient development, leading to improved feed conversion ratios and greater yields. Moreover, the interaction between nutrition and the aquatic environment is crucial for ensuring sustainable aquaculture. By providing fish with high-quality diets that are well used, the discharge of nitrogen and phosphorus is minimized, hence lowering the potential for water contamination. The undeniable correlation between a well-balanced diet and the health of fish highlights the significance of a strategic feeding method that takes into account the unique requirements of various fish species, their developmental stages, and the conditions in which they are farmed. This approach ultimately strengthens the fish's immune system, maximizes their growth and overall welfare, and guarantees a successful aquaculture business.

4.1 Diet Composition

The nutritional composition of the diet supplied to fish in aquaculture plays a crucial role in their health and well-being. It is required to provide them with a balanced combination of macro- and micronutrients that are needed for their survival, development, reproduction, and illness prevention (Gatlin, 2010). Proteins play a vital role in the construction and restoration of tissues, while lipids, including important fatty acids such as omega-3 s, provide energy and maintain the integrity of cell membranes. Although carbohydrate sources are not as important sources of energy for fish as for other living organisms, they assist in the usage of proteins and fats. Vitamins and minerals have important functions in the body. They operate as catalysts in biochemical events and help regulate water balance and immunological responses. Deficiencies or excesses of these micronutrients may cause health problems and make fish more susceptible to diseases. A proper diet enhances the immune system by supplying immunostimulant elements like vitamins C and E, as well as selenium, which aid in sustaining a strong immunological response (Lim & Webster, 2001). Accurate diet formulation is necessary to provide a proper balance between the amount of energy and protein in the diet, ensuring that protein is used efficiently for development rather than being converted into energy (Bulut et al., 2014). The nutritional needs of fish differ based on their life stage and species (Khalili Tilami & Sampels, 2018). Juveniles need greater quantities of protein to support development, while adults benefit from diets specifically designed to promote optimum reproduction or maintain good health. An optimally balanced diet facilitates efficient feed conversion, enabling the effective transformation of nutrients into body mass while minimizing waste production (Kestemont & Henrotte, 2015). Essentially, the formulation of fish feed is crucial for sustaining robust and efficient aquaculture systems. A well-balanced diet, including essential nutrients in appropriate ratios, enhances the immune system, fosters robust development, and ensures the general well-being of fish populations. Otherwise, homeostasis is broken because metabolic requirements cannot be met, and sensitivity to metabolic stress and related diseases develops.

4.2 Feeding Practices

Proper and consistent feeding methods are essential in aquaculture to maintain the health and well-being of fish. These methods directly impact the efficiency of the immune system, growth rates, reproductive success, and general vitality of the fish. Every fish species has distinct nutritional prerequisites that vary based on age, size, and life cycle stage. Therefore, formulating feeding methods that address these individual requirements guarantees a well-rounded consumption of proteins, lipids, vitamins, and minerals (Alava, 2002). Establishing consistent feeding schedules that replicate the natural eating patterns of fish is crucial for maintaining a stable and healthy metabolism. Additionally, synchronizing feeding times with the biological cycles of the fish enhances feed conversion ratios and promotes efficient development (Yakuputiyage, 2013). Ensuring the appropriate quantity of feed is crucial to prevent excessive feeding and wastage, which may contribute to water pollution and the spread of diseases, while also avoiding insufficient feeding, which can cause stunted development and a compromised immune system. Optimal feed, distinguished by a precise combination of vital nutrients, freshness, ease of digestion, and absence of impurities, aids in averting health issues and enhancing immunity and disease resistance (Hasan & New, 2013). In recent years, the advancement of technology has enabled fish farming to achieve more effective feeding via real-time monitoring and dynamically adjusted feeding programs. Automated feeding systems effectively control the time and quantity of feed distribution, preventing unnecessary waste and guaranteeing a consistent supply. Feeding procedures should include the environmental variables of the aquaculture system, including water temperature and quality, since these factors may substantially impact the fish's metabolism and nutritional needs. Monitoring fish development, health, and behavior enables farmers to modify feeding tactics and help refine them for the best health results. Overall, achieving an optimal state of fish health in aquaculture requires careful management of feed frequency, amount, and quality, as well as diligent monitoring and environmental control.

4.3 Feed Additives

In aquaculture, maintaining fish health is not always possible with a diet that only meets the basic nutritional needs of the fish. Feeds must maintain their freshness and quality until they are offered to the fish. For this purpose, additives that support shelf life are used in fish feeds (Lee, 2015). The main lipid sources in fish feeds contain high amounts of polyunsaturated fatty acids, and these fatty acids are highly sensitive to oxidation. For this reason, many natural additives (Kesbiç et al., 2023; Kesbiç, 2023; Hasdemir et al., 2023), especially tocopherol groups with antioxidant activity, are added to fish feeds (Hernández et al., 2014). Studies on this subject have increased in recent years. However, feeds, especially in intensive aquaculture systems, should provide quality nutrition and support fish's immune system under stress. Essential vitamins and minerals are crucial for various biological processes such as growth, reproduction, and protection of the immune system. B-complex vitamins, as well as essential vitamins such as A, C, D, E, and K, are vital for enzymatic reactions, bone development, skin integrity, and cellular function. Similarly, minerals such as calcium, phosphorus, and magnesium and trace elements such as iron, zinc, and selenium are important for osmoregulation, energy metabolism, and antioxidant defense mechanisms (Trichet et al., 2015; Khalili Tilami & Sampels, 2018; Lall & Kaushik, 2021; El-Sayed & Izquierdo, 2022). Supplementation becomes particularly critical in high-density farming, where natural diets are replaced by formulated feeds that may lack the full spectrum of wild nutrients. By supplementing feeds with appropriate vitamins and minerals, farmers can prevent nutritional deficiencies that can impair fish health, welfare, and productivity. In addition, recent research has investigated many alternative additives to manage the stress of intensive rearing and to improve the growth and welfare of fish. Most of these additives represent bioactive ingredients extracted from plant sources by different methods (Acar et al., 2019, 2021; Kesbic et al., 2020, 2022). Supplementation strategies must be carefully tailored to species-specific needs and life stage requirements to ensure that fish receive a complete and balanced diet, leading to optimal growth, improved health, and enhanced disease resistance. These practices support sustainable aquaculture operations and contribute to the industry's overall success.

5 Disease Management in Aquaculture

Effective disease management in aquaculture is essential for maintaining the health and productivity of fish and shellfish populations. Prevention is widely regarded as the most effective method for disease control. Key biosecurity measures include quarantine protocols, disinfection procedures, regulating farm traffic (controlling the entry and exit of personnel, animals, and equipment), water treatment, monitoring water quality parameters, using clean feeds, and proper disposal of diseased animals. Additionally, biological and chemical strategies, such as probiotics, prebiotics, herbal products (including medicinal plants, essential oils, and extracts), and appropriate antibiotics, are employed to manage diseases. However, antibiotics are used with caution due to potential adverse effects, such as the emergence of multidrug-resistant bacteria and the transfer of antibiotic resistance to aquatic microbiota. A promising alternative is phage therapy, which involves using bacteriophages to target and eliminate pathogenic bacteria and has shown effectiveness against certain bacterial species in aquaculture. Nevertheless, prevention and comprehensive management strategies remain fundamental to disease control in aquaculture systems.

5.1 Pathogen Identification

Identifying pathogens and quickly detecting disease outbreaks are crucial for effectively managing diseases in aquaculture, greatly improving the effectiveness of control methods. It is essential to identify the precise causal agents, such as bacteria, viruses, parasites, or fungi, in order to provide targeted therapies and to guide the creation of vaccines or selective breeding programs for disease resistance (Rupp et al., 2019). Gaining a comprehensive understanding of the life cycle of the disease and identifying possible vectors or reservoirs are crucial in developing efficient preventative tactics. The timely identification of disease outbreaks is crucial for their containment and control, therefore averting extensive transmission (Lotz et al., 2005; Ferreira et al., 2021). Aquaculturists generally use monitoring technologies and data analytics to promptly detect possible outbreaks. Regular health evaluations and illness monitoring systems expedite a prompt reaction, aiding in the preservation of reduced mortality and morbidity rates. Swift and suitable action in response to disease outbreaks is a suitable way to mitigate their consequences. To prevent the spread of illness and reduce economic losses, it is important to implement quarantine measures promptly, give appropriate treatments, and make necessary adjustments to management methods, such as limiting stocking density or enhancing water quality. Implementing biosecurity measures at the farm level, such as disinfecting eggs in hatcheries, using uncontaminated feed, and appropriately disposing of dead animals, is of utmost importance. Technological advancements, such as methods, improve the ability to diagnose and treat diseases in aquaculture. These technological improvements and effective health management measures are crucial for ensuring aquaculture systems' long-term viability and efficiency.

5.2 Quarantine Measures

Quarantine methods are crucial for disease management in aquaculture since they limit the transmission of infections across facilities, effectively control outbreaks, and minimize the risk of introducing new illnesses (Arthur, 1996). Nevertheless, the implementation of quarantine measures varies among different aquaculture operations. These applications may be executed with greater accuracy, particularly in recirculate aquaculture facilities like hatcheries, although they are very challenging in systems situated on the same water source (Kanchanakhan, 2005). Through the practice of separating fish that are sick or in danger, these procedures protect the well-being of aquaculture systems and enable prompt identification and action, which is vital for minimizing the economic and environmental consequences. Efficient enforcement of quarantine measures guarantees the well-being and efficiency of aquatic ecosystems, hence promoting the enduring viability of aquaculture activities. Quarantine procedures play a crucial role in avoiding the transmission of novel infections, which in turn ensures the integrity and safety of end goods. This not only enhances customer trust but also safeguards the industry's standing. Timely identification and action by isolation may effectively reduce financial damages caused by epidemics (Alavandi et al., 2015).

Effective quarantine measures need the presence of suitable infrastructure, facilities, and resources for monitoring and upkeep, in addition to well-trained personnel. Effective procedures and rapid response to disease outbreaks in aquaculture facilities need collaboration and information exchange among regulatory authorities, research institutes, and aquaculture operations. Public awareness and education play a crucial role in advocating the importance of quarantine measures for the long-term viability and integrity of aquaculture. In order to garner support for these initiatives, it is essential to provide instruction and engage all parties with a vested interest, rather than only focusing on fish farmers.

5.3 Antimicrobial Use

Antibiotics are a group of drugs that have led to great advances for mankind since their discovery and use (van den Bogaard & Stobberingh, 2000). As in all health applications, the use of antimicrobials in aquaculture serves an important purpose in disease control by effectively treating bacterial infections in fish and shellfish. This serves to improve animal health and reduce disease outbreaks (Manage, 2018). Overuse and misuse of antibiotics can lead to the emergence of microbes that are resistant to many antibiotic groups, reducing the effectiveness of antibiotics in the long term and making disease control more difficult. Moreover, antimicrobial residues from aquaculture have the potential to leach into the surrounding ecosystem, thus affecting undesirable organisms and intensifying the larger problem of antibiotic resistance (Vignesh et al., 2011).

Antibiotic-resistant pathogens in aquaculture can potentially endanger human health, especially when infected fish or shellfish are consumed (Cabello, 2006). Although bacterial infections can lead to financial setbacks in aquaculture, the strategic use of antimicrobials can help mitigate these losses and provide positive economic outcomes (Lulijwa et al., 2020). However, increasing antibiotic resistance is leading to higher costs as a result of the need for alternative treatments. To address these issues, there is an increasing focus on alternative ways to manage diseases, including immunization, probiotics, phytobiotics, quorum sensing intervention, and phage therapy. For these reasons, much of the research in recent years has served this focus.

6 Strategies for Maintaining Fish Health

To ensure the strength and longevity of aquaculture stocks, a complete plan that emphasizes proactive and strategic operations is necessary. Proactive health care is essential, including vaccines, routine health screenings, and parasite tests to swiftly detect and treat illnesses. In addition to these measures, the primary objective should be to bolster the immune systems of the fish within the system. As elaborated in previous chapters, the overarching aim is to prevent disease sources from entering the system and to cultivate disease-resistant fish, rather than merely combating diseases. Effective water quality management, which involves meticulous monitoring and modification of temperature, pH levels, dissolved oxygen, and nitrogenous waste, is crucial for maintaining an optimal environment for the well-being of fish. Conducting nutritional assessments is crucial for formulating and supplying fish feeds that fulfill their dietary requirements, enhance their immune systems, and facilitate development while reducing waste generation. Enforcing biosecurity protocols, such as segregating newly acquired livestock, sterilizing equipment, and regulating entry to farm premises, is crucial in averting the introduction and dissemination of illnesses. Implementing stress reduction strategies such as providing sufficient swimming area, reducing physical disruptions, improving the surroundings, and maintaining stable rearing circumstances may successfully minimize health problems caused by stress. Utilizing selective breeding methods may augment genetic resistance to illnesses and environmental stresses, so enhancing the general health and vigor of fish populations and lowering death rates. Thorough documentation and the capacity to track information allow for the continuous monitoring of fish health and environmental conditions. This enables the assessment of measures to safeguard health and make management choices based on evidence. These reports could possibly be used as efficient instruments for formulating proactive actions and strategic strategies for tackling challenges experienced at agricultural establishments. It is well acknowledged that every aquaculture fish farm acquires its own specialized knowledge, which includes both classic and creative practices. By using these preventative measures, fish farms safeguard their populations, bolster their ability to withstand illnesses and environmental fluctuations, and eventually attain more sustainable and lucrative aquaculture techniques.

6.1 Regular Monitoring

Regular tracking is a key element of sustaining fish health in aquaculture operations, including methodical observation and measurement of vital factors for the welfare of fish (Han et al., 2007). Water quality monitoring necessitates the ongoing assessment of temperature, pH, oxygen levels, and waste byproducts via the use of automated sensors and human testing in order to get a thorough knowledge and make appropriate modifications (Lindholm-Lehto, 2023). By observing the behavior and appearance of fish, one may get valuable insights into their health without the need for intrusive methods. Changes in swimming patterns, social connections, eating habits, or symptoms of distress might indicate environmental concerns or illness (Macaulay et al., 2021). Monitoring the development of fish is crucial for evaluating their growth rates, since stunted growth may be a sign of insufficient nutrition or sickness. This information helps in making necessary modifications to the feeding regimen and determining the effectiveness of growing programs. Implementing proactive health checks and screens for parasites and pathogens allows for efficient disease monitoring, enabling the prompt identification and isolation of diseased fish to effectively prevent the spread of diseases (Bentzon-Tilia et al., 2016). Furthermore, analyzing alterations in the blood parameters of fish is a very effective method for monitoring the adverse impacts of feeding, diseases, and environmental variables. These analyses are used in scientific research, as well as in the management, diagnosis, and monitoring of illnesses in farms (Fazio, 2019; Ahmed et al., 2020). Environmental evaluations evaluate the influence of aquaculture on adjacent water bodies and local ecosystems, encouraging the use of sustainable methods. Nevertheless, the using of water resources in common entails various risks. Specifically, using a common water source for multiple purposes heightens the likelihood of cross-contamination. Hence, controlling and managing systems that facilitate sustainable usage are extremely important (Lai et al., 2018). Periodic inspections of aquaculture procedures, equipment, and facilities guarantee optimal performance and pinpoint opportunities for improvement. Vigilant surveillance ensures the well-being of fish, enhances productivity, strengthens disease prevention measures, and guides strategic choices, eventually fostering the enduring sustainability and prosperity of aquaculture enterprises.

6.2 Vaccination Programs

Vaccination programs are most important for proactive health management in aquaculture, as they serve as a strategic defense against infectious illnesses and help decrease dependence on drugs. Vaccines activate the fish's immune system, allowing it to identify and react to certain infections, therefore improving the well-being of the fish and decreasing the likelihood of disease outbreaks (Gudding & Van Muiswinkel, 2013). Choosing the suitable vaccination according to the common illnesses in the area and the unique requirements of the farmed species is crucial for the success of the program (Shoemaker et al., 2009). The choice of vaccination technique, whether it be immersion, oral delivery, or injection, is determined by factors such as the kind of vaccine, the species of fish, their age and size, as well as practical concerns for the operation (Shefat, 2018). The timing and scheduling of vaccination are crucial in order to ensure that fish build immunity prior to being exposed to disease hazards. Post-vaccination monitoring assesses the effectiveness of the vaccine and the overall health of the fish, providing valuable information for future tactics. By combining vaccination programs with complementary health policies, such as ensuring enough nutrition, implementing effective water quality management, and enforcing biosecurity controls, the overall health and productivity may be improved (Madonia et al., 2017). By integrating vaccination programs into regular health management practices, fish farmers effectively decrease the occurrence of diseases, enhance the sustainability of fish stocks, and attain higher economic profits.

6.3 Biosecurity Measures

Implementing biosecurity controls is crucial for protecting the health of fish in aquaculture and maintaining the long-term viability of the whole aquatic ecosystem. Enforcing quarantine protocols for incoming stock is crucial in order to avoid the introduction of unfamiliar infections (Burridge et al., 2010). Fish are isolated and closely observed for illnesses before being introduced into the main aquacultural regions. Imposing limitations on the mobility of people, machinery, and animals decreases the likelihood of infections spreading, while well-designed farm arrangements further decrease the chances of cross-contamination (Reilly & Kaferstein, 1997). By using water treatment techniques like filtration, UV sterilization, or chemical treatments on incoming water, several waterborne pathogens may be completely eradicated, therefore significantly decreasing the chances of illness (Roberts & Shepherd, 1997). Of course, this situation is feasible in hatcheries where water conditions are fully controlled and in inland farming systems where water can be partially controlled. Equally crucial is the appropriate processing (filtration and other processes) of wastewater to safeguard neighboring ecosystems from pollution. Consistent disinfection of equipment, tanks, vehicles, and footwear is essential in order to avoid the transmission of disease-causing organisms (Bondad-Reantaso et al., 2005). Additionally, implementing a systematic health monitoring system enables the prompt identification of infections, facilitating quick action (Subasinghe, 2005). Vaccination programs provide proactive protection against certain illnesses, decreasing the need for medications and improving general livestock well-being (Sommerset et al., 2005). It is crucial to minimize the stress caused by handling and maintain sanitary conditions during transit in order to preserve the health of fish and reduce their vulnerability to illness (Ellis et al., 2002). Providing education and training to personnel about biosecurity procedures, illness detection, and good cleanliness promotes a culture of health consciousness and prevention within the farm (Murray & Peeler, 2005). Efficient waste management decreases the amount of disease-causing organisms and the hazards of environmental pollution (Naylor et al., 2000). Additionally, strict measures such as disinfection baths and footbaths at entrance and departure points help to further minimize the spread of pathogens (Kumar & Paul, 2022). The implementation of these comprehensive biosecurity measures serves the dual purpose of safeguarding the well-being of fish and ensuring economic stability by mitigating disease-related losses and upholding customer trust in aquaculture goods.

7 Future Perspective on Fish Health in Aquaculture

Aquaculture has made substantial progress in innovation and technology, resulting in considerable improvements in fish health and productivity. An innovation in aquaculture is the creation of sophisticated water quality monitoring systems. These systems allow operators to measure, monitor, and control important water quality parameters, including temperature, dissolved oxygen levels, and pH (Loyd & Tucker, 1998). By ensuring optimal conditions for fish health and growth, these systems contribute to the overall success of aquaculture operations. In addition, developments in fish feed composition, including the incorporation of probiotics and tailored nutritional profiles, have played a role in enhancing the immune system and bolstering disease resistance in fish that are raised in aquaculture settings (Merrifield et al., 2010). Furthermore, advancements in feed manufacturing methods have enabled the more effective preservation of the quality of raw materials during the production process while simultaneously enhancing the digestibility of the products used in fish feed (NRC, 2011). The integration of cutting-edge technology, including genetic screening for disease resistance and enhanced disease diagnostic techniques, has significantly contributed to the enhancement of fish health in aquaculture (Gjedrem et al., 2012). These advancements allow the early identification and control of any health problems, enhancing the overall quality and production of livestock. Moreover, the incorporation of digital platforms for data analysis and management enables aquaculture practitioners to make well-informed judgments using up-to-date information, resulting in more efficient disease prevention and treatment tactics (Føre et al., 2018). In the future, it is crucial to invest more in the research and development of cutting-edge technologies to further improve the influence of innovation and technology on fish health in aquaculture. The use of innovative methods such as automated feeding systems, AI-driven monitoring, and vaccine development shows significant potential for enhancing fish health and welfare in the aquaculture sector.

References

Acar, Ü., Kesbiç, O. S., İnanan, B. E., & Yılmaz, S. (2019). Effects of dietary bergamot (*Citrus bergamia*) peel oil on growth, haematology and immune response of European sea bass (*Dicentrarchus labrax*) juveniles. *Aquaculture Research, 50*(11), 3305–3312. https://doi.org/10.1111/are.14288

Acar, Ü., Kesbiç, O. S., Yılmaz, S., İnanan, B. E., Zemheri-Navruz, F., Terzi, F., et al. (2021). Effects of essential oil derived from the bitter orange (*Citrus aurantium*) on growth performance, histology and gene expression levels in common carp juveniles (*Cyprinus carpio*). *Animals, 11*(5), 1431. https://doi.org/10.3390/ani11051431

Ahmed, I., & Ahmad, I. (2020). Effect of dietary protein levels on growth performance, hematological profile and biochemical composition of fingerlings rainbow trout, *Oncorhynchus mykiss* reared in Indian himalayan region. *Aquaculture Reports, 16*, 100268.

Ahmed, I., Reshi, Q. M., & Fazio, F. (2020). The influence of the endogenous and exogenous factors on hematological parameters in different fish species: A review. *Aquaculture International, 28*, 869–899. https://doi.org/10.1007/s10499-019-00501-3

Ahmed, I., Ahmad, I., & Malla, B. A. (2024). Effects of dietary tryptophan levels on growth performance, plasma profile, intestinal antioxidant capacity and growth related genes in rainbow trout (*Oncorhynchus mykiss*) fingerlings. *Aquaculture*, 740710.

Alava, V. R. (2002). Management of feeding aquaculture species. In *Nutrition in tropical aquaculture: Essentials of fish nutrition, feeds, and feeding of tropical aquatic species* (pp. 169–208). Aquaculture Department, Southeast Asian Fisheries Development Center.

Alavandi, S. V., Rajendran, K. V., Balasubramanian, C. P., Das, S., & Vijayan, K. K. (2015). *Biosecurity and quarantine measures for aquaculture health management* (p. 45). Dr. KK Vijayan.

Arthur, J. R. (1996). Fish and shellfish quarantine: The reality for Asia-Pacific. *FAO Fisheries Technical Paper*, 11–28.

Assefa, A., & Abunna, F. (2018). Maintenance of fish health in aquaculture: Review of epidemiological approaches for prevention and control of infectious disease of fish. *Veterinary Medicine International, 2018*, 1. https://doi.org/10.1155/2018/5432497

Barton, B. A., & Iwama, G. K. (1991). Physiological changes in fish from stress in aquaculture with emphasis on the response and effects of corticosteroids. *Annual Review of Fish Diseases, 1*(1), 3–26. https://doi.org/10.1016/0959-8030(91)90019-G

Bentzon-Tilia, M., Sonnenschein, E. C., & Gram, L. (2016). Monitoring and managing microbes in aquaculture–towards a sustainable industry. *Microbial Biotechnology, 9*(5), 576–584. https://doi.org/10.1111/1751-7915.12392

Beveridge, M. C. M. (2004). *Cage aquaculture*. Blackwell Publishing. https://doi.org/10.1002/9780470995761

Bogard, J., Marks, G., Mamun, A., & Thilsted, S. (2016). Non-farmed fish contribute to greater micronutrient intakes than farmed fish: Results from an intra-household survey in rural Bangladesh. *Public Health Nutrition, 20*, 702–711. https://doi.org/10.1017/S1368980016002615

Bondad-Reantaso, M. G., Subasinghe, R. P., Arthur, J. R., Ogawa, K., Chinabut, S., Adlard, R., Tan, Z., & Shariff, M. (2005). Disease and health management in Asian aquaculture. *Veterinary Parasitology, 132*(3–4), 249–272. https://doi.org/10.1016/j.vetpar.2005.07.005

Boyd, C. E. (2017). General relationship between water quality and aquaculture performance in ponds. In *Fish diseases* (pp. 147–166). Academic Press. https://doi.org/10.1016/B978-0-12-804564-0.00006-5

Bulut, M., Yiğit, M., Ergün, S., Kesbiç, O. S., Acar, Ü., Gültepe, N., et al. (2014). Evaluation of dietary protein and lipid requirements of two-banded seabream (*Diplodus vulgaris*) cultured in a recirculating aquaculture system. *Aquaculture International, 22*, 965–973. https://doi.org/10.1007/s10499-013-9720-z

Burridge, L., Weis, J. S., Cabello, F., Pizarro, J., & Bostick, K. (2010). Chemical use in salmon aquaculture: A review of current practices and possible environmental effects. *Aquaculture, 306*(1–4), 7–23. https://doi.org/10.1016/j.aquaculture.2010.05.020

Cabello, F. C. (2006). Heavy use of prophylactic antibiotics in aquaculture: A growing problem for human and animal health and for the environment. *Environmental Microbiology, 8*(7), 1137–1144. https://doi.org/10.1111/j.1462-2920.2006.01054.x

Davidson, J., Good, C., Barrows, F. T., Welsh, C., Kenney, P. B., & Summerfelt, S. T. (2013). Comparing the effects of feeding a grain-or a fish meal-based diet on water quality, waste production, and rainbow trout *Oncorhynchus mykiss* performance within low exchange water recirculating aquaculture systems. *Aquacultural Engineering, 52*, 45–57. https://doi.org/10.1016/j.aquaeng.2012.08.001

Ellis, T., North, B., Scott, A. P., Bromage, N. R., Porter, M., & Gadd, D. (2002). The relationships between stocking density and welfare in farmed rainbow trout. *Journal of Fish Biology, 61*(3), 493–531. https://doi.org/10.1111/j.1095-8649.2002.tb00893.x

El-Sayed, A. F. M., & Izquierdo, M. (2022). The importance of vitamin E for farmed fish—A review. *Reviews in Aquaculture, 14*(2), 688–703. https://doi.org/10.1111/raq.12619

Ferreira, J. G., Taylor, N. G., Cubillo, A., Lencart-Silva, J., Pastres, R., Bergh, Ø., & Guilder, J. (2021). An integrated model for aquaculture production, pathogen interaction, and environmental effects. *Aquaculture, 536*, 736438. https://doi.org/10.1016/j.aquaculture.2021.736438

Fazio, F. (2019). Fish hematology analysis as an important tool of aquaculture: A review. *Aquaculture, 500*, 237–242. https://doi.org/10.1016/j.aquaculture.2018.10.030

Fiedler, J., Lividini, K., Drummond, E., & Thilsted, S. (2016). Strengthening the contribution of aquaculture to food and nutrition security: The potential of a vitamin A-rich, small fish in Bangladesh. *Aquaculture, 452*, 291–303. https://doi.org/10.1016/j.aquaculture.2015.11.004

Føre, M., Frank, K., Norton, T., Svendsen, E., Alfredsen, J. A., Dempster, T., et al. (2018). Precision fish farming: A new framework to improve production in aquaculture. *Biosystems Engineering, 173*, 176–193. https://doi.org/10.1016/j.biosystemseng.2017.10.014

Gatlin, D. I. (2010). *Principles of fish nutrition.*

Gjedrem, T., Robinson, N., & Rye, M. (2012). The importance of selective breeding in aquaculture to meet future demands for animal protein: A review. *Aquaculture, 350*, 117–129. https://doi.org/10.1016/j.aquaculture.2012.04.008

Gratacap, R. L., Wargelius, A., Edvardsen, R. B., & Houston, R. D. (2019). Potential of genome editing to improve aquaculture breeding and production. *Trends in Genetics, 35*(9), 672–684. https://doi.org/10.1016/j.tig.2019.06.006

Grimsrud, K. M., Nielsen, H. M., Navrud, S., & Olesen, I. (2013). Households' willingness-to-pay for improved fish welfare in breeding programs for farmed Atlantic salmon. *Aquaculture, 372*, 19–27. https://doi.org/10.1016/j.aquaculture.2012.10.009

Gudding, R., & Van Muiswinkel, W. B. (2013). A history of fish vaccination: Science-based disease prevention in aquaculture. *Fish & Shellfish Immunology, 35*(6), 1683–1688. https://doi.org/10.1016/j.fsi.2013.09.031

Han, S., Kang, Y., Park, K., & Jang, M. (2007, October). Design of environment monitoring system for aquaculture farms. In *2007 Frontiers in the convergence of bioscience and information technologies* (pp. 889–893). IEEE. https://doi.org/10.1109/FBIT.2007.77

Hasan, M. R., & New, M. B. (2013). On-farm feeding and feed management in aquaculture. FAO Fisheries and Aquaculture Technical Paper, (583).

Hasdemir, Ö., Kesbiç, O. S., Cravana, C., & Fazio, F. (2023). Antioxidant performance of *Borago officinalis* leaf essential oil and protective effect on thermal oxidation of fish oil. *Sustainability, 15*(13), 10227. https://doi.org/10.3390/su151310227

Hernández, A., García, B. G., Jordán, M. J., & Hernández, M. D. (2014). Natural antioxidants in extruded fish feed: Protection at different storage temperatures. *Animal Feed Science and Technology, 195*, 112–119. https://doi.org/10.1016/j.anifeedsci.2014.06.003

Hoque, F., Hussan, A., Das, A., & Chakraborty, P. (2002). Managing water quality and fish health in aquaculture: Farmer's traditional practices in West Bengal. *International Journal of Fisheries and Aquatic Studies, 6*(4), 31–35.

Kanchanakhan, S. (2005). Thailand's current quarantine status on aquatic animal diseases. *Bull. Fish. Res. Agen. Supplement No, 2*, 41–45.

Kesbiç, F. I., Metin, H., Fazio, F., Parrino, V., & Kesbiç, O. S. (2023). Effects of bacterioruberin-rich haloarchaeal carotenoid extract on the thermal and oxidative stabilities of fish oil. *Molecules, 28*(24), 8023. https://doi.org/10.3390/molecules28248023

Kesbiç, O. S., Acar, Ü., Hassaan, M. S., Yılmaz, S., Guerrera, M. C., & Fazio, F. (2022). Effects of tomato paste by-product extract on growth performance and blood parameters in common carp (*Cyprinus carpio*). *Animals, 12*(23), 3387. https://doi.org/10.3390/ani12233387

Kesbiç, O. S., Parrino, V., Acar, Ü., Yilmaz, S., Paro, G. L., & Fazio, F. (2020). Effects of Monterey cypress (*Cupressus macrocarpa* Hartw) leaf essential oil as a dietary supplement on growth performance and haematological and biochemical parameters of common carp (*Cyprinus carpio* L.). *Annals of Animal Science, 20*(4), 1411–1426.

Kesbiç, O. (2023). Protective effect of celeriac (*Apium graveolens*) leaf essential oil on temperature and oxygen-induced fish oil oxidation. *Kastamonu University Journal of Engineering and Sciences, 9*(1), 10–16. https://doi.org/10.55385/kastamonujes.1298104

Kestemont, P., & Henrotte, E. (2015). Nutritional requirements and feeding of broodstock and early life stages of Eurasian perch and pikeperch. In *Biology and culture of percid fishes: Principles and practices* (pp. 539–564). https://doi.org/10.1007/978-94-017-7227-3_20

Khalili Tilami, S., & Sampels, S. (2018). Nutritional value of fish: Lipids, proteins, vitamins, and minerals. *Reviews in Fisheries Science & Aquaculture, 26*(2), 243–253. https://doi.org/10.1080/23308249.2017.1399104

Kumar, S., & Paul, T. (2022). Biosafety and bio-security for sustainable aquaculture development. In *Advances in fisheries biotechnology* (pp. 453–463). Springer. https://doi.org/10.1007/978-981-16-3215-0_26

Lai, W. W. P., Lin, Y. C., Wang, Y. H., Guo, Y. L., & Lin, A. Y. C. (2018). Occurrence of emerging contaminants in aquaculture waters: Cross-contamination between aquaculture systems and surrounding waters. *Water, Air, & Soil Pollution, 229*, 1–12. https://doi.org/10.1007/s11270-018-3901-3

Lall, S. P., & Kaushik, S. J. (2021). Nutrition and metabolism of minerals in fish. *Animals, 11*(09), 2711. https://doi.org/10.3390/ani11092711

Lee, C. S. (2015). *Dietary nutrients, additives and fish health.* https://doi.org/10.1002/9781119005568

Lim, C., & Webster, C. D. (2001). *Nutrition and fish health* (pp. 163–175). Food Products Press. https://doi.org/10.1201/9781439800041

Lindholm-Lehto, P. (2023). Water quality monitoring in recirculating aquaculture systems. *Aquaculture, Fish and Fisheries, 3*(2), 113–131. https://doi.org/10.1002/aff2.102

Liu, B., Liu, Y., & Sun, G. (2017). Effects of stocking density on growth performance and welfare-related physiological parameters of Atlantic salmon *Salmo salar* L. in recirculating aquaculture system. *Aquaculture Research, 48*(5), 2133–2144. https://doi.org/10.1111/are.13050

Long, L., Zhang, H., Ni, Q., Liu, H., Wu, F., & Wang, X. (2019). Effects of stocking density on growth, stress, and immune responses of juvenile Chinese sturgeon (*Acipenser sinensis*) in a recirculating aquaculture system. *Comparative Biochemistry and Physiology Part C: Toxicology & Pharmacology, 219*, 25–34. https://doi.org/10.1016/j.cbpc.2019.02.002

Lotz, J. M., Overstreet, R. M., & Grimes, D. J. (2005). Aquaculture and animal pathogens in the marine environment with emphasis on marine shrimp viruses. In *Oceans and health: Pathogens in the marine environment* (pp. 431–451). https://doi.org/10.1007/0-387-23709-7_19

Lovell, R. T. (1991). Nutrition of aquaculture species. *Journal of Animal Science, 69*(10), 4193–4200. https://doi.org/10.2527/1991.69104193x

Loyd, C. E., & Tucker, C. S. (1998). *Pond aquaculture water quality management.* Springer. https://doi.org/10.1007/978-1-4615-5407-3

Lulijwa, R., Rupia, E. J., & Alfaro, A. C. (2020). Antibiotic use in aquaculture, policies and regulation, health and environmental risks: A review of the top 15 major producers. *Reviews in Aquaculture, 12*(2), 640–663. https://doi.org/10.1111/raq.12344

Macaulay, G., Bui, S., Oppedal, F., & Dempster, T. (2021). Challenges and benefits of applying fish behaviour to improve production and welfare in industrial aquaculture. *Reviews in Aquaculture, 13*(2), 934–948. https://doi.org/10.1111/raq.12505

Madonia, A., Melchiorri, C., Bonamano, S., Marcelli, M., Bulfon, C., Castiglione, F., Galeotti, M., Volpatti, D., Mosca, F., Tiscar, P. G., & Romano, N. (2017). Computational modeling of immune system of the fish for a more effective vaccination in aquaculture. *Bioinformatics, 33*(19), 3065–3071. https://doi.org/10.1093/bioinformatics/btx341

Manage, P. M. (2018). *Heavy use of antibiotics in aquaculture; emerging human and animal health problems–a review.*

Menozzi, D., Nguyen, T., Sogari, G., Taskov, D., Lucas, S., Castro-Rial, J. L. S., & Mora, C. (2020). Consumers' preferences and willingness to pay for fish products with health and environ-

mental labels: Evidence from five European countries. *Nutrients, 12*. https://doi.org/10.3390/nu12092650
Merrifield, D. L., Dimitroglou, A., Foey, A., Davies, S. J., Baker, R. T. M., Bøgwald, J., et al. (2010). The current status and future focus of probiotic and prebiotic applications for salmonids. *Aquaculture, 302*(1–2), 1–18. https://doi.org/10.1016/j.aquaculture.2010.02.007
Murray, A. G., & Peeler, E. J. (2005). A framework for understanding the potential for emerging diseases in aquaculture. *Preventive Veterinary Medicine, 67*(2–3), 223–235. https://doi.org/10.1016/j.prevetmed.2004.10.012
Naylor, R. L., Goldburg, R. J., Mooney, H., Beveridge, M., Clay, J., Folke, C., Lubchenco, J., Mooney, H., & Troell, M. (2000). Effect of aquaculture on world fish supplies. *Nature, 405*(6790), 1017–1024. https://doi.org/10.1038/35016500
NRC (National Research Council). (2011). *Nutrient requirements of fish and shrimp*. National Academies Press. https://doi.org/10.1007/s10499-011-9480-6
Olesen, I., Gjedrem, T., Bentsen, H. B., Gjerde, B., & Rye, M. (2003). Breeding programs for sustainable aquaculture. *Journal of Applied Aquaculture, 13*(3–4), 179–204. https://doi.org/10.1300/J028v13n03_01
Oliva-Teles, A. (2012). Nutrition and health of aquaculture fish. *Journal of Fish Diseases, 35*(2), 83–108. https://doi.org/DOI
Oliveira, J., Castilho, F., Cunha, A., & Pereira, M. J. (2012). Bacteriophage therapy as a bacterial control strategy in aquaculture. *Aquaculture International, 20*, 879–910. https://doi.org/10.1111/j.1365-2761.2011.01333.x
Reilly, A., & Kaferstein, F. K. (1997). Food safety hazards and the application of the principles of the hazard analysis and critical control point (HACCP) system for their control in aquaculture production. *Aquaculture Research, 28*(10), 735–752. https://doi.org/10.1046/j.1365-2109.1997.00939.x
Roberts, R. J., & Shepherd, C. J. (1997). *Handbook of trout and Salmon diseases*. Fishing News Books. https://doi.org/10.5555/19750821327
Rupp, M., Knüsel, R., Sindilariu, P. D., & Schmidt-Posthaus, H. (2019). Identification of important pathogens in European perch (*Perca fluviatilis*) culture in recirculating aquaculture systems. *Aquaculture International, 27*, 1045–1053. https://doi.org/10.1007/s10499-019-00382-6
Schneider, O., Schram, E., Kals, J., van der Heul, J., Kankainen, M., & van der Mheen, H. (2012). Welfare interventions in flatfish recirculation aquaculture systems and their economical implications. *Aquaculture Economics & Management, 16*(4), 399–413. https://doi.org/10.1080/13657305.2012.729252
Shefat, S. H. T. (2018). Vaccines for use in finfish aquaculture. *Acta Scientific Pharmaceutical Sciences, 2*(11), 15–19.
Shoemaker, C. A., Klesius, P. H., Evans, J. J., & Arias, C. R. (2009). Use of modified live vaccines in aquaculture. *Journal of the World Aquaculture Society, 40*(5), 573–585. https://doi.org/10.1111/j.1749-7345.2009.00279.x
Sommerset, I., Krossøy, B., Biering, E., & Frost, P. (2005). Vaccines for fish in aquaculture. *Expert Review of Vaccines, 4*(1), 89–101. https://doi.org/10.1586/14760584.4.1.89
Stentiford, G. D., Bateman, I. J., Hinchliffe, S. J., Bass, D.., Hartnell, R., Santos, E. M., Devlin, M.J., Feist, S.W., Taylor, N.G.H., Verner-Jeffreys, D.W., van Aerle, R., Peeler, E.J., Higman, W.A., Smith, L., Baines, R., Behringer, D.C., Katsiadaki, I., Froehlich, H.E., & Tyler, C. R. (2020). Sustainable aquaculture through the one health lens. *Nature Food*, *1*(8), 468–474. doi:https://doi.org/10.1038/s43016-020-0127-5.
Subasinghe, R. P. (2005). Epidemiological approach to aquatic animal health management: Opportunities and challenges for developing countries to increase aquatic production through aquaculture. *Preventive Veterinary Medicine, 67*(2–3), 117–124. https://doi.org/10.1016/j.prevetmed.2004.11.004
Svobodová, Z. (1993). *Water quality and fish health* (Vol. No. 54). Food & Agriculture Org.
Thorvaldsen, T., Kongsvik, T., Holmen, I., Størkersen, K., Salomonsen, C., Sandsund, M., & Bjelland, H. V. (2020). Occupational health, safety and work environments in Norwegian

fish farming – employee perspective. *Aquaculture, 524*, 735238. https://doi.org/10.1016/j.aquaculture.2020.735238

Trichet, V. V., Santigosa, E., Cochin, E., & Gabaudan, J. (2015). The effect of vitamin C on fish health. In *Dietary nutrients, additives, and fish health* (pp. 151–171). https://doi.org/10.1002/9781119005568.ch7

van den Bogaard, A. E., & Stobberingh, E. E. (2000). Epidemiology of resistance to antibiotics: Links between animals and humans. *International Journal of Antimicrobial Agents, 14*(4), 327–335. https://doi.org/10.1016/S0924-8579(00)00145-X

Vignesh, R., Karthikeyan, B. S., Periyasamy, N., & Devanathan, K. (2011). Antibiotics in aquaculture: An overview. *South Asian Journal of Experimental Biology, 1*(3), 114–120. https://doi.org/10.38150/sajeb.1(3)

Yakuputiyage, A. (2013). On-farm feeding and feed management strategies in tropical aquaculture. *On-Farm Feeding and Feed Management in Aquaculture, 583*, 361–376.

You, X., Shan, X., & Shi, Q. (2020). Research advances in the genomics and applications for molecular breeding of aquaculture animals. *Aquaculture, 526*, 735357. https://doi.org/10.1016/j.aquaculture.2020.735357

Chapter 8
Impacts of Climate Change on Aquaculture Nutrition and Food Security

Quseen Mushtaq Reshi, Muzamil Sidiq, Aamina Hassan, Imtiaz Ahmed, Naveed Nabi, and Ubaid Ullah Shah

1 Introduction

Aquaculture has become increasingly important to humankind's aquatic food supply (Garlock et al., 2022). Aquaculture output has grown at an exponential rate, surpassing both the production of meat from terrestrial farms and landings from capture fisheries to become the fastest-growing food production business in several nations (Ahmed et al., 2024). Aquaculture has been promoted for millennia in many developing nations, including Asia and Africa, to provide rural inhabitants with a means of improving their standard of living and escaping poverty (Edwards, 2000). Aquaculture is one of the quickest ways for the underprivileged to make a livelihood, acts as a vital source of foreign exchange for the growth of the country and has gradually become an essential component of rural livelihoods as a means of addressing issues such as increasing population pressures, environmental degradation or lack of access and declining wild fisheries harvests (Halwart et al., 2003). Between 2004 and 2011, aquaculture grew at a pace of over 2.5 million tonnes per year, whereas the worldwide wild catch decreased at a rate of more than 0.5 million tonnes annually (Cleasby et al., 2014). Due to the amplification of aquaculture,

Q. M. Reshi (✉) · M. Sidiq · A. Hassan · I. Ahmed
Fish Nutrition Research Laboratory, Department of Zoology, University of Kashmir, Srinagar, Jammu and Kashmir, India

N. Nabi
Department of Biological Sciences, Indian Institute of Science Education and Research, Mohali, Punjab, India

University Center for Research and Development (UCRD), Chandigarh University, Gharuan, Mohali, Punjab, India

U. U. Shah
Department of Library Sciences, University of Kashmir, Srinagar, Jammu and Kashmir, India

I. Ahmed, I. Ahmad (eds.), *Aquaculture: Enhancing Food Security and Nutrition*, https://doi.org/10.1007/978-3-031-92858-1_8

there is now a greater need for wide-open waters for the mass production of fish-based feedstuff components (fish oil and fish meal); additional agricultural land located far from the site of farm for the production of feed components derived from crops and other energy aspects such as (pumps, infrastructure, aeration) and suitable land that may be used for preserving quality of ponds (Henriksson et al., 2021). All things considered, intensification may decrease freshwater use while escalating some environmental issues including eutrophication, acidification and freshwater environmental toxicity (Henriksson et al., 2018).

A state in which all people, at all times, have physical, social and economic access to enough food that satisfies their dietary needs and food choices for an active and healthy life is known as food security (FAO, 2018). Aspects like sustainability and stability throughout time, in addition to food availability, access and utilisation, are included in food security (HLPE, 2020; Reshi et al., 2025). Food security could be attained when every individual can have physical, cost-effective and social accessibility to nutritious as well as adequate food that gratifies their particular dietetic requirements as well as likings for a proper health and energetic lifestyle (Leandro et al., 2020; Khan et al., 2025). The collapses of food systems, including agriculture, cattle and aquatic foods as well as commotions in food supply related to global warming, famine, floods and rain inconsistency, are all considered risks to food security posed by climate change, especially for populations already at risk of food insecurity because of lower earnings, social discrimination, dwindled access to nutrient-rich foods and various additional issues (O'Neill et al., 2022). The influence of climate change on various aspects is given in Fig. 8.1.

The definition and implementation of global food security as a concept in policy have changed over the course of several decades, making it a dynamic operational

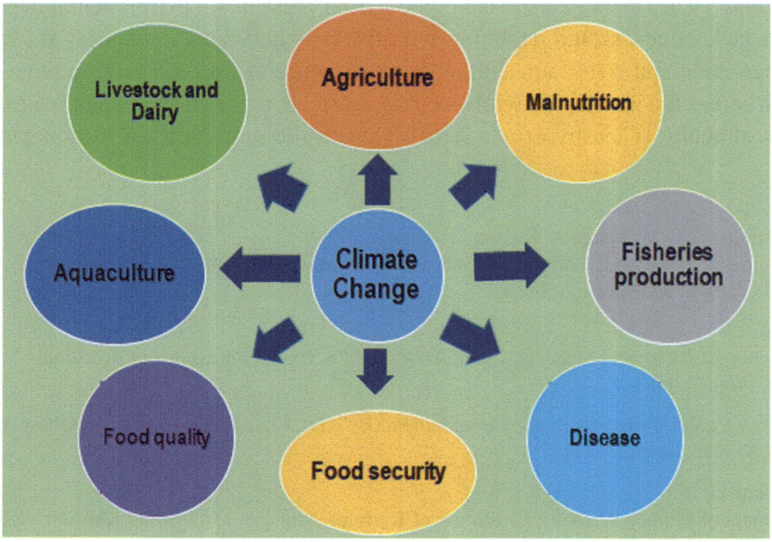

Fig. 8.1 The influence of climate change on various aspects

idea (Clay, 2011). According to Yarish et al. (2016), in order to meet present need patterns, the world would need to yield 50–70% additional foodstuff by the year 2050. The world's population is growing, making it imperative to uphold production of sustainable food, mainly in terms of maintaining the proper amount and quality to fulfil demand globally (Leandro et al., 2020). In total, 90% of the world's food supply must be met via increasing productivity per unit area on the present agricultural land; the other 10% can be met by adding more land (Garnett et al., 2013; Godfray & Garnett, 2014). Water, on the other hand, is a serious obstacle in raising agricultural productivity. It is well known that irrigation uses more than 70% of the freshwater available on earth (Madramootoo & Fyles, 2010). Systematic agriculture has led to the overuse of arable land, a reduction in the amount of freshwater available, a worsening cause of climate change and serious environmental consequences.

Global climate change is occurring in the industrial age, with many dire repercussions. Despite the fact that it is a global phenomenon, developing countries bear the brunt of its effects since they are unable to pay for the expensive mitigation of climate change. Vindication of climate change pertains to tactics and regulations that seek to lower the atmospheric greenhouse gases, either via decreasing their release or augmenting their absorption. The effects of global climate change are becoming progressively seeming and are having an influence on both marine and terrestrial habitats (Hoegh-Guldberg et al., 2018), all of which have important financial ramifications (DeFries et al., 2019). To fully realise aquaculture's potential to favourably impact the transformation of the food, it is crucial to adequately account for the environmental performance of various production system types (Bene et al., 2019; Poore & Nemecek, 2018). According to several recent studies on the topic, the three primary areas of aquaculture that have the greatest potential for ecologically sustainable expansion are marine-based finfish aquaculture, bivalve aquaculture and recirculating aquaculture systems (RASs) (Gentry et al., 2017; Stentiford et al., 2020). The first and the last areas need a great deal of outside the farm assets and are costly to run. Limited edible harvests, insufficient demand among consumers and additional challenges in processing and logistics are faced in bivalve cultivation (Belton et al., 2020; Ziegler et al., 2013).

2 Aquaculture: A Wide Spectrum of Organisms with Diverse Needs

Given that many aquaculture species have unique physiologies as well as metabolic traits, they also have varying impact on the environment profiles and efficiency under various production settings (Henriksson et al., 2021; Troell et al., 2014). While many species are filter feeders, fed animals might have dietary requirements that range from mostly herbivorous to nearly entirely carnivorous. Meanwhile, the right temperature, oxygen content and salinity levels are necessary for ideal growing conditions. Because they are cold-blooded and neutrally buoyant in the water, fish enjoy metabolic benefits over terrestrial creatures because they do not have to invest energy on regulating their body temperature, constructing supports or battling gravity (Henriksson et al., 2021). Certain species of catfish, such as the Siluriformes, have adapted to tolerate a variety of environmental situations. For instance, they can survive in water with low dissolved oxygen levels and have developed the capacity to gulp air to satisfy their oxygen demands. On the other hand, the species (like prawns) that thrive in brackish water are primarily restricted to estuary zones. Omnivorous and filter-feeding carps, bivalves, catfishes, tilapia and milkfish make up the majority of production volumes; these readily available fish are crucial for nutrition and food security (Henriksson et al., 2021). When it comes to fed fish, the most common species are those that can withstand poor water condition along with being raised on organic primary farming, farmed residues or food that is left over. Most of the time, they are offered for sale as a whole for regional or local commerce and are meant for domestic usage. If we look at live wet weight (that includes shells), bivalves account for a major portion of this group, yet they are less significant when it comes to edible yield (Edwards et al., 2019). An increasing number of non-native species (such as *Tilapia* and *Clarias* catfish) are being farmed among these widely available seafood commodities, particularly in Asia (FAO, 2021). This development might have negative effects on the environment (introduction of invasive species) and biosecurity (disease transmission) (Froehlich et al., 2017). On the other hand, a number of omnivorous finfish species and other accessible niche species—mostly other bivalves—are more often farmed and consumed locally. These species include mullets, gourami and green and blue mussels.

3 Enhanced Environmental Performance of Aquaculture

There seems to be a lot of promise for improving environmental performance and productivity in aquaculture if the performance gap is closed. The food system as a whole and the sustainability of foods derived from animals might both benefit from aquaculture, but more work is urgently needed before this promise can be fully realised (Henriksson et al., 2021). ‘Closing the yield gap’—the discrepancy with respect to achievable and realised optimum yields—or ‘sustainable intensification’, that means to obtaining better produces while utilising less resources per production output, have received a lot of attention in the field of agriculture (Willett et al., 2019). These tactics seek to increase food supply while lowering the environmental impact of food production (Willett et al., 2019). Owing to more exchanges in terms of life-cycle inputs like water, food and energy, it is more difficult to determine maximum possible yields per area in aquaculture (Henriksson et al., 2021).

3.1 Restrictions in Aquaculture and Possible Solutions

For the majority of available products as well as reachable specialised seafood, such as tilapia, carps and milkfish, a significant performance disparity still exists. A large portion of this disparity might be reduced with easily accessible, low-tech solutions including improved post-harvest handling and processing, improved cleanliness and biosecurity and improved management techniques. Simple improvements have occasionally not been realised yet because aquaculture farmers and other supply chain participants lack sufficient knowledge (Henriksson et al., 2021). Important restraints include perceived dangers and financial restrictions. This scenario shows potential for benefits through initiatives to fortify and enlarge producer organisations, as well as through training, financial assistance, extension services and training. Longer-term resource commitments would be necessary for further initiatives. These include creating genetic improvement programmes, developing vaccinations and creating innovative feeds (Farmery et al., 2021). They also involve modernising agricultural infrastructure. Encouragement to adopt more ecologically friendly practices by farmers and business might be offered by a third category of interventions. These consist of more stringent environmental laws, improved manufacturing techniques through financial incentives and better spatial planning. Given that profitability—which is mostly driven by price premiums or cost reductions—is the industry's most fundamental driver, this final set of actions is essential to pushing beyond business as usual (Sultana et al., 2023).

Despite the fact that most aquaculture systems outperform most terrestrial animal production systems in terms of environmental performance, its continuous growth will put more strain on the planet's resource systems and overall health (Naylor et al., 2021). As a result, advancements in vegetable-based and lab-based aquatic food substitutes are happening very swiftly. It is unclear how much these technologies will improve upon traditional farming in terms of the ecology and economy, and for now, these alternative goods are mostly aimed at high-income consumers and luxury aquatic food alternatives (Henriksson et al., 2021). The majority of readily available and reasonably priced fish will still be produced by farmers on small or medium scale that deal with having insufficient resources and the ability to independently optimise the benefits from their set-ups. Enhancing research and development in conjunction with extensive training initiatives as well as related services aimed towards the farmers may provide a significantly effective path ahead for increasing the accessibility of aquatic foods (Henriksson et al., 2021). Financial mechanisms that might assist farmers in managing economic risks such as contracts, and secured market places, have not been widely adopted by the aquaculture industry. This is frequently because of output volatility and confusion around the risks involved (Tisdell et al., 2010).

When we combine all of these factors, we can anticipate that over the next 10 years, worldwide aquaculture yields may rise significantly while the environmental effect per unit of output may decrease. Enhancing genetic strains, management techniques and production methods might cut feed conversion ratio and environmental effects by around 25% (Henriksson et al., 2018, 2019). Access to specially formulated commercial feed for key species (like carps) and innovative feed additives will be necessary to achieve these benefits. Promoting fish that eat diets richer at lower trophic levels will also require market access and customer acceptability. Enhancements in the post-harvest handling and dissemination process have the potential to enhance productivity and financial gain, minimise food wastage and loss and have a generally beneficial impact on the world's resource utilisation (Henriksson et al., 2019, 2021). Encouraging value-added goods will also facilitate improved by-product use and centralised processing. Achievable yields in animal production methods, such as aquaculture, are also significantly impacted by quality of feed as well as accessibility to genetically modified strains, good biosecurity and availability of medications and vaccinations. While competition between various land and water uses may restrict the potential for aquaculture to expand geographically in some areas, the obtainability of suitable feed options, scientific capabilities and social and economic factors, including need, typically present more significant obstacles to growing fed aquaculture output (Farmery et al., 2021; Oyinlola et al., 2018). Island nations should think about adopting localised, collaborative management that involves power-sharing, partaking decision-making and the amalgamation of old, common and western fisheries management tools so as to achieve suitable livelihoods as well as sound environmental governance as well as self-sufficiency of blue food and ample seafood (Cinner et al., 2012).

3.2 Empowering Small States and Islands: The Role of Aquaculture in Sustainable Development

Small states and island are predominantly reliant on food from aquatic sources and are prone to fluctuations in the rate of food on the international market as well as variation in climate. Population growth will require more food by 2050, including additional protein from the sea and other sources. For almost 20% of the world's population, the ocean provides their main source of protein (FAO, 2021). The productivity of marine fisheries is particularly important in the tropics, because despite having some of the least likely to sustain fisheries in the long run (Cheung et al., 2013; Lam et al., 2016). In the ensuing decades, climate variations in these regions are also anticipated to cause the greatest amount of reduction in aquatic fisheries production, causing high menaces to public health safety as well as the food supply (Thiault et al., 2019). Small island states, most of which are located in tropical regions, face several obstacles to sustainable development, all of which are made worse by the dangers that come with climate change (Tigchelaar et al., 2021). Small island states have limited local food systems; livelihoods are frequently hampered by a strong reliance on nutritious natural resources; and food security is susceptible to changes in global market prices (Teneva et al., 2023). Because of their remote location and heavy reliance on imports for food supply, they are vulnerable to fluctuations in the market and price of imported items. Public health concerns about nutrition, such as the high incidence of non-communicable illnesses associated with a diet heavy in imported processed foods, aggravate this economic vulnerability (Santos et al., 2019). Growing populations will raise the need for access to wholesome food, especially seafood, particularly in island nations with limited resources. According to some predictions, the world's food demand may rise from 2005 levels by as much as 100% by 2050 (Valin et al., 2014).

4 Climate Change Effects on Blue Food Security

Food security is often defined as having access to sustainable, healthy and safe food that satisfies local needs; food self-sufficiency, on the one hand, refers to a region's ability to ensure food security local scale and self-sufficiency in seafood, on the other hand, is the ability of a region to provide the sufficient quantity of sea food that its population requires locally (FAO, 2008; Teneva et al., 2023). Worrisome risks to self-sufficiency of seafood and, consequently, food security are compounded by the great dependency of tiny island nations on the ocean for their food and the reduction in the harvests of wild stocks of marine fisheries, together with the effects of climate change and expected population pressures (Golden et al., 2016; Teneva et al., 2023). The demand for seafood, or 'blue food', is predicted to almost quadruple globally by the middle of the century; therefore, supplying this need will require responsible resource management that takes into account the negative effects that improper management of the ocean has on the environment and human health (Naylor et al., 2021).

The realistic adaption scenarios recognise that yearly changes in management are not common. Rather, these scenarios take into account management that establishes, at regular intervals (5, 10 and 20 years), an economically appropriate scale of harvest according to current production and then holds onto this rate until the next evaluation. As was previously said, depending on the particular fisheries regulatory frameworks of each nation, economically optimal harvest rates may be attained by control on input/output, management or other methods (Teneva et al., 2023). In some reefs that might support fish populations, ecosystem connectivity and genetic rehabilitation throughout reef systems may increase climate flexibility; nevertheless, broad and diversified adaptation complexes are required to maximise the ability of fish and coral species to adapt (Webster et al., 2017).

In light of climate change and population growth, the countries experiencing predicted seafood deficits—that is, locally obtained seafood falling short of expected demand—will have to explore alternatives to prudent fisheries management in order to achieve more sustainable food self-sufficiency. Enhanced enforcement, international fleet fishing agreements, better national and regional regulatory environments and supply chain management through trade and market rules are a few other strategies that may be employed (Cinner et al., 2012; Teneva et al., 2023). These countries could require large expenditures in alternate food sources, such as sustainable aquaculture for both domestic and international markets. As a result, in certain regions, certain types of small-scale aquaculture may be suitable and contribute significantly to local food security while also building climate resilience through the diversification of vulnerable food systems. It is crucial to remember that depending on local preferences, costs and environmental circumstances, aquaculture—whether small or large scale—might not be suitable for all small island nations (Teneva et al., 2023).

The researchers should further investigate the prospective for affordable and prolific aquaculture opportunities for the needs of both food security and the economy. Actually, as discussed in FAO (2018), the FAO has a strategic plan that highlights the necessity of collaboration and the expansion of climate-smart small-scale fisheries and aquaculture organisation in the Caribbean to prevent fisheries misadventures and also build up this sector for local livelihood objectives and food security (Teneva et al., 2023).

It would be crucial to include 'blue food' into national food plans, which would entail specifically including marine food production in national nutrition objectives while admitting that climate change poses a danger to this source of nutrition (Tigchelaar et al., 2021). National commitments to changing the country's food systems and local knowledge of social susceptibility and flexibility related to people's ability to adapt, their access to capital, their ability to use fish gear with flexibility, the availability of substitute livelihoods and added dynamics that frequently can range to native coastal communities would drive these governance arrangements (Cinner et al., 2013; Teneva et al., 2023). In particular, lowering fisheries conflict and susceptibility from overseas marines catching too much fish in impoverished nations that depend on ocean resources can result from integrating blue food into domestic food policies and running fisheries from a food security standpoint (McClanahan et al., 2015).

Given the obvious links between ecological resources, community health, public stability and conflict, blue food policies may offer stronger benefits for handling ecological wealth with an emphasis on sustainable use. A typical illustration of how food poverty and a general lack of strong national food policy may upset social as well as administrative stability and result in enforced relocation is the Arab Spring of 2011 (Soffiantini, 2020). Moreover, by lowering the disease risk linked to insufficient nutrient intake, coordinated blue food policies that support ocean sustainability can enhance human health (Golden et al., 2021). Development of climate-suitable fisheries managing in conjunction with food planners also requires adjustment and collaboration across local and countrywide fishery management institutions, particularly for the numerous tiny island states having inadequate administration and man power. Currently serving as administration entities to promote international collaboration are Regional Fishery Management Organisations (RFMOs). For some fisheries, they can promote climate-based suitable fisheries explanations by allocating resources to monitor the effects of climate change, enabling data-sharing in situations where data is frequently scarce, boosting maritime enforcement and promoting equity in negotiations for day leases of foreign fleet vessels (Goodman et al., 2022; Pentz et al., 2018).

A number of perspectives may be used to analyse the ability of island fisheries to adapt, including those of individual fishermen, communities, countries and regions (Cinner et al., 2013; Goodman et al., 2022). Few studies have looked at the ways in which national governments can work together internationally on blue food security in addition to sustainable fisheries, despite the abundance of research on how people and communities can adapt to fishery regimes and climate change, predominantly with respect to small island settings in the Indian and Pacific Oceans (Cinner et al., 2012). The footprint of global fisheries has expanded over the past few spans due to globalised seafood markets, going far beyond coastal seas, deeper into the ocean and away from efficient enforcement management (Teneva et al., 2023). Because of this, the enormous marine resources under the national sovereignty of tiny island states with little capacity are frequently overfished as a result of illicit, unreported and unregulated fishing, which is profitable and results from the demand for seafood throughout the world (Worm et al., 2009). Countries with strong ecological connection and species having great kinesis in mature or larval stages dispersion will gain most from international collaboration. Benefits will be more restricted for isolated countries and animals that have strong ties to their habitats and/or restricted mobility. While there is a wealth of research on how communities and individuals can adapt to fishery regimes and climate change, predominantly in small islands of the Indian and Pacific Oceans, very little has been done to look at how national governments can work together internationally to promote sustainable fisheries and blue food security. The impression of whole fisheries has expanded over the past few decades due to globalised seafood markets, going past coastline, deeper into the ocean and away from efficient application regulation (Watson et al., 2015).

This suggests that the island countries with the biggest deficits have the most obstacles to becoming self-sufficient with respect to food from oceanic resources besides being most likely to become food insecure as a nation if nothing is done. As a low-carbon protein supply, aquaculture will only grow in importance, especially for tiny island nations whose local seafood demand cannot be sustained in the long run by any amount of sustainable fisheries management (Teneva et al., 2023). For several Caribbean regions and a few Pacific Ocean islands with previously historically troubled fisheries, indigenous preliminary programmes and substantial domestic and international investments in aquaculture will remain essential. Incorporating these blue crops into national food policy is going to be essential to transforming the food system and improving ocean governance in the face of climate change (Teneva et al., 2023).

5 Seaweed: A Sustainable Solution for Global Food Security

Owing to its plentiful environmental, societal and commercial benefits, seaweeds are among the utmost resourceful macroalgae of the new time. Seaweed is emerging as a crucial resource in the fight against global food insecurity. With a global seaweed market valued at US$11.8 billion, seaweed production from farming now stands at 35.8 million tonnes, or 97% of the total seaweed output (Sultana et al., 2023). Rich in nutrients, easy to cultivate and environmentally sustainable, seaweed provides an efficient way to produce food without relying on traditional farming methods. Abundant mineral concentration, little lipid concentration, fibre concentration, polyunsaturated fatty acid concentration, polysaccharide concentration, vitamins, bioactive substances and seaweeds are source of nutrient-rich diet for humans (Hamid et al., 2015). Unlike crops that require freshwater and large areas of land, seaweed grows in the ocean, making it a valuable alternative in regions facing freshwater scarcity and land degradation due to climate change. Numerous seaweed byproducts have special qualities that can be used to create a range of useful meals for the food processing sector. Seaweed farming has a low environmental impact, absorbing carbon dioxide and promoting marine biodiversity, further enhancing its role in creating a sustainable, resilient food system. Seaweed farms contribute to the mitigation of climate change by absorbing carbon, acting as a sink for CO_2 and lowering agricultural emissions by supplying the raw materials needed to produce biofuel and animal feed (Bach et al., 2021). By absorbing wave energy, protecting seashores, shifting the pH of the nearby waters, and increasing oxygenation of the waters to lessen the effects of acidification in ocean and or low oxygen on a localised level, seaweed farming systems also aid in the adaptation to climate change (Campbell et al., 2019). Aside from capturing carbon, the production of low-carbon food, the production of biofuel, the reduction of methane emissions through the addition of seaweed to cattle feed and the avoidance of negative effects by replacing inorganic fertilisers with seaweed are additional potential benefits in global climate change mitigation. With its advantages for the environment, economy and society, it is regarded as the plant of the millennium. Because of its many uses in biotechnology, food, colour, enzymes, animal feed, fertiliser, animal feed, cosmetics, textiles, animal feed and biochemical raw constituents (carrageenan, agar, agarose and alginate), seaweeds are well-known commercial marine biota. Additionally, they produce a vast array of secondary metabolic products with various biological purposes (Jeeva et al., 2012). Beyond these industrial uses, seaweeds have the capacity to control the whole benthic environment, serving as both a key food source and a haven for a wide variety of marine animals (Siddiqui et al., 2019).

Furthermore, it ensures financial stability and creates livelihood options for coastal women, both of which significantly improve their economic situation in a sustained manner. The primary obstacles to seaweed being a staple diet for millions of people are the development of appropriate seaweed species and the creation of consumer-friendly, reasonably priced, easily available and nutritionally balanced seaweed products (Forster & Radulovich, 2015). It is necessary to produce a variety of seaweed-based food items that are comparable to the meals that humans now eat. In order to develop seaweed farming, produce aquatic food items and help mitigate and adapt to the effects of climate change globally, there must be appropriate marine design so as to prevent user conflicts. Therefore, the main obstacles to women's empowerment through seaweed farming are diminished financial resources and technical know-how to set up the coastal farming system (Sultana et al., 2023).

Human civilizations will have to contend with the inconsistent consequences of climate changes and resource degradation by the middle of the twenty-first century, on top of the enormous job of feeding and supporting a population of well over 9 billion people (FAO, 2018). Seaweed farming has been promoted as an administration policy in many developing countries in an effort to reduce burden on fishing (Sievanen et al., 2005), reduce poverty, as well as provide an alternative source of income for women living in resource-poor coastal communities (Mantri et al., 2017). Seaweed agriculture has been shown to have positive economic consequences in a number of nations, including the Philippines, Tanzania, Indonesia and the Pacific Islands (Namudu & Pickering, 2006). Numerous organisations have taken action as a result of this campaign to support seaweed agri-business as a feasible unconventional source of income for the seaside people, particularly for women from fishing villages (Sultana et al., 2023).

Because seaweed biofuels often referred to as ‘Blue biofuel’ do not require freshwater, land, fertiliser or pesticides, they are not in competition with land cultivation for resources. Consequently, compared to biofuels made from terrestrial crops, these biofuels are more environmentally friendly (Duarte et al., 2013). Carbon sequestration will also be greatly increased by turning seaweed into bioethanol (Zacharia et al., 2015). In terms of preventing emissions from fossil fuels, the collected seaweed might be used entirely or in part to make biofuel with a considerable reduction of carbon dioxide potential (Duarte et al., 2017). Hargreaves et al. (2013) reported that a number of seaweed species have large concentrations of polysaccharides that, with the right technology may be converted to ethanol. Biofuel (bioethanol) might be made from seaweed leftovers from several processing businesses (Zacharia et al., 2015). There would be significant benefits to climate resilience if seaweed could be successfully converted to bioethanol on a large scale (Zacharia et al., 2015). Alvarado-Alvarado-Morales et al. (2013) showed through life cycle analysis that cultivating seaweed might also reduce agricultural emanations by increasing soil quality instead of synthetic fertiliser and by lowering methane emissions from livestock when supplementary to the feed (Duarte et al., 2017). A study conducted in vitro revealed that fermented seaweed might enhance digestion in ruminants, thereby reducing the amount of methane gas formed (Kinley et al., 2016). Moreover, a study found that incorporating a specific seaweed species into cattle feed at a rate of 2% could potentially reduce ruminant methane emission by up to 99% (Machado et al., 2016). In the lab, Roque et al. (2019) found that including just 5% of *Asparagopsis* (Rhodophyta)'s organic component into the feedstock might reduce emissions by as much as 95%. By improving soil quality, the use of extremely nutritious seaweed bio char (Zacharia et al., 2015) or compost seaweed (Cole et al., 2016) in agriculture could increase crop productivity and prevent emissions related to the production of artificial fertilisers (Smith, 2002).

Ocean acidification is thought to be avoided by using seaweeds' capacity to store carbon. In the marine environment, a low pH decreases the accessibility of the crystals that forms shells, which is crucial for a variety of bacteria, corals, molluscs and crustaceans (Gattuso & Hansson, 2011). A significant quantity of carbon may be absorbed by large quantities of seaweeds, raising the pH of the surrounding water and reducing the consequences of ocean acidification (Campbell et al., 2019). According to Langton et al. (2019), seaweed farming, particularly in the exponential growth phase of their life phase, may have contributed to an overall positive increase in oxygen and pH in the nearby agricultural activities. Significant worldwide climate change threats to nutrition and food security include those that might have an impact outside of food systems, cause millions of people to experience severe and ongoing food insecurity and malnutrition and are those to that we are insufficiently prepared to avert or effectively manage (Mirzabaev et al., 2023).

6 Climate Change and Its Effects on Cattle and Fisheries: A Growing Concern

The impact of climate change on fisheries and livestock is being studied significantly less than that of essential crops (Godde et al., 2021). However research indicates that heat stress, decreased forage availability, freshwater shortage and the spread of livestock illnesses may negatively affect livestock reproduction and productivity as a result of heatwaves and famines brought on by climate change (Mbow et al., 2019; Rojas-Downing et al., 2017). In many dry and semi-arid places, the projected effects of climate change on pasture production are predominantly negative (Rojas-Downing et al., 2017). Under high emission scenarios, extreme heat stress is likely to compromise the survival of cattle raised outdoors and harm all major species in many tropical and temperate countries (Thornton et al., 2021). Each 1 °C escalation is projected to decrease worldwide animal ocean biomass by five percent (Lotze et al., 2019) along with lessening the fishery capture potential between 5.3 and 7% by the year 2050 (Cheung et al., 2019). Also, global warming has led to an average 4% decline in marine fisheries productivity and up to 35% for some regions (Free et al., 2019). Additionally, warming is causing freshwater and marine aquatic species to move poleward, which is changing the species and quantity of fisheries in exclusive economic zones (Oremus et al., 2020; Pinsky et al., 2020) and raising the danger for low-income tropical nations (Bindoff et al., 2019). According to Weatherdon et al. (2016), poleward shifts in productivity, suitable habitat and culture species are also being seen in mariculture. These shifts are expected to continue, resulting in a decrease in species and habitat in tropical and sub-tropical regions (Froehlich et al., 2018; Oyinlola et al., 2020; Stewart-Sinclair et al., 2020).

According to projections, there will be more marine heat waves in the future, which will have an adverse effect on coastal communities that depend on fisheries and aquaculture. These effects will include reduced maximum catch potentials, increased mortality and reproductive failures (Cheung et al., 2021; Green et al., 2019; Smale et al., 2019). The productivity of shellfish aquaculture and shellfisheries is being adversely affected by ocean acidification (Doney et al., 2020; Ekstrom et al., 2015), and it is anticipated that this negative impact will persist in key production regions (Chapman et al., 2020; Des et al., 2020; Rheuban et al., 2018). Drought, flooding, coastal inundation brought on by climate change and freshwater availability are expected to continue having a detrimental impact on inland fisheries and aquaculture output (Dabbadie et al., 2018; Lebel et al., 2020).

7 Addressing Critical Threats to Food Security and Nutrition in the Face of Climate Change

The world currently has a high level of food insecurity, with between 720 million and 811 million people suffering from undernourishment and roughly 2.3 billion from malnutrition. This is significant given that Sustainable Development Goal 2 calls for achieving Zero Hunger by 2030 (FAO, 2021). The Integrated Food Security Phase Classification, which distinguishes between acute and chronic food insecurity, shows that, at the moment, 200,000 people experience catastrophic food insecurity, 32.3 million experience emergency food insecurity, 112.3 million experience crisis food insecurity and 210 million experience stressed food insecurity (The Integrated Food Security Phase Classification (IPC), 2022). First-level outcomes (alterations in food intake and maintenances), second-level results (nutritional position in addition to mortality) and causative dynamics (availability of food, access, utilisation, firmness, threats and susceptibility) are all taken into consideration when determining these IPC classification criteria (IPC Global Partners, 2021). For instance, 'households have an extreme lack of food and/or other basic needs even after full employment of coping strategies', according to the definition of catastrophic food security. It is clear that there is severe acute malnutrition, starvation, mortality and poverty (IPC, 2021). Furthermore, it is anticipated that climate change would increase the severity and probability of negative effects on nutrition and food security. Shared socioeconomic pathways (SSPs) and the amount of greenhouse gas (GHG) emissions determine when these effects occur and how well we can react to them. Up to 2050, there may be a little increase in the severity of the hazards to food security and nutrition (FSN) brought on by climate change. Nonetheless, there is evidence that the dangers associated with climate change to food safety nets are expected to intensify beyond 2050 and reach 2080 (Bezner Kerr et al., 2022). A reduction in agricultural output, decreased earnings commencing climate-sensitive livelihoods, new concerns about food safety and changes in food distribution are some of the drivers that mediate the consequences of climate change on the population at risk of hunger and malnutrition (Bezner Kerr et al., 2022). The carbon dioxide concentration in the atmosphere directly impacts how extremely climate change impends nutrition. Crop yields are expected to grow when carbon dioxide concentrations in the atmosphere rise, however this effect will be tempered by other climate risk variables (Myers et al., 2017; Zhu et al., 2018).

7.1 *Dangers of Severe Climate Change for Food Security and Nutrition*

Regarding the observed consequences of climate change on food security, while detailed works are constrained by the multifaceted causal effects of food insecurity and lack of long-term data, some indirect evidence shows that at least part of the currently reported numbers of food insecure individuals may be caused by catastrophic weather occurrences (Cooper et al., 2019; Phalkey et al., 2015). Some attributional investigations of past occurrences, such as the one conducted by Verschuur et al. (2021) on the drought in Lesotho in 2007, found strong evidence that the drought was made worse by human-induced climate change, leading to a rise in food scarcity and cost rises. An ascription analysis of the El Nịo event of 2015–2016 further demonstrates how crop yield losses from drought were exacerbated by human climate change (Funk et al., 2018). Regarding the anticipated effects of climate change on food security and nutrition, we discover that there will be significant hazards associated with the rise in the number of undernourished individuals. These risks could impact more than million people that are highly vulnerable (Shared Socioeconomic Pathways (SSPs) and especially most vulnerable populations from economically poor countries. Even locally, weather events can raise the risk of malnutrition by driving up food prices and decreasing income (Hickey & Unwin, 2020; Zhang et al., 2022). The degree of warming will determine the range of dangers, from local to worldwide, with lower warming levels creating locally significant risks to FSN and greater warming levels creating globally significant risks to FSN (O'Neill et al., 2022).

7.2 *Climate Change: A Catalyst for Infrastructure Damage, Crop Failures and Food System Conflicts*

Since climate change destroys infrastructure, causes crop failures and encourages rivalry among food production systems, extreme weather events like inundations, famines, hurricanes and others that are common and severely have the probability to substantially mess up food chains (Cottrell et al., 2019). For example, the frequency of climatic extremities is increasing across Europe, a net exporter of food (Pradhan et al., 2022). The detrimental impacts of heat waves, droughts and floods on people's food security, health and nutrition as well as on their productivity and capacity to sustain themselves are especially dangerous for those who work outside or in climate-sensitive sectors. This exposure may have a negative impact on specific social and economic groups, including low-income families, elderly, female and child populations, smallholder farmers and farmworkers, more susceptible low- and middle-income nations and also those that are dependent on rain fed agriculture systems (De Lima et al., 2021; Kuhla et al., 2021).

7.3 *Rising Temperatures, Rising Hunger: The Climate Change Impact*

Climate change is intensifying global hunger by disrupting food production and supply chains. Rising temperatures, unpredictable rainfall and extreme weather events such as floods, droughts and storms are severely affecting agriculture, reducing crop yields and leading to food shortages. Communities that rely on farming and fishing are particularly vulnerable, as their livelihoods are directly tied to climate-sensitive environments. The timing of these effects determines how serious a threat climate change is to the global hunger crisis (Mirzabaev et al., 2023). The degree of greenhouse gas emissions and the properties of the associated SSPs determine how well-equipped an entity is to adapt to these effects. Climate variation poses a risk to food security under a low vulnerability development pathway at approximately 1 °C of global warming, or when impacts are detectable and attributable to climate-related factors; over 2.5 °C, nonetheless, the likelihood of damage grows to an elevated level (prominent implications for greater proportion of region or population, though with a chance to adjust or recuperate (Hurlbert et al., 2019; Mirzabaev et al., 2023). Hasegawa et al. (2018) forecast that, in a medium vulnerability-high warming scenario which takes into consideration the CO_2-fertilization impact, there would be 24 million more undernourished people in 2050 than in a scenario without climate change. With reduced CO_2-fertilization and a low warming scenario, this figure rises by almost 78 million when the effects of mitigating measures and climate change are taken into consideration. Around 183 million more people are expected to suffer from malnutrition in low-income countries by 2050 as a result of climate change under a high vulnerability-high warming scenario (Mbow et al., 2019). Increased climate variability also contributes to food price volatility, making essential staples unaffordable for many, especially in regions already facing economic challenges. Moreover, the shifting climate is causing soil degradation, reducing the quality and availability of arable land, which further hampers food production.

By 2050, a 2 °C rise in temperature is predicted to influence the crude mortality rate of roughly 54 deaths per million people. The majority of these deaths are anticipated to be in East and South Asia, with estimates ranging from 67 to 231 deaths/million, based on the nation (Springmann et al., 2016). The effects of climate change disproportionately affect poorer regions and exacerbate existing inequalities, leaving millions of people, especially in developing countries, at greater risk of hunger and malnutrition. Addressing hunger in the face of climate change requires a concerted global effort to enhance sustainable agricultural practices, improve food distribution systems and implement policies that mitigate climate impacts. Beyond worries about the availability of calories, there is a relationship between human nutrition and climate change. By 2050, one of the biggest challenges will be providing affordable, nutrient-dense meals. When extreme events increase in frequency and severity, there is a greater likelihood of acute food insecurity and malnutrition (Bezner Kerr et al., 2022). Climate change is expected to cause under nutrition and stunting in children, under nutrition-related childhood mortality, diet-related morbidity and death and a rise in disability-adjusted life years lost, with the biggest risks in Africa and Asia (Sulser et al., 2021).

Risks from climate change to food, shelter and other non-food sectors interact closely with one another and frequently intensify one another's effects (Challinor et al., 2014). Furthermore, via its impacts on several social and economic facets such as poverty, water security, peace and mobility, biodiversity and ecological changes, infrastructure, health, climate change will have an indirect as well as direct influence on FSN (O'Neill et al., 2022). As a result, various climatic threats and sources of those risks could possibly lead to multifarious consequences (Simpson et al., 2021). Drought and conflict in Ethiopia and Yemen are making famine more likely, and the present conflict in Ukraine together with unusual weather patterns is driving up food costs worldwide (Osendarp et al., 2022). Zero Hunger is not expected to be achieved by 2030, according to Bezner Kerr et al. (2022), because of non-climatic compounding hazards to food security. Because of this, it is not possible to evaluate the risks posed by climate change to FSN in a vacuum, separate from other sources of risk. These additional socioeconomic hazards to FSN must be taken into account in adaptation strategies aimed at reducing the dangers associated with climate change.

Rise in sea level, tropical cyclones and floods raise the dangers to setup, especially in low-lying coastal locations (Koks et al., 2019). These events impair food delivery and storage, disrupt farmers' and fishermen's livelihoods and ultimately restrict access to food. Acute food insecurity is already being brought on by these cascading and/or compounding consequences in sensitive areas, such as southern Africa. They can be attributed, at least partially, to climate change brought on by humans (Verschuur et al., 2021). It is predicted that these impacts will happen more frequently as global warming grows. The influence of climate variation on food safety can further exacerbate threats to food security as well as human health. Warming will have an impact on food security, nutrition and health in both marine and inland systems due to increased bioaccumulation of contaminants like methyl mercury, greater incidence, severity and range extension of infections like Vibrios and increase in toxic algal blooms (Mohamad et al., 2019). Comparably, a wide range of adaptation techniques have been put forth for the production of cattle on land and in water, including species switching, breeding animals to withstand heat stress, bathing and shade (Bezner Kerr et al., 2022).

7.4 *Climate Change and Food Systems: Unequal Burden Across Social Groups*

Not every social group experiences the effect of climate change on food systems in the same way (Mbow et al., 2019). Certain people are more susceptible than others due to social injustices based on a variety of criteria including stage, debility, gender and race. These individuals also have varying capacity for adaptation. To provide one example, persons living in tropical as well as subtropical areas—which are also expected to experience the loss of species due to poleward migrations induced by climate marine fish resources—are more vulnerable to vitamin deficiencies associated with seafood consumption. It is becoming more widely recognised that FSN also involves important elements like sustainability and agency (HLPE, 2020). Commotions in the food markets and chains brought on by change in climate will have an impact on food prices and the fairness of food access for marginalised socioeconomic groups. These impacts will frequently intersect with additional hazards that these populations face (Mbow et al., 2019). Strengthening social protection and empowering marginalised social groups via collective action are two key components of actions to address social disparities and the uneven consequences of climate change threats on FSN. Ensuring fair access to food via social protection should not compromise individuals' autonomy.

8 Research Lacunae

Our comprehension of the severity conditions of climate change dangers to FSN is severely lacking in a number of important areas. Here, we draw attention to a number of important research gaps. Acute food insecurity, as described by IPC (2021), is very relevant to climate and weather extreme events that are expected to occur more frequently, however there have been very few studies forecasting the dangers of climate change to this phenomenon. Our ability to predict these catastrophic occurrences and their effects on food systems places a limit on this. There is still little data on how climate change may affect other significant crops (vegetables and fruits), livestock, aquaculture and fisheries; therefore, the estimates of the severe climate change threats to FSN are still mostly centred on those effects on staple food crops (Mbow et al., 2019; Mirzabaev et al., 2023). Particularly, there is a dearth of knowledge on aquaculture and inland fisheries systems in low-income nations. It is challenging to evaluate compounding and cascading risks and how these affect the severity of climate change risks to FSN because current food security risk assessments are still heavily focused on crop production and disruptions of other food security pillars like access, utilisation and stability are much less studied (Davis et al., 2021). Impact estimations sometimes fail to take into consideration the varying degrees of susceptibility among various populations.

9 Conclusion

Although many farming systems continue to have significant performance gaps and unsustainable practices, aquaculture has a lot of promise to help create more sustainable meals. The most effective interventions for addressing these issues are mostly already in place and would primarily focus on aquatic food species that are easily accessible, many of which are farmed on comparatively small-sized farms. Many of these involvements might be carried out for very little money, but they all need for a combination of short- and long-term efforts. Nonetheless, the biggest problem still lies in finding the most effective solutions for the variety of agricultural systems and providing incentives to adopt them; this calls for government assistance. In addition to investments in farm management, feed and genetics including improved documentation alongside information management, financial incentives and legislative initiatives by all farmers are required to increase aquaculture output, enhance efficiency of resources and lessen environmental effects. While they would not be able to feed the whole world's population, luxurious aquatic food stuffs from RASs and offshore systems may be able to lessen ecological effects on the global food chain if they take the place of red meat in diets. Enhancing the accessibility of aquatic foods through sustainable intensification of present systems may play a major role in helping aquaculture achieve sustainability goals. These methods would be based on increasing the dosage of proven therapies that are seldom utilised. These technologies and systems need to be more fully benchmarked using complementary frameworks and life cycle assessment in order to determine their full potential sustainability benefits. If aquatic crops are created wisely, they can reduce the environmental stressors that the current food system places on the planet, even though they alone cannot provide food security in the future. Scientific intermediations can help close the output gap in aquaculture worldwide, which will enhance profitability, resource efficiency and environmental performance overall. The necessity of food security should be given top priority, and involvements which could be commercially feasible aimed at the majority of growers, sufficiently scalable to contribute to global change.

References

Ahmed, I., Ahmad, I., & Malla, B. A. (2024). Effects of dietary tryptophan levels on growth performance, plasma profile, intestinal antioxidant capacity and growth related genes in rainbow trout (*Oncorhynchus mykiss*) fingerlings. *Aquaculture, 585*, 740710.

Alvarado-Morales, M., Boldrin, A., Karakashev, D. B., Holdt, S. L., Angelidaki, I., & Astrup, T. (2013). Life cycle assessment of biofuel production from brown seaweed in Nordic conditions. *Bioresource Technology, 129*, 92–99. https://doi.org/10.1016/j.biortech.2012.11.029

Bach, L. T., Tamsitt, V., Gower, J., Hurd, C. L., Raven, J. A., & Boyd, P. W. (2021). Testing the climate intervention potential of ocean afforestation using the great Atlantic Sargassum Belt. *Nature Communications, 12*, 1–10.

Belton, B., Little, D. C., Zhang, W., Edwards, P., Skladany, M., & Thilsted, S. H. (2020). Farming fish in the sea will not nourish the world. *Nature Communications, 11*, 1–8.

Bene, C., Oosterveer, P., Lamotte, L., Brouwer, I. D., de Haan, S., Prager, S. D., Talsma, E. F., & Khoury, C. K. (2019). When food systems meet sustainability – Current narratives and implications for actions. *World Development, 113*, 116–130.

Bezner Kerr, R., Hasegawa, T., Lasco, R., Bhatt, I., Deryng, D., Farrell, A., Gurney-Smith, H., Ju, H., Lluch-Cota, S., Meza, F., Nelson, G., Neufeldt, H., & Thornton, P. (2022). Food, fibre and other ecosystem products supplementary material. In H. O. Portner et al. (Eds.), *Climate change 2022: Impacts, adaptation, and vulnerability. Contribution of working group II to the sixth assessment report of the intergovernmental panel on climate change*.

Bindoff, N. L., Cheung, W. W. L., Kairo, J. G., Arístegui, J., Guinder, V. A., Hallberg, R., Hilmi, N., Jiao, N., Karim, M. S., Levin, L., O'Donoghue, S., Purca Cuicapusa, S. R., Rinkevich, B., Suga, T., Tagliabue, A., & Williamson, P. (2019). Changing ocean, marine ecosystems, and dependent communities. In H. O. Portner et al. (Eds.), *IPCC special report on the ocean and cryosphere in a changing climate* (pp. 447–587).

Campbell, I., Macleod, A., Sahlmann, C., Neves, L., Funderud, J., Øverland, M., Hughes, A. D., & Stanley, M. (2019). The environmental risks associated with the development of seaweed farming in Europe-prioritizing key knowledge gaps. *Frontiers in Marine Science, 6*, 107. https://doi.org/10.3389/fmars.2019.00107

Challinor, A. J., Watson, J., Lobell, D. B., Howden, S. M., Smith, D. R., & Chhetri, N. (2014). A meta-analysis of crop yield under climate change and adaptation. *Nature Climate Change, 4*, 287–291. https://doi.org/10.1038/nclimate2153

Chapman, E. J., Byron, C. J., Lasley-Rasher, R., Lipsky, C., Stevens, J. R., & Peters, R. (2020). Effects of climate change on coastal ecosystem food webs: Implications for aquaculture. *Marine Environmental Research, 162*, 1–10. https://doi.org/10.1016/j.marenvres.2020.105103

Cheung, W. W., Watson, R., & Pauly, D. (2013). Signature of ocean warming in global fisheries catch. *Nature, 497*, 365–368.

Cheung, W. W. L., Bruggeman, J., et al. (2019). Projected changes in global and national potential marine fisheries catch under climate change scenarios in the twenty-first century. In M. Barange et al. (Eds.), *Impacts of climate change on fisheries and aquaculture* (pp. 63–85). FAO.

Cheung, W. W. L., et al. (2021). Marine high temperature extremes amplify the impacts of climate change on fish and fisheries. *Science Advances, 7*, eabh0895. https://doi.org/10.1126/sciadv.abh0895

Cinner, J. E., McClanahan, T. R., Graham, N. A., Daw, T. M., Maina, J., Stead, S. M., Wamukota, A., Brown, K., & Bodin, Ö. (2012). Vulnerability of coastal communities to key impacts of climate change on coral reef fisheries. *Global Environmental Change, 22*, 12–20. https://doi.org/10.1016/j.gloenvcha.2011.09.018

Cinner, J. E., Huchery, C., Darling, E. S., Humphries, A. T., Graham, N. A., Hicks, C. C., Marshall, N., & McClanahan, T. R. (2013). Evaluating social and ecological vulnerability of coral reef fisheries to climate change. *PLoS One, 8*, e74321. https://doi.org/10.1371/journal.pone.0074321

Clay, J. (2011). Freeze the footprint of food. *Nature, 475*, 287–289. https://doi.org/10.1038/475287a

Cleasby, N., Schwarz, A. M., Phillips, M., Paul, C., Pant, J., Oeta, J., Pickering, T., Meloty, A., Laumani, M., & Kori, M. (2014). The socio-economic context for improving food security through land based aquaculture in Solomon Islands: A peri-urban case study. *Marine Policy, 45*, 89–97.

Cole, A. J., Roberts, D. A., Garside, A. L., de Nys, R., & Paul, N. A. (2016). Seaweed compost for agricultural crop production. *The Journal of Applied Psychology, 28*, 629–642. https://doi.org/10.1007/s10811-015-0544-2

Cooper, M. W., Brown, M. E., Hochrainer-Stigler, S., Pflug, G., McCallum, I., Fritz, S., & Zvoleff, A. (2019). Mapping the effects of drought on child stunting. *Proceedings of the National Academy of Sciences, 116*, 17219–17224.

Cottrell, R. S., Nash, K. L., Halpern, B. S., Remenyi, T. A., Corney, S. P., Fleming, A., Fulton, E. A., Hornborg, S., Johne, A., Watson, R. A., & Blanchard, J. L. (2019). Food production shocks across land and sea. *Nature Sustainability, 2*, 130–137. https://doi.org/10.1038/s41893-018-0210-1

Dabbadie, L., Anguilar-Manjarrez, J., Beveridge, M. C. M., Bueno, P. B., Ross, L. G., & Soto, D. (2018). Chapter 20: Effects of climate change on aquaculture: Drivers, impacts and policies. In M. Barange, T. Bahri, M. C. M. Beveridge, K. L. Cochrane, S. Funge-Smith, & F. Poulain

(Eds.), *Impacts of climate change on fisheries and aquaculture: Synthesis of current knowledge, adaptation and mitigation options* (Vol. 627, pp. 449–463). FAO.

Davis, K. F., Downs, S., & Gephart, J. A. (2021). Towards food supply chain resilience to environmental shocks. *Nature Food, 2*, 54–65. https://doi.org/10.1038/s43016-020-00196-3

De Lima, C. Z., Buzan, J. R., Moore, F. C., Baldos, U. L. C., Huber, M., & Hertel, T. W. (2021). Heat stress on agricultural workers exacerbates crop impacts of climate change. *Environmental Research Letters, 16*. https://doi.org/10.1088/1748-9326/abeb9f

DeFries, R. S., Edenhofer, O., Halliday, A. N., Heal, G. M., Lenton, T., Puma, M., Rising, J., et al. (2019). *The missing economic risks in assessments of climate change impacts* (pp. 1–15). The Earth Institute Columbia University.

Des, M., et al. (2020). How can ocean warming at the NW Iberian Peninsula affect mussel aquaculture? *Science of the Total Environment, 709*, 136117.

Doney, S. C., Busch, D. S., Cooley, S. R., & Kroeker, K. J. (2020). The impacts of ocean acidification on marine ecosystems and reliant human communities. *Annual Review of Environment and Resources, 45*, 83–112. https://doi.org/10.1146/annurev-environ-012320-083019

Duarte, C. M., Losada, I. J., Hendriks, I. E., Mazarrasa, I., et al. (2013). The role of coastal plant communities for climate change mitigation and adaptation. *Nature Climate Change, 3*, 961–968. https://doi.org/10.1038/nclimate1970

Duarte, C. M., Wu, J., Xiao, X., Bruhn, A., & Krause-Jensen, D. (2017). Can seaweed farming play a role in climate change mitigation and adaptation? *Frontiers in Marine Science, 4*, 100. https://doi.org/10.3389/fmars.2017.00100

Edwards, P. (2000). Aquaculture, poverty impacts and livelihoods. In *Natural resource management: Ecological perspectives* (Vol. 56, pp. 1–4). Overseas Development Institute.

Edwards, P., Zhang, W., Belton, B., & Little, D. C. (2019). Misunderstandings, myths and mantras in aquaculture: Its contribution to world food supplies has been systematically over reported. *Marine Policy, 106*, 103547.

Ekstrom, J. A., Suatoni, L., Cooley, S. R., Pendleton, L. H., Waldbusser, G. G., Cinner, J. E., et al. (2015). Vulnerability and adaptation of US shellfisheries to ocean acidification. *Nature Climate Change, 5*, 207. https://doi.org/10.1038/nclimate2508

FAO. (2008). *An introduction to the basic concepts of food security*. FAO.

FAO. (2018). *The state of food security and nutrition in the world: Building climate resilience for food security and nutrition*. Food and Agriculture Organization of the United Nations.

FAO. (2021). *The state of food security and nutrition in the world. Transforming food systems for food security, improved nutrition and affordable healthy diets for all*. FAO. https://doi.org/10.4060/cb4474en

Farmery, A. K., Alexander, K., Anderson, K., Blanchard, J. L., Carter, C. G., Evans, K., Fischer, M., Fleming, A., Frusher, S., Fulton, E. A., & Haas, B. (2021). Food for all: Designing sustainable and secure future seafood systems. *Reviews in Fish Biology and Fisheries, 32*, 101. https://doi.org/10.1007/s11160-021-09663-x

Forster, J., & Radulovich, R. (2015). Seaweed and food security. In B. K. Tiwari & D. J. Troy (Eds.), *Seaweed sustainability* (pp. 289–313). Academic Press.

Free, C. M., Thorson, J. T., Pinsky, M. L., Oken, K. L., Wiedenmann, J., & Jensen, O. P. (2019). Impacts of historical warming on marine fisheries production. *Science, 363*, 979–983. https://doi.org/10.1126/science.aau1758

Froehlich, H. E., Gentry, R. R., Rust, M. B., Grimm, D., & Halpern, B. S. (2017). Public perceptions of aquaculture: Evaluating spatiotemporal patterns of sentiment around the world. *PLoS One, 12*, e0169281.

Froehlich, H. E., Gentry, R. R., & Halpern, B. S. (2018). Global change in marine aquaculture production potential under climate change. *Nature Ecology & Evolution, 2*, 1745–1750. https://doi.org/10.1038/s41559-018-0669-1

Funk, C., Davenport, F., Harrison, L., Magadzire, T., Galu, G., Artan, G. A., & Nsadisa, F. D. (2018). Anthropogenic enhancement of moderate-to-strong El ni˜no events likely contributed to drought and poor harvests in southern Africa during 2016. *Bulletin of the American Meteorological Society, 99*, S91–S96.

Garlock, T., Asche, F., Anderson, J., Ceballos-Concha, A., Love, D. C., Osmundsen, T. C., & Pincinato, R. B. (2022). Aquaculture: The missing contributor in the food security agenda. *Global Food Security, 32*, 100620. https://doi.org/10.1016/j.gfs.2022.100620

Garnett, T., Appleby, M. C., Balmford, A., Bateman, I. J., Benton, T. G., Bloomer, P., Burlingame, B., Dawkins, M., Dolan, L., Fraser, D., & Herrero, M. (2013). Sustainable intensification in agriculture: Premises and policies. *Science, 341*, 33–34. https://doi.org/10.1126/science.1234485

Gattuso, J. P., & Hansson, L. (2011). Ocean acidification: Background and history. In J. P. Gattuso & L. Hansson (Eds.), *Ocean acidification* (pp. 1–20). Oxford University Press.

Gentry, R. R., Froehlich, H. E., Grimm, D., Kareiva, P., Parke, M., Rust, M., Gaines, S. D., & Halpern, B. S. (2017). Mapping the global potential for marine aquaculture. *Nature Ecology & Evolution, 1*, 1–8.

Godde, C. M., Mason-D'Croz, D., Mayberry, D. E., Thornton, P. K., & Herrero, M. (2021). Impacts of climate change on the livestock food supply chain; a review of the evidence. *Glob Food Security, 28*, 100488.

Godfray, H. C. J., & Garnett, T. (2014). Food security and sustainable intensification. *Philosophical Transactions of the Royal Society B: Biological Sciences, 369*, 20120273. https://doi.org/10.1098/rstb.2012.0273

Golden, C. D., Allison, E. H., Cheung, W. W., Dey, M. M., Halpern, B. S., McCauley, D. J., Smith, M., Vaitla, B., Zeller, D., & Myers, S. S. (2016). Nutrition: Fall in fish catch threatens human health. *Nature, 534*, 317–320. https://doi.org/10.1038/534317a

Golden, C. D., Koehn, J. Z., Shepon, A., Passarelli, S., Free, C. M., Viana, D. F., Matthey, H., Eurich, J. G., Gephart, J. A., Fluet-Chouinard, E., & Nyboer, E. A. (2021). Aquatic foods to nourish nations. *Nature, 598*, 315–320. https://doi.org/10.1038/s41586-021-03917-1

Goodman, C., Davis, R., Azmi, K., Bell, J., Galland, G. R., Gilman, E., Haas, B., Hanich, Q., Lehodey, P., Manarangi-Trott, L., & Nicol, S. (2022). Enhancing cooperative responses by regional fisheries management organisations to climate-driven redistribution of tropical Pacific tuna stocks. *Frontiers in Marine Science, 9*, 1046018. https://doi.org/10.3389/fmars.2022.104618

Green, T. J., Siboni, N., King, W. L., Labbate, M., Seymour, J. R., & Raftos, D. (2019). Simulated marine heat wave alters abundance and structure of vibrio populations associated with the Pacific oyster resulting in a mass mortality event. *Microbial Ecology, 77*, 736–747. https://doi.org/10.1007/s00248-018-1242-9

Halwart, M., Funge-Smith, S., & Moehl, J. (2003). The role of aquaculture in rural development. In *Review of the state of world aquaculture* (pp. 47–58). FAO.

Hamid, N., Ma, Q., Boulom, S., Liu, T., Zheng, Z., Balbas, J., & Robertson, J. (2015). Seaweed minor constituents. In B. K. Tiwari & D. J. Troy (Eds.), *Seaweed sustainability* (pp. 193–242). Academic Press.

Hargreaves, P. I., Barcelos, C. A., da Costa, A. C. A., & Pereira, N., Jr. (2013). Production of ethanol 3G from *Kappaphycus alvarezii*: Evaluation of different process strategies. *Bioresource Technology, 134*, 257–263.

Hasegawa, T., Fujimori, S., Havlík, P., et al. (2018). Risk of increased food insecurity under stringent global climate change mitigation policy. *Nature Climate Change, 8*, 699–703. https://doi.org/10.1038/s41558-018-0230-x

Henriksson, P. J. G., Belton, B., Jahan, K. M., & Rico, A. (2018). Measuring the potential for sustainable intensification of aquaculture in Bangladesh using life cycle assessment. *Proceedings of the National Academy of Sciences of the United States of America, 115*, 2958–2963.

Henriksson, P. J. G., Banks, L. K., Suri, S. K., Pratiwi, T. Y., Fatan, N. A., & Troell, M. (2019). Indonesian aquaculture futures – Identifying interventions for reducing environmental impacts. *Environmental Research Letters, 14*, 124062.

Henriksson, P. J. G., Troell, M., Banks, L. K., Belton, B., Beveridge, M. C. M., Klinger, D. H., Pelletier, N., Phillips, M. J., & Tran, N. (2021). Interventions for improving the productivity

and environmental performance of global aquaculture for future food security. *One Earth, 4*, 1220–1232.
Hickey, G. M., & Unwin, N. (2020). Addressing the triple burden of malnutrition in the time of COVID-19 and climate change in Small Island developing states: What role for improved local food production? *Food Security, 12*, 831–835.
HLPE. (2020). *Food security and nutrition: Building a global narrative towards 2030*, pp. 112. Available at https://www.fao.org/3/ca9731en/ca9731en.pdf
Hoegh-Guldberg, O., et al. (2018). *Impacts of 1.5 C global warming on natural and human systems*. Global warming of 1.5 C. An IPCC Special Report.
Hurlbert, M., Krishnaswamy, J., Davin, E., Johnson, F. X., Mena, C. F., Morton, J., Myeong, S., Viner, D., Warner, K., Wreford, A., Zakieldeen, S., & Zommers, Z. (2019). Risk management and decision making in relation to sustainable development. In P. R. Shukla et al. (Eds.), *Climate change and land: An IPCC special report on climate change, desertification, land degradation, sustainable land management, food security, and greenhouse gas fluxes in terrestrial ecosystems* (pp. 673–800).
Integrated Food Security Phase Classification (IPC). (2022). The IPC-CH dashboard. Available at https://www.ipcinfo.org/. Accessed on 03 Oct 2024.
IPC Global Partners. (2021). Integrated food security phase classification technical manual version 3.1. *Evidence and standards for better food security and nutrition decisions*. Rome.
Jeeva, S., Marimuthu, J., Domettila, C., Anantham, B., & Mahesh, M. (2012). Preliminary phytochemical studies on some selected seaweeds from Gulf of Mannar, India. *Asian Pacific Journal of Tropical Biomedicine, 2*, S30–S33. https://doi.org/10.1016/S2221-1691(12)60125-7
Khan, S. K., Dutta, J., Ahmad, I., Rather, M. A., Badroo, I. A., & Taj, S. (2025). Fish protein: A nutritional solution for global food security. In J. K. Sundaray, M. A. Rather, I. Ahmad, & A. Amin (Eds.), *Food security, nutrition and sustainability through aquaculture technologies*. Springer. https://doi.org/10.1007/978-3-031-75830-0_27
Kinley, R. D., de Nys, R., Vucko, M. J., Machado, L., & Tomkins, N. W. (2016). The red macroalgae *Asparagopsis taxiformis* is a potent natural antimethanogenic that reduces methane production during in vitro fermentation with rumen fluid. *Animal Production Science, 56*, 282–289. https://doi.org/10.1071/AN15576
Koks, E. E., Rozenberg, J., Zorn, C., Tariverdi, M., Vousdoukas, M., Fraser, S. A., Hall, J. W., & Hallegatte, S. (2019). A global multi-hazard risk analysis of road and railway infrastructure assets. *Nature Communications, 10*, 2677. https://doi.org/10.1038/s41467-019-10442-3
Kuhla, K., Willner, S. N., Otto, C., Wenz, L., & Levermann, A. (2021). Future heat stress to reduce people's purchasing power. *PLoS One, 16*, e0251210.
Lam, V. W., Cheung, W. W., Reygondeau, G., & Sumaila, U. R. (2016). Projected change in global fisheries revenues under climate change. *Scientific Reports, 6*, 32607.
Langton, R. W., Augyte, S., Price, N., Forster, J., Noji, T., Grebe, G., Gelais, A. S., & Byron, C. J. (2019). *An ecosystem approach to the culture of seaweed*. US Department of Commerce, National Oceanic and Atmospheric Administration, National Marine Fisheries Service.
Leandro, A., Pacheco, D., Cotas, J., Marques, J. C., Pereira, L., & Gonçalves, A. M. (2020). Seaweed's bioactive candidate compounds to food industry and global food security. *Life, 10*, 140. https://doi.org/10.3390/life10080140
Lebel, L., Lebel, P., Soe, K. M., Phuong, N. T., Navy, H., Phousavanh, P., & Lebel, B. (2020). Aquaculture farmers' perceptions of climate-related risks in the Mekong region. *Animal Production Science, 20*, 95. https://doi.org/10.1007/s10113-020-01688-5
Lotze, H. K., Tittensor, D. P., Bryndum-Buchholz, A., Eddy, T. D., Cheung, W. W. L., Galbraith, E. D., Barange, M., Barrier, N., Bianchi, D., Blanchard, J. L., Bopp, L., Büchner, M., Bulman, C. M., Carozza, D. A., Christensen, V., Coll, M., Dunne, J. P., Fulton, E. A., Jennings, S., Jones, M. C., Mackinson, S., Maury, O., Niiranen, S., Oliveros-Ramos, R., Roy, T., Fernandes, J. A., Schewe, J., Shin, Y. J., Silva, T. A. M., Steenbeek, J., Stock, C. A., Verley, P., Volkholz, J., Walker, N. D., & Worm, B. (2019). Global ensemble projections reveal trophic amplification of

ocean biomass declines with climate change. *Proceedings of the National Academy of Sciences of the United States of America, 116*, 12907–12912. https://doi.org/10.1073/pnas.1900194116

Machado, L., Magnusson, M., Paul, N. A., Kinley, R., de Nys, R., & Tomkins, N. (2016). Dose-response effects of *Asparagopsis taxiformis* and *Oedogonium* sp. on in vitro fermentation and methane production. *Journal of Applied Phycology, 28*, 1443–1452. https://doi.org/10.1007/s10811-015-0639-9

Madramootoo, C. A., & Fyles, H. (2010). Irrigation in the context of today's global food crisis. *Irrigation and Drainage, 59*, 40–52. https://doi.org/10.1002/ird.555

Mantri, V. A., Eswaran, K., Shanmugam, M., Ganesan, M., Veeragurunathan, V., Thiruppathi, S., Reddy, C. R. K., & Seth, A. (2017). An appraisal on commercial farming of *Kappaphycus alvarezii* in India: Success in diversification of livelihood and prospects. *Journal of Applied Phycology, 28*, 335–357. https://doi.org/10.1007/s10811-016-0948-7

Mbow, C., Rosenzweig, C., Barioni, L. G., Benton, T. G., Herrero, M., Krishnapillai, M., Liwenga, E., Pradhan, P., Rivera-Ferre, M. G., Sapkota, T., Tubiello, F. N., & Xu, Y. (2019). Food security. In P. R. Shukla et al. (Eds.), *Climate change and land: An IPCC special report on climate change, desertification, land degradation, sustainable land management, food security, and greenhouse gas fluxes in terrestrial ecosystems*.

McClanahan, T., Allison, E. H., & Cinner, J. E. (2015). Managing fisheries for human and food security. *Fish and Fisheries, 16*, 78–103. https://doi.org/10.1111/faf.12045

Mirzabaev, A., Kerr, R. B., Hasegawa, T., Pradhan, P., Wreford, A., von der Pahlen, M. C., & Gurney-Smith, H. (2023). Severe climate change risks to food security and nutrition. *Climate Risk Management, 39*, 100473.

Mohamad, N., Amal, M. N. A., Yasin, I. S. M., Saad, M. Z., Nasruddin, N. S., Al-saari, N., & Sawabe, T. (2019). Vibriosis in cultured marine fishes: A review. *Aquaculture, 512*, 734289. https://doi.org/10.1016/j.aquaculture.2019.734289

Myers, S. S., Smith, M. R., Guth, S., Golden, C. D., Vaitla, B., Mueller, N. D., Dangour, A. D., & Huybers, P. (2017). Climate change and global food systems: Potential impacts on food security and undernutrition. *Annual Review of Public Health, 38*, 259. https://doi.org/10.1146/annurev-publhealth-031816-044356

Namudu, M. T., & Pickering, T. D. (2006). Rapid survey technique using socio-economic indicators to assess the suitability of Pacific Island rural communities for *Kappaphycus* seaweed farming development. In R. Anderson, J. Brodie, E. Onsøyen, & A. T. Critchley (Eds.), *Eighteenth international seaweed symposium* (pp. 15–23). Springer. https://doi.org/10.1007/978-1-4020-5670-3_3

Naylor, R. L., Hardy, R. W., Buschmann, A. H., Bush, S. R., Cao, L., Klinger, D. H., Little, D. C., Lubchenco, J., Shumway, S. E., & Troell, M. (2021). A 20-year retrospective review of global aquaculture. *Nature, 591*, 551–563.

O'Neill, B. C., van Aalst, M., Zaiton Ibrahim, Z., Berrang-Ford, L., Bhadwal, S., Buhaug, H., Diaz, D., Frieler, K., Garschagen, M., Magnan, A. K., Midgley, G., Mirzabaev, A., Thomas, A., & Warren, R. (2022). Key risks across sectors and regions. In H. O. Portner et al. (Eds.), *Climate change 2022: Impacts, adaptation, and vulnerability. contribution of working group II to the sixth assessment report of the intergovernmental panel on climate change*. Cambridge University Press.

Oremus, K. L., Bone, J., Costello, C., García Molinos, J., Lee, A., Mangin, T., & Salzman, J. (2020). Governance challenges for tropical nations losing fish species due to climate change. *Nature Sustainability, 3*, 277–280. https://doi.org/10.1038/s41893-020-0476-y

Osendarp, S., Verburg, G., Bhutta, Z., Black, R. E., de Pee, S., Fabrizio, C., Headey, D., Heidenkamp, R., Laborde, D., & Ruel, M. T. (2022). Act now before Ukraine war plunges millions into malnutrition. *Nature, 604*, 620–624.

Oyinlola, M. A., Reygondeau, G., Wabnitz, C. C. C., Troell, M., & Cheung, W. W. L. (2018). Global estimation of areas with suitable environmental conditions for mariculture species. *PLoS One, 13*, e0191086.

Oyinlola, M. A., Reygondeau, G., Wabnitz, C. C. C., & Cheung, W. W. L. (2020). Projecting global mariculture diversity under climate change. *Global Change Biology, 26*, 2134. https://doi.org/10.1111/gcb.14974

Pentz, B., Klenk, N., Ogle, S., & Fisher, J. A. (2018). Can regional fisheries management organizations (RFMOs) manage resources effectively during climate change? *Marine Policy, 92*, 13–20. https://doi.org/10.1016/j.marpol.2018.01.011

Phalkey, R. K., Aranda-Jan, C., Marx, S., et al. (2015). Systematic review of current efforts to quantify the impacts of climate change on undernutrition. *Proceedings of the National Academy of Sciences of the United States of America, 112*, E4522–E4529.

Pinsky, M. L., Selden, R. L., & Kitchel, Z. J. (2020). Climate-driven shifts in marine species ranges: Scaling from organisms to communities. *Annual Review of Marine Science, 12*, 153–179.

Poore, J., & Nemecek, T. (2018). Reducing food's environmental impacts through producers and consumers. *Science, 992*, 987–992.

Pradhan, P., Seydewitz, T., Zhou, B., Lüdeke, M. K., & Kropp, J. P. (2022). Climate extremes are becoming more frequent, co-occurring, and persistent in Europe. *Anthropocene Science, 1*, 264–277.

Reshi, Q. M., Ahmad, I., Nabi, N., Ahmed, I., & Sidiq, M. (2025). Global food insecurity: Role of fishes to compensate food insecurity and nutritional demand. In J. K. Sundaray, M. A. Rather, I. Ahmad, & A. Amin (Eds.), *Food security, nutrition and sustainability through aquaculture technologies*. Springer. https://doi.org/10.1007/978-3-031-75830-0_5

Rheuban, J. E., Doney, S. C., Cooley, S. R., & Hart, D. R. (2018). Projected impacts of future climate change, ocean acidification, and management on the US Atlantic sea scallop (*Placopecten magellanicus*) fishery. *PLoS One, 13*, e0203536. https://doi.org/10.1371/journal.pone.0203536

Rojas-Downing, M. M., Nejadhashemi, A. P., Harrigan, T., & Woznicki, S. A. (2017). Climate change and livestock: Impacts, adaptation, and mitigation. *Climate Risk Management, 16*, 145–163. https://doi.org/10.1016/j.crm.2017.02.001

Roque, B. M., Salwen, J. K., Kinley, R., & Kebreab, E. (2019). Inclusion of *Asparagopsis armata* in lactating dairy cows' diet reduces enteric methane emission by over 50 percent. *Journal of Cleaner Production, 234*, 132–138. https://doi.org/10.1016/j.jclepro.2019.06.193

Santos, J. A., McKenzie, B., Trieu, K., Farnbach, S., Johnson, C., Schultz, J., Thow, A. M., Snowdon, W., Bell, C., & Webster, J. (2019). Contribution of fat, sugar and salt to diets in the Pacific Islands: A systematic review. *Public Health Nutrition., 22*, 1858–1871. https://doi.org/10.1017/S1368980018003609

Siddiqui, A. A. M., Kashem, M. A., & Islam, M. A. (2019). Commercially important seaweed cultivation and its potentials for the coastal areas of cox's bazar, Bangladesh. *International Journal of Fisheries and Aquatic Studies, 7*, 463–670.

Sievanen, L., Crawford, B., Pollnac, R., & Lowe, C. (2005). Weeding through assumptions of livelihood approaches in ICM: Seaweed farming in The Philippines and Indonesia. *Ocean and Coastal Management, 48*, 297–313. https://doi.org/10.1016/j.ocecoaman.2005.04.015

Simpson, N. P., Mach, K. J., Constable, A., Hess, J., Hogarth, R., Howden, M., Lawrence, J., Lempert, R. J., Muccione, V., Mackey, B., New, M. G., O'Neill, B., Otto, F., et al. (2021). A framework for complex climate change risk assessment. *One Earth, 4*, 489–501. https://doi.org/10.1016/j.oneear.2021.03.005

Smale, D. A., Wernberg, T., Oliver, E. C. J., Thomsen, M., Harvey, B. P., Straub, S. C., & Moore, P. J. (2019). Marine heatwaves threaten global biodiversity and the provision of ecosystem services. *Nature Climate Change, 9*, 306–312. https://doi.org/10.1038/s41558-019-0412-1

Smith, B. E. (2002). Nitrogenase reveals its inner secrets. *Science, 297*, 1654–1655.

Soffiantini, G. (2020). Food insecurity and political instability during the Arab spring. *Glob Food Security, 26*, 100400. https://doi.org/10.1016/j.gfs.2020.100400

Springmann, M., Mason-D'Croz, D., Robinson, S., Garnett, T., Godfray, H. C. J., Gollin, D., Rayner, M., Ballon, P., & Scarborough, P. (2016). Global and regional health effects of future food production under climate change: A modelling study. *Lancet, 387*, 1937–1946. https://doi.org/10.1016/S0140-6736(15)01156-3

Stentiford, G. D., Bateman, I. J., Hinchliffe, S. J., Bass, D., Hartnell, R., Santos, E. M., Devlin, M. J., Feist, S. W., Taylor, N. G. H. H., Verner-Jeffreys, D. W., et al. (2020). Sustainable aquaculture through the one health lens. *Nature Food, 1*, 468–474.

Stewart-Sinclair, P. J., Last, K. S., Payne, B. L., & Wilding, T. A. (2020). A global assessment of the vulnerability of shellfish aquaculture to climate change and ocean acidification. *Ecology and Evolution, 10*, 3518–3534.

Sulser, T. B., Beach, R. H., Wiebe, K. D., Dunston, S., & Fukagawa, N. K. (2021). Disability-adjusted life years due to chronic and hidden hunger under food system evolution with climate change and adaptation to 2050. *The American Journal of Clinical Nutrition, 114*, 550–563. https://doi.org/10.1093/ajcn/nqab101

Sultana, F., Wahab, M. A., Nahiduzzaman, M., Mohiuddin, M., Iqbal, M. Z., Shakil, A., Mamun, A. A., Khan, M. S., Wong, L., & Asaduzzaman, M. (2023). Seaweed farming for food and nutritional security, climate change mitigation and adaptation, and women empowerment: A review. *Aqua Fish, 8*, 463–480.

Teneva, L., Free, C. M., Hume, A., Agostini, V. N., Klein, C. J., Watson, R. A., & Gaines, S. D. (2023). Small Island nations can achieve food security benefits through climate-adaptive blue food governance by 2050. *Marine Policy, 151*, 105577.

Thiault, L., Mora, C., Cinner, J. E., Cheung, W. W., Graham, N. A., Januchowski-Hartley, F. A., Mouillot, D., Sumaila, U. R., & Claudet, J. (2019). Escaping the perfect storm of simultaneous climate change impacts on agriculture and marine fisheries. *Science Advances, 5*, eaaw9976. https://doi.org/10.1126/sciadv.aaw9976

Thornton, P., Nelson, G., Mayberry, D., & Herrero, M. (2021). Increases in extreme heat stress in domesticated livestock species during the twenty-first century. *Global Change Biology, 27*, 5762–5772. https://doi.org/10.1111/gcb.15825

Tigchelaar, M., Cheung, W. W., Mohammed, E. Y., Phillips, M. J., Payne, H. J., Selig, E. R., Wabnitz, C. C., Oyinlola, M. A., Frölicher, T. L., Gephart, J. A., & Golden, C. D. (2021). Compound climate risks threaten aquatic food system benefits. *Nature Food, 2*, 673–682. https://doi.org/10.1038/s43016-021-00368-9

Tisdell, C. A., Hishamunda, N., Van Anrooy, R., Pongthanapanich, T., & Upare, M. A. (2010). Investment, insurance and risk management for aquaculture development. In *Proceedings of global conference on aquaculture* (pp. 303–335).

Troell, M., Naylor, R. L., Metian, M., Beveridge, M., Tyedmers, P. H., Folke, C., Arrow, K. J., Barrett, S., et al. (2014). Does aquaculture add resilience to the global food system? *Proceedings of the National Academy of Sciences of the United States of America, 111*, 13257–13263.

Valin, H., Sands, R. D., Van der Mensbrugghe, D., Nelson, G. C., Ahammad, H., Blanc, E., Bodirsky, B., Fujimori, S., Hasegawa, T., Havlik, P., & Heyhoe, E. (2014). The future of food demand: Understanding differences in global economic models. *Agricultural Economics, 45*, 51–67. https://doi.org/10.1111/agec.12089

Verschuur, J., Li, S., Wolski, P., & Otto, F. E. L. (2021). Climate change as a driver of food insecurity in the 2007 Lesotho-South Africa drought. *Scientific Reports, 11*, 1–9. https://doi.org/10.1038/s41598-021-83375-x

Watson, R. A., Nowara, G. B., Hartmann, K., Green, B. S., Tracey, S. R., & Carter, C. G. (2015). Marine foods sourced from farther as their use of global ocean primary production increases. *Nature Communications, 6*, 7365. https://doi.org/10.1038/ncomms8365

Weatherdon, L. V., Magnan, A. K., Rogers, A. D., Sumaila, U. R., & Cheung, W. W. L. (2016). Observed and projected impacts of climate change on marine fisheries, aquaculture, coastal tourism, and human health: An update. *Frontiers in Marine Science, 3*, 48. https://doi.org/10.3339/fmars.2076.00048

Webster, M. S., Colton, M. A., Darling, E. S., Armstrong, J., Pinsky, M. L., Knowlton, N., & Schindler, D. E. (2017). Who should pick the winners of climate change? *Trends in Ecology & Evolution, 32*, 167–173. https://doi.org/10.1016/j.tree.2016.12.007

Willett, W., Rockström, J., Loken, B., Springmann, M., Lang, T., Vermeulen, S., Garnett, T., Tilman, D., DeClerck, F., Wood, A., & Jonell, M. (2019). Food in the Anthropocene: The EAT–lancet commission on healthy diets from sustainable food systems. *Lancet, 393*, 447–492.

Worm, B., Hilborn, R., Baum, J. K., Branch, T. A., Collie, J. S., Costello, C., Fogarty, M. J., Fulton, E. A., Hutchings, J. A., Jennings, S., & Jensen, O. P. (2009). Rebuilding global fisheries. *Science, 325*, 578–585. https://doi.org/10.1126/science.1173146

Yarish, C., et al. (2016). Seaweed aquaculture for food security, income generation and environmental health in tropical developing countries. *Worldbank Group*, Article 10986/24919. https://openknowledge.worldbank.org/handle/10986/24919. License: CC BY 3.0 IGO.

Zacharia, P. U., Kaladharan, P., & Rojith, G. (2015). Seaweed farming as a climate resilient strategy for Indian coastal waters. In *The international conference on integrating climate, crop, ecology–the emerging areas of agriculture, horticulture, livestock, fishery, forestry, biodiversity and policy issues* (pp. 59–62).

Zhang, T., van der Wiel, K., Wei, T., Screen, J., Yue, X., Zheng, B., & Yang, X. (2022). Increased wheat price spikes and larger economic inequality with 2° C global warming. *One Earth, 5*, 907–916.

Zhu, C., Kobayashi, K., Loladze, I., Zhu, J., Jiang, Q., Xu, X., Liu, G., Seneweera, S., Ebi, K. L., Drewnowski, A., Fukagawa, N. K., & Ziska, L. H. (2018). Carbon dioxide (CO_2) levels this century will alter the protein, micronutrients, and vitamin content of rice grains with potential health consequences for the poorest rice-dependent countries. *Science Advances, 4*, eaaq1012. https://doi.org/10.1126/sciadv.aaq1012

Ziegler, F., Winther, U., Hognes, E. S., Emanuelsson, A., Sund, V., & Ellingsen, H. (2013). The carbon footprint of Norwegian seafood products on the global seafood market. *Journal of Industrial Ecology, 17*, 103–116.

Chapter 9
Emerging Technologies and Innovations in Aquaculture Nutrition

Jebarson Solomon, Ganesh Kumar, Prachi Asgolkar, B. Veeramaruthu, and Tincy Varghese

1 Introduction

Aquaculture has grown exponentially over the past few decades, outpacing capture fisheries to become a cornerstone of global food security (FAO, 2024; Ahmed et al., 2024). This rapid expansion is largely attributed to the adoption of feed-based aquaculture systems, which account for over 70% of total aquaculture production (Tacon & Metian, 2015). Unlike natural or extensive systems, feed-based aquaculture provides fish with their required nutrients directly within controlled environments, enabling higher growth rates in shorter timeframes while minimizing spatial requirements (Li et al., 2022). However, despite these advantages, feed represents the single largest operating cost in aquaculture, often comprising 50–70% of total production expenses (Tacon & Metian, 2015). For feed manufacturers, optimizing feed formulations without compromising quality is a constant challenge, while farmers must maximize feed utilization efficiency to achieve profitability. Balancing these economic constraints with the growing demand for sustainable aquaculture has spurred the need for innovative approaches to reduce feed costs while maintaining or improving fish health and growth performance.

This book chapter explores the emerging technologies and innovations in aquaculture nutrition aimed at addressing these challenges. It begins by examining advancements in feed ingredients, which form the foundation of feed production. It then delves into the methods for assessing the quality of these ingredients, which are

J. Solomon · P. Asgolkar · T. Varghese (✉)
ICAR – Central Institute of Fisheries Education, Mumbai, Maharashtra, India
e-mail: tincy@cife.edu.in

G. Kumar
Rani Lakshmi Bai Central Agricultural University, Jhansi, Uttar Pradesh, India

B. Veeramaruthu
Faculty of Fisheries, SKUAST-K, Ganderbal, Jammu and Kashmir, India

I. Ahmed, I. Ahmad (eds.), *Aquaculture: Enhancing Food Security and Nutrition*, https://doi.org/10.1007/978-3-031-92858-1_9

crucial for ensuring their optimal selection. Building on this, the chapter discusses innovations in feed formulation that utilize the selected ingredients and additives to create nutritionally balanced and cost-effective feeds. It further addresses the recently available software for the selection of the most appropriate feed for different aquaculture systems and species to maximize efficiency. Finally, the chapter addresses the advancements in feeding processes through the use of smart feeding technologies, which are transforming the way feed is utilized on farms. By following this progression, the chapter provides a comprehensive understanding of how innovations in aquaculture nutrition can reduce feed costs while improving sustainability and productivity (Fig. 9.1).

2 Advancements in Feed Ingredients

As the demand for sustainable and cost-effective feeds continues to rise, significant efforts have been directed toward improving the nutritional quality and functionality of feed components. Innovations in ingredient processing, genetic improvements, and the use of alternative protein sources have enabled the development of feeds that are more digestible, nutritionally balanced, cost-effective, and tailored to the specific needs of aquaculture species.

2.1 Enhancing Digestibility of Fibrous Ingredients

One of the key challenges in aquaculture feed formulation is the low digestibility of fibrous ingredients. Fibers in these ingredients are often rich in lignin, a complex polymer that aquatic species struggle to break down due to the lack of appropriate enzymes in their digestive systems (Denstadli et al., 2011). This reduced digestibility limits the accessibility of essential nutrients, resulting in inefficient nutrient absorption and suboptimal growth performance in fish. To address this issue, researchers have explored various strategies to enhance the digestibility of fibrous ingredients. One promising approach involves the use of microorganisms, such as anaerobic bacteria and fungi, to degrade lignin and improve nutrient availability. For instance, white rot fungi, including *Pleurotus ostreatus*, have demonstrated significant potential in breaking down lignin, thereby increasing the protein content and digestibility of fibrous feed materials (Olagunju et al., 2023). Furthermore, certain fungal strains, such as *Aspergillus niger* and *Fusarium moniliforme*, have been effectively utilized to facilitate lignocellulosic hydrolysis (Sarsaiya et al., 2019). This process transforms complex fibers into simpler, more digestible forms, making them suitable for inclusion in aquaculture diets. These microorganisms not only degrade lignin but also convert substrate materials into high-quality biomass, enriching the nutritional profile of fibrous ingredients (Sarsaiya et al., 2019).

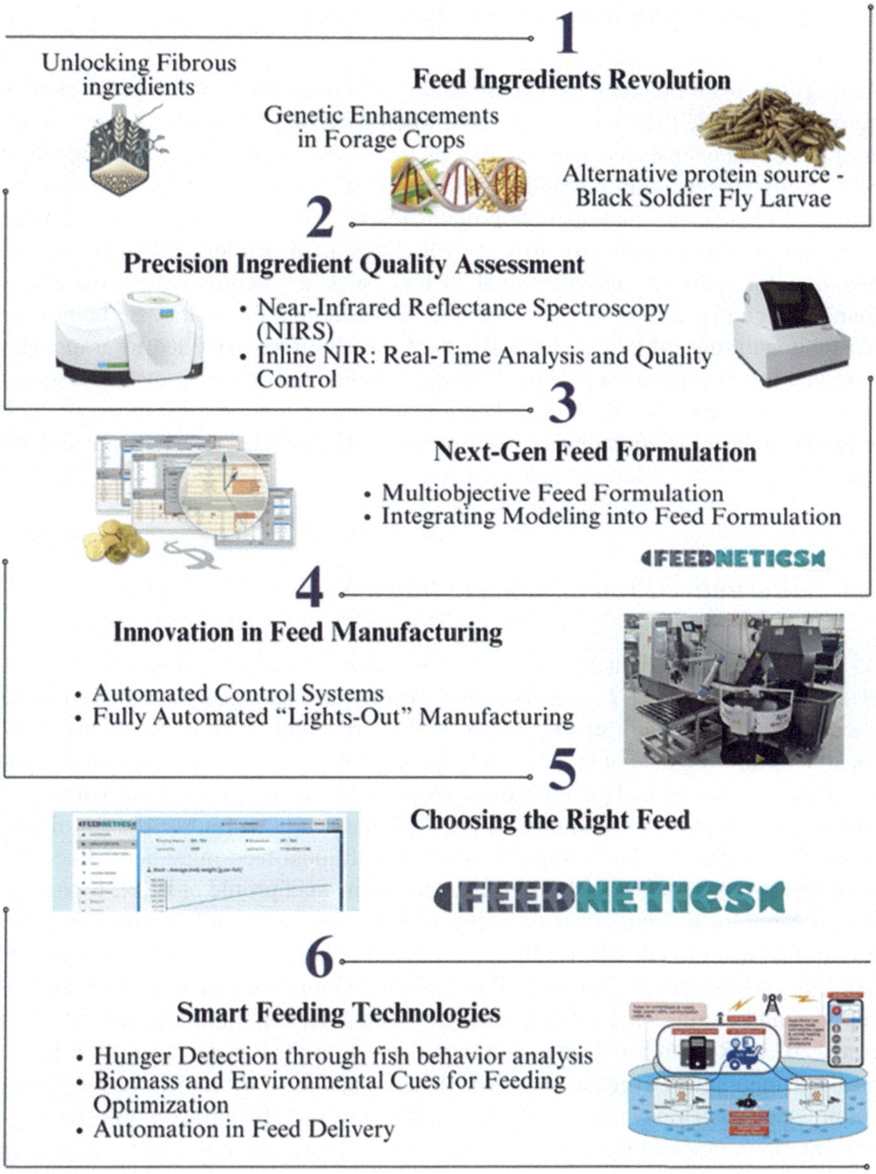

Fig. 9.1 Pioneering advancements in aquaculture nutrition

In addition to microbial fermentation, techniques like silage fermentation using enzyme additives (e.g., amylase and cellulase) have shown promise in improving the utilization of plant-based feeds (Asmare, 2014). These enzymes release sugars during the fermentation process, enhancing lactic acid production, which aids in preserving and improving the digestibility of fibrous ingredients (Sharp et al., 1992).

2.2 Genetic Enhancements in Forage Crops

Advances in plant breeding have allowed for the development of forage crops with traits that enhance their nutritional value when used in aquaculture feeds (Cottrell et al., 2020). For instance, genetically enhanced forage crops can be designed to contain higher levels of essential amino acids, which are typically low in some legumes. These crops may include proteins resistant to degradation, which ensure better nutrient availability for fish growth (Capstaff & Miller, 2018). Moreover, breeding for reduced anti-nutritional factors, such as tannins or glucosinolates found in legumes and oilseeds, has allowed for feed ingredients that are both more palatable and digestible (Chand et al., 2022). Such efforts have led to crops with modified lignin content, improving both digestibility and nutrient uptake. Research continues into creating feed ingredients that balance high-quality nutrients with reduced indigestible components, offering more efficient energy use in aquaculture species.

2.3 Alternative Protein Source: Black Soldier Fly Larvae

As the demand for sustainable and cost-effective protein sources in aquafeed grows, black soldier fly larvae (*Hermetia illucens*, BSFL) have emerged as a transformative alternative (Tippayadara et al., 2021). BSFL are highly efficient bioconverters, capable of thriving on a range of organic waste materials, such as food scraps, agricultural by-products, and plant residues. In the process, they produce nutrient-dense biomass that is rich in crude protein (32–53%), essential amino acids, and lipids, positioning them as a viable replacement for traditional feed ingredients like fishmeal and soybean meal (Lu et al., 2022). The amino acid profile of BSFL is particularly notable for its high levels of lysine, leucine, and valine, all of which are vital for optimal fish growth and health. In addition, the larvae are abundant in minerals like calcium and phosphorus, as well as bioactive compounds such as antimicrobial peptides and chitin, which may contribute to enhanced immunity in fish (Zulkifli et al., 2022). Research indicates that replacing up to 50% of fishmeal with BSFL meal can maintain or even enhance growth rates and feed efficiency in various aquaculture species (Tippayadara et al., 2021). Moreover, BSFL production offers significant environmental benefits, including reducing organic waste and minimizing greenhouse gas emissions compared to traditional protein sources (Lu et al., 2022). The scalability of BSFL production has been greatly improved by advances in insect farming technologies, which allow for large-scale and cost-effective cultivation. However, challenges such as consumer acceptance, regulatory approval, and the standardization of processing methods for consistent nutritional quality still remain (Siva Raman et al., 2022). Despite these obstacles, BSFL stands out as an innovative and sustainable solution, paving the way for the future of aquaculture nutrition while addressing global concerns about resource efficiency and environmental sustainability.

3 Advancements in Feed Ingredient Quality Assessment

Ensuring high-quality feed ingredients is essential for achieving consistent nutrient profiles and optimal health outcomes in aquaculture. Near-infrared reflectance spectroscopy (NIRS) has emerged as a pivotal tool for precise, rapid analysis of feed ingredients, allowing producers to verify and adjust nutrient levels effectively (Tahir et al., 2012). Here is how NIRS and its real-time variant, Inline NIR, have revolutionized quality checks in the feed manufacturing process.

3.1 Near-Infrared Reflectance Spectroscopy (NIRS)

NIRS uses near-infrared light, with wavelengths from 780 to 2500 nm, to analyze the chemical composition of feed ingredients based on their molecular structure (Tahir et al., 2012). This technique relies on detecting specific vibrations in hydrogen bonds, common in organic compounds. As different feed components absorb and reflect this light uniquely, NIRS can accurately identify their composition and predict nutrient levels (Berzaghi & Riovanto, 2009). Initially introduced in the 1960s to measure moisture in grains, NIRS now serves a wider range of applications across feed industries due to its ability to assess critical components like protein, moisture, fat, and fiber content (You et al., 2024).

Compared to traditional wet chemistry, NIRS is a nondestructive, rapid, and cost-effective solution that allows large-scale sampling without altering the feed samples. By enabling frequent quality checks with minimal time and expense, NIRS supports the precise formulation of feed to meet specific nutritional requirements, thus improving feed quality and consistency (Berzaghi & Riovanto, 2009).

3.2 Inline NIR: Real-Time Analysis and Quality Control

While NIRS provides accurate snapshots of feed quality, inline NIR enhances this process by allowing real-time analysis of feed ingredients as they move through the production line (You et al., 2024). Inline NIR systems continuously collect spectral data from samples, automatically assessing nutrient variations within each batch. This capability enables immediate adjustments, ensuring that the final feed product remains homogeneous and meets target specifications, even with natural fluctuations in raw materials.

In practice, inline NIR can be set up at key points in the production line to monitor ingredient consistency from the intake phase to the finished product (van der Poel et al., 2020). For instance, it can detect nutrient variability as ingredients are weighed and mixed, allowing adjustments to be made before processing continues. By integrating inline NIR technology, feed manufacturers can improve quality control while reducing the risk of nutrient imbalances that could compromise animal growth and health (Avila et al., 2020).

3.3 Challenges and Considerations for Inline NIR Implementation

Despite its advantages, implementing inline NIR comes with certain challenges, particularly in feed mills that were not originally designed to accommodate such equipment. Retrofitting older facilities may be costly and logistically complex (You et al., 2024). Additionally, the accuracy of inline NIR measurements can be affected by environmental factors, such as ambient temperature, and by probe cleanliness. Advances in sensor technology are helping to mitigate these issues, with newer systems offering greater sensitivity and resilience to environmental variations (Dixit et al., 2017).

4 Advancements in Feed Formulation

Feed formulation is a critical process in aquaculture, as it ensures that fish receive a balanced diet that meets their specific nutritional needs at each stage of growth while also optimizing cost efficiency and sustainability (Lall & Dumas, 2022). This process involves determining the right quantities of various feed ingredients to provide essential nutrients like proteins, fats, vitamins, and minerals. Key considerations in feed formulation include understanding the nutrient requirements of fish species, assessing the nutritional composition of available ingredients, evaluating ingredient cost, and ensuring ingredient availability and sustainability (Glencross et al., 2007).

4.1 Least-Cost Formulation

One of the most widely used methods in feed formulation is least-cost formulation (LCF), a strategy that combines economic and nutritional goals by matching the nutrient needs of fish with the nutritional profile of ingredients at the lowest cost (Pathumnakul et al., 2009). LCF relies on linear programming, a mathematical optimization technique that uses equations and constraints to create cost-effective feed formulations while meeting nutritional standards. Constraints might include nutritional requirements specific to fish species and stages of growth, the availability of ingredients, and sustainable sourcing practices (Ittiphalin et al., 2015).

With advancements in computing, LCF has become more accessible through specialized software, which can handle the complex calculations involved in modern feed formulation (Pathumnakul et al., 2009). These programs utilize nutrient libraries and enable rapid recalculations based on ingredient costs, nutrient profiles, and fish species requirements, streamlining the formulation process.

4.2 Multiobjective Feed Formulation in Aquaculture

Multiobjective feed formulation (MOF) is an advanced approach designed to optimize aquafeed by balancing economic viability, nutritional adequacy, and environmental sustainability (Uyeh et al., 2019). Unlike traditional least-cost feed formulations (LCF), which focus solely on minimizing costs, MOF integrates life cycle assessment (LCA) to evaluate and reduce environmental impacts, including climate change (CC), eutrophication potential, acidification, land occupation, and water dependence (Garcia-Launay et al., 2018). This makes MOF a critical tool in addressing the ecological challenges of aquaculture feed production, which contributes up to 95% of the environmental footprint of farmed aquatic species (Wilfart et al., 2023). A key advantage of MOF is its ability to significantly reduce reliance on high-impact ingredients such as fishmeal and fish oil, which are traditionally central to aquafeeds. By substituting these with more sustainable alternatives, including processed animal by-products (e.g., hydrolyzed feather meal and blood meal), plant-based proteins (e.g., pea protein concentrate), and oils (e.g., soy lecithin and poultry oil), MOF demonstrates substantial environmental benefits. These substitutions have been shown to lower greenhouse gas emissions, land use, and eutrophication potential, delivering environmental gains at both feed production and farm operation levels (Garcia-Launay et al., 2018). Recent research on rainbow trout using a multiobjective-formulated eco-diet (ECO-diet) revealed significant reductions in environmental impacts. Compared to a commercial diet (C-diet), the ECO-diet achieved a 46% decrease in CC impact, a 44% reduction in net primary production use (NPPU), and a 57% drop in nonrenewable energy demand (NRE). Additionally, the ECO-diet maintained similar digestibility for protein and energy while enhancing starch digestibility (Wilfart et al., 2023). Despite a marginal reduction in feed intake and long-term growth, the overall feed conversion ratio (FCR) remained unaffected, underscoring its potential for sustainable aquaculture practices (Wilfart et al., 2023).

The incorporation of MOF also highlights the need to avoid "pollution swapping," where reducing one environmental impact inadvertently increases another (Wilfart et al., 2023). For example, substituting fish oil with rapeseed oil can lower climate impacts but might elevate eutrophication and acidification due to land and water use. By simultaneously considering multiple objectives, MOF mitigates such trade-offs, ensuring a holistic approach to sustainability (Boissy et al., 2011). Future advancements in MOF for aquaculture include developing ingredient databases specific to aquaculture species, integrating precision feeding technologies, and adapting formulations to local resource availability and farming systems (Garcia-Launay et al., 2018). As the aquaculture industry strives to meet the growing global demand for seafood, MOF offers a scientifically robust pathway to reduce environmental impacts while supporting economic and production goals. This approach represents a vital step toward achieving a sustainable and resilient aquaculture sector.

4.3 Integrating Modeling into Feed Formulation

Mathematical and computer models are increasingly integrated into aquaculture to simulate fish responses to diet, environmental factors, and life stages, facilitating the development of cost-effective, nutritionally adequate feeds (Leishman et al., 2023). These models enable producers to optimize feed formulations by accounting for factors such as nutrient requirements, feed costs, and growth performance. By integrating predictive models with least-cost formulation (LCF), aquaculture feed mills can create customized feeds that adapt to production goals and market demands, promoting sustainability and economic efficiency (Chrystal, 2022).

Recent research highlights advancements in mathematical modeling for aquafeed formulation. For instance, Din et al. (2024) developed models tailored for red hybrid tilapia across different life stages, employing AIMMS software to design feeds that meet essential nutrient requirements (protein, lipid, carbohydrate, and fiber) while reducing costs by up to 38% compared to commercial pellets. Similarly, Aizam et al. (2018) optimized feed formulations for *Mystus nemurus* catfish using mathematical models to balance nutrient needs with production costs, yielding cost-effective combinations with superior nutrient profiles. These approaches offer significant benefits for small-scale farmers by lowering feed expenses without compromising fish health or growth performance.

Feednetics (a tool developed by SPAROS Lda.) further exemplifies the role of advanced tools in feed formulation. This decision-support software is designed to assist feed formulators in developing new feed formulations based on fish growth, body composition, feed cost, and nutrient requirements. It integrates data on diet composition and its impact on fish growth, allowing for precise adjustments to feed ingredients to optimize both nutritional content and cost efficiency (SPAROS, n.d.).

By leveraging mathematical models and tools like Feednetics, aquaculture feed producers can refine diets continually, achieving a balance between economic viability, environmental sustainability, and nutritional adequacy.

5 Advancements in Feed Manufacturing

The field of feed manufacturing for aquaculture has evolved considerably, with technological advancements making it possible to produce high-quality, cost-effective feed at an industrial scale. These innovations enhance production efficiency, reduce human error, improve biosecurity, and contribute to sustainability.

5.1 Automated Control Systems

One of the most significant advancements in feed manufacturing is the adoption of automated control systems, which streamline the complex manufacturing process by automating various stages of feed production (Coffey et al., 2016). Since their introduction in the 1970s, these systems have transformed how feed is produced, utilizing hardware and software solutions like programmable logic controllers (PLCs), distributed control systems (DCS), and computer numerical control (CNC) systems. Automated controls help manage and optimize processes, including ingredient receiving, grinding, mixing, pelleting, extrusion, and logistics (Zhang & Pierce, 2013). By automating repetitive and well-defined tasks, these systems minimize human intervention, thereby reducing labor needs, human error, and variability in feed quality. As automation technology advances, feed mills can maintain consistent product standards, meet stringent quality requirements, and enhance the overall efficiency of feed manufacturing.

5.2 Toward Fully Automated "Lights-Out" Manufacturing

"Lights-out manufacturing" represents the future of feed production facilities—a vision of fully autonomous operations where human involvement is minimal (Olivier, 2017). In this approach, production facilities could theoretically operate without lights, as no human presence would be required, aside from occasional supervision and maintenance. This high level of automation improves efficiency, reduces operating costs, and enhances safety by minimizing potential human-related risks (Schafer, 2022).

While lights-out manufacturing is an aspirational goal for most feed mills, incremental steps are already being taken toward this model. For instance, micro-scale bins for precise ingredient measurement and inline sampling systems for quality control are increasingly integrated into feed production (Lee, 2018). Coupling these technologies with machine learning algorithms enables real-time data analysis, which can predict and adjust feed formulations or processing conditions based on live feedback. For example, in grain handling, lights-out elevators use automated sampling and testing to direct ingredients to the appropriate storage based on quality results, improving both efficiency and biosecurity (Schafer, 2022).

The vision of a fully autonomous feed mill aligns with the need for reliable, high-volume production in aquaculture. As these technologies continue to advance, they could lead to the creation of highly efficient, self-sufficient feed mills that not only reduce operational costs but also enhance the quality and consistency of feed, supporting sustainable aquaculture practices worldwide.

6 Innovations in the Selection of Appropriate Feed

Advancements in software technology are revolutionizing feed selection and optimization for aquaculture, with tools like Feednetics by SPAROS emerging as game-changers for efficient feed management. Feednetics is smart, decision-support software designed to aid fish farmers, feed producers, and researchers in formulating and evaluating aquafeeds to maximize fish production under varied nutritional and environmental conditions. Using a sophisticated mathematical model, the software simulates fish growth, feed conversion, and waste output, helping users to project outcomes and make informed choices quickly (SPAROS, n.d.). One of the major benefits of this software is its ability to test different feeding scenarios in a virtual environment. Fish farmers can compare the impact of various feed compositions on fish growth and feed conversion ratio (FCR), allowing them to identify the most cost-effective strategies while minimizing environmental impact, such as nitrogen and phosphorus waste. This type of predictive modeling is especially useful in optimizing feeding regimens, with potential savings on feed costs and improvements in growth rates, as demonstrated in comparative trials. The model is calibrated for several important aquaculture species, including gilthead seabream, European seabass, rainbow trout, Atlantic salmon, and Nile tilapia. Feednetics enables users to input specific parameters such as feed composition, amino acid profiles, feeding schedules, and temperature variations, providing precise, species-specific predictions. These features allow for detailed analysis of the long-term effects of different feeds and the conditions under which they are most effective. Accessible as a web application or through a report-based service, Feednetics offers flexibility for users to either operate the tool independently or receive expert analysis from SPAROS. This versatile approach enables fish farmers, aquafeed companies, and academic researchers to refine their feeding strategies, enhance production efficiency, and contribute to sustainable aquaculture practices (SPAROS, n.d.). Through predictive tools like Feednetics, the aquaculture industry is moving toward a data-driven approach that promotes optimal growth performance, resource efficiency, and reduced environmental impact.

7 Smart Feeding Technologies

Aquaculture operations face significant challenges in feed management due to the high cost of feed and the need for labor efficiency (Edwards, 2015). Smart feeding technologies present a modern solution that automates and optimizes feed delivery based on real-time monitoring and data analysis. Smart feeding systems are designed to replace or complement manual feeding by automating feed distribution based on fish-specific needs. Unlike traditional methods, which may rely on rough estimates, smart systems make data-driven decisions on how much feed to deliver and when, enhancing feed utilization and reducing feed wastage (Vo et al., 2021). Smart feeding in aquaculture addresses two major concerns: feed waste and growth optimization. Overfeeding leads to wasted resources and increased water pollution, while underfeeding restricts fish growth. Smart feeding systems minimize these issues by delivering the precise amount of feed that fish need at any given time, ensuring sustainability and profitability (Lan et al., 2024).

7.1 Working Principle of Smart Feeding System

The working principle of a smart feeding system in aquaculture begins with continuous data collection from an array of sensors and cameras installed in the aquaculture system. These sensors and cameras monitor the fish's behavior reflecting its hunger, water quality conditions, fish density and size, and uneaten feed showing overfeeding (Alammar & Al-Ataby, 2018). These data are transmitted to a central processing unit (CPU) or cloud-based processor, where machine learning algorithms analyze patterns in behavior and environmental changes. The decision-making algorithm then determines feeding needs based on hunger cues and optimal conditions (Imai et al., 2019). When signals indicate fish are hungry and conditions are favorable, the algorithm activates the automated feeder to dispense the right amount of feed. Feedback loops allow the system to adjust feed levels dynamically, ensuring precise feeding that minimizes waste and enhances growth, all while improving resource efficiency and reducing labor needs (An et al., 2021).

7.2 Analyzing Fish Behavior: The Core of Hunger Detection

The detection and analysis of fish behavior patterns serve as fundamental indicators for optimizing aquaculture management strategies, particularly in determining optimal feeding schedules and environmental conditions (Cade et al., 2020). Behavioral modifications in fish populations manifest through various parameters, including swimming patterns, spatial distribution, and social interactions, which collectively indicate their physiological state and environmental response mechanisms (Rosa et al., 2020). Computer vision technology, particularly when enhanced by deep learning algorithms, has emerged as a predominant method for quantifying these behavioral indicators, offering advantages in both cost-effectiveness and practical implementation compared to traditional biological sensors or sonar systems (Zhao et al., 2016). The technology enables precise measurement of critical behavioral metrics, such as tail-beat frequency (TBF), which correlates strongly with swimming speed and energy expenditure, and wall-hitting rate (WHR), which serves as an indicator of environmental stress responses (Terayama et al., 2017; Xiao et al., 2015). These measurements, when analyzed collectively, provide insights into both individual and group behavioral dynamics, such as inspection behaviors, feeding responses, and escape patterns (Papadakis et al., 2012; Walberg, 2014). The technology's capability extends to differentiating between normal and abnormal behaviors, tracking individual fish trajectories within schools, and analyzing collective movement patterns using adapted Reynolds models, which were originally developed for studying flock behavior (Sadoul et al., 2014). Moreover, these behavioral analyses encompass responses to various environmental stressors, including water quality parameters, temperature fluctuations, artificial noise, and chemical pollutants, allowing for comprehensive environmental monitoring through behavioral proxies (Spampinato et al., 2010). The integration of behavioral data with environmental parameters has enabled the development of sophisticated monitoring systems that can effectively determine optimal feeding timing and quantities while simultaneously assessing environmental quality. This approach facilitates rapid detection of adverse conditions and enables prompt environmental optimization, thereby maintaining optimal growth conditions for fish populations (Xiao et al., 2015). As the field advances, the fusion of heterogeneous data sources, including behavioral indicators and physical parameters, presents opportunities for developing a more nuanced understanding of fish behavior under varying environmental conditions, particularly in relation to feeding patterns and stress responses (Papadakis et al., 2012). This integrated approach to behavioral monitoring has become instrumental in modern aquaculture systems, enabling data-driven decision-making for feed management and environmental control, ultimately contributing to more efficient and sustainable aquaculture practices (Rosa et al., 2020; Zhao et al., 2016) (Table 9.1).

7.3 *Biomass and Environmental Cues for Feeding Optimization*

Along with the fish behavior analysis, the accurate assessment of fish biomass and environmental parameters stands as a critical foundation for optimizing feeding strategies in smart aquaculture systems. The precise determination of feed quantities directly correlates with both biomass measurements and environmental conditions. These advanced systems employ various methodological approaches, including Bayesian stratification methods and deep learning architectures such as LinkNet-34 convolutional neural networks (CNNs), to establish accurate length-weight relationships (LWR) and biomass estimations (Froese et al., 2014; Konovalov et al., 2019). The implementation of stereoscopic camera systems, utilizing direct linear transformation (DLT) methods, has further enhanced the capability to perform precise three-dimensional measurements of fish morphometry while accounting for the dynamic nature of fish movement through specialized bending models (Torisawa et al., 2011; Muñoz-Benavent et al., 2018). Environmental parameters are continuously monitored through an array of sophisticated sensors: dissolved oxygen levels are measured using optical oxygen sensors (optodes) that offer high precision and minimal drift; water temperature is tracked through strategically placed

Table 9.1 Summary of methods for computer vision–based fish behavior analysis

Method	Judging index	Analysis result	References
3D computer vision	Foreground extraction accuracy (e.g., 91.2% with LB adaptive SOM)	Tracks 3D trajectories of fish; identifies abnormal behaviors effectively in real time	Tang and Liu (2015)
Clustering	Correct rate (CR) and error rate (ER)	Successfully analyzes the abnormal behavior of fish	Spampinato et al. (2010)
IMMJPDA algorithm	Abnormal trajectories recognition	Achieves a 92.3% recognition rate for abnormal behaviors using a hidden Markov model	Liu (2016)
Statistical analysis	Fish interaction (e.g., inspection and biting)	Detects behavioral changes caused by stressors; cost-effective and efficient for behavioral studies	Papadakis et al. (2012)
Optical flow algorithm	Motion vectors and mutual information	Detects abnormal behaviors with up to 99.92% accuracy but requires noise-reducing hardware	Yu et al. (2014)
Reynolds model	Total power (Ptot), difference power (Adiff), standard deviation (SD)	Sensitive, noninvasive method; applicable for behavioral assessments across various conditions	Sadoul et al. (2014)
Differential operation method	Fish mouth opening/closing states and duration	Accurately detects abnormal behavior in spotted groupers with 84% success rate	Xu and Xing (2016)
Modified kinetic energy model (KEM)	Human observation; emergent gathering and scattering	Detects special behaviors effectively using a rearing tank and monitoring platform	Zhao et al. (2016)
FIFFB and reflective image removal	Flocking index of fish feeding behavior (FIFFB)	Effectively quantifies feeding behavior changes and eliminates errors caused by surface reflections	Zhou et al. (2017)
Wall-hitting rate (WHR) and tail-beat frequency (TBF)	Number of fish and toxin concentration	Reliable for toxicity assessments and behavioral changes due to environmental stressors	Xiao et al. (2015)
Overlap coefficient adjustment	Fish feeding overlap and surface reflection correction	Measures Atlantic salmon feeding behaviors accurately, minimizing overlap issues and reflective errors	Liu et al. (2014)
T-tests and descriptive statistics	Behavior per trial observation period	Useful for farmed fish management, with significant results on behavioral changes over time	Walberg (2014)
Shoal behavior and entropy analysis	Foraging intensity using combined entropy	Enables intelligent feeding control for tilapia based on shoaling behavior and statistical modeling	Ye and Zhao (2016)

digital temperature sensors with resistance temperature detector (RTD) or thermistor technology; and light conditions are monitored via underwater photosensors that measure both intensity and spectral composition (Spoorthi et al., 2022). These environmental measurements are crucial as they directly influence fish metabolism and feeding behavior patterns. Fish feeding behavior and hunger responses vary mainly with the water conditions and their condition factor. The true innovation lies in the integration of the fish behavior, biomass levels, and water condition data through advanced machine learning algorithms that create a comprehensive understanding of the aquaculture ecosystem (O'Donncha et al., 2021; Zielinski et al., 2024). For instance, when dissolved oxygen sensors detect declining levels, the system correlates this with observed changes in swimming patterns and adjusts feeding rates accordingly, preventing overfeeding during periods of respiratory stress (Mandal & Ghosh, 2024). Similarly, temperature-dependent metabolic rates are assessed alongside behavioral cues and biomass data to optimize feed quantities, with increased distribution during periods of elevated metabolism and reduced quantities during cooler periods (Volkoff & Rønnestad, 2020). Photoperiod effects, measured through light sensors, are correlated with both behavioral patterns and growth rates, enabling feeding schedules that align with natural biological rhythms (Li et al., 2021). This three-way integration of behavioral monitoring, biomass estimation, and environmental sensing creates a robust framework for feed optimization that maximizes growth efficiency while maintaining optimal water quality and fish welfare, ultimately contributing to more sustainable and economically viable aquaculture operations.

7.4 Automation in Feed Delivery

Building upon the established framework of behavioral monitoring and integrated environmental-biomass assessment, modern aquaculture systems implement sophisticated automated feed delivery mechanisms that transcend traditional time-based feeding approaches. These advanced systems leverage real-time data integration and intelligent decision-making algorithms to optimize feed distribution with high precision (An et al., 2021). These systems employ various analytical approaches, from adaptive network-based fuzzy inference systems (ANFIS) to specialized neural networks, to process complex data patterns in real time (Zhou et al., 2018). The decision-making algorithms incorporate all the data collected from the sensors and cameras implemented in the system (AlZubi et al., 2016). Advanced systems implement negative feedback mechanisms that continuously refine feeding decisions based on post-feeding behavior analysis and feed waste assessment, enabling progressive optimization of feeding strategies (Zhao et al., 2016). To overcome challenges in data collection under suboptimal conditions, such as low light or turbid water, these systems integrate near-infrared imaging capabilities and specialized image processing algorithms that maintain operational efficiency across varying environmental conditions (Zhou et al., 2018). The automation extends beyond mere feed dispensing to include sophisticated spatial optimization, with some systems implementing designated feeding zones that help train fish populations in efficient feed utilization patterns (An et al., 2021). This comprehensive approach to automated feed delivery represents the culmination of behavioral monitoring, environmental sensing, and biomass estimation, creating a self-optimizing system that continuously refines its feeding strategies to maximize growth efficiency while minimizing waste and resource utilization. The integration of machine learning algorithms enables these systems to adapt to long-term patterns and seasonal variations, establishing a dynamic equilibrium between optimal growth conditions and resource efficiency.

8 Conclusion

The field of aquaculture nutrition is undergoing a transformative evolution driven by innovation and technological advancements. These developments not only redefine the boundaries of feed manufacturing and management but also set new benchmarks for sustainability and efficiency in the aquaculture industry. By integrating novel feed ingredients, advanced analytical techniques, optimized feed formulation models, and automated manufacturing processes, feed companies are scaling their operations to meet the increasing demands of modern aquaculture. Concurrently, the adoption of intelligent software for feed selection and smart feeding technologies enables farmers to transition from traditional practices to precision aquaculture, characterized by high yields and resource efficiency. Looking ahead, continued research is essential to advance predictive models and foster innovative solutions tailored to the diverse needs of aquaculture systems. The technologies explored in this chapter signify not just incremental progress but a fundamental reimagining of nutritional strategies, positioning aquaculture as a cornerstone of global sustainable food production. As the demand for protein surges and environmental pressures intensify, these innovations provide a clear pathway to developing resilient, efficient, and environmentally responsible aquaculture practices, ensuring the industry's role in addressing future food security challenges.

References

Ahmed, I., Ahmad, I., & Malla, B. A. (2024). Effects of dietary tryptophan levels on growth performance, plasma profile, intestinal antioxidant capacity and growth related genes in rainbow trout (*Oncorhynchus mykiss*) fingerlings. *Aquaculture*, 740710.

Aizam, N. A. H., Ibrahim, R. A., et al. (2018). Mathematical modelling for fish feed formulation of *Mystus nemurus Sp.* catfish: Optimizing growth and nutrients requirements. *Jurnal Teknologi, 80*, 10.11113/jt.v80.12488.

Alammar, M., & Al-Ataby, A. (2018). An intelligent approach of the fish feeding system. In *Computer Science & Information Technology (CS & IT). Presented at the 3rd international conference on data mining & knowledge management* (pp. 85–97). AIRCC Publication Corporation. https://doi.org/10.5121/csit.2018.81506

AlZubi, H. S., Al-Nuaimy, W., Buckley, J., & Young, I. S. (2016). An intelligent behavior-based fish feeding system. In *2016 13th international multi-conference on systems, signals & devices (SSD)*.

An, D., Hao, J., Wei, Y., Wang, Y., & Yu, X. (2021). Application of computer vision in fish intelligent feeding system – A review. *Aquaculture Research, 52*, 423–437. https://doi.org/10.1111/are.14907

Asmare, B. (2014). Biotechnological advances for animal nutrition and feed improvement. *WJAR, 2*, 115–118. https://doi.org/10.12691/wjar-2-3-5

Avila, C. R., Ferré, J., de Oliveira, R. R., de Juan, A., Sinclair, W. E., Mahdi, F. M., Hassanpour, A., Hunter, T. N., Bourne, R. A., & Muller, F. L. (2020). Process monitoring of moisture content and mass transfer rate in a fluidised bed with a low cost inline MEMS NIR sensor. *Pharmaceutical Research, 37*, 84. https://doi.org/10.1007/s11095-020-02787-y

Berzaghi, P., & Riovanto, R. (2009). Near infrared spectroscopy in animal science production: Principles and applications. *Italian Journal of Animal Science, 8*, 39–62. https://doi.org/10.4081/ijas.2009.s3.39

Boissy, J., Aubin, J., Drissi, A., van der Werf, H. M. G., Bell, G. J., & Kaushik, S. J. (2011). Environmental impacts of plant-based salmonid diets at feed and farm scales. *Aquaculture, 321*, 61–70. https://doi.org/10.1016/j.aquaculture.2011.08.033

Cade, D. E., Levenson, J. J., Cooper, R., de la Parra, R., Webb, D. H., & Dove, A. D. M. (2020). Whale sharks increase swimming effort while filter feeding, but appear to maintain high foraging efficiencies. *The Journal of Experimental Biology, 223*(Pt 11). https://doi.org/10.1242/jeb.224402

Capstaff, N. M., & Miller, A. J. (2018). Improving the yield and nutritional quality of forage crops. *Frontiers in Plant Science, 9*, 535. https://doi.org/10.3389/fpls.2018.00535

Chand, S., et al. (2022). Agronomical and breeding approaches to improve the nutritional status of forage crops for better livestock productivity. *Grass and Forage Science, 77*, 11–32. https://doi.org/10.1111/gfs.12557

Chrystal, P. (2022). Application of simulation modelling in broiler integration: Is it a necessary nutritional tool? *Proceedings of the Arkansas Nutrition Conference*, 12.

Coffey, D., Dawson, K., Ferket, P., & Connolly, A. (2016). Review of the feed industry from a historical perspective and implications for its future. *JAAN, 4*, e3. https://doi.org/10.1017/jan.2015.11

Cottrell, R. S., Blanchard, J. L., Halpern, B. S., Metian, M., & Froehlich, H. E. (2020). Global adoption of novel aquaculture feeds could substantially reduce forage fish demand by 2030. *Nature Food, 1*, 301–308. https://doi.org/10.1038/s43016-020-0078-x

Denstadli, V., Hillestad, M., Verlhac, V., Klausen, M., & Øverland, M. (2011). Enzyme pretreatment of fibrous ingredients for carnivorous fish: Effects on nutrient utilisation and technical feed quality in rainbow trout (*Oncurhynchus mykiss*). *Aquaculture, 319*, 391–397. https://doi.org/10.1016/j.aquaculture.2011.07.012

Din, N. S., Mubarak, A., Zaulir, Z., Jung, L. H., & Aizam, N. A. H. (2024). Mathematical programming for an optimised fish feed formulation: Towards sustainable red hybrid tilapia farming. *Journal of Sustainability Science and Management, 19*(4), 56–71.

Dixit, Y., Casado-Gavalda, M. P., Cama-Moncunill, R., Cama-Moncunill, X., Markiewicz-Keszycka, M., Cullen, P. J., & Sullivan, C. (2017). Developments and challenges in online NIR spectroscopy for meat processing. *Comprehensive Reviews in Food Science and Food Safety, 16*, 1172–1187. https://doi.org/10.1111/1541-4337.12295

Edwards, P. (2015). Aquaculture environment interactions: Past, present and likely future trends. *Aquaculture, 447*, S0044848615000605. https://doi.org/10.1016/j.aquaculture.2015.02.001

FAO. (2024). The state of world fisheries and aquaculture 2024—Blue transformation in action. *Rome*. https://doi.org/10.4060/cd0683en

Froese, R., Thorson, J. T., & Reyes, R. B. (2014). A Bayesian approach for estimating length-weight relationships in fishes. *Journal of Applied Ichthyology, 30*(1), 78–85. https://doi.org/10.1111/jai.12299

Garcia-Launay, F., Dusart, L., Espagnol, S., Laisse-Redoux, S., Gaudré, D., Méda, B., & Wilfart, A. (2018). Multiobjective formulation is an effective method to reduce environmental impacts of livestock feeds. *British Journal of Nutrition, 120*, 1298–1309. https://doi.org/10.1017/S0007114518002672

Glencross, B., et al. (2007). A feed is only as good as its ingredients – A review of ingredient evaluation strategies for aquaculture feeds. *Aquaculture Nutrition, 13*, 17–34. https://doi.org/10.1111/j.1365-2095.2007.00450.x

Imai, T., Arai, K., & Kobayashi, T. (2019). Smart aquaculture system: A remote feeding system with smartphones. In *2019 IEEE 23rd international symposium on consumer technologies (ISCT)* (pp. 93–96). IEEE. https://doi.org/10.1109/ISCE.2019.8901026

Ittiphalin, M., Nakrachata-Amon, T., & Pathumnakul, S. (2015). Feed formulation under raw material and production costs consideration. *AMM, 781*, 667–670. https://doi.org/10.4028/www.scientific.net/AMM.781.667

Konovalov, D. A., Saleh, A., Efremova, D. B., Domingos, J. A., & Jerry, D. R. (2019). Automatic weight estimation of harvested fish from images. In *2019 digital image computing: Techniques and applications (DICTA)* (Vol. 7, p. 7).

Lall, S. P., & Dumas, A. (2022). Chapter 3 – Nutritional requirements of cultured fish: Formulating nutritionally adequate feeds. In D. A. Davis (Ed.), *Feed and feeding practices in aquaculture. Woodhead publishing series in food science, technology and nutrition* (2nd ed., pp. 65–132). Woodhead Publishing. https://doi.org/10.1016/B978-0-12-821598-2.00005-9

Lan, H. -Y., et al. (2024). A novel process-based digital twin for intelligent fish feeding management using multi-mode sensors and smart feeding machine. https://doi.org/10.20944/preprints202401.0927.v1.

Lee, N. K. (2018). Total automation: The possibility of lights-out manufacturing in the near future. *Missouri S&T's Peer to Peer, 2*(1), 4.

Leishman, E. M., You, J., Ferreira, N. T., Adams, S. M., Tulpan, D., Zuidhof, M. J., Gous, R. M., Jacobs, M., & Ellis, J. L. (2023). Review: When worlds collide – Poultry modeling in the 'Big Data' era. animal. In *10th international workshop on modelling nutrient digestion and utilization in farm animals (MODNUT)* (Vol. 17, p. 100874). https://doi.org/10.1016/j.animal.2023.100874

Li, X., Wei, P., Liu, S., Tian, Y., Ma, H., & Liu, Y. (2021). Photoperiods affect growth, food intake and physiological metabolism of juvenile European Sea bass (*Dicentrachus labrax L.*). *Aquaculture Reports, 20*, 100656. https://doi.org/10.1016/j.aqrep.2021.100656

Li, M. H., Robinson, E. H., & Davis, D. A. (2022). Chapter 4 – Complete feeds – Intensive systems. In D. A. Davis (Ed.), *Feed and feeding practices in aquaculture. Woodhead publishing series in food science, technology and nutrition* (2nd ed., pp. 133–149). Woodhead Publishing. https://doi.org/10.1016/B978-0-12-821598-2.00007-2

Liu, S. S. (2016). *Research on the key technology of the seismic behavior analysis of fish based on 3D computer vision*. Zhejiang University of Technology.

Liu, Z., Li, X., Fan, L., Lu, H., Liu, L., & Liu, Y. (2014). Measuring feeding activity of fish in RAS using computer vision. *Aquacultural Engineering, 60*, 20–27. https://doi.org/10.1016/j.aquaeng.2014.03.005

Lu, S., Taethaisong, N., Meethip, W., Surakhunthod, J., Sinpru, B., Sroichak, T., Archa, P., Thongpea, S., Paengkoum, S., Purba, R. A. P., & Paengkoum, P. (2022). Nutritional composition of black soldier Fly larvae (*Hermetia illucens L.*) and its potential uses as alternative protein sources in animal diets: A review. *Insects, 13*, 831. https://doi.org/10.3390/insects13090831

Mandal, A., & Ghosh, A. R. (2024). Role of artificial intelligence (AI) in fish growth and health status monitoring: A review on sustainable aquaculture. *Aquaculture International, 32*, 2791–2820. https://doi.org/10.1007/s10499-023-01297-z

Muñoz-Benavent, P., García, G. A., González, J. M., Vanacloig, V. L., PuigPons, V., & Espinosa, J. V. (2018). Enhanced fish bending model for automatic tuna sizing using computer vision. *Computers and Electronics in Agriculture, 150*, 52–61.

O'Donncha, F., Stockwell, C. L., Planellas, S. R., Micallef, G., Palmes, P., Webb, C., Filgueira, R., & Grant, J. (2021). Data driven insight into fish behaviour and their use for precision aquaculture. *Frontiers in Animal Science, 2*, 695054. https://doi.org/10.3389/fanim.2021.695054

Olagunju, L. K., Isikhuemhen, O. S., Dele, P. A., Anike, F. N., Essick, B. G., Holt, N., Udombang, N. S., Ike, K. A., Shaw, Y., Brice, R. M., Orimaye, O. E., Wuaku, M., & Anele, U. Y. (2023). Pleurotus ostreatus can significantly improve the nutritive value of Lignocellulosic crop residues. *Agriculture, 13*, 1161. https://doi.org/10.3390/agriculture13061161

Olivier, L. E.. (2017). *On lights-out process control in the minerals processing industry*. Ph.D. thesis, University of Pretoria, Pretoria.

Papadakis, V. M., Papadakis, I. E., Lamprianidou, F., Glaropoulos, A., & Kentouri, M. (2012). A computer-vision system and methodology for the analysis of fish behavior. *Aquacultural Engineering, 46*, 53–59. https://doi.org/10.1016/j.aquaeng.2011.11.002

Pathumnakul, S., Piewthongngam, K., & Apichottanakul, A. (2009). A neural network approach to the selection of feed mix in the feed industry. *Computers and Electronics in Agriculture, 68*, 18–24. https://doi.org/10.1016/j.compag.2009.04.001

Rosa, L. V., Costa, F. V., Canzian, J., Borba, J. V., Quadros, V. A., & Rosemberg, D. B. (2020). Three- and bi-dimensional analyses of the shoaling behavior in zebrafish: Influence of modulators of anxiety-like responses. *Progress in Neuro-Psychopharmacology & Biological Psychiatry, 102*(10), 9957.

Sadoul, B., Evouna Mengues, P., Friggens, N. C., Prunet, P., & Colson, V. (2014). A new method for measuring group behaviours of fish shoals from recorded videos taken in near aquaculture conditions. *Aquaculture, 430*, 179–187. https://doi.org/10.1016/j.aquaculture.2014.04.008

Sarsaiya, S., Jain, A., Kumar Awasthi, S., Duan, Y., Kumar Awasthi, M., & Shi, J. (2019). RETRACTED: Microbial dynamics for lignocellulosic waste bioconversion and its importance with modern circular economy, challenges and future perspectives. *Bioresource Technology, 291*, 121905. https://doi.org/10.1016/j.biortech.2019.121905

Schafer, E. (2022). *"Lights out" feed mills improve biosecurity, reduce errors*. https://www.feedstrategy.com/feed-mill-of-the-future/lights-out-feed-mills-improve-biosecurity-reduce-errors/

Sharp, R., O'Donnell, A. G., Gilbert, H. G., & Hazlewood, G. P. (1992). Growth and survival of genetically manipulated lactobacillus plantarum in silage. *Applied and Environmental Microbiology, 58*, 2517–2522. https://doi.org/10.1128/aem.58.8.2517-2522.1992

Siva Raman, S., Stringer, L. C., Bruce, N. C., & Chong, C. S. (2022). Opportunities, challenges and solutions for black soldier fly larvae-based animal feed production. *Journal of Cleaner Production, 373*, 133802. https://doi.org/10.1016/j.jclepro.2022.133802

Spampinato, C., Giordano, D., Di Salvo, R., Chen-Burger, Y.-H. J., Fisher, R. B., & Nadarajan, G. (2010). Automatic fish classification for underwater species behavior understanding. In *Proceedings of the first ACM international workshop on analysis and retrieval of tracked events and motion in imagery streams, ARTEMIS '10* (pp. 45–50). Association for Computing Machinery. https://doi.org/10.1145/1877868.1877881

SPAROS. (n.d.). Feednetics 4.0 project. Retrieved December 2, 2024, from https://www.sparos.pt/projects/feednetics-4-0/

Spoorthi, L. E., Kashyap, N., Rajatha, B., Priyanka, M. H., & Bhargavi, K. (2022). A review on smart water quality monitoring system. *International Journal for Research in Applied Science and Engineering Technology*.

Tacon, A. G. J., & Metian, M. (2015). Feed matters: Satisfying the feed demand of aquaculture. *Reviews in Fisheries Science & Aquaculture, 23*, 1–10. https://doi.org/10.1080/23308249.2014.987209

Tahir, M., Shim, M. Y., Ward, N. E., Westerhaus, M. O., & Pesti, G. M. (2012). Evaluation of near-infrared reflectance spectroscopy (NIRS) techniques for total and phytate phosphorus of common poultry feed ingredients. *Poultry Science, 91*, 2540–2547. https://doi.org/10.3382/ps.2012-02211

Tang, Y. P., & Liu, S. S. (2015). A study of fish behavior analysis based on 3D computer vision. *High Technology Communications, 25*(3), 249–256.

Terayama, K., Hioki, H., & Sakagami, M. A. (2017). Measuring tail beat frequency and coast phase in school of fish for collective motion analysis. In *Eighth international conference on graphic and image processing (ICGIP 2016)*. International Society for Optics and Photonics.

Tippayadara, N., Dawood, M. A. O., Krutmuang, P., Hoseinifar, S. H., Doan, H. V., & Paolucci, M. (2021). Replacement of fish meal by black soldier Fly (*Hermetia illucens*) larvae meal: Effects on growth, Haematology, and skin mucus immunity of Nile tilapia, *Oreochromis niloticus*. *Animals, 11*, 193. https://doi.org/10.3390/ani11010193

Torisawa, S., Kadota, M., Komeyama, K., Suzuki, K., & Takagi, T. (2011). A digital stereo-video camera system for three-dimensional monitoringof free-swimming pacific bluefin tuna, *thunnus orientalis*, cultured in a net cage. *Aquatic Living Resources, 24*(2), 107–112.

Uyeh, D. D., Pamulapati, T., Mallipeddi, R., Park, T., Asem-Hiablie, S., Woo, S., Kim, J., Kim, Y., & Ha, Y. (2019). Precision animal feed formulation: An evolutionary multi-objective approach. *Animal Feed Science and Technology, 256*, 114211. https://doi.org/10.1016/j.anifeedsci.2019.114211

van Der Poel, A. F. B., Abdollahi, M. R., Cheng, H., Colovic, R., den Hartog, L. A., Miladinovic, D., Page, G., Sijssens, K., Smillie, J. F., Thomas, M., Wang, W., Yu, P., & Hendriks, W. H. (2020). Future directions of animal feed technology research to meet the challenges of a changing world. *Animal Feed Science and Technology, 270*, 114692. https://doi.org/10.1016/j.anifeedsci.2020.114692

Vo, T. T. E., Ko, H., Huh, J.-H., & Kim, Y. (2021). Overview of smart aquaculture system: Focusing on applications of machine learning and computer vision. *Electronics, 10*, 2882. https://doi.org/10.3390/electronics10222882

Volkoff, H., & Rønnestad, I. (2020). Effects of temperature on feeding and digestive processes in fish. *Temperature, 7*, 307–320. https://doi.org/10.1080/23328940.2020.1765950

Walberg, E. (2014). Effect of increased water temperature on warm water fish feeding behavior and habitat use. *Journal of Undergraduate Research at Minnesota State University, Mankato, 11*. https://doi.org/10.56816/2378-6949.1035

Wilfart, A., Garcia-Launay, F., Terrier, F., Soudé, E., Aguirre, P., & Skiba-Cassy, S. (2023). A step towards sustainable aquaculture: Multiobjective feed formulation reduces environmental impacts at feed and farm levels for rainbow trout. *Aquaculture, 562*, 738826. https://doi.org/10.1016/j.aquaculture.2022.738826

Xiao, G., Feng, M., Cheng, Z., Zhao, M., Mao, J., & Mirowski, L. (2015). Water quality monitoring using abnormal tail-beat frequency of crucian carp. *Ecotoxicology and Environmental Safety, 111*, 185–191. https://doi.org/10.1016/j.ecoenv.2014.09.028

Xu, S., & Xing, K. Z. (2016). A method for identifying abnormal behavior of point grouper based on machine vision. *Fishery Modernization, 43*(1), 18–23.

Ye, Z., & Zhao, J. (2016). Behavioral characteristics and statistics-based imaging techniques in the assessment and optimization of tilapia feeding in a recirculating aquaculture system. *Transactions of the ASABE, 59*(1), 345–355.

You, J., Ellis, J. L., Tulpan, D., & Malpass, M. C. (2024). Review: Recent advances and future technologies in poultry feed manufacturing. *World's Poultry Science Journal, 80*, 643–655. https://doi.org/10.1080/00439339.2024.2323536

Yu, X., Hou, X., Lu, H., Yu, X., Fan, L., & Liu, Y. (2014). Anomaly detection of fish school behavior based on features statistical and optical flow methods. *Transactions of the Chinese Society of Agricultural Engineering, 30*, 162–168.

Zhang, Q., & Pierce, F. J. (2013). *Agricultural automation: Fundamentals and practices*. CRC Press.

Zhao, J., Gu, Z., Sh, M., Lu, H., Li, J., Shen, M., Ye, Z., & Zhu, S. (2016). Spatial behavioral characteristics and statistics-based kinetic energy modeling in special behaviors detection of a shoal of fish in a recirculating aquaculture system. *Computers and Electronics in Agriculture, 127*, 271–280. https://doi.org/10.1016/j.compag.2016.06.025

Zhou, C., Zhang, B., Lin, K., Xu, D., Chen, C., Yang, X., & Sun, C. (2017). Near-infrared imaging to quantify the feeding behavior of fish in aquaculture. *Computers and Electronics in Agriculture, 135*, 233–241. https://doi.org/10.1016/j.compag.2017.02.013

Zhou, C., Lin, K., Xu, D., Chen, L., Guo, Q., Sun, C., & Yang, X. (2018). Near infrared computer vision and neuro-fuzzy model-based feeding decision system for fish in aquaculture. *Computers and Electronics in Agriculture, 146*, 114–124. https://doi.org/10.1016/j.compag.2018.02.006

Zielinski, D. P., Kerr, J., Bærum, K. M., Simmons, O. M., Silva, A. T., & Goodwin, R. A. (2024). Advancements in riverine fish movement modeling: Bridging environmental complexity and fish behavior. *Reviews in Fisheries Science & Aquaculture, 1–16*. https://doi.org/10.1080/23308249.2024.2374964

Zulkifli, N. F. N. M., Seok-Kian, A. Y., Seng, L. L., Mustafa, S., Kim, Y.-S., & Shapawi, R. (2022). Nutritional value of black soldier fly (*Hermetia illucens*) larvae processed by different methods. *PLoS One, 17*, e0263924. https://doi.org/10.1371/journal.pone.0263924

Chapter 10
Role of Fat-Soluble Vitamins in Fish Nutrition

Aamina Hassan, Imtiaz Ahmed, Quseen Mushtaq Reshi, and Parveena Hassan

1 Introduction

Providing fish with proper nutrition has long been recognized as a critical factor in promoting normal growth, health, and generation of well body composition. In addition to influencing fish health, growth, and optimum flesh quality, good nutrition also enhances the economic returns of the industry as feed accounts for a significant portion of the production cost (Ahmed et al., 2024). Regardless of the quality of the environment, an undernourished animal cannot maintain its health. It is evident that faulty nutrition reduces the productivity of fish and deteriorates their health until recognizable diseases become evident (Khan & Khan, 2020). With the development of balanced commercial diets that promote optimal fish growth and health, fish nutrition has made significant advancements in recent years (Adewumi, 2018). Fish nutrition ranks as one of the most important aspects of fish production, necessitating continuous enhancements in feed formulation and the incorporation of diverse feed ingredients. These efforts are essential for the comprehensive development and long-term sustainability of fish farming systems (Davis, 2009; Hixson, 2014). Depriving fish of essential nutrients not only hampers their growth and feed efficiency but also increases their vulnerability to diseases and prompts the display of deficiency symptoms such as altered behavior and pathology.

Among all the nutrients required by the fish, vitamins play a critical role in the nutrition and were discovered by F. G. Hopkins in 1906. Vitamins are distinct from amino acids, fats, and carbohydrates, as they are required in trace amounts from

A. Hassan · I. Ahmed (✉) · Q. M. Reshi
Fish Nutrition Research Laboratory, Department of Zoology, University of Kashmir, Srinagar, Jammu and Kashmir, India

P. Hassan
Cytogenetics and Molecular Biology Lab, Department of Botany, University of Kashmir, Srinagar, Jammu and Kashmir, India

I. Ahmed, I. Ahmad (eds.), *Aquaculture: Enhancing Food Security and Nutrition*, https://doi.org/10.1007/978-3-031-92858-1_10

external sources (diet) but are essential for maintaining normal growth, reproduction, and overall health in all animals, including fish (Shiau & Lin, 2006; NRC, 2011). Originally thought to be amines vital for existence, it was later discovered that vitamins are not related to amines but are compounds with specific functions within the body. They are essential components for many proteins and hormones and also play a crucial role in the proper functioning of various enzymes in the body, affecting their activity when the fish are supplied with an imbalanced diet (Halver, 2002; Yossa et al., 2015). Vitamins are added as supplements to fish feed to promote optimal growth and health (Barrows et al., 2008; Hemre et al., 2016). Since fish have higher vitamin requirements than other animals and cannot synthesize them in their bodies, therefore fish feed must be supplemented with vitamins (Halver, 1989; Woodward, 1994; De Silva & Anderson, 1995). Depending on their solubility, there are two groups of vitamins: water-soluble and fat-soluble vitamins. Thiamine, niacin, riboflavin, folic acid, pantothenic acid, biotin, pyridoxine, and cobalamins are the eight water-soluble vitamins known as vitamin B complex, required in relatively small amounts, and have primarily coenzyme functions. The other three water-soluble vitamins-choline, inositol, and ascorbic acid, are required in relatively large amounts and serve functions beyond acting as coenzymes.

The fat-soluble vitamins, which include vitamin A (retinol), vitamin D (calciferol), vitamin E (tocopherol), and vitamin K (phylloquinone), play indispensable roles in the health and development of aquatic organisms. These essential nutrients are crucial for a multitude of physiological functions that support the survival, growth, and reproductive success of fish. Given the expanding scope of aquaculture as a significant source of global protein, understanding the nuanced roles of these vitamins becomes not only a scientific pursuit but also a practical necessity to enhance the sustainability and efficiency of fish farming practices (Webster & Lim, 2002; NRC, 2011).

1.1 Overview

Vitamins A, D, E, and K, each with unique biochemical properties and functions, are vital for maintaining the health of fish. Vitamin A is essential for good vision, growth, and immune function, supporting skin and mucosal health in environments where pathogens are abundant (John & Ronald, 2002). Vitamin D directly influences bone strength and structural integrity through its role in calcium and phosphorus balance (Santosh, 2003). Vitamin E, known for its antioxidant properties, enhances disease resistance by protecting fish from oxidative stress (El-Sayed & Izquierdo, 2022). Vitamin K is crucial for blood clotting and bone metabolism, ensuring proper physiological functions in fish (Krossøy et al., 2011).

The significance of these vitamins extends beyond individual health, impacting the overall productivity and sustainability of aquaculture operations. Ensuring optimal levels of these nutrients in fish diets not only improves the welfare and growth rates of aquaculture species but also enhances the nutritional quality of the fish products available to consumers. This comprehensive understanding informs better feed formulation and management practices, leading to more resilient and productive aquaculture systems.

As the aquaculture industry continues to evolve, the integration of advanced nutritional research and practical feed solutions will play a pivotal role in addressing the challenges of feeding a growing global population. This chapter delves into these essential nutrients, exploring both current knowledge and future directions, thus setting the stage for ongoing innovation in aquaculture nutrition.

2 Fat-Soluble Vitamins: An Overview

Fat-soluble vitamins are essential for the normal physiological functions of aquatic organisms. These vitamins, specifically A, D, E, and K, are integral to maintaining health, enhancing growth, and ensuring the reproductive success of fish species. Each of these vitamins has unique roles and mechanisms of action, which are crucial for optimal fish nutrition and aquaculture productivity.

2.1 Vitamin A (Retinol)

Vitamin A plays a pivotal role in vision, bone growth, and epithelial integrity and is essential for immune function. In fish, vitamin A is involved in the phototransduction pathway, critical for light absorption in the retina, which is fundamental for navigating their aquatic environments. Additionally, it influences gene expression that governs cell differentiation and growth. Deficiencies in vitamin A can lead to night blindness, reduced growth rates, and increased susceptibility to infections, impacting overall aquaculture yields (Hernandez & Hardy, 2020).

2.2 Vitamin D (Cholecalciferol)

Vitamin D is vital for calcium and phosphorus homeostasis, influencing bone mineralization and structural development in fish (Karges et al., 2001; Lin et al., 2005; Tipton et al., 2007; Samuel & Sitrin, 2008; Lock et al., 2010). Unlike humans, fish are cannot synthesize vitamin D (Mazurais et al., 2009), possibly due to ultraviolet radiation absorbed by the water before it reaches the fish (Boglione et al., 2001), so it must be supplied through diet (Hamre et al., 2010). It is synthesized in the skin under exposure to sunlight and converted to its active form, calcitriol, in the liver and kidneys. This active form facilitates the intestinal absorption of calcium and phosphorus, which are critical for skeletal development, especially in rapidly growing fish species. Insufficient levels of vitamin D can lead to metabolic bone disease, manifesting as deformities and reduced mechanical strength in bones (Fenwick, 1984; Suda et al., 2003).

2.3 Vitamin E (Tocopherol)

Among the fat-soluble vitamin family, vitamin E is being utilized in the food industry for the preparation of animal feed (Shahkar et al., 2018). Vitamin E has received great interest in nutrition because of its major role in several biological and metabolic processes of fish, like enhancing growth performance, feed efficiency, and preventing muscle degeneration (El-Sayed & Izquierdo, 2022). As a powerful antioxidant, vitamin E protects cell membranes from oxidative damage, which is crucial for maintaining cellular integrity and health in fish (Bender, 2003). It plays a significant role in immune response and is involved in several metabolic processes. The antioxidant properties of vitamin E are particularly beneficial in aquaculture, where fish are often exposed to higher levels of polyunsaturated fatty acids in their diets, which can predispose them to lipid peroxidation and oxidative stress (Kumar et al., 2022).

2.4 Vitamin K

Vitamin K is essential for the synthesis of proteins required for blood coagulation and bone metabolism. In fish, it facilitates the activation of calcium-binding properties of proteins, which is critical for bone formation and repair. Vitamin K exists in two natural forms: K_1, which is found in plant sources, and K_2, which is synthesized by intestinal microflora in the gut. Deficiencies in vitamin K can lead to hemorrhaging and poor bone development, affecting overall health and viability of aquaculture operations (Tan & Mai, 2001; Jobling, 2011).

These fat-soluble vitamins are typically stored in the liver and adipose tissues and are absorbed in the intestine along with dietary fats, which necessitates the presence of lipids in the diet for efficient absorption. The dietary requirements can vary significantly among different fish species, influenced by factors such as age, species, and the specific conditions of aquaculture (Suttie, 1980; Udagawa et al., 1993; Tan & Mai, 2001; Krossøy et al., 2011). Fat-soluble vitamin A, D, E, and K requirements of some fish species are given in the Table 10.1.

3 Biochemical Roles and Mechanisms

3.1 Pathways and Interactions

3.1.1 Vitamin A Metabolism in Fish

Vitamin A, or retinol, is vital for various physiological processes in fish, including vision, growth, and reproduction. Once ingested, vitamin A is absorbed in the intestine, where it is stored primarily in the liver as retinyl esters. When needed, these esters are hydrolyzed to retinol, which is then transported by retinol-binding protein

Table 10.1 Fat-soluble vitamin A, D, E, and K requirements of some fish species

Vitamin	Fish	Requirement	Author	Journal
A	*Siniperca chuatsi*	37.32 mg/kg	Chen et al. (2024)	Fish Physiology and Biochemistry
	Cyprinus carpio communis	0.11 g/kg	Hassan et al. (2022)	Aquaculture Nutrition
	Epinephelus fuscoguttatus × *E. lanceolatus*	2688.8 IU/kg	Liang et al. (2020)	Fish & Shellfish Immunology
	Ctenopharyngodon idella	2622 IU/kg	Jiang et al. (2019)	Fish & Shellfish Immunology
	Sillago sihama Forskál.	2516 IU/kg	Huang et al. (2018)	Aquaculture Research
	Rhamdia quelen.	2610 IU/kg	Battisti et al. (2017)	Aquaculture
	Carassius auratus gibelio	2698 IU/kg	Shao et al. (2016)	Aquaculture
	Solea senegalensis	203,000 IU/kg	Fernandez et al. (2015)	Fish & Shellfish Immunology
	Cyprinus carpio var. *communis*	400 mg/kg	Jeyaraj et al. (2012)	Journal of Fisheries and Aquatic Science
	Dicentrarchus labrax.	69,245 IU/kg	Mazurais et al. (2009)	Aquaculture
	Oreochromis niloticus× *Oreochromius aureus*	1.76–2.09 mg/kg	Hu et al. (2006)	Aquaculture
D	*Ctenopharyngodon idella*	1994.80 IU/kg	Wang et al. (2024)	Fish Physiology and Biochemistry
	Epinephelus coiodes.	930 IU/kg	Lin et al. (2022)	Aquaculture Reports
	Sparus aurata	25–30 μg/kg	Sivagurunathan et al. (2022)	Aquaculture
	S. Aurata	11.6 IU/g	Dominguez et al. (2021)	Aquaculture
	Mylopharyngodon piceus	535.09 IU/kg	Wu et al. (2020)	Fish Physiology and Biochemistry
	Salmo salar	0.06–0.09 mg/kg	Prabhu et al. (2019)	Peer J

(continued)

Table 10.1 (continued)

Vitamin	Fish	Requirement	Author	Journal
E	*Chanos chanos*	205–243 mg/kg	Sivaramakrishnan et al. (2024)	Journal of Applied Aquaculture
	Labeo rohita	80.62 mg/kg	Zulfiqar et al. (2023)	Pakistan Journal of Zoology
	Caspian trout, *S. Caspius*	77.94 mg/kg	Saheli et al. (2021)	Aquaculture Reports
	Sillago sihama	100.37 mg/kg	Huang et al. (2020)	Animal Feed Science & Technology
	Oreochromis niloticus	160–320 mg/kg	Qiang et al. (2019)	Fish & Shellfish Immunology
	Pseudoplatystoma reticulatum	166.6 mg/kg	Zanon et al. (2018)	Aquaculture Research
	Ictalurus punctatus	50–100 mg/kg	He et al. (2017)	Experimental and Therapeutic Medicine
	Pelteobagrus fulvidraco	33 mg/kg	Lu et al. (2016)	Aquaculture
	Ctenopharyngodon idellus	100.36 mg/kg	Li et al. (2014)	Aquaculture
	Sciaenops ocellantus	31 mg/kg	Peng and Gatlin (2009)	Aquaculture Nutrition
	Labeo rohita	131.91 mg/kg	Sau et al. (2004)	Aquaculture
K	*Oreochromis niloticus*	2–4 mg/kg	Abdelhamid et al. (2024)	Aquaculture International
	Micropterus salmoides	9.93–15.22 mg/kg	Wei et al. (2023)	Aquaculture and Fisheries
	Sparus aurata	12 mg/kg	Dominguiz et al. (2022)	Aquaculture Nutrition
	C. carpio var. *jian*	8.93 g	Yuan et al. (2016)	Aquaculture
	Salmo salar	0.1 mg/kg	Krossoy et al. (2011)	Aquaculture Nutrition

to target tissues. In the retina, retinol is converted to retinaldehyde, which is essential for the visual cycle. This conversion is critical for the regeneration of rhodopsin, the photopigment required for vision in low-light conditions (Tanumihardjo, 2011).

3.1.2 Vitamin D Synthesis and Function

Fish synthesize vitamin D in a manner similar to other vertebrates. Upon exposure to ultraviolet light, 7-dehydrocholesterol in the skin is converted to vitamin D_3 (cholecalciferol). Vitamin D_3 is hydroxylated in the liver to 25-hydroxyvitamin D_3 and subsequently in the kidneys to the biologically active form, 1,25-dihydroxyvitamin D_3 (calcitriol). Calcitriol acts primarily to regulate calcium and phosphorus homeostasis, enhancing the intestinal absorption of these minerals and facilitating bone mineralization (Lock et al., 2010).

3.1.3 Vitamin E as an Antioxidant

Vitamin E, primarily as alpha-tocopherol, functions as a lipid-soluble antioxidant, protecting cellular membranes from oxidative damage caused by free radicals and reactive oxygen species. This protection is crucial for maintaining membrane integrity and function, especially in highly polyunsaturated fatty acids prevalent in fish membranes. Moreover, vitamin E plays a role in immune enhancement and cellular signaling processes that are vital for normal growth and development in fish (Traber & Atkinson, 2007).

3.1.4 Vitamin K Metabolic Pathways

Vitamin K is involved in the posttranslational modification of certain proteins required for blood clotting and bone health. In fish, as in other vertebrates, vitamin K acts as a cofactor for the enzyme gamma-glutamyl carboxylase, which catalyzes the carboxylation of glutamate residues on vitamin K–dependent proteins, enabling them to bind calcium ions necessary for their biological activity. This biochemical role is essential not only for coagulation but also for regulating bone mineral density and other physiological functions (Shearer & Newman, 2008).

3.2 Interactions with Other Nutrients

3.2.1 Synergies with Minerals

Fat-soluble vitamins often act synergistically with minerals to enhance their biological functions. For example, vitamin D's role in promoting calcium absorption is well known, but it also enhances phosphorus absorption, working in concert with dietary minerals to ensure proper skeletal development and maintenance in fish. Moreover, the presence of adequate vitamin K levels is crucial for utilizing calcium in the process of bone mineralization (Krossøy et al., 2011).

3.2.2 Interactions with Fatty Acids

The relationship between vitamin E and fatty acids is particularly notable. Vitamin E protects dietary and cellular polyunsaturated fatty acids from lipid peroxidation. This interaction is crucial in fish diets rich in unsaturated fats, which are susceptible to oxidation. Adequate levels of vitamin E can prevent the deleterious effects of oxidative stress on fish health, thereby improving growth performance and disease resistance (Kamal-Eldin & Appelqvist, 1996).

3.2.3 Vitamin A and Iron

Another interesting interaction occurs between vitamin A and iron. Vitamin A is known to enhance iron absorption by upregulating the expression of proteins involved in iron metabolism. This synergy is beneficial in preventing anemia and supporting the oxygen-carrying capacity of the blood, which is crucial for the metabolic demands of fish, especially during rapid growth phases or under stress conditions (Hernandez & Hardy, 2020).

4 Nutritional Requirements and Dietary Sources

4.1 Species-Specific Requirements

Understanding the nutritional requirements for fat-soluble vitamins varies significantly among different fish species. For example, trout and salmon, while both are carnivorous and thrive in cold water, exhibit distinct dietary needs that influence their health and growth.

4.1.1 Trout

Trout, particularly rainbow trout, require a balanced intake of vitamins A, D, E, and K to support their rapid growth rates and high metabolic needs. Vitamin A is crucial for their vision and epithelial health, especially given their highly visual predatory lifestyle. A study by the National Research Council suggests that the dietary vitamin A requirement for optimal growth in rainbow trout is around 2500 IU/kg of diet. Vitamin D is less of a concern due to their ability to synthesize it under natural light conditions, but supplemental sources can benefit indoor aquaculture operations (Jobling, 2011).

4.1.2 Salmon

Salmon have higher requirements for vitamin D due to their life cycle stages, which include long periods in deep or murky waters where sunlight penetration is limited. The dietary vitamin D level recommended for salmon is approximately 500 IU/kg to prevent skeletal malformations and ensure proper mineral absorption. Vitamin E requirements are also higher in salmon to combat oxidative stress associated with their high-fat diets, particularly in farmed settings where fish oil is a major dietary component (Hardy & Barrows, 2003).

4.2 Dietary Sources

4.2.1 Natural Sources

In the wild, fish obtain fat-soluble vitamins through a natural diet consisting of smaller fish, crustaceans, and aquatic plants. For instance, salmon feed on a variety of organisms like krill and small fish, which are rich in vitamins A and E. Likewise, trout consume aquatic insects and zooplankton, which provide them with necessary vitamins but may vary seasonally and geographically in concentration.

4.2.2 Supplemented Sources in Feed

In aquaculture, achieving the right vitamin balance often requires fortified feeds. These feeds are supplemented with synthetic forms of vitamins A, D, E, and K to ensure consistency and meet the specific requirements of each species. The use of stabilized forms of these vitamins is common to prevent degradation during feed processing and storage (Tacon & De Silva, 1997).

4.3 Factors Affecting Nutrient Availability in Fish Feeds

4.3.1 Preparation

The process of feed manufacture, including extrusion and pelleting, can significantly affect the stability and availability of fat-soluble vitamins. High temperatures and exposure to moisture during these processes can lead to the degradation of sensitive vitamins like vitamin E and vitamin A. It is crucial to optimize these processes to minimize nutrient losses, which can be achieved by incorporating vitamins post-extrusion or using more stable forms of the vitamins (De Silva & Anderson, 1995; Prabu et al., 2017).

4.3.2 Storage

Storage conditions such as temperature, humidity, and exposure to light play critical roles in the stability of vitamins in fish feed. Vitamins A and D are particularly susceptible to degradation under improper storage conditions. Studies have shown that storing feeds in cool, dark, and dry places can significantly enhance the shelf life and efficacy of the vitamin content within (Wang et al., 2016; McArley et al., 2021).

4.3.3 Feed Composition

The overall composition of the feed, including the presence of antioxidants and the type of fat used, can influence the stability of fat-soluble vitamins. For instance, feeds with higher levels of polyunsaturated fats require more vitamin E to prevent oxidative damage. Similarly, the presence of minerals like zinc and iron can affect the stability of vitamin A, necessitating careful formulation to avoid interactions that may impair vitamin availability (Storebakken et al., 1998).

5 Impacts of Deficiencies and Excesses

5.1 Deficiency Symptoms

Vitamin deficiencies in fish can lead to a range of physiological and health issues that impact their growth, reproduction, and survival. Each fat-soluble vitamin plays a specific role in the body, and their absence can lead to distinct symptoms:

5.1.1 Vitamin A Deficiency

Vitamin A is crucial for vision, skin integrity, and mucosal immunity. In fish, a deficiency of vitamin A can lead to xerophthalmia (dryness of the cornea and conjunctiva), night blindness, and poor growth rates. For example, juvenile salmon and trout showing signs of vitamin A deficiency often exhibit slower growth and higher susceptibility to infections due to impaired epithelial barrier functions. There can also be reproductive failures characterized by reduced egg quality and hatchability (John & Ronald, 2002). According to Kitamura et al. (1967), deficiency of vitamin A causes anemia, twisted opercula and hemorrhages in the eyes and base of fins in rainbow trout.

5.1.2 Vitamin D Deficiency

Vitamin D deficiency primarily affects bone development and mineral homeostasis. In fish, this can manifest as rickets or osteomalacia, conditions marked by softening and deformity of the bones. Young fish, especially rapidly growing species like salmon, are particularly susceptible. Symptoms include lethargy, deformities in the spine and fins, and overall stunted growth. Long-term deficiency can lead to increased mortality rates due to compromised skeletal structure (Santosh, 2003).

5.1.3 Vitamin E Deficiency

Vitamin E functions as a major antioxidant. Deficiency in fish can lead to muscular dystrophy, as observed in salmonids, where there is progressive degeneration of cardiac and skeletal muscle. Other common symptoms include anemia, poor immune response, and reproductive issues such as lower fecundity. Fish under environmental stress or high dietary polyunsaturated fatty acids are at higher risk, as vitamin E is crucial for preventing lipid peroxidation (He et al., 2017; El-Sayed & Izquierdo, 2022; Rohani et al., 2023). As per Draper and Csallany (1969), erythrocyte fragility has been used as an indicator of vitamin E status in some animals. While peroxide hemolysis of red blood cells has been used to determine vitamin E deficiency in rainbow trout and rohu (Hung et al., 1981; Sau et al., 2004).

5.1.4 Vitamin K Deficiency

Though less common, vitamin K deficiency in fish affects blood coagulation and bone metabolism. Symptoms include hemorrhaging in various tissues, including muscles and the intestinal tract. Deficient fish often show signs of blood spots on the skin and reduced bone density, which can lead to frequent fractures. Hatchlings and juvenile fish are more sensitive to vitamin K deficiency due to their rapid growth phases and high bone formation rates (Dominguez et al., 2022).

Some of the deficiency signs of fat-soluble vitamins in different fish species have been summarized in Table 10.2.

6 Practical Applications in Aquaculture

6.1 Formulation of Fish Feeds

Achieving a balanced nutritional profile in fish feed is crucial for the success of aquaculture enterprises. Ensuring the correct amounts and ratios of fat-soluble vitamins is particularly important due to their vital roles in fish health and development.

Table 10.2 Some of the deficiency signs of fat-soluble vitamins in fish

Vitamin	Deficiency	Fish
A	Ascites	Salmonids
	Dark skin coloration	Yellowtail
	Spinal jaws or snout deformities	Jain carp
	Edema	Channel catfish
E	Congested fin, skin, or bronchial mantles	Yellowtail
	Exudative diathesis	Channel catfish
	Lordosis	Common carp, grass carp
K	Hematological index changes	Salmonids
	Muscular dystrophy	Korean rockfish, Salmonids Common carp, grass carp
D	Fatty liver, lipid infiltration of liver	Rainbow trout
	Tetany	Rainbow trout

6.1.1 Balanced Vitamin Incorporation

Optimal fish health and growth are supported by a well-balanced intake of vitamins A, D, E, and K, each essential for different physiological processes. Fish feed formulations incorporate a mix of natural and synthetic sources to achieve precise nutrient profiles. Synthetic vitamin premixes are often favored for their stability and accuracy in dosing, especially crucial for heat-sensitive vitamins like A and E, which might degrade during feed processing. Incorporating these vitamins at the right stage of feed processing can significantly affect their bioavailability and effectiveness in promoting fish health (Tacon & Metian, 2015).

6.1.2 Advanced Feed Processing Techniques

Modern feed processing involves sophisticated techniques such as extrusion and pelleting, which help in maintaining the physical and nutritional quality of fish feed. The extrusion process involves cooking the feed ingredients under high pressure and temperature, which, while ensuring the destruction of anti-nutritional factors and pathogens, can also lead to the degradation of sensitive vitamins. Techniques such as encapsulation of vitamins or the addition of heat-stable forms have been developed to address these challenges. Additionally, coating the pellets with a vitamin solution post-extrusion is a common practice designed to minimize nutrient losses (Sørensen, 2012).

6.2 Quality Control and Testing

Quality assurance is integral to feed production, involving stringent testing to ensure that the feed not only meets the defined nutritional specifications but also remains consistent across batches. This includes laboratory analysis of vitamin levels and on-farm trials to observe the feed's performance in real-world conditions. Feedback from these trials is crucial for ongoing improvements to feed formulations, ensuring that they meet the evolving nutritional needs of different fish species (Bell & Waagbø, 2008).

6.3 Case Studies

6.3.1 Case Study 1: Enhanced Vitamin D Fortification in Salmon Feed

A pivotal study conducted in Norway (Jakobsen et al., 2019) explored the impact of increased vitamin D levels in the diets of farmed Atlantic salmon. Researchers investigated how varying levels of dietary vitamin D supplementation could influence the overall health and bone density of salmon. By doubling the standard dietary inclusion rates of vitamin D, significant improvements were observed in the skeletal structure of the salmon. These changes were characterized by increased bone density and reduced instances of deformities that are common during rapid growth phases in aquaculture.

This intervention not only improved the physiological health of the salmon but also brought substantial economic benefits. Enhanced bone density contributed to a reduction in losses typically associated with handling and processing, thereby improving the yield and quality of the harvest. As a result, the marketability of the farmed salmon increased, leading to better economic returns for aquaculture operations.

This case study underscores the critical role of vitamin D in aquaculture diets and highlights the potential for nutritional strategies to significantly enhance both the welfare of fish and the profitability of farming practices. It serves as a testament to the importance of targeted nutritional enhancements in aquaculture, reflecting a move toward more scientifically informed feeding practices.

6.3.2 Case Study 2: Enhanced Vitamin E Supplementation in Caspian Trout Diet

A landmark investigation (Saheli et al., 2021) into the effects of vitamin E supplementation on Caspian trout (*Salmo caspius*) has highlighted significant improvements in fish health and growth. Conducted over an eight-week period, the study evaluated six different diets, each with an increasing concentration of vitamin E, ranging from 4.9 mg to 137.0 mg per kg of diet.

The findings from this study were compelling. Trout fed with diets containing optimal levels of vitamin E, specifically around 78.8 mg per kg, demonstrated notable enhancements in growth metrics such as weight gain ratio and specific growth rate. Furthermore, these trout exhibited improved feed efficiency and protein utilization, critical factors for sustainable aquaculture operations.

Beyond growth performance, the study also revealed that higher dietary vitamin E significantly bolstered the antioxidant defence system in trout. This was evidenced by increased activities of antioxidant enzymes like superoxide dismutase and glutathione peroxidase, which play vital roles in mitigating oxidative stress.

Additionally, the immune response of the trout was strengthened. Enhanced levels of vitamin E were linked with higher concentrations of immunoglobulin M and increased activity of the alternative complement pathway, crucial components of the trout's immune system.

6.3.3 Case Study 3: Integrated Nutrient Management in Tilapia Farming

The Aquaculture Stewardship Council (ASC, 2011) developed comprehensive better management practices to advance nutrient management and overall productivity in tilapia farming. In an effort to enhance the sustainability and productivity of tilapia farming, a significant focus has been placed on implementing better management practices (BMPs). These practices are comprehensively designed to optimize nutrient management, which is crucial for improving growth rates, health, and the overall yield of tilapia farming operations.

The adoption of BMPs in tilapia farming includes precise feeding strategies that ensure nutritional adequacy without overfeeding, thereby reducing waste and preventing nutrient pollution. Additionally, these practices incorporate effective disease management protocols, which further safeguard fish health, enhancing the robustness and marketability of the farmed tilapia.

A documented example of these practices can be found in the guidelines provided by the Aquaculture Stewardship Council. Their detailed documentation outlines strategies that have been applied successfully in various aquaculture settings, demonstrating significant improvements in operational efficiency and environmental sustainability. The implementation of these BMPs not only supports the health and growth of tilapia but also aligns with global sustainability goals by minimizing the environmental impact of aquaculture practices.

7 Future Directions and Research Needs

7.1 Emerging Research

The field of aquaculture nutrition is rapidly evolving, with ongoing research continually reshaping our understanding of vitamin nutrition in fish. New studies are increasingly focusing on not only the requirements but also the optimal levels of vitamins for enhancing fish health, growth, and environmental sustainability.

7.1.1 Personalized Nutrition in Aquaculture

One promising area of research is the development of personalized nutrition strategies tailored to the specific genetic makeup and health needs of different fish species. Emerging studies are examining how genetic variations among fish affect their nutrient metabolism, which could lead to more precise dietary formulations. This approach aims to maximize feed efficiency and growth rates while minimizing waste (Munang'andu & Evensen, 2017).

7.1.2 Enhanced Bioavailability of Vitamins

Another focal point of current research is enhancing the bioavailability of vitamins in fish feeds through new forms of supplementation, such as nanoencapsulation. This technology involves encapsulating vitamins in nano-sized carriers that protect them from degradation during feed processing and release them more efficiently in the digestive tract of fish. Preliminary studies have shown promising results, particularly in the stabilization of vitamins A and E, which are prone to oxidation (Ahmed et al., 2023).

7.1.3 Role of Vitamins in Disease Resistance

Recent research is also exploring the potential role of vitamins in enhancing disease resistance in aquaculture species. Vitamins A and D, for instance, have been shown to modulate immune responses in fish, suggesting that optimizing their levels in diets could bolster fish health against infectious diseases, which are a major concern in aquaculture operations. Ongoing studies aim to determine the specific mechanisms by which these vitamins enhance immune functions and how they can be best utilized in feed formulations (Akbar Ali et al., 2022; Wang et al., 2024).

7.2 Gaps in Knowledge

Despite significant advancements, several gaps remain in our understanding of vitamin nutrition in aquaculture, necessitating further investigation.

7.2.1 Interactions Between Vitamins and Other Dietary Components

While it is known that vitamins interact with various minerals and other nutrients, the complexities of these interactions are not fully understood, especially under the diverse environmental conditions of commercial aquaculture. Research is needed to elucidate these interactions more clearly, particularly how they affect the absorption and efficacy of both vitamins and other essential nutrients (Lall & Kaushik, 2021; Zhu et al., 2024).

7.2.2 Impact of Environmental Stressors on Vitamin Requirements

The impact of environmental stressors such as water temperature, salinity, and pollution on the vitamin requirements of fish is another area that requires further exploration. Stress conditions may alter the nutritional needs of fish, and adapting feed formulations accordingly could significantly improve their resilience and overall health (Lall & Kaushik, 2021).

7.2.3 Long-Term Effects of Vitamin Supplementation

Finally, while the immediate effects of vitamin supplementation on growth and health are well-documented, there is a lack of long-term studies assessing the potential repercussions of chronic high-level intake of certain vitamins. Understanding these effects is crucial to ensuring the sustainability and safety of fortified feeds over the lifespan of aquaculture species (Selvam et al., 2022).

8 Conclusion

8.1 Summary of Key Points

This chapter has explored the critical roles of fat-soluble vitamins A, D, E, and K in fish nutrition, emphasizing their essential functions in maintaining health, growth, and vitality in aquaculture species. From ensuring proper vision and skeletal integrity with vitamins A and D to enhancing antioxidant protection and immune responses with vitamins E and K, these nutrients play foundational roles in the physiological processes of fish.

We discussed species-specific requirements, highlighting how different fish, such as trout and salmon, have varying dietary needs that must be met through carefully formulated feeds. The discussion extended into the practical applications of these vitamins in aquaculture feeds, including the strategies for balancing vitamin content and the technological advancements in feed processing that preserve vitamin potency.

8.2 *Final Thoughts*

The implications of this comprehensive overview extend beyond basic nutritional guidelines, suggesting a path toward more sustainable and efficient aquaculture practices. As the industry continues to grow, the need for precision in nutrition will become increasingly important. Optimizing vitamin intake not only improves the health and productivity of aquaculture operations but also enhances the nutritional quality of the fish products available to consumers.

Future research directions, such as personalized nutrition and the development of advanced feed technologies, promise to further refine our understanding of vitamin nutrition. Addressing the identified gaps in knowledge, particularly regarding nutrient interactions and the long-term effects of vitamin supplementation, will be crucial for advancing fish farming practices globally.

By embracing these insights and continuing to invest in research, the aquaculture industry can achieve greater sustainability and productivity, ensuring a reliable and healthy food source for the growing global population.

References

Abdelhamid, A. F., Gewida, A. G., El-Sayed, A. F. M., & Badran, M. F. (2024). Impacts of different levels of vitamin K on the growth performance, hematological parameters, and immunological response of juvenile Nile tilapia (*Oreochromis niloticus*). *Aquaculture International, 32*, 477–488.

Adewumi, A. A. (2018). The impact of nutrition on fish development, growth and health. *International Journal of Scientific and Research Publications, 8*, 147–153.

Ahmed, J., Vasagam, K. P. K., & Ramalingam, K. (2023). Nanoencapsulated Aquafeeds and current uses in fisheries/shrimps: A review. *Applied Biochemistry and Biotechnology, 195*, 7110–7131. https://doi.org/10.1007/S12010-023-04418-9/METRICS

Ahmed, I., Ahmad, I., & Malla, B. A. (2024). Effects of dietary tryptophan levels on growth performance, plasma profile, intestinal antioxidant capacity and growth related genes in rainbow trout (*Oncorhynchus mykiss*) fingerlings. *Aquaculture, 740710*, 740710.

Akbar Ali, I., Radhakrishnan, D. K., & Kumar, S. (2022). Immunostimulants and their uses in aquaculture. *Aquatic Sciences and Engineering*, 291–322. https://doi.org/10.1007/978-981-19-0817-0_11

Aquaculture Stewardship Council (ASC). (2011). Better management practices for tilapia aquaculture: A tool to assist with compliance to the International Standards for Responsible Tilapia Aquaculture.

Barrows, F. T., Gaylord, T. G., Sealey, W. M., et al. (2008). The effect of vitamin premix in extruded plant-based and fish meal based diets on growth efficiency and health of rainbow trout, *Oncorhynchus mykiss*. *Aquaculture, 283*, 148–155.

Battisti, E. K., Marasca, S., Durigon, E. G., Villes, V. S., Schneider, T. L., Uczay, J., Peixoto, N. C., & Lazzari, R. (2017). Growth and oxidative parameters of Rhamdia quelen fed dietary levels of vitamin A. *Aquaculture, 474*, 11–17.

Bell, J. G., & Waagbø, R. (2008). Safe and nutritious aquaculture produce: Benefits and risks of alternative sustainable Aquafeeds. *Aquatic Ecosystem, 185–225*. https://doi.org/10.1007/978-1-4020-6810-2_6

Bender, D. A. (2003). *Nutritional biochemistry of the vitamins*. Cambridge University Press. https://doi.org/10.1017/CBO9780511615191

Boglione, C., Gagliardi, F., Scardi, M., & Cataudella, S. (2001). Skeletal descriptors and quality assessment in larvae and post-larvae of wild-caught and hatchery-reared gilthead sea bream (*Sparus aurata* L. 1758). *Aquaculture, 192*, 1–22. https://doi.org/10.1016/S0044-8486(00)00446-4

Chen, J., Zeng, M., Liang, X. F., Peng, D., Xie, R., & Wu, D. (2024). Dietary supplementation of VA enhances growth, feed utilization, glucose and lipid metabolism, appetite, and antioxidant capacity of Chinese perch (*Siniperca chuatsi*). *Fish Physiology and Biochemistry, 50*, 225–237.

Davis, K. B. (2009). *Optimizing nutrient utilization and reducing waste through diet composition and feeding strategies*. SRAC Publication No.6003.

De Silva, S. S., & Anderson, T. A. (1995). *Fish nutrition in aquaculture*. Chapman & Hall. https://link.springer.com/book/9780412550300

Dominguez, D., Montero, D., Zamorano, M. J., Castro, P., Fontanillas, R., Prabhu, P. A. J., & Izquierdo, M. (2021). Effects of vitamin D_3 supplementation in gilthead sea bream (*Sparus aurata*) juveniles fed diets high in plant based feedstuffs. *Aquaculture, 543*, 736991.

Dominguez, D., Castro, P., Lall, S., Montero, D., Zamorano, M. J., Fontanillas, R., & Izquierdo, M. (2022). Effects of menadione sodium bisulphite (Vitamin K_3) supplementation of the diets based on plant feed ingredients on growth and bone health of gilthead seabream (*Sparus aurata*) fingerlings. *Aquaculture Nutrition, 2022*, 1613030.

Draper, H. H., & Csallany, A. S. (1969). A simplified hemolysis test for vitamin E deficiency. *The Journal of Nutrition, 98*, 390–394.

El-Sayed, A. F. M., & Izquierdo, M. (2022). The importance of vitamin E for farmed fish – A review. *Reviews in Aquaculture, 14*, 688–703. https://doi.org/10.1111/RAQ.12619

Fenwick, J. C. (1984). Effect of vitamin D_3(cholecalciferol) on plasma calcium and intestinal calcium absorption in goldfish, *Carassius auratus* L. *Canadian Journal of Zoology, 62*, 34–36. https://doi.org/10.1139/Z84-007

Fernandez, I., López-Joven, C., Andree, K. B., Roque, A., & Gisbert, E. (2015). Vitamin A supplementation enhances Senegalese sole (*Solea senegalensis*) early juvenile's immunocompetence: New insights on potential underlying pathways. *Fish & Shellfish Immunology, 46*, 703–709.

Halver, J. E. (1989). The vitamins. In J. E. Halver (Ed.), *Fish nutrition*.

Halver, J. E. (2002). The vitamins. In J. E. Halver & R. W. Hardy (Eds.), *Fish nutrition* (3rd ed., pp. 61–141). Academic.

Hamre, K., Krossøy, C., Lock, E. J., & Moren, M. (2010). Roles of lipid-soluble vitamins during ontogeny of marine fish larvae. *Aquaculture Research, 41*, 745–750.

Hardy, R. W., & Barrows, F. T. (2003). Diet formulation and manufacture. In *Fish nutrition* (pp. 505–600). Academic. https://doi.org/10.1016/B978-012319652-1/50010-0

Hassan, A., Ahmed, I., & Wani, G. B. (2022). Effect of supplementation of vitamin A on growth, Haemato-biochemical composition, and antioxidant ability in *Cyprinus carpio* var. *communis*. *Aquaculture Nutrition, 2022*, 8446092.

He, M., Wang, K., Liang, X., Fang, J., Geng, Y., Chen, Z., Pu, H., Hu, Y., Li, X., & Liu, L. (2017). Effects of dietary vitamin E on growth performance as well as intestinal structure and function of channel catfish (*Ictalurus punctatus*, Rafinesque 1818). *Experimental and Therapeutic Medicine, 14*, 5703. https://doi.org/10.3892/ETM.2017.5295

Hemre, G. I., Lock, E. J., Olsvik, P. A., Hamre, K., Espe, M., Torstensen, B. E., Silva, J., Hansen, A. C., Waagbo, R., Johansen, J. S., Sanden, M., & Sissener, N. H. (2016). Atlantic salmon (*Salmo salar*) require increased dietary levels of B-vitamins when fed diets with high inclusion of plant based ingredients. *Peer J, 4*, e2493.

Hernandez, L. H., & Hardy, R. W. (2020). Vitamin A functions and requirements in fish. *Aquaculture Research, 51*, 3061–3071. https://doi.org/10.1111/ARE.14667

Hixson, S. M. (2014). Fish nutrition and current issues in aquaculture: The balance in providing safe and nutritious seafood, in an environmentally sustainable manner. *Journal of Aquaculture Research & Development, 5*, 234.

Hu, C. J., Chen, S. M., Pan, C. H., & Huang, C. H. (2006). Effects of dietary vitamin A or β-carotene concentrations on growth of juvenile hybrid tilapia, *Oreochromis niloticus*× *O. aureus*. *Aquaculture, 253*, 602–607.

Huang, Q. C., Kwaku, A., Du, T., Tan, B. P., Chi, S. Y., Yang, Q. H., Zhang, S., Liu, H. Y., Dong, X. H., & Yang, Y. Z. (2018). Dietary vitamin A requirement of *Sillago sihama* Forskål. *Aquaculture Research, 49*, 2587–2596.

Huang, Q. C., Zhang, S., Du, T., Yang, Q. H., Chi, S. Y., Liu, H. Y., Yang, Y. Z., Dong, X. H., & Tan, B. P. (2020). Effects of dietary vitamin E on growth, immunity and oxidation resistance related to the Nrf2/Keap1 signalling pathway in juvenile *Sillago sihama*. *Animal Feed Science and Technology, 262*, 114403.

Hung, S. S., Cho, C. Y., & Slinger, S. J. (1981). Effect of oxidized fish oil, DL-α-tocopheryl acetate and ethoxyquin supplementation on the vitamin E nutrition of rainbow trout (*Salmo gairdneri*) fed practical diets. *The Journal of Nutrition, 111*, 648–657.

Jakobsen, J., Smith, C., Bysted, A., & Cashman, K. D. (2019). Vitamin D in wild and farmed Atlantic Salmon (*Salmo Salar*) – What do we know? *Nutrients, 11*, 982. https://doi.org/10.3390/NU11050982

Jeyaraj, N., Rajan, M. R., & Santhanam, P. (2012). Dietary requirement of vitamin A and biochemical composition of common carp *Cyprinus carpio* var. *communis*. *Journal of Fisheries and Aquatic Science, 7*, 65.

Jiang, W. D., Zhou, X. Q., Zhang, L., Liu, Y., Wu, P., Jiang, J., Kuang, S. Y., Tang, L., Tang, W. N., Zhang, Y. A., Shi, H. Q., & Feng, L. (2019). Vitamin A deficiency impairs intestinal physical barrier function of fish. *Fish & Shellfish Immunology, 87*, 546–558.

Jobling, M. (2011). National Research Council (NRC): Nutrient requirements of fish and shrimp. *Aquaculture International, 20*, 601–602. https://doi.org/10.1007/S10499-011-9480-6

John, E. H., & Ronald, W. H. (2002). *Fish nutrition* (3rd ed.). Academic.

Kamal-Eldin, A., & Appelqvist, L. Å. (1996). The chemistry and antioxidant properties of tocopherols and tocotrienols. *Lipids, 31*, 671–701. https://doi.org/10.1007/BF02522884

Karges, K., Baker, H. J., Brooks, J. C., Owens, F. N., Pioneer, D., & Morgan, B. (2001). Effects of supplemental vitamin D_3 on feed intake, carcass characteristics, tenderness, and muscle properties of beef steers. *Journal of Animal Science*. https://doi.org/10.2527/2001.79112844x

Khan, Y. M., & Khan, M. A. (2020). Dietary folic acid requirement of fingerling Catla, *Catla catla* (Hamilton). *Aquaculture Nutrition, 26*, 1035–1045.

Kitamura, S., Suwa, T., Ohara, S., & Nakagawa, K. (1967). Studies on vitamin requirements of rainbow trout-III. Requirement for vitamin A and deficiency symptoms. *Bulletin of the Japanese Society of Scientific Fisheries, 33*, 1126–1131.

Krossøy, C., Waagbø, R., & Ørnsrud, R. (2011). Vitamin K in fish nutrition. *Aquaculture Nutrition, 17*, 585–594. https://doi.org/10.1111/J.1365-2095.2011.00904.X

Kumar, N., Chandan, N. K., Gupta, S. K., Bhushan, S., & Patole, P. B. (2022). Omega-3 fatty acids effectively modulate growth performance, immune response, and disease resistance in fish against multiple stresses. *Aquaculture, 547*, 737506. https://doi.org/10.1016/J.AQUACULTURE.2021.737506

Lall, S. P., & Kaushik, S. J. (2021). Nutrition and metabolism of minerals in fish. *Animals, 11*, 2711. https://doi.org/10.3390/ANI11092711

Li, J., Liang, X. F., Tan, Q., Yuan, X., Liu, L., Zhou, Y., & Li, B. (2014). Effects of vitamin E on growth performance and antioxidant status in juvenile grass carp *Ctenopharyngodon idellus*. *Aquaculture, 430*, 21–27.

Liang, D., Yang, Q., Tan, B., Dong, X., Chi, S., Liu, H., & Zhang, S. (2020). Dietary vitamin A deficiency reduces growth performance, immune function of intestine, and alters tight junction proteins of intestine for juvenile hybrid grouper (*Epinephelus fuscoguttatus*♀× *Epinephelus lanceolatus*♂). *Fish & Shellfish Immunology, 107*, 346–356.

Lin, A. M. Y., Chen, K. B., & Chao, P. L. (2005). Antioxidative effect of vitamin D_3 on zinc-induced oxidative stress in CNS. *Annals of the New York Academy of Sciences, 1053*, 319–329. https://doi.org/10.1196/ANNALS.1344.028

Lin, H., Ding, M., Ray, G. W., Tan, B., & Yang, Q. (2022). Vitamin D promotes growth, feed utilization and gene expression related to lipid metabolism for juvenile orange-spotted grouper Epinephelus coioides. *Aquaculture Reports, 23*, 101074.

Lock, E. J., WaagbØ, R., Wendelaar Bonga, S., & Flik, G. (2010). The significance of vitamin D for fish: A review. *Aquaculture Nutrition, 16*, 100–116. https://doi.org/10.1111/J.1365-2095.2009.00722.X

Lu, Y., Liang, X. P., Jin, M., Sun, P., Ma, H. N., Yuan, Y., & Zhou, Q. C. (2016). Effects of dietary vitamin E on the growth performance, antioxidant status and innate immune response in juvenile yellow catfish (*Pelteobagrus fulvidraco*). *Aquaculture, 464*, 609–617.

Mazurais, D., Glynatsi, N., Darias, M. J., Christodoulopoulou, S., Cahu, C. L., Zambonino-Infante, J. L., & Koumoundouros, G. (2009). Optimal levels of dietary vitamin A for reduced deformity incidence during development of European sea bass larvae (*Dicentrarchus labrax*) depend on malformation type. *Aquaculture, 294*, 262–270. https://doi.org/10.1016/j.aquaculture.2009.06.008

McArley, T. J., Sandblom, E., & Herbert, N. A. (2021). Fish and hyperoxia – From cardiorespiratory and biochemical adjustments to aquaculture and ecophysiology implications. *Fish and Fisheries, 22*, 324–355. https://doi.org/10.1111/FAF.12522

Munang'andu, H. M., & Evensen, O. (2017). Current advances in functional genomics in aquaculture. In *Applications of RNA-Seq and omics strategies – From microorganisms to human health* (pp. 173–212). https://doi.org/10.5772/INTECHOPEN.69883

National Research Council. (2011). *Nutrient requirements of fish and shrimp* (pp. 57–101). National Academy Press.

Peng, L. I., & Gatlin Iii, D. M. (2009). Dietary vitamin E requirement of the red drum *Sciaenops ocellatus*. *Aquaculture Nutrition, 15*, 313–319.

Prabhu, P. A. J., Lock, E. J., Hemre, G. I., Hamre, K., Espe, M., Olsvik, P. A., Silva, J., Hassan, A., Johansen, J., Sissener, N. H., & Waagbø, R. (2019). Recommendations for dietary level of micro-minerals and vitamin D3 to Atlantic salmon (*Salmo salar*) parr and post-smolt when fed low fish meal diets. *Peer J, 7*, e6996.

Prabu, E., Felix, S., Felix, N., Ahilan, B., & Ruby, P. (2017). An overview on significance of fish nutrition in aquaculture industry. *International Journal of Fisheries and Aquatic Studies, 5*, 349–355.

Qiang, J., Wasipe, A., He, J., Tao, Y. F., Xu, P., Bao, J. W., Chen, D. J., & Zhu, J. H. (2019). Dietary vitamin E deficiency inhibits fat metabolism, antioxidant capacity, and immune regulation of inflammatory response in genetically improved farmed tilapia (GIFT, *Oreochromis niloticus*) fingerlings following *Streptococcus iniae* infection. *Fish & Shellfish Immunology, 92*, 395–404.

Rohani, M. F., Tarin, T., Hasan, J., Islam, S. M. M., & Shahjahan, M. (2023). Vitamin E supplementation in diet ameliorates growth of Nile tilapia by upgrading muscle health. *Saudi Journal of Biological Sciences, 30*, 103558. https://doi.org/10.1016/J.SJBS.2023.103558

Saheli, M., Rajabi Islami, H., Mohseni, M., & Soltani, M. (2021). Effects of dietary vitamin E on growth performance, body composition, antioxidant capacity, and some immune responses in Caspian trout (*Salmo caspius*). *Aquaculture Report, 21*, 2352–5134. https://doi.org/10.1016/j.aqrep.2021.100857

Samuel, S., & Sitrin, M. D. (2008). Vitamin d's role in cell proliferation and differentiation. *Nutrition Reviews, 66*(10 Suppl 2), S116. https://doi.org/10.1111/J.1753-4887.2008.00094.X

Santosh, P. L. (2003). The minerals. In J. E. Halver & R. W. Hardy (Eds.), *Fish nutrition* (3rd ed., pp. 260–301). Academic.

Sau, S. K., Paul, B. N., Mohanta, K. N., & Mohanty, S. N. (2004). Dietary vitamin E requirement, fish performance and carcass composition of rohu (*Labeo rohita*) fry. *Aquaculture, 240*, 359–368.

Selvam, C., Prabhu Philip, A. J., Lutfi, E., Sigholt, T., Norberg, B., Bæverfjord, G., Rosenlund, G., Ruyter, B., & Sissener, N. H. (2022). Long-term feeding of Atlantic salmon with varying levels of dietary EPA + DHA alters the mineral status but does not affect the stress responses after mechanical delousing stress. *British Journal of Nutrition, 128*, 2291–2307. https://doi.org/10.1017/S0007114522000514

Shahkar, E., Hamidoghli, A., Yun, H., Kim, D. J., & Bai, S. C. (2018). Effects of dietary vitamin E on hematology, tissue α-tocopherol concentration and non-specific immune responses of Japanese eel, *Anguilla japonica*. *Aquaculture, 484*, 51–57. https://doi.org/10.1016/J.AQUACULTURE.2017.10.036

Shao, L., Zhu, X., Yang, Y., Jin, J., Liu, H., Han, D., & Xie, S. (2016). Effects of dietary vitamin A on growth, hematology, digestion and lipometabolism of on-growing gibel carp (*Carassius auratus gibelio* var. CAS III). *Aquaculture, 460*, 83–89.

Shearer, M. J., & Newman, P. (2008). Metabolism and cell biology of vitamin K. *Thrombosis and Haemostasis, 100*, 530–547.

Shiau, Y. S., & Lin, Y. H. (2006). Vitamin requirements of tilapia. A review. In L. E. Cruz Suares et al. (Eds.), *Avances on nutrition Acuicola VIII. VIII. Simposium internacionalde nutricion acuicola* (pp. 15–17). Universidad Autonoma de Nuevo Leon.

Sivagurunathan, U., Dominguez, D., Tseng, Y., Eryalçın, K. M., Roo, J., Boglione, C., Antony Jesu Prabu, P., & Izquierdo, M. (2022). Effects of dietary vitamin D_3 levels on survival, mineralization, and skeletal development of gilthead sea bream (*Sparus aurata*) larvae. *Aquaculture, 560*, 738505.

Sivaramakrishnan, T., Ambasankar, K., Sathish Kumar, T., Sandeep, K. P., Bera, A., Raja, R. A., Suresh, E., Kailasam, M., & Felix, N. (2024). Effect of dietary vitamin E supplementation on growth, fatty acid composition, intestinal histology, and haemato-immune indices of milkfish, *Chanos chanos*, larvae. *Journal of Applied Aquaculture, 36*, 270–291.

Sørensen, M. (2012). A review of the effects of ingredient composition and processing conditions on the physical qualities of extruded high-energy fish feed as measured by prevailing methods. *Aquaculture Nutrition, 18*, 233–248. https://doi.org/10.1111/J.1365-2095.2011.00924.X

Storebakken, T., Shearer, K. D., & Roem, A. J. (1998). Availability of protein, phosphorus and other elements in fish meal, soy-protein concentrate and phytase-treated soy-protein-concentrate-based diets to Atlantic salmon, *Salmo salar*. *Aquaculture, 161*, 365–379. https://doi.org/10.1016/S0044-8486(97)00284-6

Suda, T., Ueno, Y., Fujii, K., & Shinki, T. (2003). Vitamin D and bone. *Journal of Cellular Biochemistry, 88*, 259–266. https://doi.org/10.1002/JCB.10331

Suttie, J. W. (1980). The metabolic role of vitamin K. *Federation Proceedings, 39*, 2730–2735.

Tacon, A. G. J., & De Silva, S. S. (1997). Feed preparation and feed management strategies within semi intensive fish farming systems in the tropics. *Aquaculture, 151*, 379–404. https://doi.org/10.1016/S0044-8486(96)01494-9

Tacon, A. G. J., & Metian, M. (2015). Feed matters: Satisfying the feed demand of aquaculture. *Reviews in Fisheries Science and Aquaculture, 23*, 1–10. https://doi.org/10.1080/23308249.2014.987209

Tan, B., & Mai, K. (2001). Effects of dietary vitamin K on survival, growth, and tissue concentrations of phylloquinone (PK) and menaquinone-4 (MK-4) for juvenile abalone, *Haliotis discus* hannai Ino. *Journal of Experimental Marine Biology and Ecology, 256*, 229–239. https://doi.org/10.1016/S0022-0981(00)00316-6

Tanumihardjo, S. A. (2011). Vitamin A: Biomarkers of nutrition for development. *The American Journal of Clinical Nutrition, 94*, 658S. https://doi.org/10.3945/AJCN.110.005777

Tipton, N. C., King, D. A., Paschal, J. C., Hale, D. S., & Savell, J. W. (2007). Effects of oral vitamin D_3 supplementation and supplement withdrawal on the accumulation of magnesium, calcium, and vitamin D in the serum, liver, and muscle tissue and subsequent carcass and meat quality of *Bos indicus* influenced cattle. *Meat Science, 75*, 150–158. https://doi.org/10.1016/J.MEATSCI.2006.06.024

Traber, M. G., & Atkinson, J. (2007). Vitamin E, antioxidant and nothing more. *Free Radical Biology and Medicine, 43*, 4–15.

Udagawa, M., Nakazoe, J. I., & Murai, T. (1993). Tissue distribution of phylloquinone and menaquinone-4 in sardine, *Sardinops melanostictus*. *Comparative Biochemistry and Physiology. B: Comparative Biochemistry, 106*, 297–230.

Wang, K., Wang, E., Qin, Z., Zhou, Z., Geng, Y., & Chen, D. (2016). Effects of dietary vitamin E deficiency on systematic pathological changes and oxidative stress in fish. *Oncotarget, 7*, 83869. https://doi.org/10.18632/ONCOTARGET.13729

Wang, Y., Liu, J., Xiao, H., Sun, H., Hu, H., Ma, X., Zhang, A., & Zhou, H. (2024). Dietary intakes of vitamin D promote growth performance and disease resistance in juvenile grass carp (*Ctenopharyngodon idella*). *Fish Physiology and Biochemistry, 50*, 1189. https://doi.org/10.1007/S10695-024-01330-9/METRICS

Webster, C. D., & Lim, C. (2002). Introduction to fish nutrition. In *Nutrient requirements and feeding of finfish for aquaculture* (pp. 1–27).

Wei, X., Hang, Y., Li, X., Hua, X., Cong, X., Yi, W., & Guo, X. (2023). Effects of dietary vitamin K_3 levels on growth, coagulation, calcium content, and antioxidant capacity in largemouth bass, *Micropterus salmoides*. *Aquaculture Fish, 8*, 159–165.

Woodward, B. (1994). Dietary vitamin requirements of cultured young fish, with emphasis on quantitative estimates for salmonids. *Aquaculture, 124*, 133–168.

Wu, C., Lu, B., Wang, Y., Jin, C., Zhang, Y., & Ye, J. (2020). Effects of dietary vitamin D_3 on growth performance, antioxidant capacities and innate immune responses in juvenile black carp *Mylopharyngodon piceus*. *Fish Physiology and Biochemistry, 46*, 2243–2256.

Yossa, R., Sarker, P. K., Mock, D. M., Lall, S. P., & Vandenberg, G. W. (2015). Current knowledge on biotin nutrition in fish and research perspectives. *Reviews in Aquaculture, 7*, 59–73.

Yuan, J., Feng, L., Jiang, W. D., Liu, Y., Jiang, J., Li, S. H., Kuang, S. Y., Tang, L., & Zhou, X. Q. (2016). Effects of dietary vitamin K levels on growth performance, enzyme activities and antioxidant status in the hepatopancreas and intestine of juvenile Jian carp (*Cyprinus carpio* var. *Jian*). *Aquaculture Nutrition, 22*, 352–366.

Zanon, R. B., Silva, T. S. D. C., Cerozi, B. D. S., & Cyrino, J. E. P. (2018). Effects of graded levels of dietary vitamin E on striped Surubim *Pseudoplatystoma reticulatum*. *Aquaculture Research, 49*, 1423–1429.

Zhu, C. B., Ren, H. C., Wu, Y. J., Yang, S., & Fei, H. (2024). Benefits and applications of vitamin C in farmed aquatic animals: An updated review. *Aquaculture International, 32*, 1295–1315.

Zulfiqar, U., Fatima, M., Shah, S. Z. H., Afzal, M., Shamas, S., Roshan, S., Batool, M., & Bilal, M. (2023). Assessment of dietary requirement of vitamin E for Rohu, *Labeo rohita* fingerlings fed practical diet. *Pakistan Journal of Zoology, 55*, 581.

Chapter 11
Role of Genomics and Nutrigenomics in Aquaculture Nutrition

Seemab Zehra, Majed Alghamdi, Asaad H. Mohamed, Y. S. Alhafedh, and A. Al-Shaikhi

1 Introduction

Aquaculture is increasingly critical in global food security, contributing over 50% of the fish consumed worldwide (FAO, 2020; Reshi et al., 2025). As wild fish stocks face depletion and environmental pressures, the reliance on farmed seafood is rising, pushing the aquaculture sector to innovate in various domains, particularly nutrition. The efficiency of feed, along with its environmental impact, is central to aquaculture's sustainability. Over 50% of aquaculture production costs are linked to feed (Bedolfe, 2015), making feed optimization a key driver of both economic and environmental outcomes in the industry.

The advent of genomic technologies, such as next-generation sequencing (NGS), has revolutionized how researchers approach aquaculture nutrition. Genomics helps uncover the genetic blueprint of farmed species, while nutrigenomics offers insights into how these organisms interact with their diets at a molecular level (Sindelar, 2016). Together, these fields support precision aquaculture, where diets can be tailored to meet the genetic and metabolic needs of specific species, improving growth rates, feed efficiency, and health outcomes (Moran et al., 2024).

Traditional aquaculture nutrition relied on general formulations designed for species groups, but genomics and nutrigenomics now provide a more precise understanding of how nutrients affect growth, metabolism, and immune responses at a

S. Zehra (✉) · M. Alghamdi · A. H. Mohamed
Kaust Beacon Development, King Abdullah University of Science and Technology, Jeddah, Saudi Arabia
e-mail: seemab.zehra@kaust.edu.sa; asaad.mohamed@kaust.edu.sa

Y. S. Alhafedh · A. Al-Shaikhi
Ministry of Environment, Water and Agriculture, Riyadh, Saudi Arabia

I. Ahmed, I. Ahmad (eds.), *Aquaculture: Enhancing Food Security and Nutrition*, https://doi.org/10.1007/978-3-031-92858-1_11

genetic level. This chapter delves into the roles of genomics and nutrigenomics in aquaculture nutrition, highlighting their potential to improve both efficiency and sustainability, supported by numerous studies and advances in the field.

2 The Role of Genomics in Aquaculture Nutrition

Genomics is the study of the entire DNA of an organism and provides a comprehensive understanding of the genetic factors that influence key traits such as growth, feed conversion, and disease resistance (Yáñez et al., 2014). In aquaculture, genomics has enabled the development of selective breeding programs aimed at enhancing desirable traits such as faster growth rates, improved feed efficiency, and greater disease resistance (Rodde, 2020). The application of genomic technologies allows breeders to identify and select for specific genetic markers that confer these traits, accelerating genetic improvement compared to traditional breeding methods (Dunham, 2023; Kajungiro et al., 2019).

3 Selective Breeding for Nutritional Traits

Selective breeding based on genomic information allows aquaculture practitioners to breed fish with superior nutritional traits. For example, studies on Atlantic salmon (*Salmo salar*) have identified key genomic regions associated with growth, feed conversion efficiency, and disease resistance (Yáñez et al., 2014). By targeting these regions in selective breeding programs, it is possible to develop strains of salmon that require less feed to achieve the same growth rates, thereby reducing feed costs and environmental impacts (Houston et al., 2020).

In tilapia (*Oreochromis niloticus*), genomic selection has been used to improve growth performance by selecting for specific genes involved in metabolism and nutrient utilization (Yáñez et al., 2020; Palaiokostas & Houston, 2018). This approach has demonstrated that fish with certain genetic profiles can utilize feed more efficiently, improving feed conversion ratios and reducing waste (Palaiokostas & Houston, 2018). Genomics-based selective breeding has thus become a critical tool in optimizing aquaculture feed efficiency and sustainability.

4 Genomic Insights into Species-Specific Nutritional Requirements

Aquaculture encompasses a wide range of species, each with its own unique nutritional requirements. Genomics provides the ability to explore these requirements at a molecular level by identifying the genes involved in key metabolic pathways (Yue & Wang, 2017). For instance, species-specific differences in amino acid and fatty acid metabolism have been revealed through genomic studies, enabling the formulation of diets tailored to the genetic makeup of each species (Zhang et al., 2020). This has been particularly valuable in optimizing feed formulations for species such as rainbow trout (*Oncorhynchus mykiss*) and gilthead seabream (*Sparus aurata*), where genomic data has provided new insights into their specific nutritional needs (Hardy & Kaushik, 2021; Munang'andu & Evensen, 2017; Tupac-Yupanqui et al., 2013).

5 Nutrigenomics: A Bridge Between Nutrition and Genomics

Nutrigenomics, the study of how nutrition influences gene expression and vice versa, plays a pivotal role in understanding the interaction between diet and an organism's genome (Fig. 11.1). Nutrigenomics allows researchers to explore how specific nutrients affect the activation or suppression of genes involved in growth, immunity, and stress responses (Panserat & Kaushik, 2010). In aquaculture, nutrigenomics is a relatively new but rapidly expanding field, with growing evidence of its potential to optimize feed formulations based on the genetic and metabolic responses of different species (Lazzarotto et al., 2016; Martin & Król, 2017).

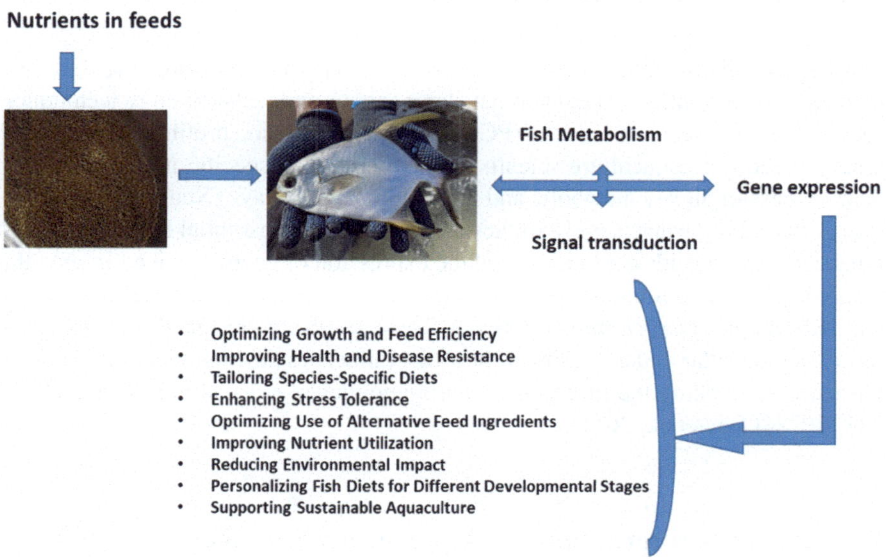

Fig. 11.1 Impact of nutrigenomics on different aspects of fish

6 The Origins of Nutrigenomics in Aquaculture

6.1 Early Observations and the Role of Nutrition in Phenotypic Expression

Long before the molecular understanding of nutrition, fish farmers and researchers observed that dietary variations could lead to significant differences in fish growth, survival, and reproduction. These early insights were based on phenotypic changes, such as body composition and feed conversion efficiency, in response to dietary interventions. Nutritional studies in aquaculture traditionally focused on macronutrient optimization, addressing the protein and energy requirements of various fish species to maximize growth.

In the mid-twentieth century, advances in biochemical techniques allowed for deeper investigations into how specific nutrients modulated metabolic pathways, laying the foundation for molecular studies in aquaculture nutrition. By understanding nutrient metabolism at the biochemical level, researchers were able to predict how different feed ingredients would affect fish physiology.

6.2 Transition to Molecular Techniques and Early Nutrigenomics

The late twentieth century marked a transition from basic nutritional research to molecular investigations, thanks to developments in molecular biology techniques such as polymerase chain reaction (PCR) and gene expression profiling. These technologies allowed aquaculture scientists to examine how specific nutrients altered gene expression in key metabolic and physiological pathways (Xue, 2022; Browdy et al., 2010). For instance, early studies demonstrated that essential fatty acids, vitamins, and amino acids could influence the expression of genes involved in growth, stress response, and immune function in species like Atlantic salmon (*Salmo salar*) and Nile tilapia (*Oreochromis niloticus*). These studies represented the first steps toward a molecular understanding of how diet influences the biological functions of farmed fish, marking the inception of nutrigenomics in aquaculture (Varga, 2022; Xue, 2022; Geay et al., 2011).

7 Gene-Diet Interactions in Aquaculture Species

Gene-diet interactions are central to nutrigenomics. Certain nutrients, such as omega-3 fatty acids or essential amino acids, can influence the expression of genes involved in lipid metabolism, immune function, and growth (Mondal & Panda, 2020). Studies have shown that altering dietary components can lead to significant changes in gene expression, which in turn affects growth performance and disease resistance (Pérez-Sánchez et al., 2018; Amri et al., 2022). For example, in European seabass (*Dicentrarchus labrax*), altering the lipid composition of the diet significantly impacted the expression of genes related to lipid metabolism and growth, demonstrating the potential of nutrigenomics to fine-tune feed formulations for optimal performance (Varga, 2022; Rimoldi et al., 2023; Geay et al., 2011).

8 Nutrigenomics in Aquaculture: Milestones and Key Discoveries

8.1 The Impact of Feed Ingredients on Gene Expression

One of the most significant milestones in the field of nutrigenomics is the discovery of how various feed ingredients affect gene expression in aquaculture species. As the demand for sustainable aquaculture practices continues to grow, researchers have increasingly focused on the nutritional composition of feeds and how they can influence the health, growth, and overall performance of aquatic organisms. Nutrigenomics, the study of how nutrition interacts with the genome, offers critical insights into the molecular mechanisms underlying these effects, allowing for more informed decisions in feed formulation and dietary management (Rimoldi et al., 2023).

The composition of feed ingredients plays a pivotal role in modulating gene expression linked to various metabolic pathways, particularly those related to lipid metabolism and growth. Traditional aquaculture diets have relied heavily on fishmeal and fish oil, which provide essential nutrients and fatty acids that promote optimal growth and health. However, the increasing pressure on marine resources has necessitated a shift toward alternative feed ingredients, especially plant-based sources. This transition is not merely an economic response to dwindling fish stocks; it is also a scientific challenge that requires a deeper understanding of how these ingredients can affect gene regulation and, consequently, the physiological processes in fish and shrimp (Serra et al., 2024; Ma & Hu, 2024).

Research has demonstrated that plant-based feed ingredients can influence the expression of genes involved in lipid metabolism. For instance, specific dietary lipids derived from plants can modulate the expression of genes that encode enzymes responsible for fatty acid synthesis and oxidation (Zou et al., 2017; Estruch et al., 2018). This modulation is crucial as it can lead to alterations in lipid deposition, growth rates, and overall health. When fish are fed plant-based diets, there is often a notable shift in their metabolic pathways. This shift can enhance their ability to utilize nontraditional feed sources, ultimately contributing to improved feed conversion ratios and growth performance (Estruch et al., 2018; Evliyaoğlu et al., 2022).

Furthermore, the incorporation of various plant ingredients in aquaculture feeds has been linked to the regulation of genes associated with growth factors, signaling pathways, and immune responses (Zhu et al., 2018). For example, certain plant proteins and carbohydrates have been shown to enhance the expression of growth hormone genes and other anabolic factors, leading to accelerated growth rates in species such as tilapia and salmon (Dhanasiri et al., 2020; Zou et al., 2017; El-Naggar et al., 2021; Katan, 2022). These findings not only highlight the potential of plant-based ingredients to support growth but also indicate their role in enhancing the health and resilience of farmed fish against diseases (Katan, 2022; Magbanua & Ragaza, 2024).

Moreover, the use of alternative feed ingredients in aquaculture has significant implications for reducing the environmental impact of fish farming. By utilizing plant-based diets, the reliance on fishmeal and fish oil can be decreased, thereby alleviating the pressure on wild fish stocks. This transition aligns with global sustainability goals and helps minimize the ecological footprint of aquaculture operations (Jannathulla et al., 2019). The insights gained from nutrigenomic studies enable researchers and aquaculture producers to design feeds that not only meet the nutritional needs of fish but also promote their health while ensuring sustainable practices (Heras, 2021).

One of the critical aspects of this research involves understanding the interactions between various feed ingredients and the genetic makeup of different aquaculture species. Different species exhibit unique responses to dietary changes, necessitating species-specific feed formulations (Bhatt & Pandey, 2024; Alhoshy et al., 2022; Puvanasundram et al., 2021; Cerda & Manchado, 2013). By leveraging genomic tools and techniques, researchers can identify specific genes and pathways that are activated or repressed in response to various feed components. This information can be used to optimize feed formulations, ensuring that they align with the nutritional requirements and metabolic capabilities of the target species (Yáñez et al., 2014).

8.2 Epigenetics in Aquaculture

Nutrigenomics, a burgeoning field at the intersection of nutrition and genomics, extends beyond the traditional focus on gene expression to encompass the fascinating realm of epigenetics. Epigenetic mechanisms, such as DNA methylation and histone modifications, play crucial roles in regulating gene activity without altering the underlying DNA sequence. In aquaculture, understanding these mechanisms has profound implications for optimizing fish performance and health through dietary interventions.

8.2.1 Understanding Epigenetics

Epigenetics refers to the chemical modifications that regulate gene expression and cellular activity, acting as a control mechanism over how genes are turned on or off in response to environmental stimuli, including dietary inputs. Unlike genetic mutations, which change the DNA sequence itself, epigenetic modifications are reversible and can be influenced by various factors, such as nutrition, stress, and environmental conditions (Granada et al., 2018; Moghadam et al., 2015; Piferrer, 2023). DNA methylation involves the addition of a methyl group to the DNA molecule, typically at cytosine bases, which can inhibit gene expression. Histone modifications, on the other hand, alter the structure of histone proteins around which DNA is wrapped, influencing the accessibility of DNA to the transcription machinery (Navarro-Martín et al., 2023).

8.2.2 Dietary Influences on Epigenetic Mechanisms

Research has shown that dietary components can significantly impact epigenetic modifications in fish. For example, essential nutrients such as folate, vitamin B12, and methionine are vital for the methylation processes that regulate gene expression (Skjærven et al., 2022; Mierziak et al., 2021). Specific dietary fats, such as omega-3 fatty acids, have also been linked to changes in histone acetylation, affecting the expression of genes involved in growth and metabolism (Lin et al., 2019).

A notable aspect of epigenetic regulation in aquaculture is the concept of early nutritional programming. Studies have demonstrated that the early dietary exposure of juvenile fish can lead to lasting epigenetic changes that influence growth, stress response, and overall fitness later in life. For instance, the early provision of high-quality protein sources has been shown to result in beneficial epigenetic modifications that enhance muscle development and stress resilience in species like Atlantic salmon (Naya-Català et al., 2023; Yang et al., 2024; Zhu et al., 2024). This phenomenon emphasizes the importance of diet during critical developmental windows and suggests that nutritional strategies can be employed to optimize growth and survival rates.

8.3 Nutrigenomics and Fish Health: Functional Feeds for Enhanced Health and Immunity

Nutrigenomics also has important implications for the development of functional feeds—diets designed to promote specific health outcomes such as improved immunity or stress tolerance. Studies have shown that certain nutrients, such as selenium or vitamin E, can influence the expression of genes involved in immune responses, enhancing disease resistance in farmed species (Martin & Król, 2017). For example, dietary selenium has been shown to upregulate genes related to antioxidant defense in rainbow trout, improving their ability to withstand environmental stressors (Keyvanshokooh, 2023). This highlights the potential of nutrigenomics to create functional feeds that not only support growth but also improve overall health and welfare in aquaculture systems. Functional feeds that target specific genetic pathways related to immune response have gained attention as a way to enhance disease resistance in aquaculture (Waagbø & Remø, 2020). Ibrahim et al. (2022) investigated the use of functional nutrients in tilapia diets and found that gene expression patterns related to immune responses were significantly influenced by these nutrients, improving the fish's resistance to diseases. This study demonstrates the potential of nutrigenomics to create diets that enhance immune function, reducing the need for antibiotics and other treatments. Earlier studies conducted on fish examined how gene-nutrient interactions affected immune system responses when trout were fed functional diets (Ibrahim et al., 2022; Zhang et al., 2020; Saba et al., 2024; Waagbø & Remø, 2020). Their findings further support the idea that nutrigenomic-based dietary formulations can strengthen immune responses, improving disease resistance in aquaculture species.

8.4 Genomics and Nutrigenomics for Sustainable Aquaculture

Sustainability is a critical issue in aquaculture, with increasing pressure to reduce the environmental impact of feed production and use. Traditional fishmeal and fish oil, which are derived from wild-caught fish, are increasingly being replaced by alternative ingredients such as plant-based proteins, insect meals, and algae. Genomics and nutrigenomics play a crucial role in evaluating the nutritional value of these alternatives and understanding how they impact the growth, health, and metabolism of farmed species (Martin & Król, 2017).

8.5 Optimizing Alternative Feeds Using Nutrigenomics

One of the major challenges in replacing fishmeal and fish oil with alternative ingredients is ensuring that these alternatives meet the nutritional needs of farmed species without compromising growth or health. Genomic studies have provided valuable insights into how different species metabolize alternative feed ingredients, allowing for the development of optimized diets that maintain performance while reducing the reliance on marine resources (Lende et al., n.d.). For instance, studies on Atlantic salmon have shown that certain plant-based proteins can replace fishmeal in diets without affecting growth, provided that they are supplemented with specific amino acids to match the fish's nutritional requirements (Dhanasiri et al., 2020; Sissener et al., 2013; Sørensen et al., 2021; Tacchi et al., 2012).

8.6 *Nutrigenomics and Sustainable Feed Formulation*

Nutrigenomics also supports sustainability by enabling more efficient use of feed ingredients. By understanding how different nutrients influence gene expression and metabolic pathways, researchers can design diets that maximize nutrient utilization and minimize waste. For example, nutrigenomic studies on rainbow trout have shown that diets rich in certain amino acids can improve protein retention and reduce nitrogen waste, contributing to more environmentally sustainable farming practices (Saha et al., 2020). The key studies in nutrigenomics in aquaculture are given in Table 11.1.

9 Nutrigenomic Variability Among Species

One of the critical challenges in aquaculture nutrition is the inherent variability in how different species, and even individuals within the same species, respond to identical diets. This variability can arise from several factors, including genetic makeup, environmental conditions, and dietary composition, all of which can significantly influence growth, feed efficiency, and overall health. Nutrigenomics, a field that examines the interactions between nutrition and gene expression, plays a pivotal role in addressing these challenges by uncovering the genetic basis of this variability. By understanding how specific genes are activated or silenced in response to dietary components, researchers can develop tailored diets that cater to the unique nutritional needs of different fish populations and species.

For instance, studies in tilapia have illustrated that genetic differences between strains lead to significant variations in the utilization of dietary lipids and proteins. In Tian et al. (2018), it was demonstrated that different tilapia strains showed distinct metabolic responses when fed diets with varying lipid profiles. This highlights the importance of considering genetic variation when formulating feeds, as a one-size-fits-all approach is unlikely to optimize growth and feed efficiency across diverse populations.

Furthermore, research by (Torsabo et al., 2024; Chandhini & Rejish Kumar, 2019; Cardinaletti et al., 2022) emphasizes that such genetic variability is not limited to tilapia but is observed across multiple aquaculture species. For example, in salmonids, genetic variation can influence how fish metabolize dietary carbohydrates and proteins, affecting overall growth rates and feed conversion ratios. Understanding these species-specific responses enables aquaculture nutritionists to design diets that not only meet the general nutritional requirements but also take into account the genetic predispositions of the fish.

The implications of nutrigenomic variability extend beyond growth and feed efficiency; they also encompass health and disease resistance. In many species, certain genetic lines may exhibit greater resilience to stressors when provided with specific dietary components. For example, in species such as Atlantic salmon,

Table 11.1 Key studies in aquaculture nutrigenomics

Study	Species	Key findings
Ma et al. (2023)	*Cyprinus carpio*	Cultivated in land-based container recirculating aquaculture systems (C-RAS); exhibit superior muscle quality compared to those raised in traditional pond systems (TP)
Machado et al. (2018)	*Sardina pilchardus*	Serves as a critical resource for future conservation efforts and comparative genomic studies among teleosts
Ngoh et al. (2015)	*Lates calcarifer*	Transcriptomic analysis revealed distinct patterns of gene expression associated with different feeds (mainly fatty acids profile)
Murray et al. (2010)	*Hippoglossus hippoglossus*	Demonstrated that soybean meal (SBM) can replace fish meal (FM) in juvenile Atlantic halibut (*Hippoglossus hippoglossus*) diets without affecting feed intake or palatability. However, it did reveal slight changes in the expression of immune genes, suggesting potential long-term impacts on immune function
Hays et al. (2023)	*Scophthalmus maximus*	Taurine supplementation significantly enhanced the expression of muscle growth-promoting genes such as *TauT*, *myoD*, *Myf5*, myogenin, and follistatin
Juárez Gutiérrez et al. (2021)	*Odontesthes bonariensis*	Larvae fed rotifers exhibited higher growth and survival, with overexpression of genes related to growth, development, and digestive enzymes. In contrast, larvae on the microdiet showed overexpression of genes associated with stress and genome instability, correlating with poorer performance
Jin et al. (2015)	*Ctenopharyngodon idella*	Increased dietary protein seems to upregulate acetyl-CoA carboxylase 1 (ACC1) in both the liver and hypothalamus, playing a crucial role in fatty acid synthesis and metabolism
Minghetti et al. (2011)	*Salmo salar*	Supplementation with cholesterol specifically influenced the expression of genes such as acyl-CoA oxidase, HMG-CoA reductase, and Elovl5 elongase
Zhou et al. (2015)	*Trachinotus ovatus*	The expression of the hepatic growth hormone gene was positively correlated with growth performance. This indicates that as dietary carbohydrate levels increased, leading to improved growth metrics, the expression of GH was also enhanced
Kortner et al. (2012)	*Salmo salar*	The combination of soybean saponins with pea protein concentrate in the diet of Atlantic salmon led to significant enteritis and major transcriptome changes, marked by acute inflammation and alterations in key metabolic and protective gene expressions

Leduc et al. (2018)	*Dicentrarchus labrax*	Dietary inclusion of protein hydrolysates from shrimp and tilapia in a low fish meal diet restored gene expression patterns in the anterior intestine of European seabass to levels similar to those observed in a control diet with 20% fish meal, with shrimp hydrolysate specifically modulating immunity-related gene pathways, while tilapia hydrolysate primarily affected nutritional metabolism
Lei et al. (2024)	*Micropterus salmoides*	A carbohydrate-rich diet in *Micropterus salmoides* influenced gene expression related to lipid metabolism and growth. Specific SNP markers were linked to growth performance, with certain genotypes (e.g., AA at rs31435313) associated with slower growth, underscoring nutrition's impact on gene expression and growth in largemouth bass
Wanna et al. (2021)	*Litopenaeus vannamei*	Supplementing *Litopenaeus vannamei* diets with the probiotic Pediococcus pentosaceus MR001 improved growth, digestive enzyme activity, immune response, and survival rates, with genetic analysis highlighting beneficial traits like bacteriocin production and stress resistance
Geay et al. (2011)	*Dicentrarchus labrax*	An all-plant diet significantly altered the hepatic transcriptome of European sea bass, with 582 differentially expressed genes, including upregulation of LC-PUFA and cholesterol biosynthesis pathways
Wong et al. (2013)	*Oncorhynchus mykiss*	The type of diet (fishmeal-based vs. grain-based) in rainbow trout (*Oncorhynchus mykiss*) significantly influenced physiological responses, suggesting that dietary composition affects gene expression related to growth and health, despite only minor changes in intestinal microbiota composition
Song et al. (2023)	*Trachinotus ovatus*	A dietary lipid level of approximately 12.12% maximized growth and improved body composition in golden pompano (*Trachinotus ovatus*). This level positively influenced muscle growth regulation, hormone response, and nutrient metabolism while lowering mRNA expression of immune response and antioxidant defense regulators
He et al. (2013)	*Oreochromis niloticus* ♀ × *Oreochromis aureus* ♂	Supplementation of hybrid tilapia (*Oreochromis niloticus* × *Oreochromis aureus*) diets with low doses of *Bacillus subtilis* C-3102 (2.5 × 10^5 and 5.0 × 10^5 CFU g − 1) altered gut bacterial communities and significantly upregulated intestinal cytokine expression (IL-1β, TGF-β, and TNF-α), while downregulating intestinal HSP70
Zhou et al. (2018)	*Ctenopharyngodon idella*	The addition of 0.06 g/kg bile acids (BA) to the high lipid diet (7 L group) significantly increased the relative expression of lipid catabolism genes in the hepatopancreas and muscle of grass carp (*Ctenopharyngodon idella*), compared to both the lower lipid diet (5 L group) and the high lipid diet without BA
Safari et al. (2020)	*Huso huso*	Supplementation of the diet with 0.1% tannins (from chestnut wood, Silvafeed®ATX) significantly upregulated the expression of growth-related genes (gh and igf-i) in juvenile beluga sturgeon (*Huso huso*), while also improving weight parameters and food conversion ratio (FCR)

(continued)

Table 11.1 (continued)

Study	Species	Key findings
Prisingkorn et al. (2017)	*Megalobrama amblycephala*	The high-carbohydrate diet (HCBD) significantly altered gene expression in blunt snout bream (*Megalobrama amblycephala*), particularly affecting genes associated with nonalcoholic fatty liver disease (NAFLD) and insulin signaling pathways
Sharawy et al. (2022)	*Litopenaeus vannamei*	Dietary supplementation with 1% nanoparticle-based *Arthrospira platensis* (NAP1%) significantly upregulated the expression of growth-related genes IGF-I and IGF-II in Pacific white shrimp (*Litopenaeus vannamei*), in a dose-dependent manner
Shu-Chien et al. (2017)	*Sagmariasus verreauxi*	The vegetable oil diet for juvenile spiny lobster (*Sagmariasus verreauxi*) resulted in higher expression of genes related to β-oxidation and fatty acid biosynthesis, indicating a preference for utilizing monounsaturated fatty acids for energy
Cai et al. (2020)	*Lateolabrax maculatus*	The 47% protein diet at low temperature (27 °C) significantly enhanced the expression of genes involved in amino acid, fatty acid, and glucose metabolism in spotted seabass (*Lateolabrax maculatus*)
Poon et al. (2023)	*Lutjanus malabaricus*	Supplementation of the diet with astaxanthin significantly enhanced red coloration in Malabar snappers (*Lutjanus malabaricus*), as evidenced by the upregulation of key pigment-related genes such as *bco1*, *bco2l*, *pax7*, *slc7a11*, *hsp70*, and *stard5*
Yuan et al. (2015)	*Cynoglossus semilaevis*	Dietary supplementation with 1.07–1.42% arachidonic acid (ARA) significantly increased the mRNA expression of COX-2 and 5-LOX in larval half-smooth tongue sole (*Cynoglossus semilaevis*), indicating their involvement in ARA metabolism
Yue et al. (2022)	*Acipenser baerii*♀ × *A. schrenckii* ♂	Glutamine supplementation in the soybean meal (SBMG diet) significantly influenced gene expression in hybrid sturgeon, as evidenced by the identification of 1564 differentially expressed unigenes (DEGs) when comparing the SBMG diet to the SBM diet

genetic variations have been linked to differences in immune response when exposed to pathogens. Nutrigenomic studies can help identify these variations, allowing for the formulation of feeds that enhance immune function and stress resistance, ultimately improving the sustainability of aquaculture operations.

10 Technological Advances Driving Nutrigenomics in Aquaculture

10.1 Genomic Tools and High-Throughput Techniques

Advances in genomic technologies have dramatically accelerated nutrigenomics research. Next-generation sequencing (NGS) allows researchers to conduct comprehensive analyses of the transcriptome, providing detailed information on how fish genes respond to different diets. These tools enable the identification of thousands of genes affected by dietary changes, offering new insights into the molecular pathways that control growth, metabolism, and health in farmed fish.

The integration of bioinformatics tools has also played a critical role in analyzing the vast datasets generated by nutrigenomic studies. With the help of computational methods, researchers can predict gene-nutrient interactions and identify novel pathways for improving feed efficiency and health.

10.2 Metabolomics and Proteomics

In addition to genomics, metabolomics and proteomics have emerged as complementary approaches to nutrigenomics. Metabolomics, the study of small metabolites produced during metabolism, helps identify the biochemical changes induced by diet, while proteomics, the study of proteins, reveals how gene expression translates into functional proteins. Together, these technologies provide a more comprehensive view of how nutrition influences the physiology of aquaculture species at the molecular level.

10.3 Challenges and Future Directions

While the integration of genomics and nutrigenomics in aquaculture nutrition holds great promise, several challenges remain. The complexity of gene-diet interactions necessitates more research to fully understand the effects of various nutrients on different species. Additionally, the cost of implementing genomic and nutrigenomic technologies can be prohibitive for small-scale farmers.

Looking forward, the future of aquaculture nutrition lies in precision feeding—developing diets that are specifically tailored to the genetic and metabolic needs of each species, population, or even individual. This will lead to more sustainable practices, with improved growth rates, better feed efficiency, and reduced environmental impact.

11 Conclusion

The application of genomics and nutrigenomics in aquaculture nutrition has transformed how the industry approaches feed formulation and sustainability. Genomic tools have enabled the development of selective breeding programs that improve growth, feed efficiency, and disease resistance, while nutrigenomics provides a deeper understanding of how diet interacts with an organism's genome. Together, these fields are paving the way for precision nutrition in aquaculture, where diets are tailored to the specific needs of different species and populations, improving performance while reducing environmental impacts. As aquaculture continues to grow, the integration of genomics and nutrigenomics will be essential in ensuring that the industry remains sustainable and efficient.

References

Alhoshy, M., Shehata, A. I., Habib, Y. J., Abdel-Latif, H. M., Wang, Y., & Zhang, Z. (2022). Nutrigenomics in crustaceans: Current status and future prospects. *Fish & Shellfish Immunology, 129*, 1–12.

Amri, A., Bouraoui, Z., Balbuena-Pecino, S., Capilla, E., Gharred, T., Haouas, Z., Guerbej, H., Hosni, K., Navarro, I., & Jebali, J. (2022). Dietary supplementation with Aloe vera induces hepatic steatosis and oxidative stress together with a disruption of cellular signaling pathways and lipid metabolism related genes' expression in gilthead sea bream (*Sparus aurata*). *Aquaculture, 559*, 738433.

Bedolfe, S. (2015). Addressing the dependence of aquaculture on wild fishery resources. *Marine Biology*.

Bhatt, D., & Pandey, A. (2024). Potential novel feed ingredients in aquaculture for future FeeD: A review. *Journal of Experimental Zoology India, 27*(1).

Browdy, C. L., Hulata, G., Liu, Z., Allan, G. L., Sommerville, C., de Andrade, T. P., Pereira, R., Yarish, C., Shpigel, M., & Chopin, T. (2010). Novel and emerging technologies: Can they contribute to improving aquaculture sustainability. In *Farming the waters for people and food. Proceedings of the global conference on aquaculture* (pp. 149–191).

Cai, L.-S., Wang, L., Song, K., Lu, K.-L., Zhang, C.-X., & Rahimnejad, S. (2020). Evaluation of protein requirement of spotted seabass (Lateolabrax maculatus) under two temperatures, and the liver transcriptome response to thermal stress. *Aquaculture, 516*, 734615. https://doi.org/10.1016/j.aquaculture.2019.734615

Cardinaletti, G., Di Marco, P., Daniso, E., Messina, M., Donadelli, V., Finoia, M. G., Petochi, T., Fava, F., Faccenda, F., & Contò, M. (2022). Growth and welfare of rainbow trout (*Oncorhynchus mykiss*) in response to graded levels of insect and poultry by-product meals in fishmeal-free diets. *Animals, 12*(13), 1698.

Cerda, J., & Manchado, M. (2013). Advances in genomics for flatfish aquaculture. *Genes & Nutrition, 8*, 5–17.

Chandhini, S., & Rejish Kumar, V. J. (2019). Transcriptomics in aquaculture: Current status and applications. *Reviews in Aquaculture, 11*(4), 1379–1397.

Dhanasiri, A. K., Johny, A., Xue, X., Berge, G. M., Bogevik, A. S., Rise, M. L., Fæste, C. K., & Fernandes, J. M. (2020). Plant-based diets induce transcriptomic changes in muscle of zebrafish and Atlantic salmon. *Frontiers in Genetics, 11*, 575237.

Dunham, R. (2023). *Aquaculture and fisheries biotechnology: Genetic approaches*. CABI.

El-Naggar, K., Mohamed, R., El-katcha, M. I., Abdo, S. E., & Soltan, M. A. (2021). Plant ingredient diet supplemented with lecithin as fish meal and fish oil alternative affects growth performance, serum biochemical, lipid metabolism and growth-related gene expression in Nile tilapia. *Aquaculture Research, 52*(12), 6308–6321.

Estruch, G., Collado, M. C., Monge-Ortiz, R., Tomás-Vidal, A., Jover-Cerdá, M., Peñaranda, D. S., Pérez Martínez, G., & Martínez-Llorens, S. (2018). Long-term feeding with high plant protein based diets in gilthead seabream (*Sparus aurata*, L.) leads to changes in the inflammatory and immune related gene expression at intestinal level. *BMC Veterinary Research, 14*, 1–15.

Evliyaoğlu, E., Kilercioğlu, S., Yılmaz, H. A., Turchini, G. M., Paolucci, M., Clark, T. D., Demirkale, İ., & Eroldoğan, O. T. (2022). Iron supplementation in plant-based aquafeed: Effects on growth performance, tissue composition, iron-related serum parameters and gene expression in rainbow trout (*Oncorhynchus mykiss*). *Aquaculture, 550*, 737884.

FAO. (2020). World fisheries and aquaculture. In *Food and agriculture organization* (Vol. 2020, pp. 1–244).

Geay, F., Ferraresso, S., Zambonino-Infante, J. L., Bargelloni, L., Quentel, C., Vandeputte, M., Kaushik, S., Cahu,C. L., & Mazurais, D. (2011). Effects of the total replacement of fish-based diet with plant-based diet on the hepatic transcriptome of two European sea bass *(Dicentrarchus labrax) half-sibfamilies showing different growth rates with the plant-based diet. BMC Genomics, 12(1), 522.* https://doi.org/10.1186/1471-2164-12-522

Granada, L., Lemos, M. F., Cabral, H. N., Bossier, P., & Novais, S. C. (2018). Epigenetics in aquaculture–the last frontier. *Reviews in Aquaculture, 10*(4), 994–1013.

Hardy, R. W., & Kaushik, S. J. (2021). *Fish nutrition*. Academic press.

Hays, H., Gu, Z., Mai, K., & Zhang, W. (2023). Transcriptome-based nutrigenomics analysis reveals the roles of dietary taurine in the muscle growth of juvenile turbot (Scophthalmus maximus). *Comparative Biochemistry and Physiology Part D: Genomics and Proteomics, 47*, 101120. https://doi.org/10.1016/j.cbd.2023.101120

Heras, J. (2021). Fish transcriptomics: Applied to our understanding of aquaculture. In *Transcriptomics from aquatic organisms to humans* (pp. 71–90). CRC Press.

He, S., Zhang, Y., Xu, L., Yang, Y., Marubashi, T., Zhou, Z., & Yao, B. (2013). Effects of dietary Bacillus subtilis C-3102 on the production, intestinal cytokine expression and autochthonous bacteria of hybrid tilapia Oreochromis niloticus ♀ × Oreochromis aureus ♂. *Aquaculture, 412–413*, 125–130. https://doi.org/10.1016/j.aquaculture.2013.06.028

Houston, R. D., Bean, T. P., Macqueen, D. J., Gundappa, M. K., Jin, Y. H., Jenkins, T. L., Selly, S. L. C., Martin, S. A., Stevens, J. R., & Santos, E. M. (2020). Harnessing genomics to fast-track genetic improvement in aquaculture. *Nature Reviews Genetics, 21*(7), 389–409.

Ibrahim, D., Arisha, A. H., Khater, S. I., Gad, W. M., Hassan, Z., Abou-Khadra, S. H., Mohamed, D. I., Ahmed Ismail, T., Gad, S. A., & Eid, S. A. (2022). Impact of omega-3 fatty acids nano-formulation on growth, antioxidant potential, fillet quality, immunity, autophagy-related genes and *Aeromonas hydrophila* resistance in Nile tilapia (*Oreochromis niloticus*). *Antioxidants, 11*(8), 1523.

Jannathulla, R., Rajaram, V., Kalanjiam, R., Ambasankar, K., Muralidhar, M., & Dayal, J. S. (2019). Fishmeal availability in the scenarios of climate change: Inevitability of fishmeal replacement in aquafeeds and approaches for the utilization of plant protein sources. *Aquaculture Research, 50*(12), 3493–3506.

Jin, Y., Tian, L., Xie, S., Guo, D., Yang, H., Liang, G., & Liu, Y. (2015). Interactions between dietary protein levels, growth performance, feed utilization, gene expression and metabolic products in juvenile grass carp (*Ctenopharyngodon idella*). *Aquaculture, 437*, 75–83. https://doi.org/10.1016/j.aquaculture.2014.11.031

Juárez Gutiérrez, M., Navarrete-Ramírez, P., Monroy, F., Llera-Herrera, R., Martínez Chávez, C., Ríos-Durán, M., Palomera-Sanchez, Z., Raggi, L., Lozano-Olvera, R., Pedroza, R., & Martinez-Palacios, C. (2021). Using nutrigenomics to evaluate microdiet performance in pike silverside larvae. *Aquaculture Nutrition.* https://doi.org/10.1111/anu.13305

Kajungiro, R., Mapenzi, L., Nyinondi, C., Haldén, A. N., Mmochi, A., Chacha, M., Mtolera, M., Lamtane, H., & de Koning, D. J. (2019). The need of a structured tilapia breeding program in Tanzania to enhance aquaculture production: A review. *Tanzania Journal of Science, 45*(3), 355–371.

Katan, T. (2022). *Impact of plant-based oils and other feed ingredients on growth, tissue composition, gene expression, and health of Atlantic salmon (Salmo salar)*. Memorial University of Newfoundland.

Keyvanshokooh, S. (2023). A review of the quantitative real-time PCR and omics approaches applied to study the effects of dietary selenium nanoparticles (nano-se) on fish. *Comparative Immunology Reports*, 200127.

Kortner, T. M., Skugor, S., Penn, M. H., Mydland, L. T., Djordjevic, B., Hillestad, M., Krasnov, A., & Krogdahl, Å. (2012). Dietary soyasaponin supplementation to pea protein concentrate reveals nutrigenomic interactions underlying enteropathy in Atlantic salmon (*Salmo salar*). *BMC Veterinary Research, 8*(1), 101. https://doi.org/10.1186/1746-6148-8-101

Lazzarotto, V., Corraze, G., Larroquet, L., Mazurais, D., & Médale, F. (2016). Does broodstock nutritional history affect the response of progeny to different first-feeding diets? A whole-body transcriptomic study of rainbow trout alevins. *British Journal of Nutrition, 115*(12), 2079–2092.

Leduc, A., Zatylny-Gaudin, C., Robert, M., Corre, E., Corguille, G. L., Castel, H., Lefevre-Scelles, A., Fournier, V., Gisbert, E., Andree, K. B., & Henry, J. (2018). Dietary aquaculture by-product hydrolysates: Impact on the transcriptomic response of the intestinal mucosa of European seabass (*Dicentrarchus labrax*) fed low fish meal diets. *BMC Genomics, 19*(1), 396. https://doi.org/10.1186/s12864-018-4780-0

Lei, C., Song, H., Song, H., Zhu, T., Du, J., & Li, S. (2024). RNA-seq and whole-genome resequencing reveal Micropterus salmoides growth-linked gene and selection signatures under carbohydrate-rich diet and varying temperature. *Scientific Reports, 14*(1), 25184. https://doi.org/10.1038/s41598-024-76685-3

Lende, S., Mahida, P., Shrivastava, V., Yusufzai, S., & Jha, A. (n.d.). Nutrigenomics: The New Era of Aquaculture Nutrition. *INDIAN FARMER*, pp. 320

Lin, G., Thevasagayam, N. M., Wan, Z. Y., Ye, B., & Yue, G. H. (2019). Transcriptome analysis identified genes for growth and omega-3/−6 ratio in saline tilapia. *Frontiers in Genetics, 10*, 244.

Ma, F., Wang, L., Huang, J., Chen, Y., Zhang, L., Zhang, M., Yu, M., Jiang, H., & Qiao, Z. (2023). Comparative study on nutritional quality and serum biochemical indices of common carp (*Cyprinus carpio*) aged 11 to 13 months aged cultured in traditional ponds and land-based container aquaculture systems. *Food Research International, 169*, 112869. https://doi.org/10.1016/j.foodres.2023.112869

Ma, M., & Hu, Q. (2024). Microalgae as feed sources and feed additives for sustainable aquaculture: Prospects and challenges. *Reviews in Aquaculture, 16*(2), 818–835.

Machado, A. M., Tørresen, O. K., Kabeya, N., Couto, A., Petersen, B., Felício, M., Campos, P. F., Fonseca, E., Bandarra, N., Lopes-Marques, M., Ferraz, R., Ruivo, R., Fonseca, M. M., Jentoft, S., Monroig, Ó., da Fonseca, R. R., Castro, C., & L. F. (2018). "Out of the can": A draft genome assembly, liver transcriptome, and nutrigenomics of the European sardine. *Sardina pilchardus. Genes, 9*(10), 485. https://doi.org/10.3390/genes9100485

Magbanua, T. O., & Ragaza, J. A. (2024). Selected dietary plant-based proteins for growth and health response of Nile tilapia *Oreochromis niloticus*. *Aquaculture and Fisheries, 9*(1), 3–19.

Martin, S. A., & Król, E. (2017). Nutrigenomics and immune function in fish: New insights from omics technologies. *Developmental & Comparative Immunology, 75*, 86–98.

Mierziak, J., Kostyn, K., Boba, A., Czemplik, M., Kulma, A., & Wojtasik, W. (2021). Influence of the bioactive diet components on the gene expression regulation. *Nutrients, 13*(11), 3673.

Minghetti, M., Leaver, M. J., & Tocher, D. R. (2011). Transcriptional control mechanisms of genes of lipid and fatty acid metabolism in the Atlantic salmon (*Salmo salar* L.) established cell line, SHK-1. *Biochim Biophys Acta (BBA) - Molecular and Cell Biology of Lipids, 1811*(3), 194–202. https://doi.org/10.1016/j.bbalip.2010.12.008

Moghadam, H., Mørkøre, T., & Robinson, N. (2015). Epigenetics – Potential for programming fish for aquaculture? *Journal of Marine Science and Engineering, 3*(2), 175–192.

Mondal, S., & Panda, D. (2020). Nutrigenomics: An interface of gene-diet-disease interaction. In *Mineral deficiencies-electrolyte disturbances, genes, diet and disease interface*. IntechOpen.

Moran, M. N., Jones, D., Jensen, S., Marcoli, R., & Jerry, D. (2024). Optimising commercial traits through gene editing in aquaculture: Strategies for accelerating genetic improvement. *Reviews in Aquaculture, 16*(3), 1127–1159.

Munang'andu, H. M., & Evensen, Ø. (2017). Current advances in functional genomics in aquaculture. In *Applications of RNA-Seq and omics strategies-from microorganisms to human health* (pp. 173–212).

Murray, H. M., Lall, S. P., Rajaselvam, R., Boutilier, L. A., Blanchard, B., Flight, R. M., Colombo, S., Mohindra, V., & Douglas, S. E. (2010). A nutrigenomic analysis of intestinal response to partial soybean meal replacement in diets for juvenile Atlantic halibut, *Hippoglossus hippoglossus*, L. *Aquaculture, 298*(3), 282–293. https://doi.org/10.1016/j.aquaculture.2009.11.001

Navarro-Martín, L., Mennigen, J. A., & Asselman, J. (2023). Transcriptional epigenetic mechanisms in aquatic species. *Epigenetics in Aquaculture*, 45–64.

Naya-Català, F., Belenguer, A., Montero, D., Torrecillas, S., Soriano, B., Calduch-Giner, J., Llorens, C., Fontanillas, R., Sarih, S., & Zamorano, M. (2023). Broodstock nutritional programming differentially affects the hepatic transcriptome and genome-wide DNA methylome of farmed gilthead sea bream (*Sparus aurata*) depending on genetic background. *BMC Genomics, 24*(1), 670.

Ngoh, S. Y., Tan, D., Shen, X., Kathiresan, P., Jiang, J., Liew, W. C., Thevasagayam, N. M., Kwan, H. Y., Saju, J. M., Prakki, S. R. S., Goh, C. H., Wong, H. C., Chan, T. T., Mézes, M., & Orbán, L. (2015). Nutrigenomic and nutritional analyses reveal the effects of pelleted feeds on Asian seabass (*Lates calcarifer*). *PLoS One, 10*(12), e0145456. https://doi.org/10.1371/journal.pone.0145456

Palaiokostas, C., & Houston, R. D. (2018). Genome-wide approaches to understanding and improving complex traits in aquaculture species. *CABI Reviews, 2017*, 1–10.

Panserat, S., & Kaushik, S. (2010). Regulation of gene expression by nutritional factors in fish. *Aquaculture Research, 41*(5), 751–762.

Pérez-Sánchez, J., Simó-Mirabet, P., Naya-Català, F., Martos-Sitcha, J. A., Perera, E., Bermejo-Nogales, A., Benedito-Palos, L., & Calduch-Giner, J. A. (2018). Somatotropic axis regulation unravels the differential effects of nutritional and environmental factors in growth performance of marine farmed fishes. *Frontiers in Endocrinology, 9*, 687.

Piferrer, F. (2023). Epigenetics in aquaculture: Knowledge gaps, challenges, and future prospects. *Epigenetics in Aquaculture*, 451–463.

Poon, Z. W. J., Shen, X., Uichanco, J. A., Terence, C., Chua, S. W. G., & Domingos, J. A. (2023). Comparative transcriptome analysis reveals factors involved in the influence of dietary astaxanthin on body colouration of Malabar snapper (*Lutjanus malabaricus*). *Aquaculture, 562*, 738874. https://doi.org/10.1016/j.aquaculture.2022.738874

Prisingkorn, W., Prathomya, P., Jakovlić, I., Liu, H., Zhao, Y.-H., & Wang, W.-M. (2017). Transcriptomics, metabolomics and histology indicate that high-carbohydrate diet negatively affects the liver health of blunt snout bream (*Megalobrama amblycephala*). *BMC Genomics, 18*(1), 856. https://doi.org/10.1186/s12864-017-4246-9

Puvanasundram, P., Chong, C. M., Sabri, S., Yusoff, M. S., & Karim, M. (2021). Multi-strain probiotics: Functions, effectiveness and formulations for aquaculture applications. *Aquaculture Reports, 21*, 100905.

Reshi, Q. M., Ahmad, I., Nabi, N., Ahmed, I., & Sidiq, M. (2025). Global food insecurity: Role of fishes to compensate food insecurity and nutritional demand. In J. K. Sundaray, M. A. Rather, I. Ahmad, & A. Amin (Eds.), *Food security, nutrition and sustainability through aquaculture technologies*. Springer. https://doi.org/10.1007/978-3-031-75830-0_5

Rimoldi, S., Montero, D., Torrecillas, S., Serradell, A., Acosta, F., Haffray, P., Hostins, B., Fontanillas, R., Allal, F., & Bajek, A. (2023). Genetically superior European sea bass (*Dicentrarchus labrax*) and nutritional innovations: Effects of functional feeds on fish immune response, disease resistance, and gut microbiota. *Aquaculture Reports, 33*, 101747.

Rodde, C. (2020). *Individual feed efficiency in fishes: Direct measurement methods and indirect predictors to develop selective breeding programs in two major aquaculture species: European sea bass Dicentrarchus labrax and Nile tilapia Oreochromis niloticus*. Université Montpellier.

Saba, K., Sofi, F., Sravani, K., Reddy, S. V. K., Aismi, O., Kumar, A., Hussain, T., Devadharshini, S., Ganesan, P., & Dar, S. A. (2024). Nutrigenomics: Boost for aquaculture research and development. In *Coldwater fisheries and aquaculture management* (pp. 369–381). Apple Academic Press.

Saha, A., Nayak, M., & Giri, S. (2020). Nutrigenomics studies in fishes. In *Fish nutrition and its relevance to human health* (pp. 215–228). CRC Press.

Safari, R., Hoseinifar, S. H., Imanpour, M. R., Mazandarani, M., Sanchouli, H., & Paolucci, M. (2020). Effects of dietary polyphenols on mucosal and humoral immune responses, antioxidant defense and growth gene expression in beluga sturgeon (*Huso huso*). *Aquaculture, 528*, 735494. https://doi.org/10.1016/j.aquaculture.2020.735494

Sharawy, Z. Z., Ashour, M., Labena, A., Alsaqufi, A. S., Mansour, A. T., & Abbas, E. M. (2022). Effects of dietary Arthrospira platensis nanoparticles on growth performance, feed utilization, and growth-related gene expression of Pacific white shrimp, Litopenaeus vannamei. *Aquaculture, 551*, 737905. https://doi.org/10.1016/j.aquaculture.2022.737905

Song, F., Qin, Y., Geng, H., He, C., Yang, P., Wang, W., & Chen, Y. (2023). Transcriptome analysis reveals the effects of dietary lipid level on growth performance and immune response in golden pompano (Trachinotus ovatus). *Aquaculture, 563*, 738959. https://doi.org/10.1016/j.aquaculture.2022.738959

Shu-Chien, A. C., Han, W.-Y., Carter, C. G., Fitzgibbon, Q. P., Simon, C. J., Kuah, M.-K., Battaglene, S. C., Codabaccus, B. M., & Ventura, T. (2017). Effect of dietary lipid source on expression of lipid metabolism genes and tissue lipid profile in juvenile spiny lobster *Sagmariasus verreauxi*. *Aquaculture, 479*, 342–351. https://doi.org/10.1016/j.aquaculture.2017.05.036

Serra, V., Pastorelli, G., Tedesco, D. E. A., Turin, L., & Guerrini, A. (2024). Alternative protein sources in aquafeed: Current scenario and future perspectives. *Veterinary and Animal Science*, 100381.

Sindelar, R. D. (2016). Genomics, other "omics" technologies, personalized medicine, and additional biotechnology-related techniques. In *Pharmaceutical biotechnology* (pp. 149–190). CRC Press.

Sissener, N., Hemre, G. I., Espe, M., Sanden, M., Torstensen, B., & Hevrøy, E. (2013). Effects of plant-based diets on glucose and amino acid metabolism, leptin, ghrelin and GH-IGF system regulation in a tlantic salmon (*Salmo Salar* L.). *Aquaculture Nutrition, 19*(3), 399–412.

Skjærven, K. H., Adam, A.-C., Takaya, S., Waagbø, R., & Espe, M. (2022). Nutritional epigenetics. In *Cellular and molecular approaches in fish biology* (pp. 161–192). Elsevier.

Sørensen, S. L., Park, Y., Gong, Y., Vasanth, G. K., Dahle, D., Korsnes, K., Phuong, T. H., Kiron, V., Øyen, S., & Pittman, K. (2021). Nutrient digestibility, growth, mucosal barrier status, and activity of leucocytes from head kidney of Atlantic salmon fed marine-or plant-derived protein and lipid sources. *Frontiers in Immunology, 11*, 623726.

Tacchi, L., Secombes, C. J., Bickerdike, R., Adler, M. A., Venegas, C., Takle, H., & Martin, S. A. (2012). Transcriptomic and physiological responses to fishmeal substitution with plant proteins in formulated feed in farmed Atlantic salmon (*Salmo salar*). *BMC Genomics, 13*, 1–21.

Tian, J., Wen, H., Lu, X., Liu, W., Wu, F., Yang, C.-G., Jiang, M., & Yu, L.-J. (2018). Dietary phosphatidylcholine impacts on growth performance and lipid metabolism in adult genetically improved farmed tilapia (GIFT) strain of Nile tilapia *Oreochromis niloticus*. *British Journal of Nutrition, 119*(1), 12–21.

Torsabo, D., Ishak, S. D., Noordin, N. M., Waiho, K., Chu, I. K. C., Abduh, M. Y., & Abol-Munafi, A. B. (2024). Optimizing reproductive performance in pangasius catfish broodstock: A review of dietary and molecular strategies. *Veterinary and Animal Science*, 100375.

Tupac-Yupanqui, I., Vasquez, L., Villarroel, M., & Dunner, S. (2013). Fatty acid food source affects expression of genes involved in the stress response in tilapia (*Oreochromis niloticus*). *Journal of Biological Research, 19*, 30.

Varga, J. (2022). Investigating the effects of synthetic peptides on aquatic pathogens and rainbow trout immunity.

Waagbø, R., & Remø, S. C. (2020). Functional diets in fish health management. In *Aquaculture health management* (pp. 187–234). Elsevier.

Wanna, W., Surachat, K., Kaitimonchai, P., & Phongdara, A. (2021). Evaluation of probiotic characteristics and whole genome analysis of *Pediococcus pentosaceus* MR001 for use as probiotic bacteria in shrimp aquaculture. *Scientific Reports, 11*(1), 18334. https://doi.org/10.1038/s41598-021-96780-z

Wong, S., Waldrop, T., Summerfelt, S., Davidson, J., Barrows, F., Kenney, P. B., Welch, T., Wiens, G. D., Snekvik, K., Rawls, J. F., & Good, C. (2013). Aquacultured rainbow trout (*Oncorhynchus mykiss*) possess a large Core intestinal microbiota that is resistant to variation in diet and rearing density. *Applied and Environmental Microbiology, 79*(16), 4974–4984. https://doi.org/10.1128/AEM.00924-13

Xue, X. (2022). *Transcriptomic studies of Atlantic salmon responses to dietary omega-3 long-chain polyunsaturated fatty acids, Piscirickettsia salmonis infection, immunogens, and functional feed*. Memorial University of Newfoundland.

Yáñez, J. M., Houston, R. D., & Newman, S. (2014). Genetics and genomics of disease resistance in salmonid species. *Frontiers in Genetics, 5*, 415.

Yáñez, J., Joshi, R., & Yoshida, G. (2020). Genomics to accelerate genetic improvement in tilapia. *Animal Genetics, 51*(5), 658–674.

Yang, Z., Fu, G., Lee, M., Yeo, S., & Yue, G. H. (2024). Genes for editing to improve economic traits in aquaculture fish species. *Aquaculture and Fisheries*.

Yue, G., & Wang, L. (2017). Current status of genome sequencing and its applications in aquaculture. *Aquaculture, 468*, 337–347.

Yue, H., Wu, J., Fu, P., Ruan, R., Ye, H., Hu, B., Chen, X., & Li, C. (2022). Effect of glutamine supplementation against soybean meal-induced growth retardation, hepatic metabolomics and transcriptome alterations in hybrid sturgeon *Acipenser baerii*♀× *a. schrenckii* ♂. *Aquaculture Reports, 24*, 101158. https://doi.org/10.1016/j.aqrep.2022.101158

Yuan, Y., Li, S., Mai, K., Xu, W., Zhang, Y., & Ai, Q. (2015). The effect of dietary arachidonic acid (ARA) on growth performance, fatty acid composition and expression of ARA metabolism-related genes in larval half-smooth tongue sole (*Cynoglossus semilaevis*). *British Journal of Nutrition, 113*(10), 1518–1530. https://doi.org/10.1017/S0007114515000781

Zhang, Y., Lu, R., Qin, C., & Nie, G. (2020). Precision nutritional regulation and aquaculture. *Aquaculture Reports, 18*, 100496.

Zhou, C., Ge, X., Niu, J., Lin, H., Huang, Z., & Tan, X. (2015). Effect of dietary carbohydrate levels on growth performance, body composition, intestinal and hepatic enzyme activities, and growth hormone gene expression of juvenile golden pompano, *Trachinotus ovatus*. *Aquaculture, 437*, 390–397. https://doi.org/10.1016/j.aquaculture.2014.12.016

Zhou, J. s., Chen, H. j., Ji, H., Shi, X. c., Li, X. x., Chen, L. q., Du, Z. y., & Yu, H. b. (2018). Effect of dietary bile acids on growth, body composition, lipid metabolism and microbiota in grass carp (*Ctenopharyngodon idella*). *Aquaculture Nutrition, 24*(2), 802–813. https://doi.org/10.1111/anu.12609

Zhu, T., Corraze, G., Plagnes-Juan, E., Quillet, E., Dupont-Nivet, M., & Skiba-Cassy, S. (2018). Regulation of genes related to cholesterol metabolism in rainbow trout (*Oncorhynchus mykiss*) fed a plant-based diet. *American Journal of Physiology-Regulatory, Integrative and Comparative Physiology, 314*(1), R58–R70.

Zhu, M., Sumana, S. L., Abdullateef, M. M., Falayi, O. C., Shui, Y., Zhang, C., Zhu, J., & Su, S. (2024). CRISPR/Cas9 technology for enhancing desirable traits of fish species in aquaculture. *International Journal of Molecular Sciences, 25*(17).

Zou, Q., Huang, Y., Cao, J., Zhao, H., Wang, G., Li, Y., & Pan, Q. (2017). Effects of four feeding stimulants in high plant-based diets on feed intake, growth performance, serum biochemical parameters, digestive enzyme activities and appetite-related genes expression of juvenile GIFT tilapia (Oreochromis sp.). *Aquaculture Nutrition, 23*(5), 1076–1085.

Chapter 12
Role of Omega-3-Fatty Acids in Aquaculture Feeds and Their Impact on Human Health

Oshin Dhillon

1 Introduction

Aquaculture forms one of the steadily expanding global food production sectors that contributes a significant portion of the world's fish and seafood supply. Advanced aquaculture systems are helpful in mitigating the growing fish demand as the primary source of high-quality protein and essential nutrients for consumers. The proper and healthy diet maintained regularly has a remarkable impact on supporting vital body functions and metabolism. Aquaculture has adapted to the needs of a growing population that relies on fish as a key source of nutrition, particularly as fish are a rich source of essential fatty acids, vitamins, and minerals. The inclusion of omega-3 polyunsaturated fatty acids (PUFAs), imparting well-established health benefits, significantly influences the nutritional profile of feed for quality farmed fish. Omega-3 fatty acids (ω-3 fatty acids; n-3 fatty acids) are polyunsaturated fatty acids (PUFAs) that play an essential role in maintaining physiological functions and growth; these play a crucial role in the prevention of cardiovascular diseases, inflammation, and cognitive decline in both aquatic and terrestrial organisms (Keservani et al., 2020). The most notable omega-3 fatty acids include alpha-linolenic acid (ALA, C18:2n-6), eicosapentaenoic acid (EPA, 20:5n-3), and docosahexaenoic acid (DHA, 22:6n-3), each contributing significantly to health and development but unable to be synthesized efficiently by most fish or humans, necessitating dietary source incorporation. The human body lacks *cis* double bond inserting enzymes, linoleoyl-CoA-desaturase (Δ6-desaturase) and Δ5-desaturase, making them incapable of synthesizing EPA and DHA from ALA, thus requiring dietary supplementation of omega-3s. These essential fatty acids are primarily derived from marine sources, such as fish and microalgae, which serve as critical components of

O. Dhillon (✉)
Division of Fish Genetics and Biotechnology, Faculty of Fisheries, SKUAST-K, Ganderbal, Jammu and Kashmir, India

I. Ahmed, I. Ahmad (eds.), *Aquaculture: Enhancing Food Security and Nutrition*, https://doi.org/10.1007/978-3-031-92858-1_12

aquaculture feeds (Tocher, 2015). The presence of long-chain PUFAs in fish and other marine organisms makes aquaculture an essential omega-3-rich food production sector. In fish farming, omega-3s are typically supplemented in feed to enhance fish health and ultimately provide omega-3-abundant seafood for human consumption.

However, the increasing need from the aquaculture industry for omega-3 fatty acids, as traditional omega-3 sources from wild fish populations are limited and often overexploited, has created problems for the aquaculture industry. Aquaculture-based strategies involving omega-3-rich feeds have the potential for direct impact on human health, since the consumption of omega-3-enriched fish can have a significant impact on human dietary intake of EPA and DHA. Omega-3 intake in humans benefits cardiovascular health and brain development and function, enhances immune response, and reduces risk of chronic diseases. The availability of marine-derived omega-3 fatty acids is constrained, making farmed fish enriched in omega-3s a sustainable means for addressing the growing global demand for beneficial fatty acids in aquaculture. This section examines the importance of omega-3 fatty acids in aquaculture diets, their effects on fish well-being and productivity, and the subsequent implications for human health, emphasizing sustainable sourcing of omega-3s to meet the growing demands of aquaculture. It also examines the current research on innovative approaches to enhance omega-3 content in aquaculture products, offering a comprehensive overview of the symbiotic interconnectedness among aquaculture practices, omega-3 fatty acids, and human health outcomes. The findings underscore the importance of continuing to optimize feed formulations, improve sustainable source strategies, and promote omega-3-rich seafood consumption as a part of a healthy diet.

1.1 Omega-3 Fatty Acids: An Overview

Omega-3 polyunsaturated fatty acids (ω-3 PUFAs) are vital fats characterized by their first double bond located at the third carbon atom from the methyl end of the fatty acid chain. These are essential for human health but cannot be synthesized by the body; hence, they must be obtained through diet. There are several primary sources of omega-3 fatty acids, including marine oils, seafood, and certain plant-based or modified sources like algae and single-cell organisms (Fig. 12.1):

- Eicosapentaenoic acid (EPA; 20:5n-3; *cis*-5,*cis*-8,*cis*-11,*cis*-14,*cis*-17-eicosapentaenoic acid): EPA has 20 carbon atoms and is found mainly in cold-water fatty fish like menhaden, salmon, tuna, sardines, mackerel, herring, pilchard, shad, freshwater eels, and freshwater fish such as shad, loach, salmon, and trout (Table 12.1). EPA is known for its anti-inflammatory properties and benefits for cardiovascular health. It also plays a role in energy metabolism and immune system modulation. Newborns receive sufficient EPA through breast milk, while adults are recommended to consume about 220 mg per day, which

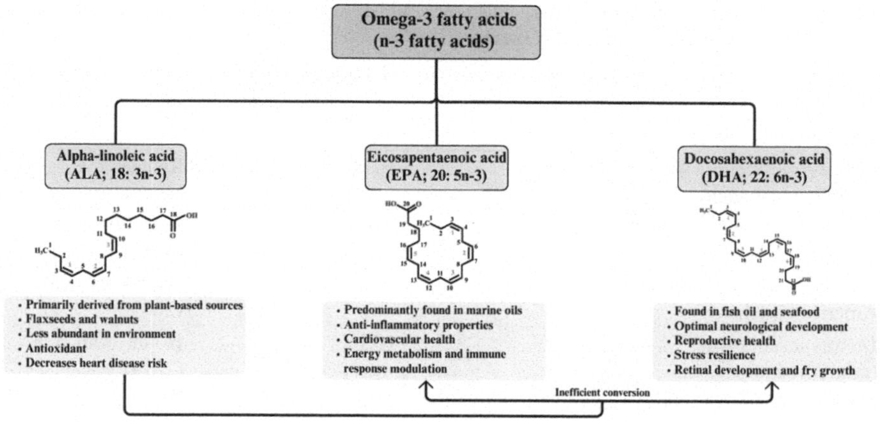

Fig. 12.1 Omega-3 fatty acid classification

can be met by eating two to three servings of fatty fish weekly, providing approximately 1250 mg of both EPA and DHA. Fish oil supplements can offer between 3000 and 4000 mg of standardized fish oils.

- Docosahexaenoic acid (DHA; 22:6n-3; *cis*-4,*cis*-7,*cis*-10,*cis*-13,*cis*-16,*cis*-19-docosahexaenoic acid): DHA, which consists of 22 carbon atoms, is primarily sourced from microalgae and fish like menhaden and trout (Table 12.1). DHA is a crucial component of cell membranes in the brain, eyes, and nervous system. In fish, it supports neurological development, reproductive health, and resilience to stress. Newborns also get enough DHA through breast milk. Adults need about 1000–2500 mg of DHA daily, generally provided by consuming two to three servings of fatty fish, which equates to about 1250 mg. Fish oil supplements provide around 3000–4000 mg of standardized fish oils, while algae can provide DHA, but not EPA, at a rate of about 200 mg per day.
- Alpha-linoleic acid (ALA; 18:3n-3; *cis*-9, *cis*-12, *cis*-15-octadecatrienoic acid): ALA, a shorter-chain omega fatty acid with 18 carbon atoms, is mainly found in plant sources like flaxseeds, walnuts, and chia seeds. Although it is less commonly found in marine environments, ALA serves as a precursor that can be converted into EPA and DHA in certain organisms, although this process is not very efficient (Thiyam-Holländer et al., 2012). Infants and children do not need ALA supplementation, while adults should aim for 1–2 grams daily, which can be achieved through sources like flaxseed, canola oil, and walnuts.

In addition to the above-mentioned omega-3s, PUFAs also include stearidonic acid (SDA; 18:4n-3) and docosapentaenoic acid (DPA, 22:5n-3), short-chain omega-3 and omega-6 fatty acids derived from plant seeds, algae, zooplankton, and smaller organisms in the form of α-linolenic acid (ALA; 18:3n-3) and linoleic acid can be converted into long-chain omega-3 fatty acids like eicosapentaenoic acid (EPA) and docosahexaenoic acid (DHA) in fish. Linoleoyl-CoA-desaturase (Δ6-desaturase) catalyzes the removal of two hydrogen atoms from α-linolenic acid and linoleic acid

Table 12.1 Omega-3 fatty acid: Eicosapentaenoic acid (EPA; 20:5n-3) and Docosahexaenoic acid (DHA; 22:6n-3) content in various fish species

Fish species	Eicosapentaenoic acid (EPA; 20:5n-3)	Docosahexaenoic acid (DHA; 22:6n-3)	References
Marine			
Brevoortia tyrannus (atlantic menhaden)	15	10.62	Rahnan et al. (1995)
Gymnura (butterfly ray)	0.03	0.09	Abd Aziz et al. (2013)
Sardinops sagax (South American pilchard)	6.6	19	Huynh and Kitts (2009)
Dussumieria acuta (rainbow sardine)	3.43	10.16	Homayooni et al. (2014)
Ethmalosa fimbriata (bonga shad)	2.11	2.25	Abd Aziz et al. (2013)
Clupea pallasi (Pacific herring)	4.99	5.76	Huynh and Kitts (2009)
Plotosus spp. (striped eel catfish)	1.46	0.89	Abd Aziz et al. (2013)
Chirocentrus dorab (dorab wolf-herring)	0.24	0.54	Abd Aziz et al. (2013)
Sardinops pilchardus (pilchard)	8.5	8.37	Huynh and Kitts (2009)
Pagellus acarne (axillary seabream)	3.19	3.41	Garcia-Moreno et al. (2013)
Trachurus mediterraneus (Mediterranean horse mackerel)	4.4	5.49	Garcia-Moreno et al. (2013)
Leiostomus xanthurus (spot croaker)	4.85	4.65	Cladis (2014)
Mugil cephalus (flathead gray mullet)	1.10	1.36	Cladis (2014)
Anoplopoma fimbria (Sablefish)	6.53	5.66	Cladis (2014)
Tenualosa macrura (longtail shad)	20.42	1.69	Abd Aziz et al. (2013)
Freshwater			
Channa punctatus (spotted snakehead)	0.32	1.54	Rahnan et al. (1995)
Hypophthalmichthys nobilis (bighead carp)	0.19	0.35	Hoseini et al. (2013)
Catla catla (catla)	0.19	0.07	Rahnan et al. (1995)
Clarias gariepinus (African sharptooth catfish)	0.08	0.14	Rahnan et al. (1995),

(continued)

Table 12.1 (continued)

Fish species	Eicosapentaenoic acid (EPA; 20:5n-3)	Docosahexaenoic acid (DHA; 22:6n-3)	References
Clarias macrocephalus (broadhead catfish)	0.08	0.48	Rahnan et al. (1995)
Anguilla anguilla (European eel)	3.48	0.89	Rahnan et al. (1995)
Pangasius pangasius (Pangas catfish)	0.08	–	Rahnan et al. (1995)
Trichogaster pectoralis (snakeskin gourami)	0.44	1.88	Rahnan et al. (1995)
Lates calcarifer (barramundi)	0.043	3.78	Rahnan et al. (1995)
Mastacembelus armatus (zig-zag eel)	–	1.99	Gulzar and Zuber (2000)
Labeo calbasu (orange-fin labeo)	–	0.89	Gulzar and Zuber (2000)
Labeo rohita (rohu)	0.58	1.27	Rahnan et al. (1995)
Oreochromis mossambicus (Mozambique tilapia)	0.42	0.35	Rahnan et al. (1995)
Alosa sapidissima (American shad)	1.33	4.70	Cladis (Cladis, 2014)
Barbatula barbatula (stone loach)	2.97	1.73	Gladyshev et al., 2018
Phoxinus czekanowskii (Czekanowski's minnow)	2.64	1.90	Gladyshev et al. (2018)
Gobio gobio (gudgeon)	2.49	1.64	Gladyshev et al. (2018)
Leuciscus leuciscus (common dace)	1.66	1.65	Gladyshev et al., 2018
Cobitis melanoleuca (Siberian spined loach)	1.40	1.68	Gladyshev et al. (2018)
Carassius carassius (Prussian carp)	0.45	0.97	Yilmaz et al. (2016)
Blicca bjoerkna (white bream)	0.40	0.80	Ahlgren et al. (2009)
Cyprinus carpio (common carp)	0.34	0.50	Rahnan et al. (1995),
Oncorhynchus keta (chum salmon)	4.00	6.00	Abramova and Kozin (2023)
Oncorhynchus mykiss (rainbow trout)	1.70	3.96	Heissenberger et al. (2010), Cladis (2014)
Salmo trutta (brown trout)	0.9	3.5	Heissenberger et al. (2010)
Coregonus macrophthalmus (gangfisch)	6.41	4.34	Gladyshev et al., 2018

using molecular oxygen as a cofactor to introduce a carbon-carbon double bond, forming octadecatetraenoic acid (18:4n-3) and γ-linolenic acid (18:3n-6), respectively. Further, elongase catalyzes the elongation of the aliphatic chain of carbons, adding two carbon units to the carboxylic end of octadecatetraenoic acid (18:4n-3) and γ-linolenic acid (18:3n-6), resulting in eicosatetraenoic acid (20:4n-3) and dihomo-γ-linolenic acid (20:3n-6), respectively. Δ5-desaturase introduces a double bond at the Δ5 position in eicosatetraenoic acid and dihomo-γ-linolenic acid, leading to the formation of eicosapentaenoic acid (20:5n-3; EPA) and arachidonic acid (20:4n-6). Elongase acts on EPA to form docosapentaenoic acid (22: 5n-3) that further undergoes a second round of elongation, forming tetracosapentaenoic acid (24: 5n-3) and resulting in the formation of nisinic acid (24: 6n-3). Nisinic acid goes through peroxisomal chain shortening, forming docosahexaenoic acid (DHA; 22:6n-3) (Scorletti & Byrne, 2018) (Fig. 12.2).

1.2 Role of Omega-3 Fatty Acids in Aquaculture Feeds

Fatty acids play a crucial role in various physiological processes in both marine and freshwater organisms, making them essential components of aquaculture feeds. Adding omega-3 fatty acids, especially EPA and DHA, to aquaculture feed helps improve fish health, ultimately leading to omega-3-rich seafood for human consumption. The aquaculture industry depends on the sustainable production of omega-3s to boost the nutritional value of fish and other aquatic species, which are vital sources of these fatty acids in our diets. However, the increasing demand has led to overfishing of traditional sources like fish meal and fish oil, creating challenges for the industry and prompting the search for alternative feed ingredients. Alternatives such as plant-based oils, microalgae, and genetically modified organisms are coming up as promising substitutes to provide omega-3s in aquaculture feeds without relying on marine resources (Senanayake & Fichtali, 2006). Nevertheless, the bioavailability and composition of omega-3s from these alternative sources may vary, which might impact the nutritional qualities of farmed fish.

As aquaculture expands globally as a key source of seafood, it is increasingly crucial to understand how omega-3 fatty acids function within aquafeeds. These essential fats play a vital role in numerous biological processes and are naturally found in the tissues of fish and shellfish that eat algae and other omega-3-rich foods. Many farmed fish, especially carnivorous species like salmon, depend on these fatty acids for optimal growth, which is why they are included in their feed to promote healthy and sustainable farming practices. Omega-3 fatty acids are not only important for growth and development, but they are also crucial for various physiological functions in fish, such as maintaining cell membranes, regulating gene expression, and modulating inflammation and immune responses. Because of their diverse roles, omega-3s are often called "essential fats," and a lack of them can greatly affect the health and performance of farmed fish (Sultana et al., 2023).

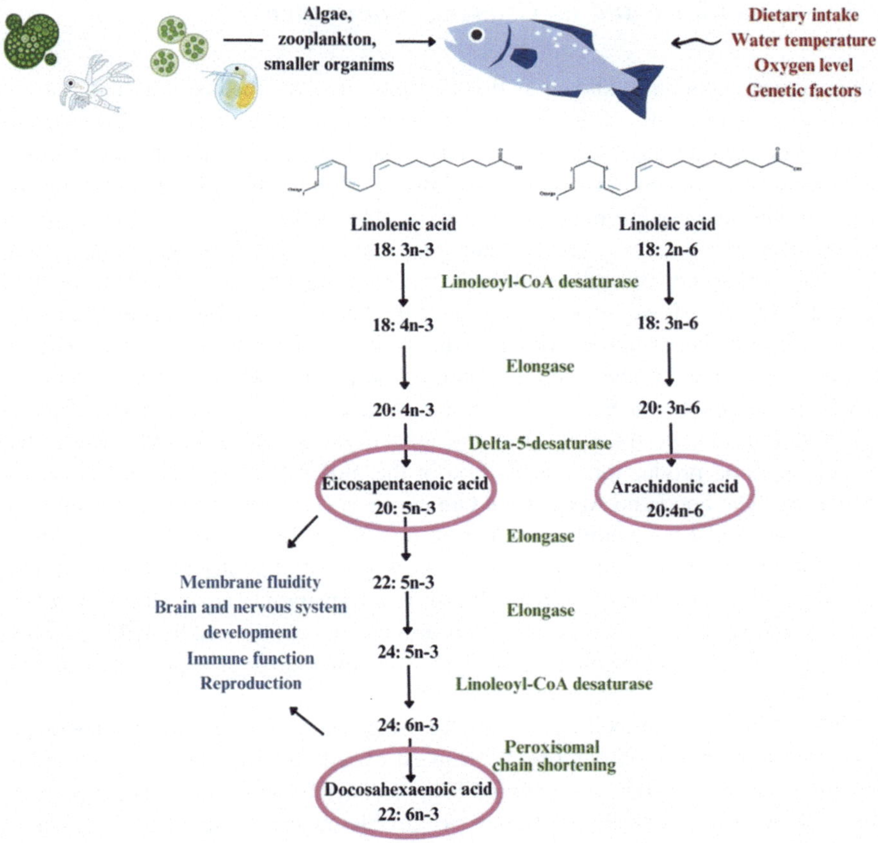

Fig. 12.2 Biosynthetic pathway of omega-3 fatty acids in fish, showing conversion of short fatty acid chains derived from dietary intake to eicosapentaenoic acid (EPA; 20: 5n-3) and docosahexaenoic acid (DHA; 22: 6n-3) that facilitate in membrane fluidity, brain and nervous system development, immune functions, and reproduction in fish

For rapidly growing species like salmon, trout, and tilapia, omega-3 fatty acids play a significant role in muscle development and fat accumulation. Research shows that these fatty acids in aquafeeds are associated with faster growth rates and improved feed efficiency. Omega-3s are key components of the brain and nervous system, aiding in nerve cell membrane formation and proper neurotransmission. During their larval and juvenile stages, omega-3 fatty acids are especially crucial for cognitive function, motor skills, and behavioral development. This is critical for fish, which rely on keen senses and quick reactions to evade predators and pursue prey. In aquaculture, where crowded environments and stress can lead to disease outbreaks, having a robust immune system is essential. Omega-3s bolster the natural defence mechanisms of fish, helping them stay healthy and resilient against common illnesses.

1.3 Aquaculture and Its Growing Importance

Aquaculture, commonly known as fish farming, involves the breeding and rearing of aquatic organisms like fish, shellfish, and other marine life in controlled environments. In recent decades, this practice has expanded significantly and has become an essential part of food production worldwide. With the global population on the rise and wild fish stocks under increasing pressure, aquaculture is seen as a crucial solution for ensuring food security and sustainability. The history of aquaculture goes back thousands of years, with early fish farming activities recorded in China around 2000 BC. However, it was not until the latter part of the twentieth century that aquaculture began to transform into the modern industry we know today. Driven by the demand for seafood, declining wild fish numbers, and the necessity to diversify food sources, advancements in technology have facilitated the large-scale farming of various aquatic species. Currently, aquaculture accounts for over 50% of the world's seafood production, and its role in the global food supply is continually increasing. The Food and Agriculture Organization (FAO) highlighted this growth, noting that global aquaculture production rose from 2.6 million tons in 1970 to more than 87.5 million tons by 2020, illustrating its vital contribution to feeding a growing global population (Bu et al., 2024). This transition to aquaculture addresses not only the need for more seafood but also the challenges faced by wild fisheries, such as overfishing, habitat loss, and the imperative to preserve marine biodiversity.

One of the most compelling reasons for the growing importance of aquaculture is its contribution to food security. With an estimated 10 billion people projected to inhabit the Earth by 2050, the demand for protein-rich food sources is expected to rise dramatically. Aquaculture offers a sustainable means of meeting this demand. Unlike terrestrial farming, aquaculture has the ability to produce large quantities of high-quality animal protein in a relatively short period of time. The nutritional benefits make seafood a vital component of balanced diets, especially in developing countries where malnutrition and micronutrient deficiencies are prevalent. In many coastal and island nations, aquaculture provides a critical source of livelihood and a local, affordable source of food. With fish and shellfish production on the rise, more people have access to the nutrients they need, which is crucial for public health. Additionally, aquaculture can help alleviate pressure on wild fish stocks, which are currently overexploited.

According to the FAO, nearly 60 million people are employed in the global aquaculture sector, a number that continues to grow as the industry expands. In many developing Southeast Asian countries, aquaculture offers a pathway out of poverty by providing stable income and livelihood opportunities. Aquaculture has allowed these nations to export seafood to global markets, boosting trade and generating significant revenue. Moreover, the development of sustainable aquaculture practices has the potential to stimulate innovation in the food industry, including the development of new fish feed formulas, improved breeding techniques, and environmental friendly farming practices. This, in turn, can create further economic opportunities in research, technology, and production.

Aquaculture has the potential to boost food production, but it also encounters various environmental issues that can lead to water pollution, loss of habitats, and disease spread among wild fish. For instance, the intensive farming of species like salmon and shrimp often results in nutrient-rich waste building up in local waters, which can negatively impact ecosystems and biodiversity. Additionally, the industry's reliance on wild-caught fish for making fishmeal and fish oil—crucial ingredients in many fish feeds—can worsen overfishing and further stress marine ecosystems. Fortunately, researchers are exploring alternative sources of protein and fats, such as algae, insects, and plant-based feeds, to lessen the industry's dependency on wild fish and enhance its environmental sustainability.

1.4 The Link Between Aquaculture, Omega-3s, and Human Health

Omega-3 fatty acids, which are important for various bodily functions, are abundant in marine life, especially in fatty fish. Recently, aquaculture has adjusted to address the rising demand for omega-3s in people's diets. Fish such as salmon, mackerel, and sardines naturally contain high levels of omega-3s because they feed on smaller fish and plankton rich in EPA and DHA. However, as the need for farmed fish has grown, a major challenge has been ensuring that these fish still provide as much omega-3 as those caught in the wild. Traditionally, farmed fish diets have included fishmeal and fish oil sourced from wild fish, which effectively boosts their omega-3 content but raises sustainability concerns regarding marine resources. Issues like overfishing, the decline of fish stocks, and the ecological effects of catching smaller fish have led the industry to look for more sustainable options. One promising approach is to enhance omega-3 levels through alternative feed sources. Algae, for instance, are a natural source of omega-3s, especially EPA and DHA. Increasingly, algal oils are being incorporated into aquaculture feeds to supplement the diets of farmed fish, providing a sustainable substitute for fish oil. Algae-based omega-3s can be grown without harming fish populations, and fish that are fed algae-enriched diets maintain high omega-3 levels in their tissues, ensuring their nutritional value remains high.

Seafood consumption rich in omega-3 fatty acids, especially from farmed fish, comes with a variety of health benefits. Regularly including omega-3s in your diet is a great way to enhance overall health and help prevent chronic illnesses. Fish such as salmon, trout, and sardines are particularly high in these essential fats, making them important for a healthy diet. One of the most recognized benefits of omega-3s is their positive impact on heart health. By incorporating omega-3-rich seafood into meals, people can lower their risk of heart disease significantly. In fact, the American Heart Association suggests consuming at least two servings of fatty fish each week to get enough omega-3s for maintaining a healthy heart. This can lead to a reduced risk of heart attacks, strokes, and various cardiovascular issues, with salmon being especially beneficial due to its omega-3 content. Additionally, omega-3s contribute to mental health; research indicates that diets high in these fats can lessen symptoms of depression and anxiety. They are also crucial for keeping our brains healthy and may help prevent cognitive decline as we age. Regular consumption of omega-3-rich seafood can enhance memory, focus, and mental clarity, making it a worthwhile addition to the diets of people of all ages. Furthermore, omega-3s possess anti-inflammatory properties that can greatly affect conditions like arthritis. Eating seafood rich in omega-3s can help ease joint pain, diminish swelling, and improve mobility, serving as a natural remedy for those with inflammatory issues. Beyond heart health, omega-3 fatty acids are linked to a reduced risk of arrhythmias, which can be serious. Research shows that long-term consumption of omega-3-rich foods, especially fish, can decrease the chances of heart attacks and strokes (Kris-Etherton et al., 2002). DHA, an important component of the brain, plays a crucial role in brain development, function, and cognitive abilities. Sufficient DHA levels are associated with better memory and learning skills, particularly in children and older adults. DHA is also essential for eye health, contributing to improved retinal function and helping to prevent age-related macular degeneration (AMD), which is a leading cause of vision loss in seniors. Eating omega-3-rich seafood can support eye health and reduce the risk of developing AMD. Additionally, omega-3s may play a key role in lowering the risk of neurodegenerative diseases, such as Alzheimer's and dementia. Studies have shown that omega-3 supplementation can lead to better cognitive function in older adults, with numerous clinical trials validating its benefits in preventing cognitive decline (Chang et al., 2024). Chronic inflammation is often at the root of several health issues, including arthritis, diabetes, and autoimmune disorders. The anti-inflammatory properties of omega-3s can help manage the immune response and prevent excessive inflammation, which is particularly important for conditions like rheumatoid arthritis (RA), where inflammation can result in pain and disability. Clinical studies support the idea that omega-3 supplementation can improve symptoms of inflammation and promote better joint health (Akbar et al., 2017).

2 Omega-3 Fatty Acids in Fish Physiology and Nutrition

Omega-3s exert a wide array of biological effects in fish, influencing various physiological and biochemical pathways. These effects are critical for maintaining overall fish health, growth, reproduction, immune function, stress resilience, and the nutritional quality of fish meat. Omega-3 fatty acids influence fish health through several key mechanisms that can be explained at the molecular, cellular, and physiological levels.

2.1 Cell Membrane Fluidity

EPA and DHA are integral parts that modulate the fluidity of cell membranes. The presence of omega-3s in membranes enhances protein function, nutrient transport, and cellular signaling, which is particularly important for immune cells, neurons, and reproductive cells.

- Incorporation into membranes: Omega-3s are incorporated into the phospholipid bilayer of cell membranes, replacing other fatty acids, particularly saturated and monosaturated fats (Alhshim, 2024). Due to their polyunsaturated nature, omega-3s increase the fluidity of membranes.
- Increase membrane flexibility: The higher fluidity of omega-3-enriched membranes enhances the function of membrane-bound proteins, including receptors, transporters, and enzymes, which are involved in cellular processes like nutrient uptake, ion exchange, and signal transduction.
- Enhanced cell signaling: Membrane fluidity also facilitates better signal transduction, particularly in immune cells (e.g., macrophages, neutrophils) and nervous tissue. This results in more efficient immune responses, faster neurotransmission, and improved tissue repair mechanisms.

Physiological impact: The fluidity of immune cell membranes affects the function of immune receptors, such as toll-like receptors (TLRs) and antigen-presenting proteins. This facilitates a quicker and more effective immune response (Aggarwal & Heber, 2014). In neurons, DHA particularly plays an important role in the central nervous system, where it helps form the structural components of synapses and promotes neurotransmission.

2.2 Pro-inflammatory Cytokine Regulation

Omega-3s, EPA, and DHA modulate the inflammatory response. Chronic inflammation in fish, as in other animals, can lead to reduced immune function, disease susceptibility, and poor growth.

- Anti-inflammatory mediators: Omega-3 fatty acids play a crucial role in producing anti-inflammatory compounds such as resolvins, protectins, and maresins. These compounds are formed when EPA and DHA undergo enzymatic conversion and work to reduce inflammation by helping clear out apoptotic cells and decreasing the levels of pro-inflammatory cytokines.
- Reduced pro-inflammatory cytokines: Additionally, omega-3s are effective at inhibiting pro-inflammatory cytokines like interleukin-1β (IL-1β), tumor necrosis factor-alpha (TNF-α), and interleukin-6 (IL-6). This is especially significant because it helps lower chronic inflammation, which can negatively affect the functionality and growth of aquatic systems.

Physiological impact: Omega-3s improve the immune response of the fish by reducing excessive inflammation, thus enhancing their ability to fight off the infections and reducing susceptibility to diseases like bacterial, viral, and parasitic infections. These can also modulate the stress response by reducing the inflammatory consequences of acute stress, preventing immunosuppression that often follows stress events in farmed fish.

2.3 Gene Expression and Protein Synthesis

Omega-3s can regulate gene expression and protein synthesis through several signaling pathways. They also modulate signaling pathways involved in cell growth, differentiation, and apoptosis by activating specific transcription factors.

- Activation of transcription factors: The activation of various transcriptional factors, such as peroxisome proliferator-activated receptor alpha (PPAR-α), PPAR-gamma (PPAR-γ), and sterol regulatory element-binding proteins (SREBPs), is mediated by omega-3s, as noted by Hedge et al. (2016). These transcription factors are essential for controlling the expression of genes that are involved in lipid metabolism, inflammation, and cellular responses to stress.
- Modulation of protein synthesis: Additionally, omega-3 fatty acids influence protein synthesis through their effect on the mechanistic target of rapamycin (mTOR) pathway, which is a crucial regulator of cell growth and metabolism. In fish, omega-3s help optimize the balance between protein synthesis and degradation, promoting muscle and tissue growth.

Physiological impact: Omega-3s enhance the efficiency of feed utilization and improve growth rates in farmed fish by regulating gene expression associated with muscle growth and lipid metabolism. These also influence the expression of heat shock proteins (HSPs) and other protective proteins that help fish cope with stress and maintain cellular homeostasis during adverse environmental conditions.

2.4 Immune System Modulation and Disease Resistance

Fish have highly responsive immune systems that react strongly to ingested changes, particularly benefiting from omega-3 fatty acids. The modulation of the immune system by omega-3s enhances fish resistance to pathogens, reduces inflammation, and promotes effective health.

- Modulation of leukocytes: Omega-3s increase both the quantity and functionality of key cells like neutrophils, macrophages, and lymphocytes, which are essential for identifying and destroying pathogens.

- Phagocytosis enhancement: Omega-3s improve the phagocytic activity of macrophages, enabling these immune cells to effectively engulf and break down pathogens and foreign particles.
- Improved antibody production: DHA and EPA also enhance the production of specific antibodies and the activation of the adaptive immune system, leading to more efficient immune responses to infections.

Physiological impact: Omega-3 fatty acids contribute to improved antibody production, specifically supporting the functions of DHA and EPA that activate the adaptive immune system. For instance, omega-3 supplementation has been shown to reduce disease incidence, such as vibriosis, furunculosis, and *Ichthyophthirius multifiliis,* in aquaculture. Omega-3s enhance the ability of fish to recover from infections by reducing the severity and duration of inflammatory responses, promoting quicker healing and regeneration of the tissue (Hegde et al., 2016).

2.5 Stress and Cortisol Regulation

Omega-3 mitigates the negative effects of environmental and physiological stressors, which are common in aquaculture systems. By regulating the release of cortisol, the primary stress hormone, omega-3s improve fish welfare and resilience.

- Reduction of cortisol levels: Omega-3 fatty acids play a role in lowering cortisol levels by affecting the hypothalamic-pituitary-adrenal (HPA) axis that controls cortisol release during stress. Specifically, EPA helps to prevent excessive cortisol production by influencing the enzymes that synthesize stress hormones.
- Regulation of stress-related gene expression: Omega-3s help regulate the expression of genes linked to stress responses, including those that contribute to cortisol production. By dampening the stress response at the molecular level, omega-3s help fish maintain homeostasis under challenging conditions.

Physiological impact: Fish fed omega-3-enriched diets show improved resistance to common stressors such as handling, crowding, temperature fluctuations, and water quality changes. Omega-3s help prevent the physiological damage often caused by chronic stress, which can lead to reduced growth, impaired immune function, and increased susceptibility to disease. By reducing cortisol levels and modulating stress responses, omega-3s contribute to lower mortality rates in farmed fish under stressful conditions.

2.6 *Reproductive Health and Fecundity*

Omega-3 fatty acids support the reproductive health of fish, particularly for broodstock and selective breeding, as highlighted by Zhang et al. (2024). DHA and EPA are known for their involvement in oocyte development and regulating reproductive hormone levels, respectively.

- Oocyte development and egg quality: DHA is abundantly found in the phospholipid membranes of oocytes. The inclusion of omega-3s in broodstock diets improves egg quality, yolk composition, and embryo development. DHA also enhances the structural integrity of eggs, ensuring better hatchability and larval survival.
- Hormonal regulation: EPA and DHA regulate the production of key reproductive hormones, including gonadotropins and prostaglandins, which are essential for ovulation, fertilization, and early embryonic development.

Physiological impact: Omega-3 supplementation increases egg production, improves fecundity, and results in healthier offspring with higher survival rates. This is particularly important in species where reproductive success is closely linked to the quality of the eggs and larvae. Larvae that are nourished by omega-3-rich eggs show improved survival and growth during the stages of life.

The mechanisms of action by which the omega-3s influence the fish's health and performance are multifaceted and complex, involving changes at the cellular, molecular, and physiological levels. From improving cell membrane function and enhancing immune responses to regulating stress and reproductive health, omega-3s sustain the overall well-being of farmed fish. By understanding these mechanisms, aquaculture systems can optimize fish health, growth, and sustainability while also strengthening the nutritional value of fish products for consumers.

3 Sources of Omega-3 Fatty Acids in Fish Feeds

In aquaculture, omega-3 fatty acids are typically included in fish feeds to enhance the nutritional value of farmed fish and to meet the demands of consumers for omega-3-rich products (Akoh, 2017). Farmed fish, such as salmon, trout, and tilapia, are often fed a mixture of plant-based ingredients, such as soy or corn, alongside fishmeal and fish oil, which are rich sources of omega-3s (Hegde et al., 2016). The primary sources of omega-3 fatty acids in aquaculture feeds include:

- Fishmeal and fish oil: These traditional sources are the most direct means of supplying EPA and DHA in fish feeds. Marine oils are an enriched source of omega-3 fatty acids, while the livers of cod, shark, whales, haddock, halibut, and tuna and their flesh are rich in DHA. Other marine life, including crustaceans and certain mollusks, also contributes to ω-3 PUFAs. Fishmeal is derived from small fish species like anchovies, herrings, and menhaden, whereas fish oil is extracted from the fatty fish tissue. These ingredients provide high concentrations of omega-3 fatty acids essential for the growth, development, and overall health of farmed fish.

- Alternative sources of omega-3s: As the supply of fish meal and fish oil becomes more limited and costly, there's a noticeable shift toward alternative omega-3 sources like algal oils, flaxseed, and microalgae. While plant oils such as flaxseed and canola are high in ALA, they have limited capability for being converted into EPA and DHA in the body. Ongoing research aims to enhance these conversion rates, making plant oils a more practical source of omega-3s for aquaculture. Algal oil is particularly notable because it serves as a sustainable source of EPA and DHA, frequently used in organic aquaculture or when environmental considerations are a priority. Derived from microalgae, algal oil provides a rich supply of both DHA and EPA and is becoming increasingly popular for use in aquaculture feeds. Microalgae and some microorganisms, including certain fungi, also yield omega-3 fatty acids. Notably, dinoflagellate microalgae, such as *Crypthecodinium cohnii*, and unicellular eukaryotes like *Schizochytrium* spp. are recognized as excellent sources of omega-3s (Senanayake & Fichtali, 2006). Additionally, insect-based proteins and oils are gaining attention as effective alternatives to traditional fishmeal and fish oil. Insects, such as black soldier fly larvae and mealworms, are packed with essential fatty acids, including omega-3s. Insect farming is a sustainable practice since insects need far less land, water, and feed compared to conventional livestock. The incorporation of insect oils and meals into aquaculture feeds is becoming increasingly appealing due to their nutritional benefits and positive environmental impact (Akoh, 2017).

4 Impact of Omega-3 Enriched Feeds on Fish Growth and Quality

With the growing global demand for aquaculture products, optimizing fish diets to improve these aspects is essential not only for increasing production efficiency but also for enhancing the nutritional quality of fish for human consumption. Omega-3-enriched feeds are known to improve growth performance in many fish species, including salmon, trout, and tilapia. These maintain cellular integrity and function, enhancing the overall metabolic efficiency of the fish and therefore forming an essential cell membrane component. Omega-3s are integral to the growth and development of fish, supporting energy metabolism, tissue formation, and cellular function. These fatty acids are essential for efficient conversion of feed into body mass, enabling fish to achieve optimum growth rates. Farmed fish that are fed omega-3-enriched diets contribute to the energy metabolism, exhibiting faster growth rates. They are involved in mitochondrial function, enhancing the efficiency of cellular energy production and improving growth performance. DHA is crucial for the development of the brain and eyes, while EPA aids in overall tissue growth and repair, affecting both skeletal and muscle development (Hegde et al., 2016). Research by Sarker et al. (2016) found that fish fed a diet high in omega-3 fatty acids showed better specific growth rates (SGR) and feed conversion ratios (FCR), indicating the efficient feed conversion into body mass leading to enhanced growth. Including omega-3s in fish diets not only provides a readily available energy source but also supports anabolic processes like protein synthesis, resulting in improved growth. Furthermore, omega-3 fatty acids can boost the expression of growth hormones, such as insulin-like growth factor 1 (IGF-1), which is essential for fish growth and development (Kumar et al., 2022). It is worth noting that the effects of omega-3-enriched feeds can vary between different fish species. Davis and Devine (2023) found that when fed omega-3-rich diets, salmon showed enhanced growth rates and increased muscle mass. Similarly, rainbow trout and tilapia fed omega-3-enriched diets exhibited increased body weight and reduced fat deposition, making them more efficient in converting energy into lean tissue (Bhardwaj et al., 2023).

Omega-3-enriched feeds can significantly enhance the fatty acid profile and nutritional value of fish. Fish, especially species like salmon, mackerel, and sardines, are well-known for their high omega-3 content, making them important dietary sources of these fatty acids for humans. When fish are supplemented with diets rich in omega-3, the levels of EPA and DHA in their tissues rise, which improves the overall quality of the meat. Not only do omega-3s boost the amount of beneficial fatty acids, but they also help create a better balance between omega-6 and omega-3 fatty acids, which is crucial for human health. An optimal omega-3 to omega-6 ratio is linked to lower risks of chronic diseases, including cardiovascular issues and inflammation. Recent research indicates that consuming trans unsaturated fats can negatively affect blood lipid levels, leading to increased LDL cholesterol, decreased HDL cholesterol, higher lipoprotein(a), elevated ApoB levels, and reduced ApoA1 levels (Oteng & Kersten, 2020; Kenneth, 2024). A study conducted by Katan (2022) found that salmon fed a diet rich in plant-based omega-3 sources like algae oil showed notable increases in EPA and DHA in their muscle tissue without raising omega-6 levels, thus enhancing the health benefits of the fish.

Omega-3 fatty acids are known for their anti-inflammatory effects, which can significantly benefit the health of farmed fish. In aquaculture environments, where fish frequently face high stocking densities, pathogens, and various environmental stresses, omega-3s are essential for supporting the immune system of the fish. Omega-3s help fish resist infections and recover from disease by modulating immune cell function and inflammatory pathways. These impact the composition of immune cell membranes, enhancing the flexibility and performance of critical immune cells like macrophages, lymphocytes, and neutrophils. Additionally, EPA and DHA help to lower the production of pro-inflammatory cytokines, which diminishes chronic inflammation and boosts resistance to diseases, including bacterial and viral infections. Xue (2022) found that omega-3-enriched feeds enhanced the immune response of Atlantic salmon, improving their resistance to *Piscirickettsia salmonis* infections, which are common in aquaculture. Chronic stress in aquaculture can lead to poor growth, increased susceptibility to diseases, and even mortality. By improving overall health and reducing disease susceptibility, omega-3-enriched diets also contribute to better welfare for farmed fish. Reduced inflammation and stress improve the quality of life for the fish, therefore contributing to the sustainability of the farming operation by decreasing the need for antibiotics and other treatments. The result is healthier fish, less environmental impact from aquaculture, and better quality fish products for human consumption. Adequate omega-3s have been shown to improve reproductive health, spawning success, and the health of offspring in many aquaculture species.

Recent studies have emphasized the beneficial effects of omega-3-enriched feeds on the growth and health of fish (Gabriel et al., 2024). Balancing omega-3s is a critical component of aquaculture feeds, they must be balanced with other essential nutrients to ensure optimal fish health and performance. Key considerations in feed formulation include:

- *Protein and amino acids:* Fish require a high-quality protein source to support growth and development. Omega-3 fatty acids must be included in a feed formulation that also provides sufficient proteins, essential amino acids, and vitamins.
- *Energy and fat content:* Omega-3-rich oils contribute to the fat content of the feed. Care must be taken to balance the energy content from omega-3 sources with the overall fat composition of the feed. Excessive fat can lead to obesity and impaired growth in fish (Gabriel et al., 2024).
- *Vitamin and mineral balance:* Omega-3-rich feeds should also include a balanced supply of vitamins and minerals, which are essential for the metabolic processes that support growth, immune function, and reproduction.

5 Sustainability of Omega-3 Production in Aquaculture

Ensuring that aquaculture feeds rich in omega-3 are sustainable is a persistent issue. The heavy dependence on fishmeal and fish oil, sourced from wild-caught fish, has sparked worries about the dwindling populations of marine fish. Consequently, there has been significant research into alternative feed ingredients, such as insect meal, plant-based oils, and algae, which offer a more sustainable and potentially cost-effective solution for omega-3 supplementation in aquaculture diets.

The sustainability of omega-3s in aquaculture is crucial in both the aquaculture industry and human health. As global fish meal and fish oil supplies become more constrained, there is an increasing shift toward finding alternative sources of omega-3s, particularly through plant-based oils and algae, which have a lower environmental footprint.

6 Challenges and Opportunities in Omega-3 Enrichment

As the global demand for aquaculture continues to rise, the necessity for efficient and sustainable practices in fish feed formulation becomes increasingly critical. The depletion of marine resources due to overfishing poses significant sustainability challenges, prompting researchers and industry stakeholders to explore alternative sources. Concurrently, the development of plant-based omega-3 substitutes offers promising opportunities to align environmental considerations with economic viability. Recent advancements in aquaculture nutrition emphasize the need for a diversified approach to feed formulation, wherein alternative raw materials are evaluated based on their nutritional complementarity rather than simply acting as substitutes. This paradigm shift highlights the importance of identifying plant-based sources of omega-3s to ensure dietary adequacy while mitigating the environmental impact associated with traditional marine ingredients (Tocher et al., 2019). Addressing these challenges not only supports the sustainability of aquaculture practices but also paves the way for economic viability in feed production.

The sustainability of fish oil and fish meal has become really important in aquaculture since they are key for adding omega-3 to fish feed. However, the excessive harvesting of wild fish populations creates serious environmental concerns, making it essential to find alternative sources. One promising strategy involves improving fish nutrition through innovative feed formulations that reduce reliance on traditional fishmeal. Additionally, enhancing fish welfare through better dietary practices not only fosters healthier aquatic populations but also supports the broader objectives of food and nutrition security. It is crucial to prioritize sustainable practices in both the production and use of fish oil and fish meal. This approach not only helps maintain ecological balance but also supports the economic health of the aquaculture industry. The environmental challenges posed by fish oil and fish meal production are significant hurdles that need to be overcome for aquaculture to become more sustainable. Overfishing has led to a depletion of wild fish stocks, primarily low-trophic-level fish used for feed, raising concerns about marine biodiversity and ecosystem health (Vianna et al., 2020). This production process not only contributes to the decline of fish populations but also exacerbates habitat destruction through unsustainable fishing practices. Furthermore, as demands for omega-3 enrichment increase, the reliance on fish meal intensifies, leading to higher costs and greater ecological strain. The production of fishmeal is associated with notable greenhouse gas emissions and negative implications for global fisheries. Consequently, exploring alternative protein sources, such as yeast cultivated from waste biomass, presents a dual opportunity to reduce ecological footprints while supporting aquaculture needs. Thus, transitioning toward plant-based and alternative proteins is essential for mitigating these environmental challenges.

Creating plant-based alternatives for aquaculture feed is an important step toward reducing our dependence on fish oil and fish meal, which are the main sources of protein in the industry these days (Gabriel et al., 2024). With approximately 70% of aquatic animal production being reliant on high-protein aquafeeds sourced from wild-caught forage fish, which is projected to strain resources significantly by 2025, stakeholders are compelled to explore alternative protein sources to meet future demands. Ingredients derived from plants like soy, peas, and various grains have great nutritional benefits and can help create a more sustainable aquaculture system. However, challenges remain in optimizing these ingredients to ensure they meet the specific nutritional needs of aquatic species. By balancing economic considerations, nutritional efficacy, and environmental impacts, the industry can leverage the potential of plant-based alternatives, creating a resilient aquaculture sector that reduces dependency on marine resources while promoting sustainability.

The nutritional efficacy of plant-based omega-3 sources presents both opportunities and challenges in the context of aquaculture feed. While these alternatives can help alleviate the sustainability issues associated with diminishing fish oil and fish meal supplies, their incorporation is not without difficulties. Plant sources like flaxseed and algae offer ALA, which needs to be converted into the more beneficial forms EPA and DHA that are crucial for the health and growth of fish. However, the conversion rate of ALA to these bioactive forms is often inadequate, raising concerns about the overall nutritional quality of plant-based feeds (Saini et al., 2021). Additionally, the fatty acid profile and the presence of anti-nutritional factors in these plant materials can make their use in aquaculture diets even more challenging. Thus, while plant-based omega-3 sources offer promising avenues for sustainability in aquaculture, their nutritional efficacy and acceptability require critical evaluation to ensure optimal fish performance and health.

The economic considerations of omega-3 enrichment in aquaculture feed are pivotal, particularly in light of the declining stocks of natural fishmeal and fish oil sources. As global demand for aquaculture products surges, driven by health-conscious consumers demanding omega-3-rich diets, producers face the challenge of maintaining profitability while ensuring sustainability. The transition from traditional fish-based feeds to plant-based alternatives presents both opportunities and obstacles. Plant-based sources of omega-3, such as algae, can reduce reliance on overfished species, yet their production and integration can impose higher initial costs on feed manufacturers and potentially compromise performance. Furthermore, the evolving regulatory landscape surrounding sustainability practices is likely to influence market dynamics, compelling producers to invest in eco-friendly options. Thus, a thorough analysis of market trends and production costs is essential for aquaculture ventures aiming to navigate these economic challenges successfully while meeting nutritional demands.

In evaluating the cost-effectiveness of traditional versus alternative omega-3 enrichment methods in aquaculture feed, it is pivotal to consider both economic feasibility and sustainability. Traditional methods often rely on fish oil and fish meal, posing significant sustainability challenges due to overfishing and high ecological impact (Warwas, 2023). Conversely, alternative methods, such as the incorporation of plant-based sources or microbial oils, present a promising avenue for reducing reliance on marine resources. Although initial investment costs for developing and refining plant-based alternatives can be substantial, long-term savings are anticipated through lower raw material costs and improved feed conversion ratios.

Moreover, the environmental benefits associated with these alternatives may offset their higher initial costs, making them viable in the context of increasing regulatory and consumer pressures for sustainable aquaculture practices. Thus, a comprehensive cost analysis highlights the potential for shifting toward more sustainable omega-3 enrichment strategies.

7 Human Health Benefits of Omega-3-Rich Fish

With aquaculture's recognized significant role in satisfying the global seafood demand, it's important to understand how omega-3-enriched farmed fish can affect our health. Recent research is exploring the molecular mechanisms behind omega-3s and their potential in preventing or managing chronic diseases. Omega-3s are essential for cell function, maintaining membrane integrity, and managing inflammation. There are three main types of omega-3 fatty acids important for human health: ALA, which comes from plant sources like flaxseeds and walnuts, and EPA and DHA, primarily found in fish and algae. Omega-3s are best known for their effects on cardiovascular health, but emerging evidence underscores their broader impact on cognitive, immune, and metabolic functions.

7.1 Cardiovascular Health

Omega-3 fatty acids are widely recognized for their ability to lower the risk of cardiovascular diseases. Similarly, a case study by Rawat et al. (2024) found that consumption of omega-3s from fish oil regularly can help reduce blood triglycerides, lower blood pressure, and decrease the chances of heart attacks and strokes. For example, research by Shen et al. (2022) found that higher levels of EPA and DHA intake significantly lowered heart disease rates among individuals with high cholesterol.

- Impact on lipid profile: Studies indicate that taking EPA and DHA can have a significant positive impact on lipid levels, particularly by lowering triglycerides. This happens because these fatty acids help to reduce the liver's release of very low-density lipoprotein (VLDL). When taken in high doses, around 2–4 grams daily, these supplements can lead to a 20–50% decrease in triglyceride levels. However, the effects of omega-3s on cholesterol are more mixed. While some people might see a slight rise in low-density lipoprotein cholesterol (LDL-C), they also tend to experience an increase in high-density lipoprotein cholesterol (HDL-C), which is good for heart health. The overall impact is often favorable when balancing the lipid profile.
- Blood pressure: Multiple meta-analyses have demonstrated that omega-3 fatty acids can effectively lower blood pressure, especially in those with hypertension.

For instance, a study conducted by Kelsey and Pagidipati (2023) found that individuals consuming omega-3s experienced a drop in systolic blood pressure of about 4.51 mmHg and a decrease of 3.05 mmHg in diastolic blood pressure. Kris-Etherton et al. (2002) reviewed multiple clinical traits and concluded that long-term omega-3 supplementation (3–4 grams per day) could reduce cardiovascular events by up to 30%.

- Anti-inflammatory and antithrombic effect: Omega-3 fatty acids have anti-inflammatory properties, helping to reduce the production of harmful cytokines and eicosanoids. They also play a role in preventing blood clots by inhibiting platelet aggregation, which can lower the risks of heart attacks and strokes.

7.2 Cognitive Function and Mental Health

DHA is crucial for brain health, as it forms a significant part of the brain's gray matter and is important for neural functions, synaptic plasticity, and cognitive abilities. DHA-rich fish consumption has been linked to better cognitive performance, especially among older adults. Recent research indicates that regularly consuming omega-3 fatty acids can help prevent cognitive decline, including conditions like Alzheimer's disease and dementia. A study by Raji et al. (2014) found that individuals with higher omega-3 levels in their diets tend to have improved cognitive abilities and experience a slower decline in cognitive function as they age. Omega-3s are thought to influence mood regulation by modulating neurotransmitter systems and inflammatory responses. Various randomized controlled trials (RCTs) have shown positive effects of omega-3 supplements on symptoms of depression and anxiety. A meta-analysis by Bhardwaj et al. (2023) found that omega-3 supplementation, especially EPA, significantly alleviated depression symptoms in people diagnosed with major depressive disorder. Hudson (2015) found that omega-3s might also reduce anxiety symptoms, particularly in individuals with clinical levels of anxiety. Recent studies indicate that omega-3 fatty acids might help enhance behavior and social skills in people with autism spectrum disorder (ASD) (de Andrade Wobido et al., 2022).

7.3 Inflammation and Immune Function

Omega-3 fatty acids, especially EPA, are known for their strong anti-inflammatory properties, making them beneficial for managing chronic inflammatory conditions like RA, inflammatory bowel disease (IBD), and psoriasis. They help regulate the immune system and may lower the risk of chronic diseases such as arthritis and diabetes. Omega-3s are recognized for their role in promoting a healthy inflammatory response by adjusting the production of pro-inflammatory cytokines. A meta-analysis by Wei et al. (2021) revealed that omega-3 supplementation can significantly decrease inflammation markers in patients suffering from chronic inflammatory diseases. They work by lowering the production of pro-inflammatory substances like eicosanoids and cytokines, including tumor necrosis factor-alpha (TNF-α) and interleukin-6 (IL-6), which are linked to chronic inflammation. Clinical trials have indicated that supplementing with omega-3s can lead to reduced levels of inflammation markers such as C-reactive protein (CRP) and IL-6. Additionally, research suggests that omega-3s may alleviate symptoms of RA, including joint pain and stiffness. While findings are somewhat inconsistent, omega-3s appear to have an impact on gut inflammation in patients with IBD. MacDonald (2006) suggested that EPA and DHA supplementation might reduce flare-ups in ulcerative colitis but have fewer effects on Crohn's disease.

7.4 Pregnancy and Early Development

Pregnancy and early development are critical periods in a human's life, where optimal nutrition is paramount for the health of both the mother and the developing fetus. Omega-3 fatty acids are among the many nutrients that are crucial during pregnancy, primarily because of their significant role in fetal development and maternal well-being. These polyunsaturated fats are essential for brain, eye, and nervous system development. Despite the clear benefits of omega-3s during pregnancy, the consumption of fish during pregnancy raises concerns regarding mercury exposure. It's important to have higher amounts of EPA and DHA to support the rapid growth and development of the fetus. While DHA is predominantly incorporated into the brain and retina, EPA has important anti-inflammatory effects that can benefit both maternal health and the development of the fetus. DHA is essential for the neurodevelopment of infants, comprising approximately 30% of the structural lipids in the brain and retina. Maternal omega-3 supplementation has been linked to improved outcomes in child development, including better cognitive and motor skills.

Research has consistently demonstrated that DHA plays a crucial role in the proper development of neurons and synapses in the brain. Adequate DHA levels in the mother's diet have been linked to improved cognitive function in the offspring. Research suggests that DHA supplementation during pregnancy can enhance brain development, particularly in terms of cognitive performance, attention, and memory in early childhood. A landmark study by Ramakrishnan et al. (2016) found that infants whose mothers had higher levels of DHA during pregnancy scored higher on tests of cognitive development and visual attention at an early age. DHA is also critical for the development of the retina, where it helps form the photoreceptor cells necessary for vision. It is believed that maternal DHA intake directly influences the visual acuity of infants during the first year of life. Research by Hoffman et al. (1993) indicated that infants whose mothers had higher DHA intake during pregnancy showed better visual development and healthier eyes, as reflected in quicker visual maturation. Additionally, omega-3 fatty acids are linked to a lower risk of preterm birth and complications like preeclampsia. In a study by Jones et al. (2014), it was proposed that omega-3 supplements may help reduce the risk of preterm birth by influencing inflammatory processes and maintaining the health of the cell membranes in the uterus and placenta. Omega-3 fatty acids, especially EPA and DHA, possess anti-inflammatory effects that could help mitigate the inflammation that triggers premature labor. Furthermore, maternal consumption of omega-3s has been correlated with a decreased likelihood of preeclampsia, which is marked by elevated blood pressure and potential organ damage during pregnancy. A meta-analysis by Rajati et al. (2024) revealed that omega-3 supplementation could reduce the occurrence of preeclampsia by about 20% in pregnancies deemed high-risk.

Fish is a fantastic source of DHA and EPA, along with other vital nutrients for pregnancy like protein, vitamins D and B_{12}, and minerals such as selenium and iodine. These nutrients are crucial for maternal health and the growth and development of the fetus. Regular fish consumption during pregnancy has been shown to contribute to higher maternal levels of DHA and EPA, which are associated with better pregnancy outcomes, including healthier birth weights and longer gestational periods. Despite the nutritional benefits of fish, there are concerns about the potential for mercury contamination, particularly in larger fish that are higher up in the food chain, such as tuna. Mercury is a neurotoxin that can negatively affect fetal brain development, leading to cognitive impairments, motor dysfunction, and other developmental delays. To address this, the US Food and Drug Administration (FDA) and the Environmental Protection Agency (EPA) recommend that pregnant women limit their intake of high-mercury fish while promoting the consumption of those with low mercury levels. Some low-mercury fish options include salmon, sardines, anchovies, and trout, which are also great sources of omega-3 fatty acids. To enjoy the benefits of omega-3 fatty acids while minimizing the risks of mercury exposure, health experts suggest that pregnant women eat two to three servings of low-mercury fish each week. This can provide about 200–300 mg of DHA daily, which is essential for fetal development. Although fish is a great source of omega-3s, not everyone may get enough through their diet, especially if they do not eat fish regularly. In those situations, omega-3 supplements, like fish oil or algal oil, can be beneficial for ensuring sufficient DHA and EPA intake. Algal oils, in particular, offer a plant-based alternative that is ideal for vegetarians and those who have sustainability concerns regarding fish-derived omega-3s. Research indicates that omega-3 supplementation during pregnancy can enhance the health of both mothers and babies. Specifically, taking DHA in doses between 600 and 1000 mg daily has been linked to better cognitive outcomes for infants and a lower risk of premature birth (Carlson & Colombo, 2021). The significance of omega-3s does not stop after pregnancy, as DHA is transferred through breast milk from mother to child. Omega-3s play a vital role in the ongoing development of an infant's brain and visual system in the initial months of life. Mothers who have higher intakes of omega-3s during breastfeeding tend to have higher levels of DHA in their milk, which can support their infants' cognitive and visual growth.

7.5 *Cancer Prevention*

Recent studies indicate that omega-3 fatty acids could be beneficial in preventing specific cancers, such as those of the breast, prostate, and colon. Evidence suggests that these fatty acids might aid in cancer prevention by affecting cell growth, inducing cell death (apoptosis), and limiting the spread of cancer cells (metastasis) (Kong, 2016). EPA and DHA are believed to have an impact on the tumor environment and can decrease cancer cell growth, especially in breast, prostate, and colon cancers. Additionally, a study conducted by Lui and Ma (2014) found that a high intake of omega-3 fatty acids correlates with a reduced risk of breast cancer, possibly by influencing hormone-related mechanisms and inflammatory responses.

The optimal dosage of omega-3s varies depending on the health condition being treated. General recommendations for cardiovascular health are typically around 1–3 grams of EPA and DHA daily. Higher doses (4–5 grams) may be required for certain conditions like hypertriglyceridemia, while doses of 2–3 grams may be effective for mood disorders. Omega-3 fatty acids offer a wide range of health benefits, including heart protection, brain support, and preventing chronic diseases. There is increasing evidence that omega-3s are essential not only for managing chronic diseases but also for improving overall health and well-being. However, challenges such as sourcing sustainable omega-3s and optimizing dosages for specific conditions remain. Continued research into the molecular mechanisms underlying the health benefits of omega-3 fatty acids will help refine clinical guidelines and maximize their therapeutic potential.

8 Future Perspectives and Sustainability of Omega-3s

There is growing interest in finding sustainable omega-3s owing to their importance in fish health and aquaculture. Alternative sources such as microalgae, plant-based oils, and insects are being explored with the demand for fishmeal and fish oil on the rise. These options hold great potential for fulfilling the omega-3 requirements of farmed fish while also minimizing the environmental impact of aquaculture. However, relying on fishmeal and fish oil in feeds raises important questions about the sustainability of our marine resources. Wild-caught fish used in these products are often sourced from fisheries that may be overexploited, leading to reduced biodiversity and ecosystem degradation. This has prompted the aquaculture industry to seek sustainable alternatives. As discussed earlier, algae and plant-based sources of omega-3s offer promising alternatives to traditional fish-derived omega-3s. Algal oil, in particular, is gaining attention due to its sustainability, as it does not depend on wild fish stocks. Additionally, growing algae can take place in land-based setups, which helps lessen the environmental footprint. By turning to alternative and more sustainable sources of omega-3s, like algae and plant oils, we can meet the global appetite for omega-3-rich seafood without putting too much pressure on marine resources. Research into more efficient fish feed formulas and genetic advancements in aquaculture will further optimize the production of omega-3s in farmed fish, allowing for greater yields of high-quality seafood.

Ongoing research is exploring innovative solutions to improve omega-3 production and availability in aquaculture (Bhardwaj et al., 2023). These include the genetic engineering of algae and plants to produce higher levels of omega-3s, the development of sustainable fish feed technologies, and the use of insect-based ingredients. Another promising avenue is the use of fermentation technologies to produce omega-3-rich oils from microorganisms. This could provide a scalable and sustainable method for producing omega-3s without relying on marine resources. These initiatives have helped raise awareness of the importance of sustainable practices and are encouraging producers to adopt more eco-friendly approaches. The future also holds the potential for innovations in biotechnology, which may enable the development of fish that can naturally produce higher levels of omega-3s from plant-based diets, reducing the need for fish oil altogether. This would provide a more sustainable and ethical solution to meeting the global demand for omega-3s in aquaculture.

9 Conclusion

Aquaculture has become a key component of the global food system, offering vital protein sources, economic opportunities, and a way to lessen the strain on wild fish populations. As the industry progresses, enriching farmed fish with omega-3 fatty acids is proving crucial for feeding the growing world while also enhancing human health. The relationship among aquaculture, omega-3s, and health is clear. Omega-3s, which are plentiful in farmed fish, provide various health advantages, such as improving heart health, boosting brain function, and reducing inflammation. Adding omega-3s to aquaculture feeds not only promotes the growth and well-being of farmed fish but also enhances the nutritional value of the seafood consumed by humans. As the demand for sustainable and nutritious aquaculture products increases, the use of omega-3-enriched feeds will be essential in fulfilling consumer dietary needs and addressing the environmental issues the industry faces. With advancements in biotechnology, genetic research, and sustainable farming methods, the industry has the potential to become even more efficient and environmentally friendly. Innovations in feed formulation, including alternative omega-3 sources like algae, are leading to more sustainable aquaculture methods. By adopting these practices, the future of omega-3-rich seafood looks bright, ensuring that these important fatty acids continue to be a crucial part of a nutritious diet for years to come.

References

Abd Aziz, N., Azlan, A., Ismail, A., Mohd Alinafiah, S., & Razman, M. R. (2013). Quantitative determination of fatty acids in marine fish and shellfish from warm water of straits of Malacca for nutraceutical purposes. *BioMed Research International, 2013*(1), 1.

Abramova, L. S., & Kozin, A. V. (2023). Assessment of the nutrient and metabolic profile of the chum Salmon (*Oncorhynchus keta*). *Applied Biochemistry and Microbiology, 59*(6), 986–997.

Aggarwal, B. B., & Heber, D. (Eds.). (2014). *Immunonutrition: Interactions of diet, genetics, and inflammation*. CRC Press.

Ahlgren, G., Vrede, T., & Goedkoop, W. (2009). Fatty acid ratios in freshwater fish, zooplankton and zoobenthos–are there specific optima? In *Lipids in aquatic ecosystems* (pp. 147–178).

Akbar, U., Yang, M., Kurian, D., & Mohan, C. (2017). Omega-3 fatty acids in rheumatic diseases: A critical review. *Journal of Clinical Rheumatology, 23*(6), 330–339.

Akoh, C. C. (2017). *Food lipids: Chemistry, nutrition, and biotechnology*. CRC press.

Alhshim, E. (2024). *Exploring the effects of N, N-dimethylbiguanide and ω-3 PUFAs on muscle stem cell differentiation*. Doctoral dissertation. University of Arkansas

Bhardwaj, S., Thakur, K., Sharma, A. K., Sharma, D., Brar, B., Mahajan, D., Kumar, S., & Kumar, R. (2023). Regulation of omega-3 fatty acids production by different genes in freshwater fish species: A review. *Fish Physiology and Biochemistry, 49*(5), 1005–1016.

Bu, X., Li, Y., Lai, W., Yao, C., Liu, Y., Wang, Z., Zhao, Z., Han, S., Du, J., Yao, X., & Mai, K. (2024). Innovation and development of the aquaculture nutrition research and feed industry in China. *Reviews in Aquaculture, 16*(2), 759–774.

Carlson, S. E., & Colombo, J. (2021). DHA and cognitive development. *The Journal of Nutrition, 151*(11), 3265–3266.

Chang, J., Liu, M., Liu, C., Zhou, S., Jiao, Y., Sun, H., & Ji, Y. (2024). Effects of vitamins and polyunsaturated fatty acids on cognitive function in older adults with mild cognitive impairment: A meta-analysis of randomized controlled trials. *European Journal of Nutrition, 63*, 1–20.

Cladis, D. P. (2014). *Fatty acids and mercury in seventy seven species of commercially available finfish in the United States*. Master's thesis. Purdue University.

Davis, B. A., & Devine, M. D. (2023). Evaluation of long-chain omega-3 canola oil on Atlantic salmon growth, performance, and essential fatty acid tissue accretion across the life cycle: A review. *Aquaculture International, 31*(5), 2559–2579.

de Andrade Wobido, K., et al. (2022). Non-specific effect of omega-3 fatty acid supplementation on autistic spectrum disorder: Systematic review and meta-analysis. *Nutritional Neuroscience, 25*(9), 1995–2007.

Gabriel, N. N., Abasubong, K. P., Erasmus, V. N., & Kamble, M. T. (2024). *Sustainable feed ingredients and additives for aquaculture farming*. Springer.

García-Moreno, P. J., Pérez-Gálvez, R., Morales-Medina, R., Guadix, A., & Guadix, E. M. (2013). Discarded species in the West Mediterranean Sea as sources of omega-3 PUFA. *European Journal of Lipid Science and Technology, 115*(9), 982–989.

Gladyshev, M. I., Sushchik, N. N., Tolomeev, A. P., & Dgebuadze, Y. Y. (2018). Meta-analysis of factors associated with omega-3 fatty acid contents of wild fish. *Reviews in Fish Biology and Fisheries, 28*(2), 277–299.

Gulzar, S. A. B. A., & Zuber, M. U. H. A. M. M. A. D. (2000). Determination of Omega-3 fatty acid composition in fresh water fish. *International Journal of Agriculture and Biology, 2*, 342–343.

Hegde, M. V., Zanwar, A. A., & Adekar, S. P. (2016). *Omega-3 fatty acids*. Springer International.

Heissenberger, M., Watzke, J., & Kainz, M. J. (2010). Effect of nutrition on fatty acid profiles of riverine, lacustrine, and aquaculture-raised salmonids of pre-alpine habitats. *Hydrobiologia, 650*, 243–254.

Hoffman, D. R., Birch, E. E., Birch, D. G., & Uauy, R. D. (1993). Effects of supplementation with ω3 long-chain polyunsaturated fatty acids on retinal and cortical development in premature infants. *The American Journal of Clinical Nutrition, 57*(5), 807S–812S.

Homayooni, B., Sahari, M. A., & Barzegar, M. (2014). Concentrations of omega-3 fatty acids from rainbow sardine fish oil by various methods. *International Food Research Journal, 21*(2), 743.

Hoseini, M., Baboli, M. J., & Sary, A. A. (2013). Chemical composition and fatty acids profile of farmed big head carp (*Hypophthalmichthys nobilis*) and grass carp (*Ctenopharyngodon idella*) føilet. *Aquaculture, Aquarium, Conservation & Legislation, 6*(3), 202–210.

Hudson, T. (2015). Omega-3s and mental health. *Townsend Letter, 387*, 106–110.

Huynh, M. D., & Kitts, D. D. (2009). Evaluating nutritional quality of pacific fish species from fatty acid signatures. *Food Chemistry, 114*(3), 912–918.

Jones, M. L., Mark, P. J., & Waddell, B. J. (2014). Maternal dietary omega-3 fatty acids and placental function. *Reproduction, 147*(5), R143–R152.

Katan, T. (2022). *Impact of plant-based oils and other feed ingredients on growth, tissue composition, gene expression, and health of Atlantic salmon (Salmo salar)*. Doctoral dissertation. Memorial University of Newfoundland.

Kelsey, M. D., & Pagidipati, N. J. (2023). Should we "RESPECT EPA" more now? EPA and DHA for cardiovascular risk reduction. *Current Cardiology Reports, 25*(11), 1601–1609.

Kenneth, R. (2024). The effect of diet on cardiovascular disease and lipid and lipoprotein levels. *Endotext*.

Keservani, R. K., Sharma, A. K., & Kesharwani, R. K. (Eds.). (2020). *Nutraceuticals and dietary supplements: Applications in health improvement and disease management*. CRC Press.

Kong, A. N. T. (2016). *Inflammation, oxidative stress, and cancer: Dietary approaches for cancer prevention*. CRC Press.

Kris-Etherton, P. M., Harris, W. S., & Appel, L. J. (2002). Fish consumption, fish oil, omega-3 fatty acids, and cardiovascular disease. *Circulation, 106*(21), 2747–2757.

Kumar, N., Chandan, N. K., Gupta, S. K., Bhushan, S., & Patole, P. B. (2022). Omega-3 fatty acids effectively modulate growth performance, immune response, and disease resistance in fish against multiple stresses. *Aquaculture, 547*, 737506.

Liu, J., & Ma, D. W. (2014). The role of n-3 polyunsaturated fatty acids in the prevention and treatment of breast cancer. *Nutrients, 6*(11), 5184–5223.

MacDonald, A. (2006). Omega-3 fatty acids as adjunctive therapy in Crohns disease. *Gastroenterology Nursing, 29*(4), 295–301.

Oteng, A. B., & Kersten, S. (2020). Mechanisms of action of trans fatty acids. *Advances in Nutrition, 11*(3), 697–708.

Rahnan, S. A., Huah, T. S., Nassan, O., & Daud, N. M. (1995). Fatty acid composition of some Malaysian freshwater fish. *Food Chemistry, 54*(1), 45–49.

Rajati, M., Rajati, F., Chegeni, M., Rasulehvandi, R., Rezaei, M., Ganjabi, M., & Kazeminia, M. (2024). The effect of Omega-3 supplementation and fish oil on preeclampsia: A systematic review and meta-analysis. *Clinical Nutrition ESPEN*.

Raji, C. A., Erickson, K. I., Lopez, O. L., Kuller, L. H., Gach, H. M., Thompson, P. M., Riverol, M., & Becker, J. T. (2014). Regular fish consumption and age-related brain gray matter loss. *American Journal of Preventive Medicine, 47*(4), 444–451.

Ramakrishnan, U., Gonzalez-Casanova, I., Schnaas, L., DiGirolamo, A., Quezada, A. D., Pallo, B. C., Hao, W., Neufeld, L. M., Rivera, J. A., Stein, A. D., & Martorell, R. (2016). Prenatal supplementation with DHA improves attention at 5 y of age: A randomized controlled trial. *The American Journal of Clinical Nutrition, 104*(4), 1075–1082.

Rawat, A., Khare, A., Suradkar, N. G., Kumar, D. K., & Rajmohan, H. U. (2024). A study on omega-3 fish oil safety and efficacy for cardiac patients. *Frontiers in Health Informatics, 13*(4), 826–837.

Saini, R. K., Prasad, P., Sreedhar, R. V., Akhilender Naidu, K., Shang, X., & Keum, Y. S. (2021). Omega− 3 polyunsaturated fatty acids (PUFAs): Emerging plant and microbial sources, oxidative stability, bioavailability, and health benefits – A review. *Antioxidants, 10*(10), 1627.

Sarker, P. K., Kapuscinski, A. R., Lanois, A. J., Livesey, E. D., Bernhard, K. P., & Coley, M. L. (2016). Towards sustainable aquafeeds: Complete substitution of fish oil with marine

microalga Schizochytrium sp. improves growth and fatty acid deposition in juvenile Nile tilapia (*Oreochromis niloticus*). *PLoS One, 11*(6), e0156684.
Scorletti, E., & Byrne, C. D. (2018). Omega-3 fatty acids and non-alcoholic fatty liver disease: Evidence of efficacy and mechanism of action. *Molecular Aspects of Medicine, 64*, 135–146.
Senanayake, S. P. J. N., & Fichtali, J. (2006). Single-cell oils as sources of nutraceutical and specialty lipids: Processing technologies and applications. In *Nutraceutical and specialty lipids and their co-products* (pp. 265–294). CRC Press.
Shen, S., Gong, C., Jin, K., Zhou, L., Xiao, Y., & Ma, L. (2022). Omega-3 fatty acid supplementation and coronary heart disease risks: A meta-analysis of randomized controlled clinical trials. *Frontiers in Nutrition, 9*, 809311.
Sultana, N., Mohammad, M., Lisa, S. A., & Saha, B. (2023). Effects of dietary inclusion of seaweed in regular commercial fish feed on proximate composition, minerals and fatty acid profiling of Nile tilapia *Oreochromis niloticus* (Linnaeus, 1758). *Aquatic Food Studies*, 2.
Thiyam-Holländer, U., Eskin, N. M., & Matthäus, B. (Eds.). (2012). *Canola and rapeseed: Production, processing, food quality, and nutrition*. CRC Press.
Tocher, D. R. (2015). Omega-3 long-chain polyunsaturated fatty acids and aquaculture in perspective. *Aquaculture, 449*, 94–107.
Tocher, D. R., Betancor, M. B., Sprague, M., Olsen, R. E., & Napier, J. A. (2019). Omega-3 long-chain polyunsaturated fatty acids, EPA and DHA: Bridging the gap between supply and demand. *Nutrients, 11*(1), 89.
Vianna, G. M., Zeller, D., & Pauly, D. (2020). Fisheries and policy implications for human nutrition. *Current Environmental Health Reports, 7*, 161–169.
Warwas, N. (2023). *Novel marine ingredients for aquaculture-fish nutrition, physiology and intestinal health.*
Wei, Y., Meng, Y., Li, N., Wang, Q., & Chen, L. (2021). The effects of low-ratio n-6/n-3 PUFA on biomarkers of inflammation: A systematic review and meta-analysis. *Food & Function, 12*(1), 30–40.
Xue, X. (2022). *Transcriptomic studies of Atlantic salmon responses to dietary omega-3 long-chain polyunsaturated fatty acids, Piscirickettsia salmonis infection, immunogens, and functional feed.* Doctoral dissertation. Memorial University of Newfoundland.
Yilmaz, S. T., Cakli, S., Dincer, T., Sargin, H., & Ucok, C. (2016). Evaluation of nutritional value of *Carassius gibelio* (Bloch, 1782). *Archiv für Lebensmittelhygiene, 67*(1), 2.
Zhang, Z., Miar, Y., Huyben, D., & Colombo, S. M. (2024). Omega-3 long-chain polyunsaturated fatty acids in Atlantic salmon: Functions, requirements, sources, de novo biosynthesis and selective breeding strategies. *Reviews in Aquaculture, 16*(3), 1030–1041.

Chapter 13
An Overview of Functional Genomics in Aquaculture

Yusuf Jaafar, Peerzada Zubair Ahmad, Mohd Ashraf Rather, and Ishtiyaq Ahmad

1 Introduction

Functional genomics is a field of molecular biology that attempts to describe gene (and protein) functions and interactions. The term 'genome' itself is more than 75 years old and refers to the entire genetic material of an organism, or its complete set of genes located on chromosomes (Hieter & Boguski, 1997). In 1986, 'genomics' was coined by Thomas Roderick to describe the scientific discipline of mapping, sequencing and analysing genomes. At the inception, it stated, and data was analysed; subsequently, different specific genomic techniques were developed with the rapid advances in technology, particularly the advances in PCR and sequencing technologies, a series of highly efficient approaches for genomic studies were developed. As a result of scientific demand and technological advances, a very specific branch of science evolved that is now called Genomics. In order to understand functional genomics, it is important to first define function. Define function in two possible ways. These are 'selected effect' and 'causal role'. The 'selected effect'

Y. Jaafar (✉)
Department of Water Resources, Aquaculture and Fisheries Technology, Federal University of Technology, Minna, Nigeria

Division of Fish Genetics and Biotechnology, Faculty of Fisheries, Sher-e- Kashmir University of Agricultural Science and Technology, Srinagar, Jammu and Kashmir, India
e-mail: j.yusuf@futminna.edu.ng

P. Z. Ahmad · M. A. Rather
Division of Fish Genetics and Biotechnology, Faculty of Fisheries, Sher-e- Kashmir University of Agricultural Science and Technology, Srinagar, Jammu and Kashmir, India

I. Ahmad
Division of Fish Genetics and Biotechnology, Faculty of Fisheries, Sher-e- Kashmir University of Agricultural Science and Technology, Srinagar, Jammu and Kashmir, India

Department of Zoology, University of Kashmir, Jammu and Kashmir, Srinagar, India

I. Ahmed, I. Ahmad (eds.), *Aquaculture: Enhancing Food Security and Nutrition*, https://doi.org/10.1007/978-3-031-92858-1_13

Table 13.1 Overview of scope of functional genomics in aquaculture

Applications	References
Understanding the genetic basis of desirable traits such as growth rate, disease resistance, feed efficiency	Houston et al. (2014)
Identifying genetic markers for selective breeding programmes	Sonesson et al. (2013)
Impact of genetic modification on aquatic organisms	Pott et al. (2018)
Enhanced growth, improving nutritional efficiency and feed formulation	Houston et al. (2014), Gheyas et al. (2018)
Enhancing stress tolerance and environmental adaptability	MacKenzie et al. (2017)
Developing genomics-informed breeding programmes	Gheyas et al. (2018)
Current Status of Genome Sequence and its Application in Aquaculture	
Genomics breeding programmes for environmental adaptability, sustainability, reducing environmental impact and improving water quality through bioremediation, genetic selection and enhancing biodiversity through genetic conservation	MacKenzie et al. (2017)
Enhancing nutritional value through genomics-informed breeding	Yue & Wang, 2017

function refers to the function for which a trait (DNA, RNA, protein, etc.) is selected for. The 'causal role' function refers to the function that a trait is sufficient and necessary for. Functional genomics usually tests the 'causal role' definition of function.

In aquaculture, functional genomics refers to the study of the structure, function and regulation of genes and their products in aquatic organisms, with the goal of improving aquaculture production, nutrient utilization, disease resistance and sustainability (Gheyas et al., 2018) (Table 13.1).

2 Methodologies in Functional Genomics

2.1 Transcriptomics

Transcriptomics including microarray and RNA-sequencing (RNA-seq) are the cheap, fast and easy methods to detect the differential expressed genes. Transcriptional profiling using microarrays has developed into the most prominent tool for functional genomics because expression microarrays were developed in the 1990s and marked the beginning of the genomic era and has convincingly demonstrated how information from raw sequence data can be converted into a broad understanding of gene function. After the invention of the next-generation sequencing (NGS) method in 2005, it has led to drastic changes in the way RNA-seq data is generated and analysed. Using the RNA-seq approach, it is easy to quantify the expression of known isoforms or transcripts and for unknown transcripts. Usually, transcriptomics studies provide information about what seems to be happening in an organism under specific biological circumstances (Lowe et al., 2017). The relative abundance of RNA transcripts offers a better view of active cell expression than the genomic approach (Ye et al., 2018).

Microarrays are based on hybridization, whereas RNAseq utilizes new ultrahigh throughput sequencing that became available in the recent years. The hybridization-based approaches typically involve incubating fluorescently-labelled complementary DNA (cDNA) with pre-defined sequences, such as PCR products or long oligonucleotides (mostly 60 mers), densely spotted onto a solid modified glass surface. In contrast to microarray methods, sequence-based approaches determine gene expression levels by directly sequencing cDNAs. Both approaches generate a relative abundance of mRNAs, which reflect gene expression levels. The outcome of transcriptomic studies strongly depends on sequence availability, computational methods for gene annotation and gene set enrichment. The overall goal of transcriptomics is not to identify single genes that may be altered, but to define which biological pathways are being altered in a more holistic approach. Although the technical aspects of both technologies differ considerably (as described below), they both generate lists of differentially expressed genes, and the biological interpretation of these genes is central to the biological interpretation of the experiment.

A microarray slide is typically a glass or silicon chip containing thousands of tiny spots. The microarray slide size is generally (approximately 75 mm 3 26 mm). Each DNA spot in the array contains about 10,212 moles (picomoles) of a single gene or known DNA sequence (uniquely cloned cDNA or oligonucleotide), called an oligo or probe or reporter. Probes are a small fragment of transcripts or genes that are used to hybridize a targeted cDNA sample under high-stringency conditions. Microarray not only shows the order of magnitude of transcripts but also gives the benefit of genes studied that are not influenced, whereas, on the other hand, RT-PCR (reverse transcriptase polymerase chain reaction) and northern technique blot limit the validation of a few genes per experiment (Allanach et al., 2008). TMAs are high-throughput technology that provides a comprehensive assessment of the expression. TMAs are used to analyse the protein expression at the same time in multiple individual cells and tissue samples on a single slide (Govindarajan et al., 2012).

Microarrays were first used in fish studies during the late 1990s (reviewed in Gracey & Cossins, 2003; Goetz & MacKenzie, 2008), based on the sequencing of expressed sequence tags (ESTs). Expressed sequence tags are single-pass sequences of random complementary DNA (cDNA) clones from cDNA libraries. The ESTs were often generated from cDNA libraries that were enriched for genes associated with infection or developmental stages (O'Farrell et al., 2002; Taggart et al., 2008).

The principle of microarray expression profiling study is based on the hybridization of a single-strand nucleic acid fragment to its complementary single strand with high specificity (Schena et al., 1995). The target cDNA is first labelled using fluorescent dyes and then hybridized to the array surface. The retained labelled target will then be subjected to stringent washes, capture and quantification of the fluorescent signals, followed by the data analysis process. Two platforms that are commonly in use are: cDNA microarrays and oligonucleotide microarrays (Schena et al., 1995). The cDNA microarrays contain a collection of probes generated by PCR amplification from cDNA libraries, expressed sequence tag clones or long genome cloned fragments. Oligonucleotide microarrays commonly contain short oligonucleotides (25–30 nt) or long oligonucleotides (50–80 nt). In both platforms, the short DNA segments are printed onto solid supports, usually microscopic slides, by direct contact mode (mechanical robotic spotting) or non-contact mode (ink jet technique). Different platforms each have their advantages and limitations, but in common, all have been shown reliably to capture gene expression signatures on a genomic scale (Schena et al., 1995).

2.2 Collection and Analysis of Microarray Data

For the collection of data, a microarray scanner is used comprising a computer, a camera and a laser. Microarray technique uses relative quantitation where a laser beam is passed and excites fluorescence, thus emitting the intensity from a particular spot under various conditions, which is matched to the intensity of the same spot. The excited fluorescence generates the signals when the laser beam scans the microarray, the camera captures and records the images (the pattern of fluorescence emission) produced, and this is stored as data in computer and finally analysed. The gene character of each spot is determined by the varying intensities of its colours.

EST analysis is an efficient approach for gene discovery and gene identification. For instance, between 2001 and 2007, catfish ESTs increased from 10,000 to 44,000 and the putative genes number increased from 5905 to 25,000. In the Pacific oyster (*Crassostrea gigas*), 40,845 high-quality ESTs represented 29,745 unique transcribed sequences (Fleury et al., 2009); in gilthead sea bream (*Sparus auratus*), 30,000 ESTs represented 18,196 putative unigenes (Louro et al., 2010). Currently, there are over 180 aquaculture species having more than 100 ESTs in dbEST, with approximately a dozen species having more than 10,000 ESTsPAI.

EST analysis can provide comparisons of gene expression profiling in different tissues and conditions. For instance, in a recent study with rainbow trout (*Oncorhynchus mykiss*), Kondo et al. (2011) sequenced over 30,000 ESTs from rainbow trout adipose tissue. These ESTs were used to search adipokine-related genes. The result showed that none of them encoded adipokine and PPARγgene, which play important roles in mammalian adipocytes. Further qRT-PCR results confirmed EST analysis results, that is, rainbow trout adiponectin transcripts were weakly detected in adipose tissue but strongly detected in muscle, suggesting the difference of energy metabolism between fish and mammals (Kondo et al., 2011). Chini et al. (2008) constructed normalized cDNA libraries from liver, ovary and testis in bluefin tuna (*Thunnus thynnus*), identifying several sequences with known functions in other organisms, but not previously described in this species. Also, sequences were described being expressed in one, two or more tissue libraries. Similarly, Zou et al. (2011) constructed normalized cDNA libraries from testis, ovary and mixed organs of mud crab (*Scylla paramamosain*). Through EST analysis, sex-specific transcripts were identified.

EST analysis is an effective genomic approach for rapid identification of expressed genes and has been widely used in genome-wide gene expression studies in various tissues, developmental stages and under different environmental conditions. In addition, the availability of cDNA sequences has accelerated further molecular characterization of genes of interest and provided sequence information for microarray construction and genome annotation (Rise et al., 2004).

2.3 RNA-seq Technology

RNA-seq is one of the important applications of NGS technologies to study RNA molecules, such as tRNA (transfer RNA), mRNA (messenger mRNA), rRNA (ribosomal RNA) and ncRNA (noncoding RNA), which are present in the cells. RNA-seq provides information on tens to hundreds of millions of transcripts and data on billions of individual bases of an organism. It is the study of a whole set of RNAs that are transcribed in a cell or tissue and their quantity for certain physiological and developmental stages (Gedil et al., 2016). Using the NGS approach, it is easy to identify the quantity and presence of RNA in a biological sample at a specific condition and found a more sensitive and accurate way to study the differential expression analysis, which drastically overcomes the lacuna of microarray (Voelckel et al., 2017). RNA-seq played the important role in the field of transcriptomes of aquatic animals to identify the specific genes. Moreover, it can be carried out without the prior information of the reference genome and that is very informative for a non-model organism whose genome is not yet sequenced (Voelckel et al., 2017).

RNA is polymeric in nature involving many biological processes of an organism, such as gene expression, regulation of genes, coding and decoding (Yang & Kim, 2015). Traditional methods based on cloning-based techniques, such as suppression subtractive hybridization or cDNA libraries, are more laborious and complex, which need lots of time to do, whereas on the other side RNA-seq is very easy and straight forward. RNA-seq plays an important role in gene expression, coexpression analysis, genic identification, genic variants, alternative splicing and miRNA target identifications (Yang & Kim, 2015) (Fig. 13.1).

3 Bioinformatics Analysis for RNA-seq Technology

A typical RNA-seq data analysis can be summarized as follows: (i) Perform quality control for raw RNA-seq data. Low-quality sequence tags produced from library construction or sequencing process are trimmed away by the software provided by

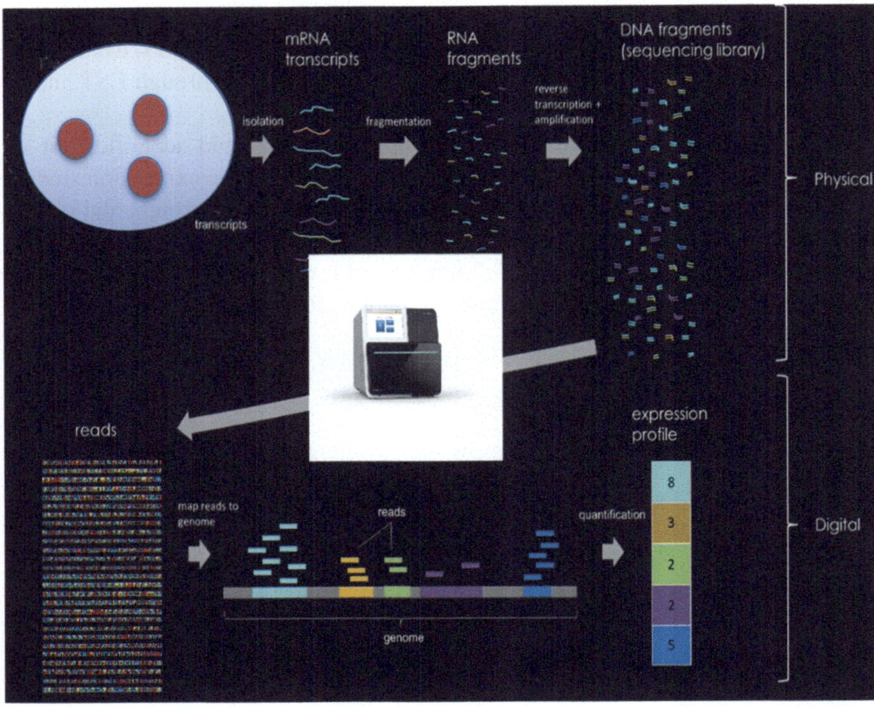

Fig. 13.1 An overview of RNA-sequence

the sequencing platform. (ii) For cases with a reference genome, map millions of short reads to the reference genome, determine the position of each RNA transcript in the genome, calculate the expression level of each transcript, and then find differentially expressed genes across the samples. All the above processes are carried out by corresponding pipelines, including representatives of Bowtie to map reads to the reference genome, Tophat to identify splice junctions, cufflinks to test for differential expression and so on. For cases without a reference genome, de novo transcriptome assembly is performed from short RNA-seq reads and then all assembled contigs are subjected to functional annotation, which requires extensive computer resources. Further specific analysis will be performed to answer certain questions involved in transcriptomics, such as analysis of RNA editing and ncRNAs, discovery of novel transcripts and correlation of transcriptome data to available genomic or epigenetic data.

4 Metabolomics

Metabolomics, a relatively new field introduced by Professor Jeremy Nicholson of Imperial College London in the mid-1990s, emerged from the investigation of the metabolome. The term 'metabolome' refers to all small molecules with a molecular weight less than 1500 Da present during a specific physiological period in a cell, organ, tissue, organism, or non-living things like food and feed (Qiu et al., 2023). Metabolomics involves the qualitative and quantitative, or semi-quantitative, study of these small molecules to examine metabolic pathways and link metabolome changes to physiological processes in living organisms. Metabolomics analysis technology is valuable in nutritional research as it reveals how external substances affect an organism's internal processes, such as growth, development, metabolism and reproduction. Unlike other omics fields, which focus on gene expression, metabolomics studies the end products of these processes, the metabolites generated by metabolic pathways. The results of metabolomics research are closely related to the phenotype of organisms. Fiehn (2002) noted that this research helps us understand how organisms change and respond to environmental stimuli, accurately describing their physiological states. Activities of the cells are reflected by the metabolome changes that can identify complex biologically essential changes. Subtle differences in genome and proteome expression can translate into metabolite differences, making them easier to identify. Unlike genomics and transcriptomics, which require high-throughput sequencing and large databases, metabolome analysis is more intricate yet simpler. Metabolites have significantly lower molecular masses than genes and proteins and are independent of genetics or species (Taylor et al., 2002). It acts as a complementary platform and is often used in accordance with transcriptomics and proteomics since metabolomics is a downstream result of gene and protein expression.

Metabolomics workflow started with the extraction of samples by chemical solvent, solid phase extraction or hydrodistillation method. In aquaculture, metabolomics analyses are sometimes used to describe the effects of fish diet and pollutants on fish (Ussery et al., 2018). In addition, metabolomics analyses are also extended to resolve issues related to fish production, such as in farmed Mandarin fish, *Siniperca chuatsi* (Xiao et al., 2020), as well as post-harvest quality control, such as in the gilthead sea bream, *Sparus aurata* (Melis et al., 2017). Moreover, metabolomics analyses can be used to investigate diseases, such as the metabolic profiling of Atlantic salmon infected with *A. salmonicida*. Metabolomics has also been used to characterize fish feed and is positioned to reveal the physiological effects of dietary manipulations. In order to improve the nutrition and quality of aquaculture goods, metabolomics is utilized to investigate the metabolites that give aquaculture products their flavour and aroma (Diez-Simon et al., 2019). Compared to other omics, researchers and scientists rarely explore the applications of metabolomics in aquaculture research. The temperature (thermal tolerance) has effect on the metabolomics as low temperatures can induce changes in cellular structure, function, metabolism, protein folding, assembly, activity and stability. Elevated temperatures can also depress metabolism, alter oxidative regulation and affect immunity. The effects of chronic elevated temperatures on *O. mykiss* were investigated through ^{1}H-NMR profiling.Metabolomics showed that a higher temperature (20 °C vs. 15 °C) induced thermal stress and downregulated levels of PCr and ATP in muscle, and glycogen levels in the liver. In a comparable study, Kullgren, Jutfelt, Fontanillas, Sundell, Samuelsson, Wiklander, Kling, Koppe, Larsson, Bjornsson and Jonsson assessed metabolic effects of elevated temperature in *S. salar*. ^{1}H-NMR plasma metabolomics revealed that *S. salar* kept at elevated temperature (8°, 12° and 18 °C) for 3 months had lower levels of glutamine, tyrosine and phenylalanine involved in energy metabolism.

The most common metabolomics tools used are mass spectrometry (MS) and nuclear magnetic resonance (NMR). Furthermore, MS technology is often combined with chromatography such as gas chromatography-mass spectrometry (GC-MS), liquid chromatography-mass spectrometry (LC-MS) or capillary electrophoresis coupled with mass spectrometry (CEMS) for the analysis of metabolites and compounds (Jorge et al., 2016). Although the NMR-based approach has dominated as the analytical platform of metabolomics studies, the number of metabolomics studies using MS-based techniques has been on the rise because of its high sensitivity and accuracy during the detection of metabolites in a sample.

NMR interrogates metabolites based on the behaviour of their atoms (frequently protons [1 H] or carbon [13C]) following exposure of a sample to a strong external magnetic field.Due to differences in molecular composition and conformation of metabolites, the nuclei of their atoms will interact and behave differently with magnetic fields to produce unique signatures that can be detected to enable molecular identification. NMR is an efficient and popular platform for structural assessment and absolute quantification of metabolites.

MS-based platforms employ molecular ionization to detect and identify metabolites in samples. By measuring the mass-to-charge ratio (m/z) of ionized molecules or their fragments, and comparing with a reference library of known compounds, metabolite identity can be established. MS-based platforms characterize the molecular weights of metabolites with high accuracy and can quantify their concentrations in the nanomolar (nM) to picomolar (pM) range. This allows the detection of additional metabolites compared with NMR.37. Metabolomics workflow started with the extraction of samples by chemical solvent, solid phase extraction or hydrodistillation method. The sample was then analysis using selected analytical platform such as NMR or MS tools, before data analysis was conducted by multi-variate analysis and statistical analysis.

5 Proteomics

Proteomics in aquaculture is one of the growing fields over the past decade, contributing to aquaculture towards reaching its primary goal of large-scale production with a quality and sustainable product. Variation in proteomics occurred during the different life stages of fish species, depicting the significance of signalling events that take place during development in a rapidly changing environment. Proteomics analysis of fishes could help in discovering proteomic changes occurring due to disease or other environmental factors such as toxins, pollutants and fluctuation in temperature that can affect normal metabolism in the body (Tomanek, 2011). Using proteomics at the molecular level study can provide vital clues regarding the transcription factors and signalling proteins regulating the growth pattern in fishes. Indeed, proteomics can be a powerful tool to investigate the physiological adaptations, evolution and biodiversity of fish species. Proteomics has been used for exploring various issues in aquaculture related to welfare, safety, quality, health and disease (Rodrigues et al., 2012). Many studies on fish have focused on proteome characterizations in the liver, blood or skin while evaluating the response to hormones (Dietrich et al., 2020), stress (Naderi et al., 2018), pollutants and pathogen infection (Saleh et al., 2018). Furthermore, studies on whole larvae (Sveinsdóttir et al., 2022), liver (Martin et al., 2007; heart (Lee et al., 2006), kidney (Martin et al., 2007) and muscle have been reported using 2-DE based proteome analyses. In the case of shellfish, proteome techniques have been used occasionally to study the changes and variety in protein expression between shellfish populations from different environments, including bio-toxins contamination (Lopez et al., 2002; Wang et al., 2008; Robalino et al., 2009).

5.1 Proteomic Techniques Used in Aquaculture

Proteomics deals with two major areas; expression and functional proteomics. It is a known fact that most of the biological process are influence by complex of proteins whose interaction might be the determinant of particular biological function (Monti et al., 2009). The proteins of interest along with their interacting counterpart can be fished out from a whole protein extract using various methods. Affinity-based methods or immune-precipitation using ligand binding properties methods can be employed (Monti et al., 2009). AQUA known as absolute quantification is another technique that uses an isotope-labelled internal standard peptides (ILISP) and liquid chromatography-mass spectrometry (LC-MS) to quantify protein and its peptides.

Nowadays, the most popular technologies in protein identification are based on peptide mass fingerprinting using mass spectrometry (MS) analysis, such as matrix-assisted laser desorption ionization-time of flight mass spectrometry (MALDI-TOF) and electrospray ionization mass spectrometry ESI-MS (Boliccio et al., 2016). These technologies involve further separation by labelling approaches using stable isotopes such as isotope-coded affinity tags (ICAT), stable isotope labelling with amino acids in cell culture (SILAC) and isobaric tags for relative and absolute quantification (iTRAQ). While for the three-dimensional structures study of the protein, the x-ray crystallography and nuclear magnetic resonance (NMR) spectroscopy technologies are two widely used techniques that might be helpful to understand its biological mechanism (Aslam et al., 2017). Proteomic workflow usually starts with the total protein extraction before the proteins are separated either using the gel-based or shotgun approaches. In the gel-based approach, several methods can be used for identifying and quantifying expressed proteins such as 2D gel electrophoresis, protein spot selection and protein digestion. Meanwhile, in shotgun approach, methods such as peptide clean-up and mass spectrometry were usually used.

Various proteomic techniques used for expression profiling include separation methods such as two-dimensional gel electrophoresis (2DE), protein microarray, chromatographic methods and mass spectrometry (MS). A typical proteomic workflow involves several steps of sample preparation, various protein separation techniques, different methods for protein digestion and identification approaches. Any step in a proteomic analysis can be selected depending upon the type of sample and aim of the expression.

As one of the most powerful tools, the proteomics approach has been increasingly used over the last decade for drug discovery research, biomarker identification, understanding pathogenicity mechanisms and altering protein expression patterns in response to different diseases (Boliccio et al., 2016; Aslam et al., 2017). Proteomics also contributes to understand intra- and extracellular processes, where differential protein expression such as protein abundance, protein posttranslational modifications and protein–protein interactions can be observed at certain times and in different environments (Aslam et al., 2017).

6 Nutrigenomics

Nutrigenomic in aquaculture, nutrigenomics or 'nutritional genomics' is concerned with studying the impacts of nutrients and food ingredients on gene expressions and comprehending the interactions that likely occur between nutrients and dietary bioactive ingredients with the genome and cellular molecules of the treated aquatic fauna at the molecular levels, which will ultimately result to mediation of gene expression. Studying the nutrigenomics in different aquaculture species holds significant. Nutrigenomics has rapidly grown in addressing several aspects related to the influences of nutrients on aquaculture species. Several researchers have studied the relationships between numerous functional genes and their expression profiles with several physiological functions. In aquaculture, nutrigenomics can be applied to improve the health, growth, reproduction and nutritional quality of farmed fish and shellfish.

Different scientists have studied the genomic changes influenced by various nutrients in different fish species. Król et al. (2016) compared the differential response of the Atlantic salmon gut to soybean meal (SBM) and fish meal (FM) as positive and negative controls for enteritis, respectively. They registered that SBM change the histology of the gut and induced extensive transcriptomic changes linked to the underlying mechanisms of SBM-induced enteropathy. Their study discovered 18 enriched pathways connected to inflammation and immune responses induced by SBM enteropathy.

Part of the pathway were the NF-kB and IL-8 signalling pathways known to induce the synthesis of various pro-inflammatory cytokines. Phagocytic pathways such as the Fcγ receptor mediated phagocytosis and monocyte pathways were highly enriched. The study of (Torrecillas et al., 2015) depicts downregulation of TCRβ, COX-2, TNFα, IL-8, IL-6, IL-10, TGFβ and IgM when MHC-II was upregulated in European seabass fed with Soya-bean oil (SBO). Combination of MOS and SBO diets reduced the inimical effects associated with SBO diets by optimizing the downregulation of GALT-related genes. Therefore, these observations show the significant of moderating feed formulation in order to produce balanced diets that will ensure preservation of the GALT-immune homeostasis.

Apart from soyabean, nutrigenomics has also been employed to examine the effects of other nutrients in fish diets. Azeredo et al. (2015) reported that the immune status of the European seabass was reduced when fed with arginine dietary supplements with downregulation of different cell-mediated immune markers in fish fed 1–2% arginine diets. Leukocytes collected from fish fed arginine diets depict a low respiratory burst compared to control fish.

Microarray and RNA-seq. RNA-seq are the Omics technologies commonly used for nutrigenomics analyses in aquaculture and are used to study the impact of different diets in various fish species including Atlantic salmon (Núñez-Acuña et al., 2016), rainbow trout, channel catfish (*Iactalurus punctatus*) (Zhao et al., 2015) blue catfish (*Ictalurus furcatus*) (Li et al., 2014) and zebrafish (Rurangwa et al., 2015), Atlantic cod (*G. morhua*) and Gilthead sea bream (*S. aurata*). However, the use of RNA-seq and microarray leads to several challenges that include the need for large data processing software as well as the need of bioinformatics tools required for differential gene expression, network pathway, alternative splicing and gene duplication analyses.

References

Allanach, K., Mengel, M., Einecke, G., Sis, B., Hidalgo, L. G., Mueller, T., et al. (2008). Comparing microarray versus RT-PCR assessment of renal allograft biopsies: Similar performance despite different dynamic ranges. *American Journal of Transplantation, 8*, 1006–1015. https://doi.org/10.1111/j.1600-6143.2008.02199.x

Aslam, B., Basit, M., Nisar, M. A., Khurshid, M., & Rasool, M. H. (2017). Proteomics: Technologies and their applications. *Journal of Chromatographic Science, 55*, 182–196. https://doi.org/10.1093/chromsci/bmw167

Azeredo, R., Pérez-Sánchez, J., Sitjà-Bobadilla, A., Fouz, B., Tort, L., Aragão, C., Oliva-Teles, A., & Costas, B. (2015). European sea bass (*Dicentrarchus labrax*) immune status and disease resistance are impaired by arginine dietary supplementation. *PLoS One, 10*(10), e0139967.

Boliccio, P. D., Rossi, C. M., Cicalini, I., Sacchetta, P., & Pieragostino, D. (2016). Integration of metabolomics and proteomics in multiple sclerosis: From biomarkers discovery to personalized medicine. *Proteomics. Clinical Applications, 10*, 470–484. https://doi.org/10.1002/prca.201500083

Chini, V., Cattaneo, A. G., Rossi, F., Bernardini, G., Terova, G., Saroglia, M., & Gornati, R. (2008). Genes expressed in blue fin tuna (*Thunnus thynnus*) liver and gonads. *Gene, 410*(1), 207–213. https://doi.org/10.1016/j.gene.2007.12.012

Dietrich, M. A., Irnazarow, I., Inglot, M., Słowińska, M., Judycka, S., & Ciereszko, A. (2020). Hormonal stimulation of carp (*Cyprinus carpio* L.) males triggers changes in the sperm proteome. *Aquaculture, 530*, 735791. https://doi.org/10.1016/j.aquaculture.2020.735791

Diez-Simon, C., Mumm, R., & Hall, R. D. (2019). Mass spectrometry-based metabolomics of volatiles as a new tool for understanding aroma and flavour chemistry in processed food products. *Metabolomics, 15*(3), 41. https://doi.org/10.1007/s11306-019-1493-6. PMID: 30868334; PMCID: PMC6476848.

Fiehn O (2002). *Max-Planck Institu te of molecular plant physiology*. 4424 Potsdam

Fleury, E., Huvet, A., Lelong, C., de Lorgeril, J., Boulo, V., Gueguen, Y., Bachère, E., Tanguy, A., Moraga, D., Fabioux, C., Lindeque, P., Shaw, J., Reinhardt, R., Prunet, P., Davey, G., Lapègue, S., Sauvage, C., Corporeau, C., Moal, J., Gavory, F., et al. (2009). Generation and analysis of a 29,745 unique expressed sequence tags from the Pacific oyster (*Crassostrea gigas*) assembled into a publicly accessible database: The GigasDatabase. *BMC Genomics, 10*, 341. https://doi.org/10.1186/1471-2164-10-341

Gedil, M., Ferguson, M., Girma, G., Gisel, A., Stavolone, L., & Rabbi, I. (2016). Perspectives on the application of next-generation sequencing to the improvement of Africa's staple food crops. *INTECH*. https://doi.org/10.5772/61665

Gheyas, A., Trujillo, A. V., Kebede, A., Al-Jumaili, A., Tadelle, D., Sparks, N., et al. (2018). Genomic analysis of African local chickens to elucidate tropical adaptation. In *Plant and animal genome XXVI conference*. PAG.

Goetz, F. W., & MacKenzie, S. (2008). Functional genomics with microarrays in fish biology and fisheries. *Fish and Fisheries, 9*(4), 378–395.

Govindarajan, R., Duraiyan, J., Kaliyappan, K., & Palanisamy, M. (2012). Microarray and its applications. *Journal of Pharmacy and Bioallied Sciences, 4*(Suppl 2), S310–S312.

Gracey, A. Y., & Cossins, A. R. (2003). Application of microarray technology in environmental and comparative physiology. *Annual Review of Physiology, 65*(1), 231–259.

Hieter, P., & Boguski, M. (1997). Functional genomics: It's all how you read it. *Science, 278*, 601–602.

Houston, R. D., Taggart, J. B., Cézard, T., et al. (2014). Development and validation of a high density SNP genotyping array for Atlantic salmon (*Salmo salar*). *BMC Genomics, 15*, 1–13.

Jorge, T. F., Rodrigues, J. A., Caldana, C., Schmidt, R., van Dongen, J. T., Thomas-Oates, J., & António, C. (2016). Mass spectrometry-based plant metabolomics: Metabolite responses to abiotic stress. *Mass Spectrometry Reviews, 35*(5), 620–649.

Kondo, H., Suga, R., Suda, S., Nozaki, R., Hirono, I., Nagasaka, R., Kaneko, G., Ushio, H., & Watabe, S. (2011). EST analysis on adipose tissue of rainbow trout *Oncorhynchus mykiss* and tissue distribution of adiponectin. *Gene, 485*, 40–45.

Król, E., Douglas, A., Tocher, D. R., Crampton, V. O., Speakman, J. R., Secombes, C. J., & Martin, S. A. (2016). Differential responses of the gut transcriptome to plant protein diets in farmed Atlantic salmon. *BMC Genomics, 17*(1), 156.

Lee, J., Valkova, N., White, M. P., & K ultz D. (2006). Proteomic identification of processes and pathways characteristic ¨ of osmoregulatory tissues in spiny dogfish shark (*Squalus acanthias*). *Comparative Biochemistry and Physiology Part D, 1*, 328–343.

Li, C., Beck, B. H., & Peatman, E. (2014). Nutritional impacts on gene expression in the surface mucosa of blue catfish (*Ictalurus furcatus*). *Developmental & Comparative Immunology, 44*(1), 226–234.

Liu, P. F., Liu, Q. H., Wu, Y., & Huang, J. (2017). Increased nucleosidediphosphate kinase activity induces white spot syndromevirus infection in *Litopenaeus vannamei*. *PLoS One, 12*(5), e0175741.

Lopez, J. L., Marina, A., et al. (2002). A proteomic approach to the study of marine mussels *Mytilus edulis* and *M. galloprovincialis*. *Marine Biology, 141*, 217–223.

Louro, B., Passos, A. L., Souche, E. L., Tsigenopoulos, C., Beck, A., Lagnel, J., Bonhomme, F., Cancela, L., Cerdà, J., Clark, M. S., Lubzens, E., Magoulas, A., Planas, J. V., Volckaert, F. A., Reinhardt, R., & Canario, A. V. (2010). Gilthead Sea bream (*Sparus auratus*) and European sea bass (*Dicentrarchus labrax*) expressed sequence tags: Characterization, tissue-specific expression and gene markers. *Marine Genomics, 3*(3–4), 179–191. https://doi.org/10.1016/j.margen.2010.09.005

Lowe, R., Shirley, N., Bleackley, M., Dolan, S., & Shafee, T. (2017). Transcriptomics technologies. *PLoS Computational Biology, 13*(5), e1005457.

Mackenzie, K. J., Carroll, P., Martin, C. A., Murina, O., Fluteau, A., Simpson, D. J., Olova, N., Sutcliffe, H., Rainger, J. K., Leitch, A., & Osborn, R. T. (2017). cGAS surveillance of micronuclei links genome instability to innate immunity. *Nature, 548*(7668), 461–465.

Martin, S. A., Mohanty, B. P., Cash, P., Houlihan, D. F., & Secombes, C. J. (2007). Proteome analysis of the Atlantic salmon (*Salmo salar*) cell line SHK-1 following recombinant IFN-gamma stimulation. *Proteomics, 7*, 2275–2286.

Melis, R., Sanna, R., Braca, A., Bonaglini, E., Cappuccinelli, R., & Slawski, H. (2017). Molecular details on gilthead sea bream (*Sparus aurata*) sensitivity to low water temperatures from 1H NMR metabolomics. *Comparative Biochemistry and Physiology Part A: Molecular & Integrative Physiology, 204*, 129–136. https://doi.org/10.1016/j.cbpa.2016.11.010

Monti, M., Cozzolino, M., Cozzolino, F., Vitiello, G., Tedesco, R., Flagiello, A., et al. (2009). Puzzle of protein complexes in vivo: A present and future challenge for functional proteomics. *Expert Review of Proteomics, 6*, 159–169.

Naderi, M., Keyvanshokooh, S., Ghaedi, A., & Salati, A. P. (2018). Effect of acute crowding stress on rainbow trout (*Oncorhynchus mykiss*): A proteomics study. *Aquaculture, 495*, 106–114.

Núñez-Acuña, G., Gonçalves, A. T., Valenzuela-Muñoz, V., Pino-Marambio, J., Wadsworth, S., Gallardo-Escárate, C., et al. (2016). Correction: Exploring early micronutrient deficiencies in rainbow trout (*Oncorhynchus mykiss*) by next-generation sequencing Technology from black box to functional genomics. *PLoS One, 11*(5), e0156668.

O'Farrell, C., Pickford, F., Vink, L., McGowan, E., & Cookson, M. R. (2002). Sequence conservation between mouse and human synphilin-1. *Neuroscience Letters, 322*(1), 9–12.

Pott, A., Otto, M., & Schulz, R. (2018). Impact of genetically modified organisms on aquatic environments: Review of available data for the risk assessment. *Science of the Total Environment, 635*, 687–698. https://doi.org/10.1016/j.scitotenv.2018.04.013

Qiu, S., Cai, Y., Yao, H., et al. (2023). Small molecule metabolites: Discovery of biomarkers and therapeutic targets. *Signal Transduction and Targeted Therapy, 8*, 132. https://doi.org/10.1038/s41392-023-01399-3

Rise, M. L., von Schalburg, K. R., Brown, G. D., Mawer, M. A., Devlin, R. H., Kuipers, N., Busby, M., Beetz-Sargent, M., Alberto, R., Gibbs, A. R., & Hunt, P. (2004). Development and application of a salmonid EST database and cDNA microarray: Data mining and interspecific hybridization characteristics. *Genome Research, 14*(3), 478–490.

Robalino, J., Carnegie, R. B., O'Leary, N., Ouvry-Patat, S. A., de la Vega, E., et al. (2009). Contributions of functional genomics and proteomics to the study of immune responses in the Pacific white leg shrimp *Litopenaeus vannamei*. *Veterinary Immunology and Immunopathology, 128*, 110–118.

Rodrigues, P. M., Silva, T. S., Dias, J., & Jessen, F. (2012). PROTEOMICS in aquaculture: Applications and trends. *Journal of Proteomics, 75*(14), 4325–4345. https://doi.org/10.1016/j.jprot.2012.03.042

Rurangwa, E., Sipkema, D., Kals, J., ter Veld, M., Forlenza, M., Bacanu, G. M., Smidt, H., & Palstra, A. P. (2015). Impact of a novel protein meal on the gastrointestinal microbiota and the host transcriptome of larval zebrafish *Danio rerio*. *Frontiers in Physiology, 6*, 133.

Saleh, M., Kumar, G., Abdel-Baki, A. A., Dkhil, M. A., El-Matbouli, M., & Al-Quraishy, S. (2018). Quantitative shotgun proteomics distinguishes wound-healing biomarker signatures in common carp skin mucus in response to *Ichthyophthirius multifiliis*. *Veterinary Research, 49*, 1–1.

Schena, M., Shalon, D., Davis, R. W., & Brown, P. O. (1995). Quantitative monitoring ofgene expression patterns with a complementary DNA microarray. *Science, 270*(5235), 467–470.

Sonesson, A. K., & Meuwissen, T. H. (2013). Testing strategies for genomic selection in aquaculture breeding programs. *Genetics Selection Evolution, 41*, 1–9.

Sveinsdóttir, H. S., Decker, A., Christensen, C., Lucena, P. B., Porsteinsson, H., Richert, E., et al. (2022). Motility phenotype in a zebrafish vmat2 mutant. *PLoS One, 17*, e0259753.

Taggart, J. B., Bron, J. E., Martin, S. A., Seear, P. J., Høyheim, B., Talbot, R., Carmichael, S. N., Villeneuve, L. A., Sweeney, G. E., Houlihan, D. F., & Secombes, C. J. (2008). A description of the origins, design and performance of the TRAITS–SGP Atlantic salmon *Salmo salar* L. cDNA microarray. *Journal of Fish Biology, 72*(9), 2071–2094.

Taylor, J., King, R., et al. (2002). Application of metabolomics to plant genotype discrimination using statistics and machine learning. *Bioinformatics, 18 supple_2*, S241–S248. https://doi.org/10.1093/bioinformatics/18.suppl_2.s241

Tomanek, L. (2011). Environmental proteomics: Changes in the proteome of marine organisms in response to environmental stress, pollutants, infection, symbiosis, and development. *Annual Review of Marine Science, 3*, 373–399.

Torrecillas, S., Montero, D., Caballero, M. J., Pittman, K. A., Custódio, M., Campo, A., Sweetman, J., & Izquierdo, M. (2015). Dietary mannan oligosaccharides: Counteracting the side effects of soybean meal oil inclusion on european sea bass (*Dicentrarchus labrax*) gut health and skin mucosa mucus production? *Frontiers in Immunology, 6*, 397. https://doi.org/10.3389/fimmu.2015.00397

Ussery, E., Bridges, K. N., Pandelides, Z., Kirkwood, A. E., Bonetta, D., Venables, B. J., et al. (2018). Effects of environmentally relevant metformin exposure on Japanese medaka (*Oryzias latipes*). *Aquatic Toxicology, 205*, 58–65. https://doi.org/10.1016/j.aquatox.2018.10.003

Voelckel, C., Gruenheit, N., & Lockhart, P. (2017). Evolutionary Transcriptomics and proteomics: Insight into plant adaptation. *Trends in Plant Science, 22*(6), 462–471. https://doi.org/10.1016/j.tplants.2017.03.001

Wang, J., Wei, Y., Wang, D., Chan, L. L., & Dai, J. (2008). Proteomic study of the effects of complex environmental stresses in the livers of goldfish (*Carassius auratus*) that inhabit Gaobeidian Lake in Beijing, China. *Ecotoxicology, 17*, 213–220.

Xiao, Z. C., Ge, C. R., Zhou, G. H., Zhang, W. G., & Liao, G. Z. (2020). 1H NMR-based metabolic characterization of Chinese Wuding chicken meat. *Food Chemistry., 274*, 574–582. https://doi.org/10.1016/j.foodchem.2018.09.008

Yang, I., & Kim, S. (2015). Analysis of whole transcriptome sequencing data: Workflow and software. *Genomics Inform, 13*(4), 119–125. https://doi.org/10.5808/GI.2015.13.4.119. Epub 2015 Dec 31. PMID: 26865842; PMCID: PMC4742321.

Ye, C., Long, Y., Ji, G., Li, Q. Q., & Wu, X. (2018). APAtrap: Identification and quantification of alternative polyadenylation sites from RNA-seq data. *Bioinformatics, 34*, 1841–1849. https://doi.org/10.1093/bioinformatics/bty029

Yue, G. H., & Wang, L. (2017). Current status of genome sequencing and its applications in aquaculture. *Aquaculture, 468*, 337–347. https://doi.org/10.1016/j.aquaculture.2016.10.036

Zhao, H., Li, C., Beck, B. H., Zhang, R., Thongda, W., Davis, D. A., & Peatman, E. (2015). Impact of feed additives on surface mucosal health and columnaris susceptibility in channel catfish fingerlings, *Ictalurus punctatus*. *Fish & Shellfish Immunology, 46*(2), 624–637.

Zou, Z., Zhang, Z., Wang, Y., Han, K., Fu, M., Lin, P., & Xiwei, J. (2011). EST (2011) analysis on the gonad development related organs and microarray screen for differentially expressed genes in mature ovary and testis of Scylla paramamosain. *Comparative Biochemistry and Physiology Part D, 6*, 150–157.

Chapter 14
CRISPR-Cas and Nanotech in Aquaculture: Pioneering Genetic Advancements for Food Security

Suvadip Ghara, Sayan Biswas, Ranjit Singh Patel, Mohd Ashraf Rather, and Ishtiyaq Ahmad

1 Introduction

The rapid growth of the world population and demand for sustainable food supply stresses traditional agricultural and animal production to the extreme. In contrast to terrestrial animal husbandry, aquaculture (i.e. production of aquatic species) has grown into a large industry to address this issue. As a result, it provides a promising means of improving the food supply at a smaller environmental cost. Though aquaculture (or the production of aquatic species) is a relative newcomer to the rise of animal husbandry, it is starting to gain popularity in comparison as we are also learning how to nourish these animals with foods other than just their original source, the sea. This has promise as a potential way of increasing food supply, while reducing its environmental effect. Technical solutions to conquer these limitations and realise the full potential of aquaculture must be based on technological advancements in genetic enhancement.

According to FAO (2022), the demand for aquatic food products will double by 2050. Creative ways to increase the sustainability and efficiency of aquaculture production are needed to meet this need. Traditional breeding methods are time-consuming and often fail to introduce the desirable traits. Genome editing technologies, most importantly the clustered regularly interspaced short

S. Ghara (✉) · R. S. Patel · M. A. Rather
Division of Fish Genetics and Biotechnology, Faculty of Fisheries, SKUAST-K, Srinagar, Jammu & Kashmir, India
e-mail: suvadipghara@gmail.com

S. Biswas
Department of Aquaculture, College of Fisheries Science, Veraval, Kamdhenu University, Gandhinagar, Gujarat, India

I. Ahmad
Department of Zoology, University of Kashmir, Jammu and Kashmir, Srinagar, India

I. Ahmed, I. Ahmad (eds.), *Aquaculture: Enhancing Food Security and Nutrition*, https://doi.org/10.1007/978-3-031-92858-1_14

palindromic repeats (CRISPR)-Cas systems, have permitted precise and successful genetic modification of aquaculture species (Okoli et al., 2022). CRISPR-Cas is an adaptive immune system in bacteria and archaea, a super powerful gene editing method from which incredibly specific targeted DNA modifications can be made to organisms. Resistance to diseases, faster development rate and higher nutritional value (Ferdous et al., 2022) can be introduced by this method.

Nanotechnology which does manipulation of matter at nanoscale has unique ability by which we can improve the drug delivery, effective gene delivery and sophisticated disease diagnostics. Thus, it improves the application of CRISPR-Cas technology in aquaculture. The CRISPR-Cas components can be delivered to aquaculture species using nanocarriers including liposomes, nanoparticles and nanoemulsions (Fajardo et al., 2022). They can be engineered to target specific tissues or organs and protect the genetic material from deterioration and can improve cellular absorption (Panyam & Labhasetwar, 2003). This targeted delivery improves the practical application of aquaculture, improves the effectiveness of gene editing and also reduces off-target effects.

By combining two existing technologies, nanotechnology and CRISPR-Cas, we create a powerful synergy to address problems in aquaculture. Through work combining the precision of CRISPR-Cas with the deployment of nanomaterials, researchers can deliver beneficial genetic modifications effectively to aquaculture species, enabling increased production and sustainability (Zhu et al., 2024). Nanotechnology is very useful for aquaculture, enabling scope for fast and effective diagnosis, thus in turn subsidising disease epidemics early. Sensors based on nanotechnology can provide real-time monitoring of water quality and aquaculture species health, and they can detect traces of pathogens or environmental pollutants (Fajardo et al., 2022). This proactive approach to disease management is less likely to produce antibiotic resistance, and lowers losses and promotes prudent use of antibiotics.

2 CRISPR-Cas Technology: Fundamentals and Mechanisms

First discovered to be adaptive immune systems in bacteria and archaea, CRISPR-Cas systems have become very powerful gene editing instruments. CRISPR-associated (Cas) proteins function in concert with CRISPR RNAs (crRNAs) as the essential basis of such systems (Barrangou, 2013). In particular, CRISPR arrays consist of short, partially palindromic, repeat sequences spaced with unique spacer sequences encoded from previous viral or plasmid invaders (Zhu et al., 2024). After transcription and processing, these spacer sequences become converted to crRNAs, which direct Cas proteins to attack complementary DNA sequences. As Burmistrz et al. (2020) point out, the crRNA complexes with a Cas protein, and directs it to the target DNA sequence through base pairing between the crRNA and the target DNA.

Cas9 endonuclease is the most widely used variant of the CRISPR-Cas system. It senses the target, and when it recognises it, it induces a double-strand break

(DSB) at the specified site. It is the 'blunt-end double-strand break (DSB) at three base pairs upstream of the protospacer adjacent motif (PAM), a short DNA sequence necessary for Cas9 to bind and cleave its target DNA' (Karvelis et al., 2015). The cell's DNA repair mechanisms are activated using this double strand break (DSB) that chooses one of the two outcomes: non-homologous end joining (NHEJ) or homology-directed repair (HDR). Error-prone NHEJ can introduce insertions, or deletions, disrupting the target gene. HDR, however, uses a donor DNA template and can do precisely insertions, deletions and substitutions (Wimberger, 2023).

Discovery of additional Cas enzymes that perform different functions beyond Cas9 has only increased the spectrum of uses for CRISPR technology. One example is Cas12a (Cpf1), which recognises a T-rich PAM sequence and introduces a staggered double-strand break, providing specific advantages for particular applications (Ansori et al., 2023). In addition, Cas13 enzymes can function in RNA knockdown and editing by binding to RNA rather than to DNA (Zhu et al., 2024). The ease of use and programmability of CRISPR-Cas systems have made them indispensable tools for genetic modification of organisms broadly relevant to aquaculture.

2.1 Origin and Evolution of CRISPR-Cas Technology

The CRISPR-Cas technologies not only have arisen from but also are derived from, the adaptive immune systems of some prokaryotes including bacteria and some archaea. This system offers protection against recurrent attacks of the same pathogens, should they be able to invade the bacterial cells, which then remember the viral invaders and can identify them. CRISPR sequences carry information about 'previous' viral contacts (Heler, 2019). That occurs when the CRISPR system directs related Cas (Cas-CRISPR-associated) proteins to find and cleave the DNA of the invader virus—neutralising the threat (van Beljouw et al., 2023).

However, these defence mechanisms have evolved over millions of years and have evolved several types of CRISPR-Cas systems specially tailored to address the need of disparate microbial or environmental conditions (Nidhi et al., 2021). This diversity provides scientists with a technology box full of choices: good accurate and adaptable gene editing. These feats also allow us to deepen our knowledge of microbial defence and genetic engineering—CRISPR-Cas has revolutionised genetic engineering, opening the door to target specific DNA sequences for altering them in very specific ways throughout any organism. The CRISPR-Cas technology has now transformed the hand-held as an advanced tool for the purpose of gene editing and the trait improvement at the industries like aquaculture, agriculture and medicine (Ferdous et al., 2022).

CRISPR-Cas technology has been developed over several large years. *Escherichia coli*'s unusual repeating DNA sequences were noticed first by Japanese researchers in 1987 (Ishino et al., 2018), eventually becoming the bases of CRISPR sequences. And it wasn't until 2002 that scientists figured out what the word CRISPR would mean (Greely, 2022). By 2005, researchers had identified these sequences as part of microbial immunity and as a store for the viral DNA chunks, which they sounded the alarm on potential invaders.

But in the years since, Emmanuelle Charpentier and Jennifer Doudna's (2012) demonstration that CRISPR-Cas9 could be harnessed to target specific DNA sequences marked a big step forward in the use of the system in genome editing across a broad range of animals. Important years that prepared the groundwork for the development of CRISPR-Cas technology as a revolutionary genetic engineering and biotechnology tool are 1987, 2002, 2005, 2012, and 2013, when Feng Zhang and others were able to alter mammalian cell gene for the first time using CRISPR-Cas9.

2.2 Mechanisms of CRISPR-Cas

CRISPR-Cas systems—borrowed from bacterial and archaeal adaptive immune mechanisms—are a completely new toolbox for genome editing, because they provide a programmable, effective and adaptable resource. These systems function to target and modify particular DNA regions that essentially require coordinated action of CRISPR-associated (Cas) proteins and CRISPR RNA (crRNA) (Van Der Oost et al., 2014). Despite this, CRISPR-Cas systems with various features and working principles all share a common theme that uses RNA-guided DNA targeting (Nishimasu & Nureki, 2017).

Streptococcus pyogenes's type II CRISPR-Cas9 is by far the most widely used genome editing system. The Cas9 protein was identified by Doudna et al. (Doudna & Charpentier, 2014) as 'RNA-guided DNA endonuclease that produces DSBs at target sites next to a crRNA and a trans-activating crRNA (tracrRNA)'. Independently, these two RNAs can be combined to make a single guide RNA (sgRNA), simplifying the experimental set-up. That the target DNA sequence must lie within a protospacer adjacent motif (PAM) is required to target specificity. According to Bhattacharya et al. (Bhattacharya & Satpati, 2024), 'The PAM sequence is the necessary one for Cas9 binding and/or cleavage' and the PAM sequence varies depending on the Cas protein's origin, but 5'-NGG-3′ is the canonical PAM of *S. pyogenes* Cas9.

Once the Cas9-sgRNA complex attaches to the purine adenine (PAM)-containing target DNA sequence, the Cas9 protein causes a DSB. If it cannot fix this DSB, then the cell will try to do that through homology directed repair (HDR) or nonhomologous end joining (NHEJ). Ceccaldi et al. (2016) explained that 'NHEJ is an error-prone pathway that frequently gives rise to small insertions or deletions (indels) at the DSB site and thus sometimes results in gene disruption, rendering NHEJ a powerful tool for gene knockout research'. However, HDR repairs the break by replacing the break with a 'donor' DNA template, which is homologous to the break flanking regions. In 'Genome Engineering with CRISPR-Cas9' (Singh & Dhar, 2020), the mechanisms under which this route allows for precise insertion, replacement or correction of genes are described. Researchers can inject a donor template with the desired sequence changes to specifically mutate the target site or insert new genetic material at that target site (Wang et al., 2022).

FLOW CHART OF USING CRISPR – Cas9 FOR GENOME EDITING

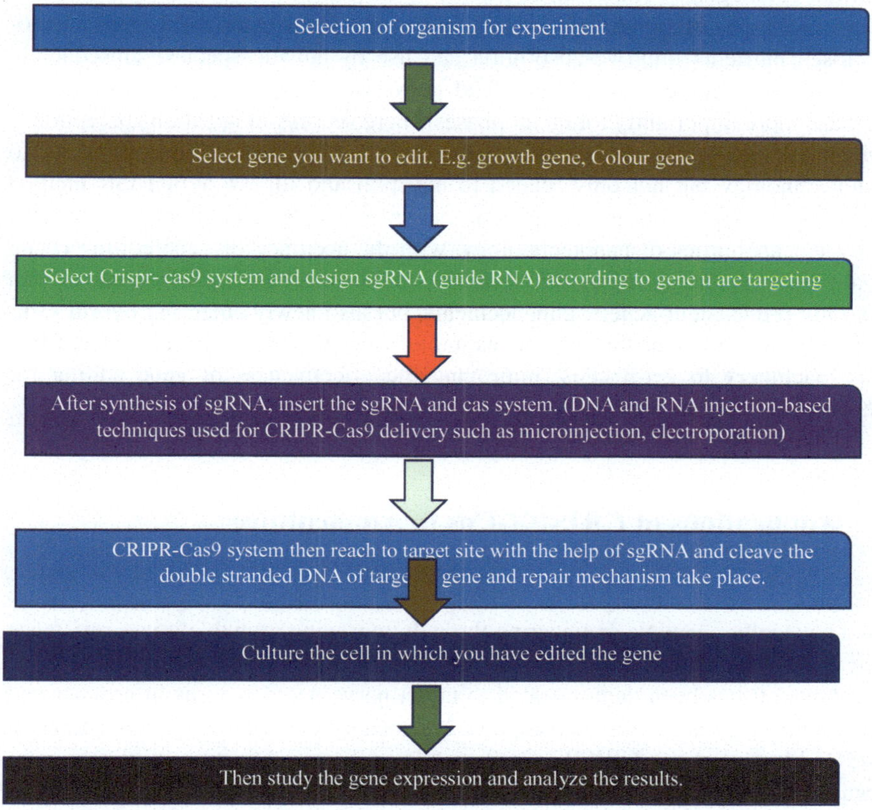

2.3 *Types of CRISPR-Cas Systems Relevant to Aquaculture*

Using genetic modification, disease control and nutrition, nanotechnology offers a range of tools to address aquaculture's problems. In fish aquaculture, a number of different nanomaterials demonstrate promise in boosting several facets of the industry. One well-known use is targeted medication and vaccine delivery using nanoparticles. Nanocarriers can encapsulate and protect therapeutic molecules to enhance the bioavailability and minimise the toxicity to non-target organisms (Huang et al., 2015). The focus of this strategy is to maximise therapeutic effectiveness while reducing environmental impact. This focused strategy maximises therapeutic effectiveness while reducing environmental impact. Chitosan nanoparticles have been studied for their biocompatibility and capacity to administer vaccinations and immunostimulants, hence improving fish illness resistance (Hossam-Elden et al., 2024). Furthermore, nanoparticles can be functionalised with ligands that specifically bind to receptors on the surface of pathogens, leading to targeted delivery of antimicrobial agents (Yun et al., 2013).

Furthermore, nanomaterials are required for improved nutrient delivery. Nano-enclosed nutrients improve absorption and use by aquatic species, subsequently resulting in increased growth and feed conversion rates (Ahmed et al., 2023). Perhaps more importantly for larval phases where successful nutrition absorption is critical for growth and survival, this is important. Genetic manipulation is related to nanotechnology but not only related to nutrition and illness. While this method opens up new possibilities for genetic enhancement by combining the targeted delivery capabilities of nanotechnology, with the accuracy of gene editing (Feng et al., 2023), it also opens up new risks due to their increased amalgamations with not only pre-existent genetic enhancements, but also newly emerging hybrid technologies. One way to do that is to, for example, use nanoparticles to deliver CRISPR-Cas machinery to germ cells, improving the effectiveness of gene editing for desired traits.

3 Applications of CRISPR-Cas in Aquaculture

The precision gene editing of the CRISPR-Cas technology has significantly benefited aquaculture genetic engineering through improving growth, disease resistance and nutritional values. CRISPR-Cas9 has been demonstrated to improve disease resistance. It has been demonstrated to target nerve necrosis virus in grouper and disrupt the grass carp Toll-like receptor 22 to enhance infection survival (Chaumont, 2024). Moreover, the CRISPR-Cas9 was used for modification of the growth-regulating genes such as myostatin (*MSTN*), in Atlantic salmon and red sea bream, resulting in faster growth rates supporting shorter production cycles and improved profitability (Wang et al., 2024). Also, CRISPR-Cas is under investigation to increase omega-3 fatty acids in fish whilst also improving the fish protein nutritional profile for human health. There are still problems too, such as minimising off-target effects and delivery of CRISPR components. Public acceptance and ethical considerations also drive the use of genetically modified organisms (GMOs) in aquaculture (Table 14.1).

3.1 Genetic Improvements for Growth and Yield

With the combination of nanotechnology and CRISPR-Cas technology, it is possible to improve growth and yield, thus improving sustainable aquaculture and food security. Unlike conventional breeding processes, it focuses on producing rapid and

Table 14.1 Outline of CRISPR-Cas genome editing observed in cultured fish

Targeted trait	Gene	Species	References
Aquatic toxicity	ahr2a and ahr2b	*Fundulus heteroclitus*	Aluru et al. (2015)
Disease resistance	Toll-like receptor TLR22 Junctional adhesion molecule-A (JAM-A)	Rohu (*Labeo rohita*)	Chakrapani et al. (2016)
Growth	Myostatin (mstn)	Red sea bream (*Pagrus major*) Common carp (*Cyprinus carpio*)	Kishimoto et al. (2018), Zhong et al. (2016)
Meiosis	aldh1a2	Nile tilapia (*Oreochromis niloticus*)	Zhong et al. (2016)
Diseases	gcjam-a	Grass carp (*Ctenopharyngodon idella*)	Ma et al. (2018)
	Cathelicidin gene	Channel catfish (*Ictalurus punctatus*)	Simora et al. (2020)
	Maf1	*Paralichthys olivaceus*	Kim et al. (2021)

focused genetic gain. By the help of CRISPR-Cas systems, fish growth–related genes can be directly altered. For instance, the CRISPR-Cas9 gene knockout of the myostatin (MSTN) gene in common carp and Atlantic salmon increased growth rate and muscle mass in these fish. It decreases manufacturing cycle and increases the fillet yield, making the process more productive and profitable. In addition, CRISPR is accurate enough to have precise changes to pathogen growth hormone (GH) genes, outperforming transgenic methods for a more reliable growth outcome (Schultz III, 2023). Genes related to increased feed conversion efficiency, disease resistance and improved stress tolerance are currently under study. Increasing delivery efficiency reduces labour-intensive and stressful traditional methods which can be achieved by the combination of nanotechnology and CRISPR-Cas. As CRISPR components can be encapsulated into liposomes and nanoparticles, the degradation is prevented, and it can be targeted for delivery into fish cells (Rani & Prabhu, 2022). Nanocarrier-mediated transport is particularly appealing for tiny eggs or larvae, where it offers a scalable and less-intrusive alternative. Nanoparticles functionalised may be able to specifically target particular tissue like muscle cells allowing localised gene editing (Majumder et al., 2019).

3.2 *Enhancing Disease Resistance*

Traditional disease management, such as antibiotics and vaccines, is often linked with high costs, limited efficacy and environmental concerns. CRISPER-Cas technology has become a boon to aquaculture enabling precise genetic modifications to enhance disease resistance in farmed species. ‘Disease outbreaks, particularly viral and bacterial infections, pose a major threat to the sustainability of aquaculture, leading to substantial economic losses and affecting the productivity of aquaculture systems' (Combe et al., 2023).

CRISPR-Cas9 has been successfully applied to improve disease resistance by targeting specific genes that pathogens exploit to infect host species. 'In orange-spotted grouper (*Epinephelus coioides*), CRISPR-Cas9-mediated knockout of genes involved in susceptibility to nervous necrosis virus (NNV) significantly increased survival rates after infection', indicating the potential of CRISPR to target disease susceptibility at the molecular level (Wang et al., 2023). This targeted approach allows for the disruption of viral entry points or replication pathways, thereby providing a robust defence mechanism against viral infections without relying on chemical interventions (Hilleman, 2004).

Similarly, CRISPR-Cas9 technology has been employed to combat bacterial infections in aquaculture species. For example, the disruption of Toll-like receptor 22 (*TLR22*), a gene critical for bacterial recognition and immune response in grass carp (*Ctenopharyngodon idella*), resulted in enhanced resistance to bacterial pathogens, highlighting the gene's vital role in host immunity (Liao et al., 2017). This genetic modification minimises the reliance on antibiotics, addressing a major environmental and health concern associated with antibiotic overuse in aquaculture (Serrano, 2005).

Moreover, the use of CRISPR-Cas9 in aquaculture is considered beneficial for sustainability. By reducing the need for chemical treatments and enhancing natural resistance, CRISPR technology supports the development of healthier and more resilient fish populations, contributing to long-term sustainability in the industry (Zhu et al., 2024). The versatility of CRISPR-Cas also enables tailored approaches to disease resistance, making it adaptable to a wide range of pathogens and aquaculture species, furthering its appeal for industry applications (Robinson et al., 2023).

3.3 Environmental Adaptation and Resilience

CRISPR-Cas technology holds significant potential for advancing environmental adaptation and resilience in aquaculture species. As climate change and environmental stressors like temperature fluctuations, salinity changes and pollution increase, aquaculture species face challenges that impact growth, survival and productivity. By enabling precise, targeted gene editing, CRISPR-Cas technology can enhance the adaptability of these species, ensuring they are better suited to withstand and thrive in changing environments (KhokharVoytas et al., 2023).

One application of CRISPR-Cas is in modifying genes that influence temperature tolerance, allowing aquaculture species to withstand higher or lower temperatures without compromising their physiological functions. For instance, CRISPR has been used to edit genes linked to thermal resilience in some fish species, paving the way for breeding strains that can survive in warmer waters, thus supporting sustainable aquaculture even in regions affected by rising temperatures (Yang et al.,

2024). Another key area is salinity tolerance. Species that can tolerate a broader range of salinity levels are particularly advantageous for aquaculture operations in coastal and estuarine environments, where salinity can vary significantly. CRISPR-Cas has been used to target genes associated with osmoregulation, helping to develop strains that can better manage changes in salinity, as shown in studies on tilapia and other brackish-water species (Kim, 2022).

In addition to temperature and salinity adaptation, CRISPR-Cas can be applied to improve resilience to environmental pollutants. Aquaculture species are often exposed to contaminants, including heavy metals and other toxins. CRISPR can be used to enhance detoxification pathways or metabolic processes, enabling fish and shellfish to more effectively neutralise or excrete harmful substances. This has been explored in research focusing on improving antioxidant responses and metal resistance, as seen in studies on oysters and other shellfish (Goud & Lanka, 2020).

3.4 *Ethical and Environmental Considerations of Gene Editing in Aquaculture*

Though CRISPR-Cas technology offers various possibilities, it brings up various ethical and environmental concerns, which include off-target mutations, harmful traits affecting the health of farmed species and their offspring. Ethically, one of the primary concerns is the potential for unintended genetic consequences. Thus, comprehensive testing and monitoring of edited organisms are crucial to mitigate these risks (Manghwar et al., 2020). Escapement of genetically modified organisms (GMOs) into the wild will contaminate and affect the natural biodiversity of the ecosystem which raises another ethical issue. For example, genetically modified fish with enhanced traits, such as faster growth rates or disease resistance, might outcompete native species, thereby altering the balance of local ecosystems. Such risks necessitate stringent containment measures and regulatory oversight to prevent the unintended spread of GMOs into natural habitats (Muir, 2004).

Environmental concerns also focus on the broader ecological impacts of gene-edited species. The introduction of modified organisms with altered growth, stress tolerance or disease resistance could disrupt local food webs and interactions between species. While CRISPR-Cas technology has the potential to help species adapt to environmental challenges, it also poses risks to the stability of ecosystems if these edited species were to escape or proliferate beyond controlled environments (Piergentili et al., 2021).

In addition, animal welfare is an important ethical consideration. Gene editing could improve the growth rate and disease resistance, but on the other hand, it may impact the natural behaviour and health of the farmed species, so long-term impacts should be studied and evaluated. Ethical concerns related to animal welfare must be integrated into the decision-making processes surrounding the use of CRISPR-Cas in aquaculture (Anne & Xu, 2021). Though some places have developed their own regulatory measures, some are yet to do so. Moreover, regulatory measures are still under evaluation and study. On the other hand, public perception of the same is also under early stages of study. Public acceptance will depend on transparent communication about the safety, ecological impacts and ethical considerations of genetically modified aquaculture species. Engaging with the public and establishing clear, responsible guidelines will be critical to the successful adoption of CRISPR-Cas technology in aquaculture (Okoli et al. 2022).

4 Fundamentals of Nanotechnology in Aquaculture

Nanotechnology that deals with materials at the nanoscale is rapidly being utilised by aquaculture to enhance sustainability and efficiency. This is a technique that exploits the high surface area and reactivity of nanoparticles to provide solution to important industry problems. Nanotechnology is also being used in feed distribution systems, a large area of aquaculture (Sarkar et al., 2022). Nanoencapsulation precisely delivers nutrients, reducing waste and improving the feed-to-weight ratio. Nanoparticles are also used to treat diseases such as nanovaccines, which is a focused and effective means to immunise against pathogens (Pati et al., 2018). Real-time monitoring of environmental parameters in water quality management can only be achieved by nanosensors, giving the aquatic life the proper conditions in the water (Vikesland, 2018). Additionally, nanomaterials can help clean water of impurities, and allow for a stronger ecology (Yaqoob et al., 2020).

As aquaculture becomes the answer to the world's food needs, nanotechnology offers a creative way to increase production, sustainability and resilience.

4.1 Introduction to Nanotechnology

The potential impact of nanotechnology to transform many industries, including aquaculture, has enormous potential to manipulate matter at the nanoscale (1–100 nanometres). Here, materials exhibit distinct physicochemical features relative to their bulk counterparts and new tools and processes with enhanced capabilities can be developed at this scale (Khan et al., 2022). As fish consumption worldwide continues to rise, so do sustainable aquaculture methods become increasingly important. Nanotechnology offers creative responses in aquaculture to problems such as growth promotion, disease prevention and delivery of effective contaminants in the future all of which contribute to food security (Nasr-Eldahan et al., 2021).

Nanomaterials possess unusual optical, magnetic and electrical properties due to high surface area-to-volume ratios, quantum effects and other factors. Nanomaterials have unique qualities, such as high reactivity, large surface area and tunable optical and magnetic properties, and therefore have applicability in a variety of areas (Baig et al., 2021). The combination of these special qualities allows such nanosensors for real-time water quality monitoring, nanocarriers for targeted delivery of drugs and genes, and antimicrobial nanomaterials for disease prevention. To use nanotechnology in aquaculture, one of the important ways is in creating effective drug delivery systems. Gaps include the risk of environmental contamination, non-specific distribution and drug degradation. Nanocarriers such as liposomes, nanoparticles and nanoemulsions provide more regulated and focused strategy. Nanocarriers can encapsulate therapeutic agents, protecting them in degradation while providing controlled release at the target site (Ding & Li, 2017). It provides controlled administration with improved therapeutic efficacy that reduces the required dosage, while reducing negative effects. Nanoparticles can also be functionalised with specific ligands to target specific cells or tissues in order to increase the accuracy of medication administration (Ding & Li, 2017).

Nanotechnology is also critical to disease detection and prevention in aquaculture. Rapid and sensitive pathogen detection is essential to effective disease management (Sloots et al., 2015). Nanosensors use the unique optical and electronics properties of nanomaterials to detect disease at very low concentration, providing means for early diagnosis before epidemics can occur. Pathogen detection was studied using nanosensors based on quantum dots, gold nanoparticles and carbon nanotubes (Kumar et al., 2020). By quick, on-site detection, these nanosensors save lives and stop disease by preventing illness and helping tackle it quickly.

Despite great potential for applications of nanotechnology in aquaculture, the use of nanomaterials involves potential hazards, which must be addressed. The environmental fate and toxicity of nanomaterials should be carefully evaluated in order to minimise the risk of unsupportable applications (Corsi et al., 2018). Research of nanomaterials on aquatic environments is important to help reduce possible negative consequences over the long term. It has been known that the safe and successful application of nanotechnology in aquaculture requires the development of standardised protocols for assessment of nanotoxicity, as well as uniform regulations of the practices (Sarkar et al., 2022). To effectively maximise the benefits of nanotechnology in sustainable aquaculture and support global food security, a comprehensive strategy that considers both the direct and indirect portable power technology hazards is necessary.

4.2 Types of Nanotechnology Applications Relevant to Aquaculture

Nanotechnology offers a broad range of tools for solving many of the most important aquaculture problems—from disease control to better nutrition, improved breeding or environmental monitoring. This section examines the main uses of nanotechnology related to transforming aquaculture methods and improving global food security.

4.2.1 Nanocarrier-Based Drug and Vaccine Delivery

Though traditional drug and vaccine administration in aquaculture often fails to deliver due to rapid clearance, poor specificity and environmental contamination, nanocarriers offer a smart alternative through their ability to encapsulate medicinal compounds and deliver to therapeutic areas with increased efficacy and fewer adverse effects. The nanocarriers can protect drugs from degradation, increase their solubility and bioavailability and allow for the delivery of the drug to the appropriate cells or tissues (Din et al., 2017). This particular distribution helps minimise the need for a large dosage and in turn cuts costs and such negative effects on the environment. One of the findings of the research is how you can use various different nanomaterials including 'liposomes, polymeric nanoparticles, chitosan nanoparticles' to deliver immunostimulants, vaccination and antibiotics. Chitosan nanoparticles 'have been shown to be effective in delivering vaccines against viral and bacterial pathogens in fish (Kumar et al., 2008)'. Nanoparticles as carriers for oral vaccination are a cost-effective and less stressful alternative to injection-based vaccination (Bezbaruah et al., 2022).

4.2.2 Nanosensors for Water Quality Monitoring and Disease Diagnostics

Productive aquaculture requires sustained ideal water quality. Live monitoring of key water parameters can be performed with high sensitivity using nanosensors. These sensors can also identify pathogens early during infection allowing quick diagnosis as well as breaking large-scale outbreaks. Creation of nanosensors for rapid detection of bacterial and viral pathogens in aquaculture systems is an important tool for proactive disease management (Foo et al., 2024). These developments, which reduce economic losses from disease outbreaks and minimise the need for preventative antibiotic usage, could help sustainability by environmental and economic means.

4.2.3 Nanomaterials for Enhanced Nutrition and Growth

But nanotechnology can improve nutrient absorption and enhance the nutritional content of fish feed. However, the same nanomaterials also may be used to encapsulate and deliver essential nutrients such as vitamins, minerals and fatty acids to improve the bioavailability and absorption (Gorantla et al., 2021). This targeted nutrient delivery can improve immune system performance, develop quicker and improve fish health as a whole. It also promotes more sustainable aquaculture pathways involving more efficient use of feeds and less waste, using nanosized particles of feed material. In the text, it also mentions the usage of nanominerals like nano-selenium and nano-zinc that has shown promising results in growing fish and improving fish immunity.

4.2.4 Nanomaterials for Bioremediation of Aquaculture Wastewater

Aquaculture wastewater contains high organic materials, nutrients and possibly hazardous substances. Nanomaterials like titanium dioxide and zinc oxide nanoparticles are used for photocatalytic degradation of organic pollutants and removal of heavy metals from aquaculture wastewater (Foo et al., 2024). All the while, these nanoparticles effectively remove contaminants which can improve water quality and reduce the aquaculture operations' negative environmental impacts.

4.2.5 Nanomaterials for Genetic Manipulation

Although CRISPR-Cas systems are revolutionising gene editing, nanotechnology can be used in tandem to bring these genetic tools into aquaculture species. As carriers of gene editing tools such as CRISPR-Cas9, nanomaterials also allow introduction of the tools into fish cells, enhancing the efficiency of gene editing (Zhu et al., 2024). This focused delivery can speed the creation of better aquaculture breeds by increasing the accuracy and effectiveness of genetic modification. Despite being in its infancy, however, the potential for future developments in aquaculture biotechnology is immense.

4.3 Safety and Regulatory Considerations

There are many advantages to using nanotechnology in aquaculture, but serious safety and legal issues have to be thought through carefully. One of the main safety concerns is the possible toxicity of nanoparticles to aquatic life, to people and to the environment. Small (< 1000 nm) and highly reactive nanoparticles can accumulate in tissues and can negatively impact health of both the individual and those raised for food (Buzea et al., 2007). The long-term effects of nanoparticle exposure need to be evaluated before risk can be evaluated, something which include bioaccumulation and biomagnification in food chains.

Another important factor is environmental safety. The intentional or accidental releases of nanoparticles into aquatic environments may have consequence for organisms that are not targeted and can disturb natural equilibrium (Tran et al., 2023). Environmental persistence and degradation processes of nanomaterials need to be assessed to reduce possible hazards. Talking about regulations, food safety frameworks and aquaculture, the existing rules often do not contain directives on nanotechnology. The regulatory agencies must establish clear guidelines for the characterisation of nanoparticles, acceptable exposure levels and monitoring needs. Labelling and risk communication are carried out transparently and are also approved by the consumer. With international regulatory harmonisation, strong safety procedures and the use of nanotechnology in aquaculture, its use will encourage innovation while sustaining aquatics ecosystems, human health and food security.

5 Synergies and Integrated Applications of CRISPR-Cas and Nanotechnology

Combined with CRISPR-Cas gene editing, nanotechnology opens up hitherto unimagined avenues for improving aquaculture and improving global food security. Nanomaterials go beyond the limits of conventional delivery techniques including microinjection and electroporation to provide significant flexibility for the safe and targeted administration of CRISPR-Cas components to aquaculture species. This synergy allows precise genetic alterations in farmed fish and shellfish resulting in better features, such as disease resistance, growth rate and nutritional value.

One of their main advantages is that nanomaterials can enhance cellular absorption and protect CRISPR-Cas components from being deteriorated. Mujtaba et al. (2021) demonstrate how nanocarriers enhance stability and bioavailability of the CRISPR-Cas components in vivo by preventing enzymatic degradation of the components. This defence is particularly important in a complex aquatic environment vulnerable to destruction of nucleases and other degrading agents that can severely degrade the efficacy of CRISPR-Cas systems. A variety of nanocarrier types have been investigated for this goal including lipid nanoparticles (LNPs) and polymeric nanoparticles, as well as inorganic nanoparticles. Among delivery platforms for CRISPR-Cas components, lipid nanoparticles (LNPs) have emerged as a promising valve, particularly for delivery of Cas nucleases and guide RNAs encoded in mRNA (Wilbie et al., 2019). CRISPR-Cas delivery in aquaculture is ideally suited to LNPs because they have inherent biocompatibility and ability to efficiently internalise cells.

Furthermore, nanomaterials enable the targeted distribution to certain tissues and cells of gene editing with less off-target damage. Functionalising nanocarriers with targeting ligands such as antibodies or peptides can enable targeted delivery of CRISPR-Cas components to specific cells or tissues within an organism (Song et al., 2021). Systemic delivery methods can be difficult to use and can lead to unintended alterations in non-target tissues, making this focused strategy of particular importance in aquaculture. For example, functionalised nanoparticles can be engineered to target specific immune cells in fish, permitting manipulation of the immune response and making disease resistant (Ji et al., 2015).

The combination CRISPR-Cas and nanotechnology makes aquaculture easier to create new disease prevention and treatment plans. Nanomaterials can be used to deliver CRISPR-Cas components aimed to attack viral DNA or RNA triggering viral replication and preventing disease spread (Dubey et al., 2022). The strategy described here should be seen as a promising alternative to traditional approaches like vaccinations and antibiotics, since they can have limitations in terms of their efficacy and may have environmental effects as well.

CRISPR-Cas has the potential beyond treating diseases for the improvement of the nutritional profiles and growth of farmed aquatic species using nanotechnology. CRISPR-Cas-mediated gene editing can be used to increase growth rate by targeting genes involved in growth hormone regulation, while nanocarriers can facilitate delivery of these gene editing tools (Mujtaba et al., 2021). This focused strategy to growth improvement leads to sustainability and financial sustainability of aquaculture operations, as well as drastically booms productivity of an operation and reduces the time needed to reach market size. Similarly, genetic modifications targeting genes related to lipid metabolism could improve the nutritional value of farmed fish with improvements in omega 3 fatty acid and other beneficial nutrient concentration, and this could be delivered precisely using nanocarrier with precise delivery of these genetic modifications (Okoli et al., 2022).

Yet there are still some problems about using nanotechnology and CRISPR-Cas on a large scale for aquaculture. Optimisation of nanocarrier design and formulation is critical to improve delivery efficiency while limiting possible toxicity to aquatic organisms (Luiz et al., 2021). More investigation is needed to assess the long-term impacts of nanomaterials on aquatic ecosystems and to ensure that genetically modified organisms are safe for human consumption. Issues of ethics arise when gene editing technology is used in aquaculture because it might have unintended effects on the ecosystem and biodiversity. Despite these hurdles, the nanotechnology and CRISPR-Cas combined potential offers great potential to revolutionise aquaculture. Given the world's population increasing, it is imperative to continue looking into this field in order to effectively and sustainably conduct premium aquatic food product production which ultimately contributes to global food security.

5.1 Complementary Advantages of CRISPR-Cas and Nanotechnology

By combining nanotechnology with CRISPR-Cas gene editing, we have developed a potent synergistic strategy to create aquaculture and address critical challenges to global food security. While CRISPR-Cas delivers the tools with which to precisely edit genetic control strings, nanotechnology provides practical solutions for the effective circulation, targeted application and augmented power of these gene editing tools in this intricate aquatic milieu. This synergy counterbalances the disadvantages of every technology when used separately, and enables unprecedented genetic development opportunities for aquaculture species.

A major hurdle for the use of CRISPR-Cas in aquaculture is effective transport of gene editing components into target cells. Microinjection is not the ideal technique for many aquatic species, particularly younger life stages, but it is inconvenient, intrusive and not appropriate. Nanotechnology offers an appealing alternative, by providing a different form, in the form of nanocarriers for encapsulating and delivering CRISPR-Cas components. Nanoparticles offer advantages for gene delivery, including protection of genetic material from degradation, targeted delivery to specific cells or tissues, and controlled release of payload (Mirza & Karim, 2021). Within this thesis, the biocompatibility, customisable size and surface characteristics, and the capacity to promote cellular uptake of lipid-based nanoparticles, polymeric nanoparticles and inorganic nanoparticles such as gold nanoparticles are investigated. One example is chitosan nanoparticles have been proven to deliver Cas9/sgRNA complexes in fish with improved transfection efficiency and gene disruption (Lin & Wang, 2023).

In addition, nanocarriers improve precision targeting and regulated release of the off-target effects and potential toxicity occurring with use of CRISPR-Cas reagents. We need to make sure gene editing works, and is safe too, in aquaculture—this is essential. 'Encapsulating CRISPR-Cas components within nanoparticles allows the liberation of them to be spatially and temporally controlled, giving gene editing greater precision and mitigating adverse effects' (Rostami et al., 2024). This focused strategy is particularly germane to germline editing, where reducing off-target alterations is paramount for producing heritable genomic alterations.

Unlike delivery, making nanotech improvements to CRISPR-Cas system effectiveness is critical (Demirer et al., 2021). Nanomaterials can be functionalised with targeting ligands that recognise specific cell surface receptors allowing selective delivery of CRISPR-Cas components to the desired cell types. This focused delivery is particularly useful in aquaculture, where modification of specific cell types, such as germ cells or immune cells, can greatly enhance breeding programs or resistance against disease. Nanoparticles conjugated to antibodies against a specific pathogen can be used to deliver CRISPR-Cas systems to infection cells, allowing for the targeted disruption of viral or bacterial genes (Dubey et al., 2022). However, as we show in this approach, this can provide new avenues for fighting infectious diseases in aquaculture, a major challenge.

In the aquatic environment, nanotechnology will furthermore enhance survival and long operation time of CRISPR-Cas components. Severe conditions include variations of temperature, salinity and presence of nucleases that can weaken the CRISPR-Cas machinery and reduce its effectiveness. Encapsulation within nanocarriers protects CRISPR-Cas components from degradation and maintains their stability in the aquatic environment, extending CRISPR component activity and increasing probability for successful gene editing (Feng et al., 2023). This improved stability is particularly critical for in vivo applications where the aquatic environment needs to be exposed for extended periods of time.

Secondly, the coupling of CRISPR-Cas and nanotechnology allows development of new aquaculture diagnostic tools. Nanomaterials are an important tool to develop highly sensitive and specific biosensor for detecting or monitoring pathogens or expression of aquaculture-related genes (Pires et al., 2021). When used together with CRISPR-Cas-based diagnostics, this speeds diagnostic, genetic modification monitoring, and simplifies the process of disease control and selective breeding.

Finally, CRISPR-Cas and nanotechnology have the potential to revolutionise aquaculture by synergistic integration. The combination of inaccurate and ineffective CRISPR-Cas delivery, high labour costs, low efficiency and genetic instability in aquatic animals is addressed by nanotechnology, which allows for gene editing on a more accurate and more effective basis. With this potent combination, this starts the development of fish that are more nutrient dense, faster growing and more resistant to disease, providing room for improvement both in sustainable aquaculture methods and in global food security. However, to fully realise the potential of these remarkable technologies in shaping aquaculture's future, more study is needed on how these technologies work together to achieve their value.

5.2 Development of Precision Aquaculture

The marriage of nanotechnology and CRISPR-Cas gene editing leads to previously unimaginable prospects for developing precision aquaculture that brings about successful and sustainable food production. This synergistic approach and advanced delivery and monitoring systems are used to develop tailored genetic therapies that will later improve fish welfare, disease resistance and production.

5.2.1 Targeted Genetic Enhancements

CRISPR-Cas systems can now be used to apply precise genetic alterations to aquaculture species, solving problems with product quality, growth and disease susceptibility. CRISPR-Cas9 has quickly become a highly effective gene editing tool for targeted editing in many types of aquaculture production organisms, leading to the development of more productive, more disease-resistant and more stress-tolerant strains (Zhu et al., 2024). By focusing researchers on genes in growth hormones that regulate growth rates, production times have been significantly reduced, and feed conversion ratios are put on the optimal side.

Improving disease resistance of farmed fish is also a huge potential use of CRISPR-Cas technology. Disease outbreaks represent a major challenge to the sustainability of aquaculture, often with high mortality rates and large economic losses (Ferdous et al., 2022). Another use of CRISPR-Cas is to introduce specific mutations into genes that may make it susceptible to the pathogen, a process that can create disease-resistant strains. However, compared to conventional illness management techniques such as the use of antibiotics, which can actually worsen environmental degradation and antimicrobial resistance, this focused approach offers a more sustainable alternative.

5.2.2 Nanotechnology-Enabled Delivery and Monitoring

For use in aquaculture, nanotechnology is necessary for enhancing the precision and utility of the CRISPR-Cas technology. Nanomaterials provide unique delivery advantages for delivery of CRISPR-Cas components into aquaculture species due to improved cellular uptake, targeted delivery and reduced off-target effects (Feng et al., 2023). An alternative is to encapsulate CRISPR-Cas components with nanoparticles to keep them from degrading and increase the ability of the particles to get into target cells. This focused delivery achieves this by reducing off-target effects and improving gene editing accuracy and safety.

But there are also cutting-edge nanotechnology instruments beyond distribution for in-the-moment fish and environmental health monitoring. Nanosensors can deliver real-time water quality parameter data including temperature, pH and dissolved oxygen, for early incipience of environmental stressors and disease outbreak detection (Sarkar et al., 2022). In this document, we explore the use of nanosensors to detect pathogens and monitor water quality in aquaculture systems. This real-time monitoring decreases the diseased outbreak and environmental change-induced effects on fish health and productivity through enabling preventative actions. Each fish can also be tracked using nanosensors to record about the fish's physiological condition, which is a very important measure of health and wellbeing.

5.2.3 Integration for Precision Aquaculture

The combination of CRISPR-Cas and nanotechnology allows for a data-driven approach to fish farming, and creates a potent precision aquaculture platform. By fusing target genetic modifications with new delivery and monitoring technologies, aquaculture procedures can be optimised to achieve the best productivity and sustainability. This combination of CRISPR-Cas and nanotechnology holds enormous potential to entirely change aquaculture by allowing precise control over fish genetics, health and the entire environment surrounding the fish (Roy et al., 2022). Finally, this integrated method increases fish welfare and production efficiency by facilitating the creation of customised feeding plans, ideal climatic conditions and individual health management procedures.

5.2.4 Challenges and Future Directions

However, combining CRISPR-Cas and nanotechnology has the potential to greatly benefit precision aquaculture; however, some issues still need to be resolved. Specifically, they include optimising gene editing procedures for a range of aquaculture species, developing more effective and biocompatible CRISPR-Cas delivery nanocarriers, and addressing any potential ethical or legal issues that will naturally arise when applying genetic modification technologies for use in food production. That is why careful research and development are necessary to overcome these obstacles and realise full potential of the combined approach in order to produce both the sustainable and secure food production from aquaculture.

6 Potential Challenges and Limitations

Yet, the combined use of CRISPR-Cas and nanotechnology has huge potential to transform aquaculture; however, there are many challenges and limitations that must be overcome to harness these technologies to enhance food security. Such challenges are biological, ethical, regulatory and technical.

6.1 Technical Challenges

A major technical problem is the effective transport of CRISPR-Cas components into target cells. Delivering CRISPR reagents to the desired cells or tissues within the aquatic organism in an efficient way is necessary for targeted gene editing (Feng et al., 2023). This is difficult, particularly for species with hard shells or a complicated life cycle. Nanocarriers hold significant potential, but the major caveat remains the ability to tailor these devices for specific aquatic species. Variables that need careful consideration in improving delivery efficiency include nanoparticle size, surface charge and stability in the aqueous environment (Mitchell et al., 2021). The potential toxicity and environmental impact of some nanomaterials also raise concerns as to their safety of consumption of genetically modified seafood (Madkour, 2023). It thus implies that nanomaterials must be subjected to elaborate toxicity testing before they become generally acceptable in use.

The other difficulty is that off-target consequences are possible. To minimise off-target activity, small details in the CRISPR-Cas system and guide RNAs need to be optimised. However, while advances in CRISPR technology (such as high-fidelity Cas enzymes, and more theoretically precise guide RNA design algorithms) are being developed to address this issue, validating the absence of off-target mutations is a complex and lengthy process; especially, when we consider the genomes of species such as fish, whose genomes are, generally, complex with vast repeat domains (Zhang, 2019). Advances in CRISPR technology, such as high-fidelity Cas enzymes and improved guide RNA design algorithms, are being developed to address this issue (Zhang, 2019). Yet, validating the absence of off-target mutations remains a complex and time-consuming process, especially in species with complex genomes (Chen et al., 2020).

6.2 Biological Challenges

Aquatic species biology is too complex to easily work around. Species with diverse genetic backgrounds or physiological characteristics require species-specific optimisation of gene editing protocols, and a better understanding of the complex interplay of genes and environment is needed to predict the phenotypic outcome of gene editing (Ferdous et al., 2022). In view of this, a clear knowledge of the physiology, the genetics and the interactions between the target species and its environment is required. The long generation times of some commercially important species also restrict progress of research and development. However, mosaicism, the ability for edited and unedited cells to coexist within an organism, would make it more difficult to interpret experimental data and reduce the ability of gene editing for trait enhancement. This is necessary for the development of stable genetic lines through germline transmission; however, defining settings to reliably germline transmit edits can be difficult in certain species (Lewens, 2020).

6.3 Ethical and Regulatory Considerations

The use of nanotechnology and CRISPR-Cas in aquaculture raises a number of ethical questions. The success of these technologies hinges on how the public perceives and will accept genetically modified seafood. Things to worry about include possible ecological effects, like gene flow out to wild populations. All of this should be taken carefully into account. To minimise the risk of gene editing from being used irresponsibly in aquaculture, risk assessment frameworks and regulatory guidelines are needed (Robinson et al., 2024). Lacking uniform rules around the world, there are barriers to international trade of genetically modified fish. The regulatory landscape for gene-edited organisms is evolving, but there is still no clarity of how to classify and approve gene-edited aquaculture products (Brookes & Smyth, 2024). In the field of clean tech, this uncertainty is also easily eroding investment and innovation.

7 Future Directions

We need the researchers, the regulators and the sector to work together to overcome these obstacles. More research is needed to increase the effectiveness and precision of CRISPR-Cas systems, to optimise nanocarrier design for targeted gene delivery and to develop reliable methods to evaluate off-target effects. This prompted the need to develop standardised protocols and guidelines for gene editing in aquaculture as a strategy to promote transparency and reproducibility (Tourapi et al., 2024). Public involvement and education are necessary in addressing ethical issues and public acceptance of genetically modified fish. Finally, transparent and uniform CRISPR-Cas and nanotechnology regulatory frameworks will enable the responsible and sustainable development of CRISPR-Cas and nanotechnology for aquaculture and extend global food security. The future of aquaculture lies in finding solutions to these problems and harnessing those gargantuan technologies to the fullest.

8 Conclusion

In brief, CRISPR-Cas in conjunction with nanotechnology holds great potential for a variety of interdisciplinary research studies that will not just reshape aquaculture methods but will also contribute to addressing problems of global food security. It is hoped that by focusing on multiplex editing, controlled release, targeted distribution and responsible deployment, researchers will fully realise how these technologies can contribute to the development of ethical and sustainable aquaculture. However, to fully realise the potential of this potent technological convergence, material scientists, geneticists, aquaculture experts and ethicists must continue to collaborate.

References

Ahmed, J., Vasagam, K. K., & Ramalingam, K. (2023). Nanoencapsulated aquafeeds and current uses in fisheries/shrimps: A review. *Applied Biochemistry and Biotechnology, 195*(11), 7110–7131.

Aluru, N., Karchner, S. I., Franks, D. G., Nacci, D., Champlin, D., & Hahn, M. E. (2015). Targeted mutagenesis of aryl hydrocarbon receptor 2a and 2b genes in Atlantic killifish (*Fundulus heteroclitus*). *Aquatic Toxicology, 158*, 192–201.

Anne, A. S. O. T. B, & Xu, I. M. W. (2021). Sustainable use of CRISPR/Cas in fish aquaculture: The biosafety perspective.

Ansori, A. N., Antonius, Y., Susilo, R. J., Hayaza, S., Kharisma, V. D., Parikesit, A. A., et al. (2023). Application of CRISPR-Cas9 genome editing technology in various fields: A review. *Narra J, 3*(2).

Baig, N., Kammakakam, I., & Falath, W. (2021). Nanomaterials: A review of synthesis methods, properties, recent progress, and challenges. *Materials Advances, 2*(6), 1821–1871.

Barrangou, R. (2013). CRISPR-Cas systems and RNA-guided interference. *Wiley Interdisciplinary Reviews: RNA, 4*(3), 267–278.

Bezbaruah, R., Chavda, V. P., Nongrang, L., Alom, S., Deka, K., Kalita, T., et al. (2022). Nanoparticle-based delivery systems for vaccines. *Vaccine, 10*(11), 1946.

Bhattacharya, S., & Satpati, P. (2024). Why does the E1219V mutation expand T-rich PAM recognition in Cas9 from streptococcus pyogenes? *Journal of Chemical Information and Modeling, 64*(8), 3237–3247.

Brookes, G., & Smyth, S. J. (2024). Risk-appropriate regulations for gene-editing technologies. *GM Crops & Food, 15*(1), 1–14.

Burmistrz, M., Krakowski, K., & Krawczyk-Balska, A. (2020). RNA-targeting CRISPR–Cas systems and their applications. *International Journal of Molecular Sciences, 21*(3), 1122.

Buzea, C., Pacheco, I. I., & Robbie, K. (2007). Nanomaterials and nanoparticles: Sources and toxicity. *Biointerphases, 2*(4), MR17-MR71.

Ceccaldi, R., Rondinelli, B., & D'Andrea, A. D. (2016). Repair pathway choices and consequences at the double-strand break. *Trends in Cell Biology, 26*(1), 52–64.

Chakrapani, V., Patra, S. K., Panda, R. P., Rasal, K. D., Jayasankar, P., & Barman, H. K. (2016). Establishing targeted carp TLR22 gene disruption via homologous recombination using CRISPR/Cas9. *Developmental and Comparative Immunology, 61*, 242–247. https://doi.org/10.1016/j.dci.2016.04.009

Chaumont, L. (2024). *Functional study of key fish interferon-stimulated genes using an in vitro knock-out approach in fish cell lines: from comparative immunology to interest for vaccine production*. Doctoral dissertation. Université Paris-Saclay.

Chen, S., Yao, Y., Zhang, Y., & Fan, G. (2020). CRISPR system: Discovery, development and off-target detection. *Cellular Signalling, 70*, 109577.

Combe, M., Reverter, M., Caruso, D., Pepey, E., & Gozlan, R. E. (2023). Impact of global warming on the severity of viral diseases: A potentially alarming threat to sustainable aquaculture worldwide. *Microorganisms, 11*(4), 1049.

Corsi, I., Winther-Nielsen, M., Sethi, R., Punta, C., Della Torre, C., Libralato, G., et al. (2018). Ecofriendly nanotechnologies and nanomaterials for environmental applications: Key issue and consensus recommendations for sustainable and ecosafe nanoremediation. *Ecotoxicology and Environmental Safety, 154*, 237–244.

Demirer, G. S., Silva, T. N., Jackson, C. T., Thomas, J. B., Ehrhardt, W., et al. (2021). Nanotechnology to advance CRISPR–Cas genetic engineering of plants. *Nature Nanotechnology, 16*(3), 243–250.

Din, F. U., Aman, W., Ullah, I., Qureshi, O. S., Mustapha, O., Shafique, S., & Zeb, A. (2017). Effective use of nanocarriers as drug delivery systems for the treatment of selected tumors. *International Journal of Nanomedicine*, 7291–7309.

Ding, C., & Li, Z. (2017). A review of drug release mechanisms from nanocarrier systems. *Materials Science and Engineering: C, 76*, 1440–1453.

Doudna, J. A., & Charpentier, E. (2014). The new frontier of genome engineering with CRISPR-Cas9. *Science, 346*(6213), 1258096.

Dubey, A. K., Kumar Gupta, V., Kujawska, M., Orive, G., Kim, N. Y., Li, C. Z., et al. (2022). Exploring nano-enabled CRISPR-Cas-powered strategies for efficient diagnostics and treatment of infectious diseases. *Journal of Nanostructure in Chemistry, 12*(5), 833–864.

Fajardo, C., Martinez-Rodriguez, G., Blasco, J., Mancera, J. M., Thomas, B., & De Donato, M. (2022). Nanotechnology in aquaculture: Applications, perspectives and regulatory challenges. *Aquaculture and Fisheries, 7*(2), 185–200.

FAO. 2022. *The State of World Fisheries and Aquaculture 2022. Towards Blue Transformation.* Rome, FAO. https://doi.org/10.4060/cc0461en

Feng, S., Xie, X., Liu, J., Li, A., Wang, Q., Guo, D., et al. (2023). A potential paradigm in CRISPR/Cas systems delivery: At the crossroad of microalgal gene editing and algal-mediated nanoparticles. *Journal of Nanobiotechnology, 21*(1), 370.

Ferdous, M. A., Islam, S. I., Habib, N., Almehmadi, M., Allahyani, M., Alsaiari, A. A., & Shafie, A. (2022). CRISPR-Cas genome editing technique for fish disease management: Current study and future perspective. *Microorganisms, 10*(10), 2012.

Foo, W. H., Chia, W. Y., Ende, S., Chia, S. R., & Chew, K. W. (2024). Nanomaterials in aquaculture disinfection, water quality monitoring, and wastewater remediation. *Journal of Environmental Chemical Engineering*, 113947.

Gorantla, S., Wadhwa, G., Jain, S., Sankar, S., Nuwal, K., Mahmood, A., et al. (2021). Recent advances in nanocarriers for nutrient delivery. In *Drug delivery and translational research* (pp. 1–26).

Goud, S., & Lanka, S. (2020). Chitosan from crustacean shell waste in heavy metal removal from industrial effluents. In *Biotechnology for sustainable development*. Immortal publications.

Greely, H. T. (2022). *CRISPR people: The science and ethics of editing humans*. MIT Press.

Heler, R. (2019). *Generation of memory of infection during the CRISPR-Cas9 immune response*.

Hilleman, M. R. (2004). Strategies and mechanisms for host and pathogen survival in acute and persistent viral infections. *Proceedings of the National Academy of Sciences, 101*(suppl_2), 14560–14566.

Hossam-Elden, N., Abu-Elala, N. M., AbuBakr, H. O., Luo, Z., Aljuaydi, S. H., Khattab, M., et al. (2024). Dietary chitosan nanoparticles enhance growth, antioxidant defenses, immunity, and *Aeromonas veronii* biovar sobria resistance in Nile tilapia *Oreochromis niloticus*. *Fishes, 9*(10), 388.

Huang, S., Wang, L., Liu, L., Hou, Y., & Li, L. (2015). Nanotechnology in agriculture, livestock, and aquaculture in China. A review. *Agronomy for Sustainable Development, 35*, 369–400.

Ishino, Y., Krupovic, M., & Forterre, P. (2018). History of CRISPR-Cas from encounter with a mysterious repeated sequence to genome editing technology. *Journal of Bacteriology, 200*(7), 10–1128.

Ji, J., Torrealba, D., Ruyra, À., & Roher, N. (2015). Nanodelivery systems as new tools for immunostimulant or vaccine administration: Targeting the fish immune system. *Biology, 4*(4), 664–696.

Karvelis, T., Gasiunas, G., Young, J., Bigelyte, G., Silanskas, A., Cigan, M., & Siksnys, V. (2015). Rapid characterization of CRISPR-Cas9 protospacer adjacent motif sequence elements. *Genome Biology, 16*, 1–13.

Khan, Y., Sadia, H., Ali Shah, S. Z., Khan, M. N., Shah, A. A., Ullah, N., et al. (2022). Classification, synthetic, and characterization approaches to nanoparticles, and their applications in various fields of nanotechnology: A review. *Catalysts, 12*(11), 1386.

KhokharVoytas, A., Shahbaz, M., Maqsood, M. F., Zulfiqar, U., Naz, N., Iqbal, U. Z., et al. (2023). Genetic modification strategies for enhancing plant resilience to abiotic stresses in the context of climate change. *Functional & Integrative Genomics, 23*(3), 283.

Kim, C. (2022). *Investigating transcriptional regulatory mechanisms of euryhaline fish in response to salinity stress using a tilapia cell line*. University of California.

Kim, J., Cho, J. Y., Kim, J.-W., et al. (2021). Molecular characterization of Paralichthys olivaceus maf1 and its potential role as an anti-viral hemorrhagic septicaemia virus factor in hirame natural embryo cells. *Intern J Mol Sci., 22*(3), 1353.

Kishimoto, K., Washio, Y., Yoshiura, Y., Toyoda, A., Ueno, T., Fukuyama, H., et al. (2018). Production of a breed of Red Sea bream *Pagrus major* with an increase of skeletal muscle muss and reduced body length by genome editing with CRISPR/Cas9. *Aquaculture, 495*, 415–427. https://doi.org/10.1016/j.aquaculture.2018.05.055

Kumar, S. R., Ahmed, V. I., Parameswaran, V., Sudhakaran, R., Babu, V. S., & Hameed, A. S. (2008). Potential use of chitosan nanoparticles for oral delivery of DNA vaccine in Asian sea bass (*Lates calcarifer*) to protect from vibrio (Listonella) anguillarum. *Fish & Shellfish Immunology, 25*(1–2), 47–56.

Kumar, H., Kuča, K., Bhatia, S. K., Saini, K., Kaushal, A., Verma, R., et al. (2020). Applications of nanotechnology in sensor-based detection of foodborne pathogens. *Sensors, 20*(7), 1966.

Lewens, T. (2020). Blurring the germline: Genome editing and transgenerational epigenetic inheritance. *Bioethics, 34*(1), 7–15.

Liao, Z., Wan, Q., Su, H., Wu, C., & Su, J. (2017). Pattern recognition receptors in grass carp *Ctenopharyngodon idella*: I. Organization and expression analysis of TLRs and RLRs. *Developmental & Comparative Immunology, 76*, 93–104.

Lin, M., & Wang, X. (2023). Natural biopolymer-based delivery of CRISPR/Cas9 for cancer treatment. *Pharmaceutics, 16*(1), 62.

Luiz, M. T., Viegas, J. S. R., Abriata, J. P., Viegas, F., de Carvalho Vicentini, F. T. M., Bentley, M. V. L. B., et al. (2021). Design of experiments (DoE) to develop and to optimize nanoparticles as drug delivery systems. *European Journal of Pharmaceutics and Biopharmaceutics, 165*, 127–148.

Ma, J., Fan, Y. D., Zhou, Y., et al. (2018). Efficient resistance to grass carp reovirus infection in JAM-A knockout cells using CRISPR/Cas9. *Fish & Shellfish Immunology, 76*, 206–215.

Madkour, L. H. (2023). Eco-friendly green biosynthesized metallic nanoparticles and biotechnological applications in pharmaceuticals sciences. *Materials Science and Engineering B, 13*, 1–69.

Majumder, J., Taratula, O., & Minko, T. (2019). Nanocarrier-based systems for targeted and site specific therapeutic delivery. *Advanced Drug Delivery Reviews, 144*, 57–77.

Manghwar, H., Li, B., Ding, X., Hussain, A., Lindsey, K., Zhang, X., & Jin, S. (2020). CRISPR/Cas systems in genome editing: Methodologies and tools for sgRNA design, off-target evaluation, and strategies to mitigate off-target effects. *Advanced Science, 7*(6), 1902312.

Mirza, Z., & Karim, S. (2021). Nanoparticles-based drug delivery and gene therapy for breast cancer: Recent advancements and future challenges. In *Seminars in cancer biology* (Vol. 69, pp. 226–237). Academic Press.

Mitchell, M. J., Billingsley, M. M., Haley, R. M., Wechsler, M. E., Peppas, N. A., & Langer, R. (2021). Engineering precision nanoparticles for drug delivery. *Nature Reviews Drug Discovery, 20*(2), 101–124.

Muir, W. M. (2004). The threats and benefits of GM fish: A generally accepted model for assessing the environmental risk of GM organisms would not only help regulators but also address fears about this technology. *EMBO Reports, 5*(7), 654–659.

Mujtaba, M., Wang, D., Carvalho, L. B., Oliveira, J. L., et al. (2021). Nanocarrier-mediated delivery of miRNA, RNAi, and CRISPR-Cas for plant protection: Current trends and future directions. *ACS Agricultural Science & Technology, 1*(5), 417–435.

Nasr-Eldahan, S., Nabil-Adam, A., Shreadah, M. A., Maher, A. M., & El-Sayed Ali, T. (2021). A review article on nanotechnology in aquaculture sustainability as a novel tool in fish disease control. *Aquaculture International, 29*, 1459–1480.

Nidhi, S., Anand, U., Oleksak, P., Tripathi, P., Lal, J. A., Thomas, G., et al. (2021). Novel CRISPR–Cas systems: An updated review of the current achievements, applications, and future research perspectives. *International Journal of Molecular Sciences, 22*(7), 3327.

Nishimasu, H., & Nureki, O. (2017). Structures and mechanisms of CRISPR RNA-guided effector nucleases. *Current Opinion in Structural Biology, 43*, 68–78.

Okoli, A. S., Blix, T., Myhr, A. I., Xu, W., & Xu, X. (2022). Sustainable use of CRISPR/Cas in fish aquaculture: The biosafety perspective. *Transgenic Research, 31*(1), 1–21.

Panyam, J., & Labhasetwar, V. (2003). Biodegradable nanoparticles for drug and gene delivery to cells and tissue. *Advanced Drug Delivery Reviews, 55*(3), 329–347.

Pati, R., Shevtsov, M., & Sonawane, A. (2018). Nanoparticle vaccines against infectious diseases. *Frontiers in Immunology, 9*, 2224.

Piergentili, R., Del Rio, A., Signore, F., Umani Ronchi, F., Marinelli, E., & Zaami, S. (2021). CRISPR-Cas and its wide-ranging applications: From human genome editing to environmental implications, technical limitations, hazards and bioethical issues. *Cells, 10*(5), 969.

Pires, N. M., Dong, T., Yang, Z., & da Silva, L. F. (2021). Recent methods and biosensors for foodborne pathogen detection in fish: Progress and future prospects to sustainable aquaculture systems. *Critical Reviews in Food Science and Nutrition, 61*(11), 1852–1876.

Rani, V., & Prabhu, A. (2022). CRISPR-Cas9 based non-viral approaches in nanoparticle elicited therapeutic delivery. *Journal of Drug Delivery Science and Technology, 76*, 103737.

Robinson, N. A., Robledo, D., Sveen, L., Daniels, R. R., Krasnov, A., Coates, A., et al. (2023). Applying genetic technologies to combat infectious diseases in aquaculture. *Reviews in Aquaculture, 15*(2), 491–535.

Robinson, N. A., Østbye, T. K. K., Kettunen, A. H., Coates, A., Barrett, L. T., Robledo, D., & Dempster, T. (2024). A guide to assess the use of gene editing in aquaculture. *Reviews in Aquaculture, 16*(2), 775–784.

Rostami, N., Gomari, M. M., Choupani, E., Abkhiz, S., Fadaie, M., Eslami, S. S., et al. (2024). Exploring advanced CRISPR delivery technologies for therapeutic genome editing. *Small Science, 4*(10), 2400192.

Roy, S., Kumar, V., Behera, B. K., Parhi, J., Mohapatra, S., Chakraborty, T., & Das, B. K. (2022). CRISPR/Cas genome editing – Can it become a game changer in future fisheries sector? *Frontiers in Marine Science, 9*, 924475.

Sarkar, B., Mahanty, A., Gupta, S. K., Choudhury, A. R., Daware, A., & Bhattacharjee, S. (2022). Nanotechnology: A next-generation tool for sustainable aquaculture. *Aquaculture, 546*, 737330.

Schultz III, R. B. (2023). *Utilization of biotechnology, including CRISPR/Cas9, to improve production efficiency in swine*. Doctoral dissertation. Iowa State University.

Serrano, P. H. (2005). *Responsible use of antibiotics in aquaculture* (Vol. 469). Food and Agriculture Organization.

Simora, R. M. C., Xing, D., Bangs, X. R., et al. (2020). CRISPR/Cas9-mediated knock-in of alligator cathelicidin gene in a non-coding region of channel catfish genome. *Scientific Reports, 10*, 22271.

Singh, V., & Dhar, P. K. (Eds.). (2020). *Genome engineering via CRISPR-Cas9 system*. Academic Press.

Sloots, T. P., Nissen, M. D., Ginn, A. N., & Iredell, J. R. (2015). Rapid identification of pathogens using molecular techniques. *Pathology-Journal of the RCPA, 47*(3), 191–198.

Song, X., Liu, C., Wang, N., Huang, H., He, S., Gong, C., & Wei, Y. (2021). Delivery of CRISPR/Cas systems for cancer gene therapy and immunotherapy. *Advanced Drug Delivery Reviews, 168*, 158–180.

Tourapi, C., Christoforou, E., Gaudêncio, S. P., & Vasquez, M. I. (2024). Aquatic biomaterial repositories: Comprehensive guidelines, recommendations, and best practices for their development, establishment, and sustainable operation. *Marine Drugs, 22*(9), 427.

Tran, T. K., Nguyen, M. K., Lin, C., Hoang, T. D., Nguyen, T. C., Lone, A. M., et al. (2023). Review on fate, transport, toxicity and health risk of nanoparticles in natural ecosystems: Emerging challenges in the modern age and solutions toward a sustainable environment. *Science of the Total Environment, 169331*.

van Beljouw, S. P., Sanders, J., Rodríguez-Molina, A., & Brouns, S. J. (2023). RNA-targeting CRISPR–Cas systems. *Nature Reviews Microbiology, 21*(1), 21–34.

Van Der Oost, J., Westra, E. R., Jackson, R. N., & Wiedenheft, B. (2014). Unravelling the structural and mechanistic basis of CRISPR–Cas systems. *Nature Reviews Microbiology, 12*(7), 479–492.

Vikesland, P. J. (2018). Nanosensors for water quality monitoring. *Nature Nanotechnology, 13*(8), 651–660.

Wang, J., He, Z., Wang, G., Zhang, R., Duan, J., Gao, P., et al. (2022). Efficient targeted insertion of large DNA fragments without DNA donors. *Nature Methods, 19*(3), 331–340.

Wang, J., Liu, Z., Zhang, Y., Li, C., Dunham, R. A., & Liu, H. (2023). CRISPR/Cas9 microinjection of transgenic embryos enhances the dual-gene integration efficiency of antimicrobial peptide genes for bacterial resistance in channel catfish, *Ictalurus punctatus*. *Aquaculture, 575, 739725*. https://doi.org/10.1016/j.aquaculture.2023.739725

Wang, J., Torres, I. M., Shang, M., Al-Armanazi, J., Dilawar, H., Hettiarachchi, D. U., et al. (2024). One-step knock-in of two antimicrobial peptide transgenes at multiple loci of catfish by CRISPR/Cas9-mediated multiplex genome engineering. *International Journal of Biological Macromolecules, 260*, 129384.

Wilbie, D., Walther, J., & Mastrobattista, E. (2019). Delivery aspects of CRISPR/Cas for in vivo genome editing. *Accounts of Chemical Research, 52*(6), 1555–1564.

Wimberger, S. (2023). Improving integration efficiency and precision of CRISPR/Cas9-mediated genome editing.

Yang, Z., Fu, G., Lee, M., Yeo, S., & Yue, G. H. (2024). Genes for editing to improve economic traits in aquaculture fish species. *Aquaculture and Fisheries*.

Yaqoob, A. A., Parveen, T., Umar, K., & Mohamad Ibrahim, M. N. (2020). Role of nanomaterials in the treatment of wastewater: A review. *Water, 12*(2), 495.

Yun, Y., Cho, Y. W., & Park, K. (2013). Nanoparticles for oral delivery: Targeted nanoparticles with peptidic ligands for oral protein delivery. *Advanced Drug Delivery Reviews, 65*(6), 822–832.

Zhang, F. (2019). Development of CRISPR-Cas systems for genome editing and beyond. *Quarterly Reviews of Biophysics, 52*, e6.

Zhong, Z., Niu, P., Wang, M., Huang, G., Xu, S., Sun, Y., et al. (2016). Targeted disruption of Sp7 and Myostatin with CRISPR-Cas9 results in severe bone defects and more muscular cells in common carp. *Scientific Reports, 6*, 1–14. https://doi.org/10.1038/srep22953

Zhu, M., Sumana, S. L., Abdullateef, M. M., Falayi, O. C., Shui, Y., Zhang, C., et al. (2024). CRISPR/Cas9 technology for enhancing desirable traits of fish species in aquaculture. *International Journal of Molecular Sciences, 25*(17).

Chapter 15
Impact of Climate Change on Shellfish and Its Implication in Posing Risk to Global Food Security

Raman Jasrotia, Menakshi Dhar, Dheeraj Sharma, Sourabh Sharma, and Seema Langer

1 Introduction

Shellfish is a prime constituent of seafood production across the globe. The term shellfish is used for the exoskeleton-bearing invertebrates that inhabit either fresh or marine aquatic ecosystems and includes crabs, lobsters, shrimp, mussels, oysters, clams, crayfish, squid, abalone, cuttlefish, octopus, snails, scallops, and others. A considerable amount of essential amino acids, polyunsaturated fatty acids, and various vitamins and minerals such as copper, zinc, sodium, potassium, and iodine are present in the shellfish, and thus, they are considered as good for health Venugopal and Gopakumar (2017). Certain functional components are present in the shellfish that are absent in the terrestrial organisms (Hosomi et al., 2012).

2 Contribution of Shellfish in Aquaculture Practices

At present, aquaculture is a rapidly growing food-producing sector throughout the world and is considered a sustainable alternative for procuring food security (Azra et al., 2021). Due to the continuous increase in population, it is prophesied that the demand for aquatic food products will surge in the near future, which will exert great pressure on the food producers (Godfray et al., 2010). Across the world,

R. Jasrotia (✉)
Department of Zoology, University of Jammu, Jammu, Jammu and Kashmir, India

Department of Zoology, Cluster University of Jammu, Jammu, Jammu and Kashmir, India
e-mail: drraman.jasrotia@jammuuniversity.ac.in

M. Dhar · D. Sharma · S. Sharma · S. Langer
Department of Zoology, University of Jammu, Jammu, Jammu and Kashmir, India

I. Ahmed, I. Ahmad (eds.), *Aquaculture: Enhancing Food Security and Nutrition*, https://doi.org/10.1007/978-3-031-92858-1_15

several initiatives have been taken to meet the growing food demands of the population, and among these, fishery resources have made a major contribution (Garcia & Rosenberg, 2010). In 2014, the worldwide availability of seafood increased to 167.2 million metric tons (Khan & Liu, 2019a, b). According to the Food and Agriculture Organization, 2014, across the globe, the annual catch of cephalopods is approximately 4.8 million tons, and cephalopod protein assures 2% of the global protein consumption, worldwide. The aquaculture sector has a significant contribution towards food security, especially in developing countries. In 2015, 76.65 million tons of production was made through aquaculture globally, out of which 9.6% were farmed shellfish. Aquaculture accounts for 45% of the animal food obtained from the aquatic ecosystem that is utilized by humans and thus provides remarkable benefits to humans (De Silva & Soto, 2009; Bostock et al., 2010).

3 Food Security from Shellfish

The global food composition database includes data on various species of shellfish, both in raw and cooked forms. These shellfish are marketed in dried, fresh, chilled, and frozen forms (Kugino et al., 1993; Paredi & Crupkin, 1997; Hurtado et al., 2001). The proximate composition of various shellfish has been given in Table 15.1. FAO/INFOODS (2016) has recorded the nutrient composition of crabs and lobsters and observed elevated content of vitamins A, B_{12}, and C, and β-carotene in the body parts, although little content of those vitamins is present in the legs and claws. Muscle composition of mollusks varies according to their environment as far as omega-3 fatty acids and minerals are concerned, the nutritive value of shellfish is higher than the terrestrial animal food sources such as lamb, hogget, mutton, chicken, pork, and beef (Wright et al., 2018). Being an excellent source of various nutrients, mollusks may be utilized in functional foods and nutricosmetics (Tabakaeva et al., 2018). Various components of the shellfish are recorded in the following subheadings:

3.1 Fatty Acids

Only 0.6–1.4% of fats have been found in the crab meat, *Portunuspelagicus* (Musaiger & Al-Rumaidh, 2005). Kathrivel et al. (2014) recorded six different kinds of fatty acids in the body and appendages of the box crab, *Calappalophus*. The white shrimp, *Litopenaeusschimitti*, has excellent nutritional value. Studies revealed high levels of omega-3 polyunsaturated fatty acids in the muscles of shrimp, particularly decosahexaenoic acid and eicosapentaneoic acid, that are considered good as far as nutritional status is concerned. Clams and snails belonging to the Phylum Mollusca are abundant sources of essential polyunsaturated fatty acids. Merline and Chitra (2020) recorded more lipid content in lobsters as compared to

Table 15.1 Proximate composition of various shellfish

Shellfish	Proteins	Lipids	Ash	Moisture	Carbohydrates	References
Metanephrops challengeri	15.2 ± 1.7	1.1 ± 0.4	–	66.4 ± 4.4	–	Meynier et al. (2008)
Thenus orientalis	18.3 ± 0.01	0.09 ± 0.01	1.80 ± 007	77.0 ± 1.00	–	Merline and Chitra (2020)
Penaeus semisulcatus	21.6 ± 0.10	0.15 ± 0.02	2.40 ± 0.07	74.4 ± 0.047	–	Merline and Chitra (2020)
Penaeus monodon	20.3 ± 0.87	5.53 ± 0.43	0.87 + 0.06	72.3 ± 2.54	0.91 ± 0.03	Islam et al. (2017)
Macrobrachium rosenbergii	25.8 ± 2.31	1.70 ± 0.11	0.81 ± 0.09	71.4 ± 1.43	0.17 ± 0.05	Islam et al. (2017)
Crassostrea madrasensis	9.41 ± 0.85	3.25 ± 0.32	1.01 ± 0.06	82.64 ± 1.31	3.2 ± 0.13	Asha et al. (2014)
Uroteuthis duvauceli	17.5 ± 0.06	0.52 ± 0.08	1.31 ± 0.15	80.40 ± 0.23	–	Remyakumari et al. (2018)
Notodarus sloani	18.9 ± 1.2	5.6 ± 1.9	–	72.7 ± 2.3	–	Meynier et al. (2008)
Enteroctopus zealandicus	13.6 ± 1.3	1.8 ± 0.6	–	80.2 ± 1.3	–	Meynier et al. (2008)
Munida gregaria	13.7	3.84	2.19	77.6	–	Varisco et al. (2020)
Palinurus elephas	19.31 ± 0.56	0.38 ± 0.06	2.13 ± 0.59	76.68 ± 2.56	–	Kampouris et al. (2021)
Penaeus aztecus	22.30 ± 0.27	0.51 ± 0.04	1.73 ± 0.07	74.39 ± 2.88	–	Kampouris et al. (2021)
Perna viridis	36.1 ± 1	19.72 ± 2	–	–	24.54 ± 1	Saritha et al. (2015)

prawns. Lipids in the crayfish, *Astacusleptodactylus*, have less amounts of saturated fatty acids and are rich in polyunsaturated fatty acids such as arachidonic acid, docosahexaenoic acid, and eicosapentaenoic acid (Öksüz & Mazlum, 2016). Studies done by Moniruzzaman et al. (2021) indicated that the bivalves inhabiting marine waters are rich sources of EPA, DHA, and high-quality proteins whereas the freshwater snails and mussels are great sources of arachidonic acid, linolenic acid, and proteins but lack EPA and DHA. Cuttlefish contains significant amounts of omega-3 polyunsaturated fatty acids that are required for the development of the body (Ozyurt et al., 2006). Significant levels of essential fatty acids like EPA, DHA, and arachidonic acid were observed by Remyakumari et al. (2018) in Indian squid, *Uroteuthisduvauceli*. High levels of polyunsaturated fatty acids and amino acids have been reported in the cuttlefish by Nurjanah et al. (2012) which are considered essential for good human health.

3.2 Proteins

Shrimps are excellent sources of energy, proteins, and minerals. The shells of tiger shrimp, white leg shrimp, and Indian white shrimp can act as an alternative source of proteins that can be consumed by humans and also used as a source of feed for animals (Ali et al., 2017). Aquatic crustaceans have excellent nutritional value and chemical composition and so can be used as a protein source for the whole world's population especially in the developing nations where animal protein is high-priced and cannot be afforded by impoverished populations (Bello-Olusoji & Oke, 2005). Studies done by Reddy and Reddy (2014) suggested that a prawn, *Macrobrachiumrosenbergii*, contains large amounts of protein and thus acts as a candidate species for consumption by the human population. Results concluded by Kiin-Kabari et al. (2017) revealed that the proteins and minerals are present in good concentration in mollusks and consuming such foods aids in mollifying malnutrition problems. Crabmeat of *Portunuspelagicus* contains approximately 17.5–18.8% of protein (Musaiger & Al-Rumaidh, 2005). Twenty-one amino acids were reported to be present in the *Callapalophus*, a box crab, of which 11 were essential and the remaining ten were nonessential amino acids (Kathirvel et al., 2014). Varisco et al. (2020) studied the proximate composition of the meat of *Munidagregaria*, a squat lobster, and concluded that glutamic acid, leucine, and aspartic acid are the main amino acids found in the meat thus providing high-quality protein which is beneficial for human health. High levels of various amino acids such as aspartic acid, glycine, leucine, lysine, alanine, and glutamate are present in the edible species of bivalve mollusks. *P. viridis*, a marine bivalve mollusk, has been reported as a useful source of food for humans due to the presence of balanced nutrients and high-quality protein (Saritha et al., 2015). Asha et al. (2014) postulated that an oyster, *C. madrasensis*, contains 9.41% protein and 3.25% fats, and is rich in macrominerals. Bano et al. (1992) reported high protein content in the squid meat that showed the presence of various amino acids such as aspartic acid, lysine, leucine, cysteine, serine, arginine, histidine, glutamic acid, and tryptophan. Studies done by Ahmed et al. (Ahmad et al., 2013) revealed the presence of a high concentration of proteins in the crayfish.

3.3 Carbohydrates

Carbohydrates are present in very small amounts in the tissues of shellfish. Islam et al., 2017 revealed that the level of carbohydrates in shellfish viz., *Macrobrachiumrosenbergii* and *Penaeusmonodon* lies in the range of 0.17–0.91%. George and Mathew (1996) reported that the glycogen content decreases with an increase in the age of a bivalve. In clam meat, they reported ribose and fructose that too in minimal concentration. Rosa and Nunes (2004) evaluated the glycogen content in *Aristeusantennatus, Parapenaeuslongirostris,* and *Nephropsnorvegicus* and found that it varies from 1.0% to 1.2%. *Meretrixlusoria*, a marine bivalve had a carbohydrate content of 7.9% (Karnjanapratum et al., 2013). Edible mollusks have less content of carbohydrates and total and saturated fats.

3.4 Vitamins and Minerals

Seven different minerals viz., calcium, magnesium, zinc, iron, copper, sodium, and potassium have been observed in the *Callapalophus*, a box crab. Common mussels, Venus calm, and Pacific oysters are good sources of iron. The nutritional content of the different shellfish varies and provides additional value to the diet (King et al., 1990; Kathirvel et al., 2014). Abulude et al. (2006) reported the presence of various minerals such as calcium, magnesium, phosphorus, manganese, and chlorine in the species of prawn. Essential minerals such as copper, magnesium, zinc, nickel, iron, potassium, phosphorus, calcium, and iron were found to be in good amounts in a shellfish, *Sudanonautes africanusafricanus* (Adeyeye, 2002). Mollusks are useful sources of essential minerals and B-complex vitamins. Proximate analysis of various spiny lobsters revealed the presence of calcium, magnesium, potassium, and sodium. According to the reports of Bender (1992), mineral content is present in a higher proportion of snail meat. Various macrominerals such as magnesium, potassium, phosphorus, and sodium, and microminerals like nickel, copper, zinc, essential fatty acids, and vitamin E have been reported in various species of squids across the globe (Torrinha et al., 2014). Bogard et al. (2015) recorded 0.06 μg/100 g of Vitamin D_3 in the shrimp. Molluscan shellfish are a perfect source of choline, iron, zinc, selenium, and vitamin B_{12} as prescribed by the Nutrient Database for Standard Reference published by the Department of Agriculture, United States.

4 Consumption of Shellfish

During the last 40 years, a sea change has been seen in the consumption of shellfish. The consumption of seafood products has grown expeditiously across the globe. As a result of which the shellfish have been adulated for their high nutritional value and are the valuable sources of proteins and minerals required by humans. Mollusks are a rich source of polyunsaturated fatty acids and so utilization in the human diet is elevating year after year (Khan & Liu, 2019a, b). The presence of essential nutrients in mollusk and their positive impacts on the health of humans have led to their harvesting and culturing on a commercial scale (Surm et al., 2015). Consumption of shellfish in everyday diet can prove beneficial in keeping away various diseases as they obtain essential nutrients and various active secondary metabolites and thus have the potential to augment immune response (Khan & Liu, 2019a, b). Demand for shellfish, oysters, prawns, and lobsters is mushrooming across the globe due to rising health consciousness. Ilam et al. (2017) suggested that both prawns and shrimps can be used as a food source by humans as they are rich in animal protein like other aquatic food. Due to the delicacy of crustaceans and their high protein content, they are preferred foodstuffs for consumption by humans. Mollusks are well-liked by people across the world due to their nutritive value and palatable taste. The meat of mollusks is nutritionally characterized by low fat and high protein content (Chen et al., 2012; Banu et al., 2016).

5 Health Benefits of Shellfish Consumption

Consumption of a diet rich in omega-3 polyunsaturated fatty acids such as DHA and EPA lessens the death rate that occurs due to cardiovascular diseases. PUFA present in the shellfish helps to reduce blood clot formation, inflammation and widens the blood vessels thus aid in averting certain diseases such as high blood pressure and elevated levels of cholesterol in the blood as reported by Marchioli et al. (2002). Besides being an excellent source of protein, *Macrobrachiumrosenbergii*, contains a good amount of fiber and thus helps in reducing constipation and other problems in humans (Devi et al., 2015). Yuan et al. (2001) conducted a study among 18,244 healthy men in China and found that those who consume more than 200 g of shellfish (rich in omega-3) every week were unlikely to die from myocardial infarction than those who consume less than 50 g weekly. Eating nutrients present in shellfish are good both for heart and brain health. Shellfish forms an important source of nutrition for humans and is considered a healthy food as far as minerals, proteins, and unsaturated fatty acids are concerned. Adequate atherogenic index, polyene index, thrombotic index, hypocholesterolemic/hypercholesterolic ratio make the meat of *Munidagregaria* beneficial for consumption by humans as it lessens the cardiovascular disease risk (Bugel et al., 2001).

Fats of aquatic animals are excellent sources of fatty acids that the human body is not able to synthesize. Mollusks act as the source of B-complex vitamins, niacinamide, and thus play an important role in human nutrition. The meat of snails can be utilized for the treatment of diseases like high blood pressure, heart attack, stroke, anemia, and fat-related diseases. It is also used as a remedy for pain during labor, loss of blood during delivery, and pus-formed blisters formed due to smallpox disease (Srilatha et al., 2013; Akinnusi, 2002; Wosu, 2003). Oysters are rich in nutrients and have various health benefits (Maurya, 2021). They are helpful in increasing the concentration of vitamin B_{12} in the body. Moreover, they have high magnesium and potassium content, and thus their consumption is helpful in maintaining blood pressure, improving circulation of blood, balancing lipid concentration in the body, strengthening immunity, healing wounds, and enhancing sex hormones. Studies conducted by Swenor et al. (2010) supported the fact that the intake of shellfish (crabs and oysters) provide protection against advanced age-related macular degeneration. Consumption of clams helps in increasing vitality and reinstates the sexual edge of men (Benjamas, 2020). Consumption of 300 g of shrimp regularly increases the level of good cholesterol by 12% and decreases triglycerides by approximately 13%. Shellfish consumption reduces the risk of cardiovascular diseases in people suffering from Type-2 diabetes in the age group of 40–64 years (De Silva et al., 1996; Kim et al., 2015).

Shellfish such as crayfish contain nine essential amino acids in equal amounts that are utilized by the body for building and repair mechanisms. This kind of seafood can be utilized by the body to fight off protein deficiency diseases such as kwashiorkor and marasmus (Ibironke et al., 2018). All the essential amino acids required by the human body for the growth and maintenance of the body can be provided by the shellfish (Friedman, 1996). The hydrolysates of the mitre squid (*Uroteuthischinensis*) have high nutritive value and have the potential of being utilized as a nutritional supplement in various dietary products (Wu et al., 2015). Mussel lipids consisting of polyunsaturated fatty acids are acclaimed for their role in the prevention and cure of rheumatoid arthritis (Grienke et al., 2014). Lipid extracts and frozen powder of mussels act beneficial for human health and have the potential to lower the inflammation (Miller et al., 2014; Ahmad et al., 2018). WHO (2018) recommends that high saturated fatty acids should be replaced with polyunsaturated fatty acids or monounsaturated fatty acids, ideally from the seafood origin so that the ratio of ω-6/ω-3 decreases, which will lessen the risk of cardiovascular diseases. Research studies conducted by Panayotova et al. (2020) revealed that the consumption of marine bivalves is propitious for human health as they are easily digestible, low in calories, and have high nutritive values. Eating 100 g of shrimp per day provides approximately ten minerals and ten vitamins. Various essential vitamins such as vitamin D, A, E, B_{12}, and B_3 are present in the shrimp. So, the presence of various vital nutrients makes it a healthy food for humans as per the findings of Dayal et al. (2013). Shellfish protein is rich in essential amino acids and is easy to digest due to the absence of connective tissue. The presence of various polyunsaturated fatty acids such as eicosapentaenoic acid and *decosahexaenoicacid* in shellfish decreases the risk of stroke and cardiovascular diseases. Diet rich in omega-3 fatty acids lowers cholesterol levels in serum which is helpful for cardiac health (Gogus & Smith, 2010; Hamed et al., 2015; Weichselbaum et al., 2013). Haryono et al. (2015) conducted a study to assess the nutritional status of five spiny lobsters. Results revealed that the *Panulirushomarus* has 6.18% lipid content, 55–68% carbohydrate content thus suggesting the high nutritional value of shellfish.

6 Effects of Climate Change on Aquaculture

Increase in anthropogenic activities is driving climate change on a global level at a great magnitude Solomon et al. (2009) and is considered the most serious global issue. Yazdi and Shakouri (2010) defined climate change as the alteration in the qualitative weather distribution over a prolonged period of time, from millions of years. According to the United Nations, continuous alteration in the temperature and weather patterns leads to climate change. UNFCCC (United Nations Framework Convention on Climate Change, 1994) defined climate change as the change in climate due to human activities that modify the composition of gases in the atmosphere. This changing climate poses a serious risk to food production across the globe and acts as a serious challenge to the quantity and quality of the production (Hamdan et al., 2012; Myers et al., 2017). Kandu (2017) reported that access to various essential nutrients through aquaculture is jeopardized by the prognosticated impacts of climate change. In an aquatic ecosystem, climatic change brings changes in temperature, pH, water level, rainfall, wind patterns, salinity, and ocean acidity that are known to affect biodiversity and ultimately have a disastrous impact on the communities and industries (Poloczanska et al., 2007; Brierley & Kingsford, 2009; Hobday et al., 2008). Studies done by Fleming et al., 2014; Bueno & Soto, 2017; Dabbadie et al., 2018 predicted that the entire aquaculture system is at risk of climate change effects. Changing climatic conditions are expected to severely affect the aquaculture systems across the globe due to their effect on the physiology, growth rate, reproductive performance, and susceptibility to disease in cultured shellfish (Handisyde et al., 2006; Brander, 2007; Cochrane et al., 2009; De Silva, 2012).

Changing climatic conditions may have a great influence on aquaculture as they directly affect the supply of shellfish or indirectly the price of shellfish resources required by the fishermen will get accrued. According to IPCC (2019), from the year 2007–2016; forestry, agriculture, and other land activities related to land use have led to the release of 82% nitrous oxide, 44% methane, and 13% carbondioxide in the atmosphere. De Silva and Soto (2009) recorded that the main greenhouse gas released from the aquaculture sector is carbon dioxide, although it is much less as compared to the other food production sectors. A case study carried out by Asiedu et al. (2019) in Ghana (Africa) revealed that the market value of the products obtained from aquaculture lowers due to change in climate, which instigates poverty among the people dwelling in rural areas.

Despite being good sources of essential nutrients such as vitamins, minerals, and omega-3 polyunsaturated fatty acids, it is believed that due to the rise in anthropogenic activities, the pollution of water bodies will also increase, which will make the shellfish unfit for human consumption due to assemblage of toxic products in their body tissues. Studies done by Perello et al., 2008; Storelli et al., 2012 revealed that there is a strong tendency to accumulate toxic metals such as lead, mercury, and cadmium in cephalopods, which serve as a potential threat for humans consuming shellfish.

6.1 Pollution Load Due to Heavy Metals

Heavy metal contamination is a key environmental concern in marine and freshwater ecosystems. Accumulation of cadmium in squid, mercury, and arsenic in other shellfish is a rising problem (Mouritsen & Styrbæk, 2018). These heavy metals get accumulated in the tissues and innards of shellfish, which makes them unfit for human consumption. Studies on a load of heavy metals such as arsenic, lead, cadmium, and mercury on the most consumed shellfish species viz., octopus, cuttlefish, and squid were done by Storelli et al. (2006, 2010). The results of their study revealed that octopus was heavily loaded with toxic elements and large amounts of these toxins were present in the hepatopancreatic tissue. Arsenic gets amassed in the cuttlefish. Thus, the presence of these heavy metals in the water bodies is increasing at a faster rate due to the growth of industries which has increased the safety issues for consumers feeding on these shellfish.

6.2 Rise in Air and Water Temperature

According to IPCC (2019), by the end of the twenty-first century, it is expected that the global mean temperature of the surface will rise by 0.3–4.8 °C. The growth and development of an organism depend upon the temperature Ngao (2018). An increase in temperature lowers the energy reserve and augments the death rate of adult oysters (Matoo et al., 2013). Studies conducted by Bastos et al. (2018) showed that an increase in temperature decreases the pH and dissolved oxygen of the aquatic ecosystem which has a negative impact on the survival and growth of juveniles of *Macrobrachiumamazonicum*. Temperature influences the concentration of omega-3 and omega-6 fatty acids as observed in *Portunuspelagicus* by Munafi et al. (2020). Pecl and Jackson (2008), put forward that climate change likely induces behavioral, physiological, and biological changes in squid that will have long-term effects on the survival rate of the species. An increase in temperature accelerates the growth, physiology, and metabolism, and behavior of feeding in shellfish species (Lemasson et al., 2018; Akegbejo-Samsons, 2009).

6.3 Ocean Acidification

It is a major threat to the calcified species of shellfish. There is a continuous increase in carbon dioxide production due to various human activities. The carbon dioxide produced reacts with the water and forms carbonic acid that lowers down the pH of an aquatic ecosystem and decreases the concentration of carbonate ions, which has an adverse impact on aquatic life. Rhien et al. (Rhein et al., 2013) revealed that the pH of oceans will decrease by −0.0014 to −0.0024 every year due to the dissolution of atmospheric carbon dioxide in the oceans and this will have a significant impact on the aquafauna that possess calcium carbonate shells. Shell hardness decreases due to reduced pH levels Ivanina et al. (2013). Larval stages are more susceptible to ocean acidification than adults (Talmage & Gobler, 2011). Reduced levels of carbonate ions and increased levels of carbon dioxide are known to increase the death rate of juvenile molluscs (*Mercenariamercenaria* and *Mya arenaria*). Raised CO_2 levels decrease the deposition of calcium in the shells of mussels and oysters and even reduce the growth of mussels (Green et al., 2004, 2009; Berge et al., 2006; Gazeau et al., 2007). Arnold et al. (2009) observed that due to the acidification of the aquatic ecosystem the carapace mass of the lobster, *Homarusgammarus* decreased at the time of completion of larval development. Studies conducted by the National Oceanic and Atmospheric Administration proposed that with the increase in CO_2 levels, it will be difficult for the shellfish to develop their shell and skeleton. Orr et al. (2005) suggested that since the industrial revolution that began in the eighteenth century, the pH levels in oceans have decreased by 0.01 pH units due to carbondioxide absorbance.

6.4 Diseases, Algal Blooms, and Marine Biotoxins

With the surge in daily temperature, the growth of harmful aquatic algae continues to rise which is responsible for the production of biotoxins. When molluscan shellfish feed on these toxin-releasing algae, the toxic compounds get accumulated in their tissues, and through the food web, they enter into other living creatures. Climate change has increased the threat of mycotoxins, as it reduces immunity and fecundity in shrimps. Changing climate increases the disease susceptibility in shrimps (Islam et al., 2016) and leads to deformities in the body and lesions on the skin. The rise in temperature and acidification of water bodies have put huge stress on aquatic life and have enhanced the occurrence of disease in shellfish. Climate change alters the susceptibility of the host towards the disease-causing microorganisms. Stewart et al. (Stewart et al., 1969; Stewart, 1980) observed that the incidence of gaffkemia proves fatal for the lobsters at a temperature above 15 °C. Fluctuation in the temperature and salinity leads to the outbreak of White Spot Disease in shrimp farms (Rebecca et al., 2021). Various human enteric viruses present in the polluted waterbodies contaminate the bivalve shellfish (Leroi, 2014;

Jennings et al., 2016). Various parasites associated with the body of shellfish contaminate shellfish harvesting water bodies. Climate change declines the abundance and increases the occurrence of diseases in various species of lobsters (Caputi et al., 2013).

6.5 Rise in Sea Level

Due to thermal expansion and melting of glaciers and ice caps, a significant increase in the sea level has been recorded. From the year 1961 to 2003, the average sea level rose at the rate of 1.8 mm per year. This rise in sea level disturbs the breeding ground of a wide variety of aquatic fauna (Mohanty, 2018). The rise in sea level is expected to influence aquaculture production such as cages, tanks, and ponds due to the intrusion of saline water. Salinization of aquaculture systems proves harmful for shellfish. Rising sea levels will change the abundance and distribution of organism productivity in aquatic ecosystems, the composition of species that endangers aquaculture productivity (Kibria et al., 2017; Doney et al., 2012).

6.6 pH and Salinity

The aquaculture system has been greatly affected by the salinity levels. Any fluctuation in the salinity level brings alteration in the defense mechanisms, growth, oxygen use, and the rate of filtration in bivalves (Pourmozaffar et al., 2019). Low pH and low salinity levels slow down the growth of larvae in Pacific oysters before metamorphosis. Ko et al. (2014). In oysters changing salinity affects their immunity and this way they become more prone to bacterial diseases at low saline conditions. The productivity of prawns in aquaculture decreases at high salinity levels (Rodrick, 2008; Ahmed, 2013).

6.7 Precipitation

It is predicted that climate change will alter the pattern of rainfall, which can even lead to a prolonged drought period and overall reduce the availability of water. This decrease will increase the water temperature and induce drastic changes in the lifespan of an organism (Oberdorff et al., 2015). Changing rainfall patterns affect the maturity and gonadal development of shellfish during the breeding season. Change in the intensity of rainfall has a direct impact on shrimp farming (Mohanty, 2018). Shortage and deteriorating water quality lead to drought-like conditions, thus having a negative impact on the production of aquaculture (Hambal et al., 1994).

7 Conclusion

The literature surveyed has indicated that the aquaculture sector has a positive contribution toward global food production. Changing climate poses a serious threat to the aquaculture sector across the globe which in turn threatens the global food system that is already floundering due to the rising demands of the growing population Aquaculture being dependent upon the environmental conditions, is greatly affected by climate change. Several steps have been recommended in order to lessen the impact of climate change on shellfish aquaculture:

- Approaches are required to manage the aquaculture systems.
- Overfishing in the water bodies should be reduced.
- Adoption of environment-friendly aquaculture practices.
- Monitoring of shellfish stocks in the aquatic ecosystem is necessary so that changes due to climate are readily observed.
- Plantation of tree barriers across the coastal areas against the rise in sea level.
- Efforts should be made to develop the blue economy.
- Valuing the undervalued aquatic resources should be done by diversifying the value chains.

References

Abulude, F. O., Lawal, L. O., Ehikhamen, G., Adesanya, W. O., & Ashafa, S. L. (2006). Chemicalcomposition and functional properties of some prawns from the coastal area of Ondo state, Nigeria. *Electron Journal of Environment, Agriculture and Food Chemistry, 5*(1), 1235–1240. https://doi.org/10.12691/ajrd-2-1-1

Adeyeye, E. I. (2002). Determination of the chemical composition of the nutritionally valuable parts of male and female common west African fresh water crab *Sudananautes africanus africanus*. *International Journal of Food Sciences and Nutrition, 53*(3), 189–196. https://doi.org/10.1080/09637480220132805

Ahmad, M., Birnin-Yauri, A., Bagudo, U., & Sahabi, M. (2013). Comparative analysis on the nutritional values of crayfish and some insects. *African Journal of Food Science and Technology, 4*(1), 9–12.

Ahmad, T. B., Liu, L., Kotiw, M., & Benkendorff, K. (2018). Review of anti-inflammatory, immune-modulatory and wound healing properties of molluscs. *Journal of Ethnopharmacology, 210*, 156–178. https://doi.org/10.1016/j.jep.2017.08.008

Ahmed, N. (2013). Linking prawn and shrimp farming towards a green economy in Bangladesh: Confronting climate change. *Ocean & Coastal Management, 75*, 33–42. https://doi.org/10.1016/j.ocecoaman.2013.01.002

Akegbejo-Samsons, Y. (2009). Climate change impacts on fisheries production in land-water interface. *IOP Conference Series. Earth and Environmental Science, 6*(29). https://doi.org/10.1088/1755-1307/6/29/292022

Akinnusi, O. (2002). Introduction to snails and snail farming. *Real Solutions Computers, 23*.

Ali, S. S. R., Ramachandran, M., Chakma, S. K., & Sheriff, M. A. (2017). Proximate composition of commercially important marine fishes and shrimps from the Chennai coast, India. *International Journal of Fisheries and Aquatic Studies, 5*(5), 113–119.

Arnold, K., Findlay, H., Spicer, J., & Boothroyd, D. (2009). Effect of CO_2-related acidification on aspects of the larval development of the European lobster, *Homarus gammarus* (L.). *Biogeosciences, 6*, 1747–1754.

Asha, K. K., Anandan, R., Mathew, S. W., & Lakshmanan, P. T. (2014). Biochemical profile of oyster *Crassostrea madrasensis* and its nutritional attributes. *The Egyptian Journal of Aquatic Research, 40*(1), 35–41.

Asiedu, B., Malcolm, D., & Iddrisu, S. (2019). Assessing the economic impact of climate change in the small-scale aquaculture industry of Ghana, West Africa. *AAS Open Res, 1*, 26. https://doi.org/10.12688/aasopenres.12911.2

Banu, S. K. S., Hareesh, K., & Reddy, M. S. (2016). Evaluation of nutritional status of Penaeid prawns through proximate composition studies. *International Journal of Fisheries and Aquatic Studies, 4*(1), 13–19.

Bastos, A. M., Lima, J. F., & Tavares-Dias, M. (2018). Effect of increase in temperature on the survival and growth of *Macrobrachium amazonicum* (Palaemonidae) in the Amazon. *Aquatic Living Resources, 31*(21), 1–6.

Bello-Olusoji, A. O., & Oke, A. (2005). Chemical index and organoleptic assessment of freshness of frozen and oven dried African river prawn *M. vollenhovenii* during storage. *Journal of Tropical Biosciences, 5*(1), 154–157.

Bender, A. (1992). Meat and meat products in human nutrition in developing countries. *FAO Food and Nutrition Paper, 53*, 1–91.

Benjamas, D. (2020). Value added product development for oyster farmers' group in Kung Krabaen Bay Royal Development Study Center, Chanthaburi. *Interdisciplinary Science Reviews, 15*(1), 54–57.

Berge, J. A., Bjerkeng, B., Pettersen, O., Schaanning, M. T., & Øxnevad, S. (2006). Effects of increased sea water concentrations of CO_2 on growth of the bivalve *Mytilus edulis* L. *Chemosphere, 62*(4), 681–687.

Bogard, J. R., Thilsted, S. H., Marks, G. C., Wahab, M. A., Hossain, M. A. R., Jakobsen, J., & Stangoulis, J. (2015). Nutrient composition of important fish species in Bangladesh and potential contribution to recommended nutrient intakes. *Journal of Food Composition and Analysis, 42*, 120–133. https://doi.org/10.1016/j.jfca.2015.03.002

Bostock, J., McAndrew, B., Richards, R., Jauncey, K., et al. (2010). Aquaculture: Global status and trends. *Philosophical Transactions of the Royal Society B, 365*, 2897–2912.

Brander, K. M. (2007). Global fish production and climate change. *Proceedings of the National Academy of Sciences of the United States of America, 104*, 19709–19714.

Brierley, A. S., & Kingsford, M. J. (2009). Impacts of climate change on marine organisms and ecosystems. *Current Biology, 19*, R602–R614.

Bueno, P. B., & Soto, D. (2017). *Adaptation strategies of the aquaculture sector to the impacts of climate change*. FAO.

Bugel, S. H., Sandstrom, B., & Larsen, E. H. (2001). Absorption and retention of selenium from shrimps in man. *Journal of Trace Elements in Medicine Biology, 14*, 198–204.

Caputi, N., de Lestang, S., Frusher, S., & Wahle, R. A. (2013). The impact of climate change on exploited lobster stocks. In *Lobsters: Biology, management, aquaculture and fisheries*. https://doi.org/10.1002/9781118517444.ch4

Chen, D. W., Su, J., Liu, X. L., Yan, D. M., Lin, Y., Jiang, W. M., & Chen, X. H. (2012). Amino acid profiles of bivalve mollusks from Beibu gulf, China. *Journal of Aquatic Food Product Technology, 21*, 369–379. https://doi.org/10.1080/10498850.2011.604820

Cochrane, K., De Young, C., Soto, D., & Bahri, T. (2009). *Climate change implications for fisheries and aquaculture: overview of current scientific knowledge*. FAO Fisheries and Aquaculture Technical Paper, No. 530. Food and Agricultural Organisation of the United Nations, Rome.

Dabbadie, L., Aguilar-Manjarrez, J. J., Beveridge, M. C. M., Bueno, P. B., Ross, L. G., & Soto, D. (2018). *Chapter 20: Effects of climate change on aquaculture: Drivers, impacts and policies*. FAO.

Dayal, J. S., Ponniah, A. G., Khan, H. I., MadhuBabu, E. P. M., Ambasankar, K., & Vasagam, K. P. K. (2013). Shrimps – A nutritional perspective. *Current Science, 104*(11), 1487–1491.

De Silva, S. S. (2012). Climate change impacts: Challenges for aquaculture. In R. P. Subasinghe, J. R. Arthur, D. M. Bartley, S. S. De Silva, et al. (Eds.), *Farming the waters for people and food. Proceedings of the global conference on aquaculture* (pp. 75–110). FAO.

De Silva, S. S., & Soto, D. (2009). Climate change and aquaculture: Potential impacts, adaptation and mitigation. In K. Cochrane, C. De Young, D. Soto, & T. Bahri (Eds.), *Climate change implications for fisheries and aquaculture: Overview of current scientific knowledge*. FAO Fisheries and Aquaculture Technical Paper, No. 530 (pp. 151–212). FAO.

De Silva, E. R., Seidman, C. E., Tian, J. J., Hudgins, L. C., Sacks, F. M., & Breslow, J. L. (1996). Effects of shrimp consumption on plasma lipoproteins. *The American Journal of Clinical Nutrition, 64*(5), 712–717. https://doi.org/10.1093/ajcn/64.5.712

Devi, D. V. P., Hareesh, K., et al. (2015). Studies on the proximate composition of tropical freshwater prawn *Macrobrachium rosenbergii*. *International Journal of Fisheries and Aquatic Studies, 3*(1), 329–336.

Doney, S. C., Ruckelshaus, M., Duffy, J. E., Barry, J. P., Chan, F., English, C. A., Galindo, H. M., Grebmeier, J. M., Hollowed, A. B., Knowlton, N., Polovina, J., Rabalais, N. N., Sydeman, W. J., & Talley, L. D. (2012). Climate change impacts on marine ecosystems. *Annual Review of Marine Science, 4*, 11–37. https://doi.org/10.1146/annurev-marine-041911-111611

Fleming, A., Hobday, A. J., Farmery, A., van Putten, E. I., Pecl, G. T., & Green, B. S. (2014). Climate change risks and adaptation options across Australian seafood supply chains–a preliminary assessment. *Climate Risk Management, 1*, 39–50. https://doi.org/10.1016/j.crm.2013.12.003

Food and Agriculture Organization (FAO)/INFOODS. (2016). *Global food composition database for fish and shellfish*.Version10-uFiSh10. https://www.fao.org/3/i6655e/i6655e.pdf

Friedman, K. (1996). Nutritive value of proteins from different food sources a review. *Journal of Agricultural and Food Chemistry, 44*, 6–29.

Garcia, S. M., & Rosenberg, A. A. (2010). Food security and marine capture fisheries: Characteristics, trends, drivers and future perspectives. *Philosophical Transactions of the Royal Society B, 365*, 2869–2880.

Gazeau, F., Quiblier, C., Jansen, J. M., Gattuso, J. P., Middelburg, J. J., & Heip, C. H. R. (2007). Impact of elevated CO_2 on shellfish calcification. *Geophysical Research Letters, 34*(7), 1–5. https://doi.org/10.1029/2006GL028554

George, C., & Mathew, P. T. (1996). Biochemical and nutritional evaluation of yellow clams. *Fishery Technology, 33*(1), 41–45.

Godfray, H. C. J., Beddington, J. R., Crute, I. R., Haddad, L., et al. (2010). Food security: The challenge of feeding 9 billion people. *Science, 327*, 812–818.

Gogus, U., & Smith, C. (2010). N-3 omega fatty acids: A review of currentknowledge. *International Journal of Food Science & Technology, 45*, 417–436.

Green, M. A., Jones, M. E., Boudreau, C. L., Moore, R. L., & Westman, B. A. (2004). Dissolution mortality of juvenile bivalves in coastal marine deposits. *Limnology and Oceanography, 49*, 727–734.

Green, M. A., Waldbusser, G. G., Reilly, S. L., Emerson, K., & O'Donnell, S. (2009). Death by dissolution: Sediment saturation state as a mortality factor for juvenile bivalves. *Limnology and Oceanography, 54*, 1037–1047.

Grienke, U., Silke, J., & Tasdemir, D. (2014). Bioactive compounds from marine mussels and their effects on human health. *Food Chemistry, 142*, 48–60. https://doi.org/10.1016/j.foodchem.2013.07.027

Hambal, H., Akhir, A. M., & Saniah, Y. (1994). *Environmental issues on aquaculture development in Malaysia. Working papers of the fisheries research institute*. Department of Fisheries, Kuala Lumpur.

Hamdan, R., Kari, F., Othman, A., & Samsi, S. M. (2012). Climate change, socio-economic and production linkages in East Malaysia aquaculture sector. In *2012 international conference on future environment and energy IPCBEE* (Vol. 28). IACSIT Press.

Hamed, I., Ozogul, F., Ozogul, Y., & Regenstein, J. M. (2015). Marine bioactivecompounds and their health benefits: A review. *Comprehensive Reviews in Food Science and Food Safety, 14*, 446–465.

Handisyde, N. T., Ross, L. G., Badjeck, M. C., & Allison, E. H. (2006). *The effects of climate change on world aquaculture: a global perspective*. Final technical report, Department for International Development, Stirling Institute of Aquaculture, Stirling.

Haryono, F. E. D., Hutabarat, S., & Hutabarat, J. (2015). Nutritional value of spiny lobsters (*Panulirus sp.*) from southern coast of Java. *AIP Conference Proceedings, 1699*(1), 030016. https://doi.org/10.1063/1.4938301

Hobday, A. J., Poloczanska, E. S., & Matear, R. (2008). *Implications of climate change for Australian fisheries and aquaculture: A preliminary assessment*. Report to the Department of Climate Change.

Hosomi, R., Yoshida, M., & Fukunaga, K. (2012). Seafood consumption and components for health. *Global Journal of Health Science, 4*(3), 72–86. https://doi.org/10.5539/gjhs.v4n3p72

Hurtado, J. L., Montero, P., & Borderías, J. (2001). Behavior of octopus muscle (*Octopus vulgaris*) under a process of pressuretime-temperature combinations. *Revista de Agaroquimicay Tecnologia de Alimentos, 7*(3), 259–267.

Ibironke, S. I., Adepeju, A. B., Otutu, O., Oyedele, D. S., & Esan, Y. O. (2018). Nutritional evaluation and comparison study of seafoods such as fish and crayfish supplement dietary. *MOJ Food Process Technol, 6*(1), 73–76. https://doi.org/10.15406/mojfpt.2018.06.00147

IPCC. (2019). *Special report on climate change, desertification, land degradation, sustainable land management, food security, and greenhouse gas fluxes in terrestrial ecosystems, summary for policymakers approved draft*. IPCC.

Islam, M. A., Islam, M. S., & Wahab, M. A. (2016). Impacts of climate change on shrimp farming in the south-west coastal region of Bangladesh. *Research in Agriculture Livestock and Fisheries, 3*(1), 227–239.

Islam, A., Mondal, S., Bhowmik, S., Islam, S., & Begum, M. (2017). A comparative analysis of the proximate composition of wild and cultured prawn (*Macrobrachium rosenbergii*) and shrimp (*Penaeus monodon*). *International Journal of Fisheries and Aquatic Studies, 5*(4), 59–62.

Ivanina, A. V., Dickinson, G. H., Matoo, O. B., Bagwe, R., Dickinson, A., Beniash, E., & Sokolova, I. M. (2013). Interactive effects of elevated temperature and CO_2 levels on energy metabolism and biomineralization of marine bivalves *Crassostrea virginica* and *Mercenaria mercenaria*. *Comparative Biochemistry and Physiology Part A: Molecular & Integrative Physiology, 166*(1), 101–111. https://doi.org/10.1016/j.cbpa.2013.05.016

Jennings, S., Stentiford, J. D., Leocadio, A. M., Jeffery, K. R., MetCalfe, J. D., Katsiadaki, J., Auchterlonie, N. A., Managi, S. C., Pinnegar, J. K., Ellis, T., Peeler, E. J., Luisetti, P., Baker-Austin, C., Brown, M., Catachpole, T. L., Clyne, F. J., Dye, S. R., Edmonds, N. J., Hyder, K., Lee, J., Lees, D. N., Morgan, O. C., O'Brien, C. M., Oldtmann, B., Posen, P. E., Santos, A. R., Taylor, N. G. H., Turner, A. D., Townhill, B. L., & Verner-Jeffreys, D. W. (2016). Aquatic food security: Insights into challenges and solutions from an analysis of interactions between fisheries aquaculture, food safety, human health, fish and human welfare economy and environment. *Fish Fisheries, 17*, 893–838.

Kampouris, T. E., Asimaki, A., Klaoudatos, D., Exadactylos, A., Karapanagiotidis, I. T., & Batjakas, I. E. (2021). Nutritional quality of the European spiny lobster *Palinurus elephas* (J.C. Fabricius, 1787) (Achelata, Palinuridae) and the non-indigenous northern Brown shrimp *Penaeus aztecus* Ives, 1891 (Dendrobranchiata, Penaeidae). *Food, 10*(10), 2480. https://doi.org/10.3390/foods10102480

Kandu, P. (2017). Papua New Guinea impacts of climate variations on local fisheries and aquaculture resources in PNG. In E. J. Ramos (Ed.), *Ecological risk assessment of impacts of cli-*

mate change on fisheries and aquaculture resources (pp. 45–49). APEC Ocean and Fisheries Working Group.

Karnjanapratum, S., Benjakul, S., Kishimura, H., & Tsar, V.-H. (2013). Chemical composition and nutritional value of Asian hard clam *Meretrix lusoria* fromthe coast of Andaman Sea. *Food Chemistry, 141*, 4138–4145.

Kathirvel, K., Eswar, A., Manikandarajan, T., Ramamoorthy, K., Sankar, G., & Anbarasu, R. (2014). Proximate composition, amino acid, fatty acid and mineral analysis of box crab, *Calappa lophus* (Herbst, 1782) from Parangipettai, southeast coast of India. *IOSR Journal of Environmental Science, Toxicology and Food Technology (IOSR-JESTFT), 8*(5), 50–57.

Khan, B. M., & Liu, Y. (2019a). Marine mollusks: Food with benefits. *Comprehensive Reviews in Food Science and Food Safety, 18*(2), 548–564. https://doi.org/10.1111/1541-4337.12429

Khan, B. M., & Liu, Y. (2019b). Marine mollusks: Food with benefits. *Comprehensive Reviews in Food Science and Food Safety, 18*, 548–564. https://doi.org/10.1111/1541-4337.12429

Kibria, G., Haroon, Y. A. K., & Dayanthi, N. (2017). Climate change impacts on tropical and temperate fisheries, aquaculture, and seafood security and implications–a review. *Livestock Research for Rural Development, 29*(1), 1–16.

Kiin-Kabari, D. B., Hart, A. D., & Nyeche, P. T. (2017). Nutritional composition of selected shellfish consumed in Rivers state, Nigeria. *American Journal of Food and Nutrition, 5*(4), 142–146. https://doi.org/10.12691/ajfn-5-4-5

Kim, H., Park, S., Yang, H., Choi, Y. J., Huh, K. B., & Chang, N. (2015). Association between fish and shellfish, and omega-3 PUFAs intake and CVD risk factors in middle-aged female patients with type 2 diabetes. *Nutrition Research and Practice, 9*(5), 496–502.

King, I., Childs, M. T., Dorsett, C., Ostrander, J. G., & Monsen, E. R. (1990). Shellfish: Proximate composition, minerals, fatty acids, and sterols. *Journal of the American Dietetic Association, 90*(5), 677–685. https://doi.org/10.1016/S0002-8223(21)01602-3

Kugino, M., Kugino, K., & Wu, Z. H. (1993). Rheological properties of dried squid mantle change on softening. *Journal of Food Science, 58*(2), 321–324.

Lemasson, A. J., Hall-Spencer, J. M., Fletcher, S., Provstgaard-Morys, S., & Knights, A. M. (2018). Indications of future performance of native and non-native adult oysters under acidification and warming. *Marine Environmental Research, 142*, 178–189. https://doi.org/10.1016/j.marenvres.2018.10.003

Leroi, F. (2014). Role of bacteria in seafood products. In S. K. Kim (Ed.), *Seafood science: Advances in chemistry technology and applications* (pp. 397–432). CRC Press.

Marchioli, R., Barzi, F., Bomba, E., Chieffo, C., Di Gregorio, D., Di Mascio, R., Franzosi, M. G., Geraci, E., Levantesi, G., Maggioni, A. P., Mantini, L., Marfisi, R. M., Mastrogiuseppe, G., Mininni, N., Nicolosi, G. L., Santini, M., Schweiger, C., Tavazzi, L., Tognoni, G., Tucci, C., & Valagussa, F. (2002). GISSI-Prevenzione investigators. Early protection against sudden death by n-3 polyunsaturated fatty acids after myocardial infarction: Time-course analysis of the results of the GruppoItaliano per lo studio dellaSopravvivenzanell'InfartoMiocardico (GISSI)-Prevenzione. *Circulation, 105*(16), 1897–1903. https://doi.org/10.1161/01.cir.0000014682.14181.f2

Matoo, O. B., Ivanina, A. V., Ullstad, C., Beniash, E., & Sokolova, I. M. (2013). Interactive effects of elevated temperature and CO(2) levels on metabolism and oxidative stress in two common marine bivalves (*Crassostrea virginica* and *Mercenari amercenaria*). *Comp Biochem Physiol A MolIntegr Physiol., 164*(4), 545–553. https://doi.org/10.1016/j.cbpa.2012.12.025

Maurya, N. K. (2021). Nutraceutical potential of oyster. Research & reviews. *Journal of Food Science & Technology, 10*(1), 1–6.

Merline, X., & Chitra, G. (2020). Comparative study on the proximate composition of prawn, lobster and puffer fish from Pamban, Rameswaram Island, south east coast of India. *International Journal of Creative Research Thoughts, 8*(5), 1278–1282.

Meynier, L., et al. (2008). Proximate composition, energy content, and fatty acid composition of marine species from Campbell plateau, New Zealand. *New Zealand Journal of Marine and Freshwater Research, 42*(4), 425–437. https://doi.org/10.1080/00288330809509971

Miller, M. R., Pearce, L., & Bettjeman, B. I. (2014). Detailed distribution of lipids in Green shell mussel (*Perna canaliculus*). *Nutrients, 6*(4), 1454–1474.

Mohanty, A. K. (2018). Impact of climate change: A curse to the shrimp farming in India. *International Journal of Advance Research, Ideas and Innovations in Technology, 4*(3), 1151–1158.

Mouritsen, O. G., & Styrbæk, K. (2018). Cephalopod gastronomy – A promise for the future. *Frontiers in Communication, 3*, 38. https://doi.org/10.3389/fcomm.2018.00038

Munafi, A. B. A., Ikhwanuddin, M., & Azra, M. N. (2020). Effects of temperature on the whole body fatty acid composition and histological changes of the gills in blue swimmer crabs, *Portunus pelagicus*. *Aquaculture Reports, 16*, 100270. https://doi.org/10.1016/j.aqrep.2019.100270

Musaiger, A. O., & Al-Rumaidh, M. J. (2005). Proximate and mineral composition of crab meat consumed in Bahrain. *International Journal of Food Sciences and Nutrition, 56*(4), 231–235. https://doi.org/10.1080/09637480500146945

Myers, S. S., Smith, M. R., Guth, S., Golden, C. D., Vaitla, B., Mueller, N. D., Dangour, A. D., & Huybers, P. (2017). Climate change and global food systems: Potential impacts on food security and undernutrition. *Annual Review of Public Health, 38*, 259–277. https://doi.org/10.1146/annurev-publhealth-031816-044356

Oberdorff, T., Jézéquel, C., Campero, M., Carvajal-Vallejos, F., Cornu, J. F., Dias, M. S., Duponchelle, F., Maldonado-Ocampo, J. A., Ortega, H., Renno, J. F., & Tedesco, P. A. (2015). Opinion paper: How vulnerable are Amazonian freshwater fishes to ongoing climate change? *Journal of Applied Ichthyology, 31*, 4–9.

Öksüz, A., & Mazlum, Y. (2016). Determination of proximate composition and fatty acid profiles of *Astacus leptodactylus* Eschscholtz, 1823 in Turkish freshwater resources. *Crustaceana, 89*(10), 1135–1147.

Orr, J. C., Fabry, V. J., Aumont, O., Bopp, L., Doney, S. C., Feely, R. A., Gnanadesikan, A., Gruber, N., Ishida, A., et al. (2005). Anthropogenic Ocean acidification over the twenty-first century and its impact on calcifying organisms. *Nature, 437*, 681–686.

Ozyurt, G., Duysak, Ö., Akamca, E., & Tureli, C. (2006). Seasonal changes of fatty acids of cuttlefish sepia *officinalis* L. (Mollusca: Cephalopoda) in the north eastern Mediterranean Sea. *Food Chemistry, 95*(3), 382–385.

Panayotova, V., Merdzhanova, A., Dobreva, D. A., Bratoeva, K., & Makedonski, L. (2020). Nutritional composition, bioactive compounds and health-beneficial properties of Black Sea shellfish. *Journal of IMAB, 26*(3), 3293–3297. https://doi.org/10.5272/jimab.2020263.3293

Paredi, M. E., & Crupkin, M. (1997). Biochemical properties of actomyosin from frozen stored mantles of squid (*Illex argentinus*) at different sexual maturation stages. *Journal of Agricultural and Food Chemistry, 45*(5), 1629–1632.

Pecl, G. T., & Jackson, G. D. (2008). The potential impacts of climate change on inshore squid: Biology, ecology and fisheries. *Reviews in Fish Biology and Fisheries, 18*, 373–385. https://doi.org/10.1007/s11160-007-9077-3

Perello, G., Marti-Cid, R., Llobet, J. M., & Domingo, J. L. (2008). Effects of various cooking processes on the concentrations of arsenic, cadmium, mercury, and lead in foods. *Journal of Agricultural and Food Chemistry, 56*(23), 11262–11269.

Poloczanska, E. S., Babcock, R. C., Butler, A., Hobday, A. J., Hoegh-Guldberg, K. T. J., Meater, R. J., Milton, D. A., Okey, T. A., & Richardson, A. J. (2007). Climate change and Australian marine life. *Oceanography and Marine Biology: An Annual Review, 45*, 407–478.

Pourmozaffar, S., Jahromi, S. T., Rameshi, H., Sadeghi, A., Bagheri, T., Behzadi, S., Gozari, M., Zahedi, M. R., et al. (2019). The role of salinity in physiological responses of bivalves. *Reviews in Aquaculture, 12*(3), 1548–1566. https://doi.org/10.1111/raq.12397

Reddy, B. S., & Reddy, K. V. S. (2014). Proximate composition of the fresh water prawn *Macrobrachium rosenbergii* in cultured and frozen stage from Nellore Coast, India. *International Food Research Journal, 21*(4), 1707–1710.

Remyakumari, K. R., Ginson, J., Ajeeshkumar, K. K., Vishnu, K. V., Asha, K. K., & Suseela, M. (2018). Biochemical profile and nutritional quality of Indian squid, *Uroteuthis duvauceli*. *International Journal of Fisheries and Aquatic Studies, 6*(3), 187–192.

Rhein, M., Rintoul, S. R., Aoki, S., Campos, E., Chambers, D., Feely, R. A., & Wang, F. (2013). Observations: Ocean. In T. F. Stocker et al. (Eds.), *Climate change 2013: The physical science basis. Contribution of working group I to the fifth assessment report of the intergovernmental panel on climate change*. Cambridge University Press.

Rodrick, G. B. (2008). Effect of temperature, salinity, and pesticides on oyster hemocyte activity. *Florida Water Resources Journal, 86*, 4–14.

Rosa, R., & Nunes, M. L. (2004). Nutritional quality of red shrimp *Aristeus antennatus* Risso, pink shrimp *Parapenaeus longirostris* Lucas and Norwaylobster *Nephrops norvegicus* Linnaeus. *Journal of the Science of Food and Agriculture, 8*, 89–94.

Saritha, K., Mary, D., & Patterson, J. (2015). Nutritional status of green mussel Perna viridis at Tamil Nadu, southwest coast of India. *Journal of Nutrition & Food Sciences, 14*(003).

Solomon, S., Plattner, G. K., Knutti, R., & Friedlingstein, P. (2009). Irreversible climate change due to carbon dioxide emissions. *Proceedings of the National Academy of Sciences of the United States of America, 106*, 1704–1709.

Srilatha, G., Chamundeeswari, K., Ramamoorthy, K., Sankar, G., & Varadharajan, D. (2013). Proximate, amino acid, fatty acid and mineral analysis of clam, Meretrix casta (Chemnitz) from Cuddalore and Parangipettai coast, south east coast of India. *Journal of Marine Biology & Oceanography, 2*(2), 2324–2386.

Stewart, J. E. (1980). Diseases. In J. S. Cobb & B. F. Phillips (Eds.), *The biology and management of lobsters* (pp. 301–342). Academic Press.

Stewart, J. E., Cornick, J. W., & Zwicker, B. M. (1969). Influence of temperature on gaffkemia, a bacterial disease of the lobster *Homarus americanus*. *Journal of the Fisheries Board of Canada, 26*(9), 2503–2510.

Storelli, M. M., Giacominelli-Stuffler, R., Storelli, A., & Marcotrigiano, G. O. (2006). Cadmium and mercury in cephalopod molluscs: Estimated weekly intake. *Food Additives and Contaminants, 23*, 25–30. https://doi.org/10.1080/02652030500242023

Storelli, M. M., Garofalo, R., Giungato, D., & Giacominelli-Stuffler, R. (2010). Intake of essential and non-essential elements from consumption of octopus, cuttlefish and squid. *Food Additives and Contaminants: Part B, 3*, 14–18. https://doi.org/10.1080/19440040903552390

Storelli, M. M., Normanno, G., Barone, G., Dambrosio, A., Errico, L., Garofalo, R., & Giacominelli-Stuffler, R. (2012). Toxic metals (hg, cd, and Pb) in fishery products imported into Italy: Suitability for human consumption. *Journal of Food Protection, 75*, 189–194.

Surm, J. M., Prentis, P. J., & Pavasovic, A. (2015). Comparative analysis and distribution of omega-3 lcPUFA biosynthesis genes in marine molluscs. *PLoS One, 10*(8), e0136301.

Swenor, B. K., Bressler, S., Caulfield, L., & West, S. K. (2010). The impact of fish and shellfish consumption on age-related macular degeneration. *Ophthalmology, 117*(12), 2395–2401.

Tabakaeva, O. V., Tabakaev, A. V., & Piekoszewski, W. (2018). Nutritional composition and total collagen content of two commercially important edible bivalve molluscs from the sea of Japan coast. *Journal of Food Science and Technology, 55*, 4877–4886.

Talmage, S. C., & Gobler, C. J. (2011). Effects of elevated temperature and carbon dioxide on the growth and survival of larvae and juveniles of three species of Northwest Atlantic bivalves. *PLoS One, 6*(10), e26941. https://doi.org/10.1371/journal.pone.0026941

Torrinha, A., Gomes, F., Oliveira, M., Cruz, R., Mendes, E., Delerue-Matos, C., Casal, S., & Morais, S. (2014). Commercial squids: Characterization, assessment of potential health benefits/risks and discrimination based on mineral, lipid and vitamin E concentrations. *Food and Chemical Toxicology, l67*, 44–56. https://doi.org/10.1016/j.fct.2014.02.014

United Nations Framework Convention on Climate Change:Convention (UNFCCC). (1994). Article 1 – Paragraph 2. http://unfccc.int/essential_background/convention/background/items/2536.php

Variscoa, M., Crovetto, C., Colomboa, J., Vinuesaa, J., & Rissob, S. (2020). Proximate composition and nutritional quality of the meat of the squat lobster *Munida gregaria* (Fabricius 1973). *Journal of Aquatic Food Product Technology, 29*(3), 229–237.

Venugopal, V., & Gopakumar, K. (2017). Shellfish: Nutritive value, health benefits, and consumer safety. *Comprehensive Reviews in Food Science and Food Safety, 16*(6), 1219–1242. https://doi.org/10.1111/1541-4337.12312

Weichselbaum, E., Coe, S., Buttriss, J., & Stanner, S. (2013). Fish in the diet: Areview. *Nutrition Bulletin, 38*, 128–177.

WHO. (2018). Better food and nutrition in Europe: A progress report monitoring policy implementation in the WHO European Region, pp. 1–58.

Wosu, L. O. (2003). *Commercial snail farming in West Africa–a guide*. AP Express Publishers Ltd.

Wright, A. C., Fan, Y., & Baker, G. L. (2018). Nutritional value and food safety of bivalve molluscan shellfish. *Journal of Shellfish Research, 37*(4), 695–708.

Wu, Y., Li, W., & Jin, T. (2015). Preparation and characterization of protein hydrolysates from little Loligo squid (*Uroteuthis chinensis*). *Journal of Aquatic Food Product Technology, 24*(1), 42–51. https://doi.org/10.1080/10498850.2012.753563

Yazdi, S. K., & Shakouri, B. (2010). The effects of climate change on aquaculture. *International Journal of Environmental Science and Development, 1*(5), 378–382.

Yuan, J. M., Ross, R. K., Gao, Y. T., & Yu, M. C. (2001). Fish and shellfish consumption in relation to death from myocardial infarction among men in Shanghai, China. *American Journal of Epidemiology, 154*(9), 809–816. https://doi.org/10.1093/aje/154.9.809

Chapter 16
Fish Pathogenesis Interactions: Effect on Protein Make-Up and Biometallic Absorption with Parasite Distribution Ecology in Garhwal Himalayan and Gangetic Plains Fish

Sandeep K. Malhotra and Anita Yadav

1 Introduction

Ecological, environmental and physiological study of a fish-parasite system illustrates the complete picture of that particular model. An attempt has been made in this investigation to study simultaneously all the above-mentioned aspects of freshwater, hill-stream riverine and marine ecosystems. The studies on population biology of fish parasites have been conducted by many earlier workers (Kennedy, 1968; Gairola, 1990; Malhotra, 1990a, b). This study covers certain major aspects of fish-parasite system, viz. effects of the environment, influence of seasonality on growth and establishment of parasites and their developmental stages, their optimal requirements of temperature, PH, dissolved oxygen, conductivity, total dissolved solids, tolerance to metal toxicity parameters and other water quality indices and effects of pathogenicity on fish health.

Fish pathogens are important as health hazards to man and animals and as handicaps to process of fish production. Fish are vulnerable to most types of infectious organisms which affects mammals. In countries where fish farming is well developed, workers report health problems with viral, bacterial, fungal, parasitic and protozoan infections. Although many diseases reported abroad have been found to occur in India, emphasis is being given only to the study of the systematics of the

S. K. Malhotra (✉)
Department of Zoology, University of Allahabad, Prayagraj, Uttar Pradesh, India

A. Yadav
Department of Zoology, CMP PG College (a Constituent College of the University of Allahabad), Prayagraj, Uttar Pradesh, India

I. Ahmed, I. Ahmad (eds.), *Aquaculture: Enhancing Food Security and Nutrition*, https://doi.org/10.1007/978-3-031-92858-1_16

causative agents and hardly any efforts have been made to study the diseases in all their perspectives (Annon, 1979); besides, fishes are some of the commonest intermediate hosts of a variety of helminth parasites. The evidence of this has been gathered in the present study to demonstrate the development of larval *Proteocephalus ritae* (Gairola & Malhotra, 1988) and black spot disease organisms, *Neascus vetestai* (Malhotra & Chaubey, 1988), harbouring gall bladder and skin of Indian fishes, respectively, in an experimental homeothermic host, *i.e. Mus musculus*. Under natural environmental conditions, aquatic fauna succumbs to infections, parasites, toxins, malnutrition and neoplasms, but predators and scavengers so quickly eliminate the week and consume the dead that most losses are inapparent to man.

In contrast, fish farming generally involves the confinement and concentration of fish into more or less unnatural environments free of predators so that health problems become apparent, often revealing serious losses which require diagnosis and control. When fish are free to seek an ideal environment for the species, they migrate with changing seasons and seek conditions providing nutrients essential for good health, optimal temperature range, optimal oxygen and water free of pollutants with toxic or noxious substances—all vital for survival, growth, reproduction and resistance to disease. Oxygen in optimal amount is not only essential for life but also an important factor in resistance to disease.

Pollution not only introduces materials which are toxic and noxious for aquatic life, but it often destroys the natural food supply. Since organic pollutants tend to absorb oxygen, they may reduce it to a fatal, especially in winter. Sewage may also expose fauna to infectious organisms and lower resistance to infection. White perch from heavily polluted surface waters of chesapeake Bay have been shown to have antibodies against bacterial pathogens of man including *Pasteurella, Salmonella, Shigella, Escherichia coli* and *Aeromonas*. Similar white perch from clean water were free of such antibodies. Crowding of aquatic fauna poses the same hazards of increased opportunity for disease concentration and transmission as does crowding of animals. Crowding also lowers resistance and contributes to pollutant additives.

Fish pathology has been one of the vital aspects for the fishery scientists with wider commercial implications. The term 'disease' has been defined by Mawdesley-Thomas (1972) as an end result of an interaction between a noxious stimulus and a biological system. Disease concerns all aspects of the biology of that species. The concept of multifactorial aetiology in the complex host-parasite-environment system (Snieszko, 1974) to assist establishment of parasite communities appeared to be operating in this investigation. Some parasitic diseases of fishes of hill-stream and Gangetic plains of India are reported in this study.

Parasites are the major aetiological agents and are the cause of diseases in fishes. Fish parasites harm their hosts in a variety of ways. Besides general blocking of alimentary canal, tissue inflammation, irritation, injury, etc., these infections damage the host's blood physiology and metabolism. They weaken the immune response of hosts, as well as act as carriers or vectors of other pathogens. Therefore, these pathogens have the potential to inflict serious economic losses. Snieszko (1984) asserted that the differences in environmental factors, physiological condition, genetic resistance, physical stress as well as age and sex also play a part in determining the susceptibility of fish to disease. Due emphasis has been laid on most of these factors in this study. Acknowledging the importance of fish, pathogen and environment as significant constituents of a host-parasite system, the study of fish environment, i.e. chemical analysis of water of the riverine ecosystem, was undertaken as an essential aspect of this investigation. The relationship of diseases to environment was discussed in detail by Bauer in the year as early as 1959. Fishes acclimate constantly to the chemical and physical variability and latitude of aquatic environment (Wedemeyer et al., 1976).

Instances or synergistic effects of pollutant interactions and parasite stress on fish hosts have received scant attention from fishery parasitologists in India. The differences in aquatic environmental factors like temperature, pressure, oxygen exert stress on fishes and such ecosystem threats put the mechanism of homeostasis into action (Margalef, 1981). It has been observed that in a toxic-waste-polluted lake ecosystem, the food chain is shortened with a reduction in top predators. Two main arguments are put forward to explain such a situation:

1. The small organisms complete the larger ones in threatened habitats both under conditions of enrichment and toxic stress.
2. Capable feeders, when resources are scarce, are subjected to bioaccumulation of toxic pollutants and in turn have vulnerable developmental stages (Dobson, 1974). Nevertheless, they are more sensitive to the disturbances than the small organisms (Reddy, 1988).

2 Material and Methods

2.1 *Biochemical Estimations*

2.1.1 Total Protein Estimation

The total protein of the dried worms and tissues was precipitated in 14% trichloroacetic acid (TCA) and purified by the method after Oser (1965). Purified proteins were dissolved in a known volume of 0.4 sodium hydroxide and estimated by the Folin-Phenol method of Lowry et al. (1951). Protein reacts with Folin-Phenol reagent to give a coloured complex. The colour formed is due to the reaction of alkaline copper with proteins and Folin-Phenol reagent by tyrosine and tryptophan present in the protein.

2.2 Precipitation and Purification of Proteins

About 20 mg of dried worms/tissues was taken and homogenized in 2–3 ml of 0.9% saline. Protein was precipitated by adding 2–3 ml of 14% TCA to homogenate and the whole centrifuged. Precipitated protein was purified as follows:

It was washed twice in 14% TCA; washed in methanol: chloroform (1:1 v/v) at 55 °C to remove phospholipids, washed in ether to remove remaining fats, treated with 7% TCA at 90 °C for 20 min to remove nucleic acids, finally washed again twice in 70% TCA and dried in acetone followed by ether.

2.3 Protein Estimation

Purified protein was dissolved in an appropriate volume of 0.1 N NaOH. 1 ml of aliquot was pipetted in a test tube, and 5 ml of freshly prepared alkaline CuSO4 solution was added to it. The contents were left for 5 minutes. 0.5 ml of freshly diluted Folin-Phenol reagent was then added and immediately shaken and incubated at 37 °C for 30 min. The optical density of the colour developed was read at 750 nm after setting ELICO spectrophotometer to zero, using a Blank of 1 ml of 0.1 N NaOH which was processed as in the test sample.

The standard graph was prepared by using different concentrations of bovine serum albumin and processing the standard samples as in the case of the test sample.

The protein content of the test samples was calculated with reference to the standard graph and expressed as mg/100 mg of dry weight.

2.4 Paper Chromatography

At least 0.2 mg and more than 1.0 mg of hot helminth parasites and infected and non-infected tissues were taken and hydrolysed for 24 and 36 h in 6 N HCl in vacuo at 120 °C (Malhotra, 1982a). Solid material remaining in the hydrolysate was removed by filtration through a Buchner funnel in the presence of Hyflo Super Cel, and the acidity of the clear filtrate was adjusted to pH 5 by repeated addition and evaporation, by gentle heat, of distilled water. Amino acids recovered in this manner were finally suspended in distilled water; this solution constituted the sample to be analysed.

A drop of the sample was placed at a spot exactly 9 cm from the edge of a sheet of Whatman No. 1 chromatographic paper. After thorough air drying, additional drops were added to the same spot and allowed to dry until the desired concentration of the sample was obtained on the paper. One edge of the paper was immersed in 'n-butanol: acetic acid: water' (4:1:1 v/v) solvent which was allowed to flow along the longitudinal axis of the solvent and allowed to thoroughly air-dry. After thorough air drying, the paper was developed in 0.2% ninhydrin in acetone (Levy & Chung, 1953), allowed to dry and baked in an oven at 60-70 °C for 5 min. Linear distances of each spot in the chromatographic pattern were compared to known amino acids run simultaneously on the same chromatographic paper as one-dimensional chromatography. Quantitative determination of amino acids was carried out with paper chromatography and photochemical colorimetry technique of Thomson and Morris (1959).

2.5 Thin-Layer Chromatography

The experimental results were confirmed by thin-layer chromatography (Stahl, 1969). The presence of detected amino acids detected was confirmed via bidimensional paper chromatography, using the solvent system 'butanol: acetic acid: water' (4:1:1) and 'phenol: water' (4:1) and by comparing with the standard chromatogram of known amino acids developed under identical conditions.

3 Interrelationship of Gonadosomatic Index and Condition Factor with Parasitic Infections

Since the gonadal development in female fishes is known to be greatly and more frequently affected by environmental and other genetic factors in females than males and the differences in condition factor (CF) are more pronounced in the female fishes than in males (Lincoln, 1981a, b), the variations in monthly values of CF and gonadosomatic index (GSI) in female fishes were worked out in the present study.

The decline in mean intensity of *Eustrongylides himalayii* and *Acanthosentis armatus* in *Channa punctatus* at Prayagraj in January, May and October (Table 16.1) illustrated weight reduction in fish body mainly because of greater contribution by fish to the production of reproductive entities, during the aforementioned months of the year that were the periods of breeding season of the fishes (Jhingran, 1975). A review of nematode weight-to-body weight relationship in Garhwal mahseer *Tor tor* (Ham.) was published earlier by Malhotra (1982b). Chauhan et al. (1981b) also summarized an account of parasitization index in relation to ecology of distribution of tapeworms in certain hill-stream fish from Garhwal Himalayas. The pathways of transmission of tapeworm, nematodes and other worms were traced and reviewed in Garhwal Himalayan fish by Malhotra (1981).

The facts analysed corroborated enhanced infection intensity because of greater susceptibility to infestation by *Clinostomum complanatum, P. ritae, Rhabdochona (Filochona) nayari, E. himalayii, Pallisentis allahabadii* and *A. armatus* to larger sized fish (Rae, 1963; Gopalakrishnan & Pal, 1964; Dunsmore & Dudzinski, 1968). However, the level of infestation by *E. himalayii* and *A. armatus* in *C. punctatus* exhibited inverse interrelationships (Kennedy & Lie, 1976). The distributional abundance of tapeworms and nematodes in hill-stream fish was comprehensively presented by Chauhan et al. (1981).

Summarily, as revealed by the linear regression patterns (Banerjee, 1987) (Table 16.1) the mean intensity by *C. complanatum, P. ritae, R.(F.) nayari, E. himalayii, P. allahabadii* and *A. armatus* was correlated positively with the size of respective fish except *E. himalayii* and *A. armatus* of *C. punctatus* at Prayagraj, in which the distribution of worms exhibited negative correlation with the size of the two respective fish (Table 16.1), and of *C. punctatus* at Garhwal in January, April and October provided conclusive evidence of more pronounced effects of parasitism on weight of fishes during these months, while fish length increased disproportionately to its weight immediately prior to or after the breeding season. Arme and Owen (1964) demonstrated in this regard that the development of sex cells proceeded no further than the earliest maturation stages and no infected fish was capable of breeding.

The findings of this study indicate that in catfishes, *C. punctatus*, and snow trout, *Schizothorax richardsonii*, the gonads are less severely affected than in *Sciaena coitor*, and their reduction in size was less marked than in *S. coitor*. The results are supported by a uniformly declining trend in GSI in all the infected fishes during the year 1987 (Table 16.2) in *S. coitor*, which reflected poor growth and development of gonads in this fish, particularly during the infection period. On the other hand, higher values of GSI were recorded in the infected *C. punctatus* at Allahabad and at Garhwal in the month of May than the non-infected catfishes. This evidently indicates that the weight of gonads was appreciably high despite infection in May, while the gains of growth were masked by consistent infections of fishes in all other months of the year resulting in lower GSI values in the infected than non-infected catfishes. The highest GSI values in the infected *S. richardsonii* in June and December than non-infected fishes also demonstrated pronounced growth of gonads during these months in relation to body weight. These findings thus corroborate the conclusions of Arme (1968) and Arme and Owen (1964) that the growth retardation may occur under the influence of parasitic infections as well as due to a derangement of the normal hormonal control of gonadal development during certain periods of the year.

This investigation further illustrated that the weight of non-parasitized *S. richardsonii* increased in direct proportion during at least 7 months of the year, 1987 (Table 16.1), i.e. January, February, March, May, July, August and November. The disproportionate increase in fish size relative to weight of fish in *S. coitor* was only evident in May, and otherwise, this fish showed a normal increase in weight vs cube

Table 16.1 Linear regression trends depicting correlation of infection intensity by various parasites with the size of their host fishes

Male 1986–87	Female	Total	Male 1987–88	Female	Total
Clinostomum complanatum $Y = -6.37 + 0.78X$ $r = 0.51, P < 0.1^S$	$Y = -8.01 + 0.79X$ $r = 0.52, P < 0.1^S$	$Y = -11.95 + 1.18X$ $r = 0.86, P < 0.01^S$	$Y = -2.97 + 0.34X$ $r = 0.63, P < 0.1^S$	$Y = -0.87 + 0.16X$ $r = 0.48, P < 0.1^S$	$Y = 1.55 + 0.02X$ $r = 0.08, P > 0.5^{ns}$
Proteocephalus ritae $Y = -10.65 + 1.87X$ $r = 0.51, P < 0.1^S$	$Y = 15.98 + 2.13X$ $r = 0.60, P < 0.1^S$	$Y = -8.85 + 1.21X$ $r = 0.64, P < 0.1^S$	$Y = 4.59 + 0.16X$ $r = 0.31, P < 0.1^S$	$Y = 2.68 + 0.48X$ $r = 0.31, P < 0.1^S$	$Y = 3.50 + 0.35X$ $r = 0.30, P < 0.1^S$
1987			1988		
Rhabdochona (Filochona) nayari $Y = 10.24 + 0.23X$ $r = 0.41, <0.1^S$	$Y = 15.12 + 0.71X$ $r = 0.5, P < 0.1^S$	$Y = 38.12 + 0.89X$ $r = 0.52, P < 0.1^S$	$Y = 11.84 + 1.04X$ $r = 0.52, P < 0.1^S$	$Y = 2.87 + 4.12X$ $r = 0.44, P < 0.1^S$	$Y = 1.46–0.03X$ $r = -0.30, P < 0.1S$
Eustrongylides himalayii $Y = 0.92{+}{+}0.01X$ $r = 0.04, P > 0.5^{ng}$	$Y = 2.35–0.08X$ $r = 0.58, P < 0.1^S$	$Y = 1.71–0.03X$ $r = 0.23, P > 0.5^{ns}$	$Y = 1.44–0.06X$ $r = -0.2, P > 0.5^{ns}$	$Y = 1.76–0.04X$ $r = -0.43, P < 0.1^S$	$Y = 1.46 + 0.03X$ $r = 0.30, P < 0.1^S$
1986–87			1987–88		
Pallisentis allahabadii $Y = 0.04 + 0.2X$ $r = 0.4, P < 0.1^S$	$Y = 0.69 + 0.22X$ $r = 0.33, P < 0.1^S$	$Y = 1.29 + 0.15X$ $r = 0.27$, P 0.5 ns	$Y = 3.27–0.13X$ $r = -0.34, P > 0.5^{ns}$	$Y = 2.16 + 0.02X$ $r = 0.003, P > 0.5^{ns}$	$Y = 3.61–0.09X$ $r = -0.37, P > 0.5^{ns}$
1987			1988		
Acanthosentis armatus $Y = 9.74–0.56X$ $r = 0.84, P < 0.1^S$	$Y = 9.74–0.55X$ $r = 0.84, P < 0.1^S$	$Y = 5.46–0.28X$ $r = -0.69, P < 0.1^S$	$Y = 1.74 + 0.03X$ $r = 0.09, P > 0.5^{ns}$	$Y = 2.18 + 2.00X$ $r = 0.46, P < 0.1^S$	$Y = 5.46–0.28X$ $r = -0.69, P < 0.1^S$

S significant, *ns* non-significant

Table 16.2 Month-wise distribution of condition factor and gastro-somatic index in various hill-stream and other fish

Months 1987	*Schizothorax richardsonii*				*Sciaena coitor*				*Channa punctatus* Prayagraj				*Channa punctatus* (Garhwal)			
	Non-infected		Infected		Non-infected		Infected		Non-infected		Infected		Non-infected		Infected	
	GSI	CF	GSI	CF	GSI	CF	GSI	CF	GSI	CF	GSI	CF	GSI	CF	GSI	
January	1.71	25.12	1.73	82.8	3.03	193.45	1.69	246.56	2.27	204.99	1.01	1.85.19	2.12	201.75	0.98	
February	1.69	24.42	1.63	9.2	3.70	71.24	1.69	437.38	2.44	155.56	1.27	375.66	2.08	148.28	1.38	
March	1.59	24.12	1.55	85.3	5.26	152.42	1.69	192.0	1.96	79.69	1.20	345.19	1.82	82.73	1.04	
April	1.35	555.56	0.33	185.2	3.45	182.90	0.84	1803.16	0.17	133.57	0.12	144.34	0.29	121.75	0.22	
May	0.86	91.03	0.36	1.24.46	3.45	197.22	2.58	193.06	0.84	226.16	4.84	166.4	0.59	2.18.12	4.02	
June	0.72	288.95	0.05	118.97	5.56	173.10	5.43	122.2	0.06	464.46	0.07	740.33	0.11	334.28	0.09	
July	1.21	1.46.74	0.34	277.78	4.11	115.97	–	–	1.85	281.6	1.64	531.55	1.56	2.67.29	1.34	
August	0.56	208.33	0.25	339.99	2.74	43.53	1.27	118.52	6.23	97.34	1.75	185.6	5.34	92.47	1.52	
September	2.46	800.0	2.42	533.46	2.66	221.91	1.82	350.0	3.30	63.48	3.00	196.81	2.89	55.14	2.54	
October	2.82	185.73	1.35	1406.7	4.17	732.42	3.45	888.89	3.45	562.5	2.46	205.01	4.87	531.28	2.97	
November	0.67	101.48	0.46	147.36	2.74	281.74	2.66	337.5	1.48	167.85	1.27	255.10	1.32	153.12	1.18	
December	0.76	520.0	2.04	156.25	4.58	81.29	2.04	180.84	2.5	233.27	1.94	612.33	2.14	214.12	1.58	

GSI gonadosomatic index, *CF* condition factor

of length during as many as remaining 11 months of the year (Table 16.2). Hence, this could be concluded that the relative growth in length of fish to its weight was more evenly balanced in *S. coitor* followed by *S. richardsonii* and lastly in catfishes of the two zoogeographically different regions.

Finally, a closer look at the monthly fluctuations in CF and GSI of the non-infected fishes would reveal that the influence on these factors in the catfishes of the two different regions closely resembled the situation in *S. coitor.* All these three fishes exhibited an elevated CF in the post-spawning period, i.e. October. On the contrary, *S. richardsonii* exhibited an elevated CF prior to spawning, i.e. April. However, higher values of GSI were seen in all the fishes in the month of October that was the post-spawning period of these Indian fishes. Thus, it is evidently clear that the reduction in gonadal development observed in the remaining months of a year, except pre- and post-spawning period, may be beneficial to aquaculture, as somatic growth is expected to be enhanced. It is also clear that the effects of gonadal development were far greater in female fishes.

4 Biochemical Correlates to Infections

The significant increase in protein content of intestinal tissues of infected fishes, viz. *S. coitor*, *C. punctatus* and *S. richardsonii* in this study (Table 16.3), corroborated with the findings of Taliaferro and Sarles (1939), Barker (1973), Symons (1965, 1976), Jones and Symons (1982), and Symons and Jones (1983). A greater enteric loss of plasma occurred in *Trichostrongylus colubriformis*-infected sheep (Barker, 1973). In addition, increased loss of other proteins such as exfoliated epithelial cells and mucoproteins also occurred, but much of this endogenous protein is reabsorption of amino acid flux absorbed in infected lambs (Jones & Symons, 1982). It was also emphasized by the latter authors that protein synthesis associated with developing immunological response is also likely to increase. Hence, a specific effect on protein metabolism due to parasitism, but not attributable to anorexia, was illustrated by these workers. Symons (1965) demonstrated an increased rate of turnover and proliferation of intestinal epithelial cells associated with villous atrophy similar to those reported in rats infected with *Nippostrongylus brasiliensis.* In the nematode infections by *Nippostrongylus*, the protein synthesis is also known to be influenced by inflammation of lamina propria (Taliaferro & Sarles, 1939). Although malabsorption is induced in the zone of infections, compensatory rates of protein absorption in the parasite-free ileum enhanced the total protein content in the infected tissues (Symons & Jones, 1983).

Hypercholesterolaemia occurring in parasitized *C. punctatus* and *S. richardsonii* (Table 16.3) resembled the reported increase of 28% in duodenum of *Clarius batrachus* infected by *Lytocestus indicus* under Indian conditions (Rasheed & Simha, 1982) and in various organs of infected fishes abroad (Canedo et al., 1976), and differential argumentation rates in liver and intestine of Indian poikilotherm, *Calotes versicolor* (Kameshwari et al., 1980), parasitized by trematode, cestode and nematode organisms. On the other hand, cholesterol depletion in the infected fishes has been attributed to the breakdown of cholesterol to fulfil the energy requirements of the hosts after depletion in the infected fishes has been attributed to the breakdown of cholesterol to fulfil the energy requirements of the hosts after depletion of carbohydrate reserve. In this investigation, a significant decline in the intestinal cholesterol content in the infected *S. coitor* by 44.8% ($P < 0.10$, Table 16.3) is thus explained accordingly.

General depletion in glycogen content of infected host tissues (Read et al., 1963) to a significant level in most of the fishes except in the intestine of *C. punctatus* was a noticeable event (Table 16.3). Parasitic invasions have an adverse influence on the activity of liver thereby disturbing carbohydrate depletion, which has been reported earlier in Indian fishes in single infections (Joshi, 1979; Kadav & Agarwal, 1981). Infected fishes parasitized by *Ligula* (Kosareva, 1961) and by *S. solidus* (Arme & Owen, 1964) have been observed to have depleted glycogen content than non-infected fishes. The significant increase in glycogen content by 52.96% in the infected *C. punctatus* than the non-infected ones was correlated with a generally lower level of cestode, nematode and acanthocephalan infections (Table 16.3) in the study by Banerjee and Malhotra (1990). Kadav and Agarwal (1981) also reported in parallel studies that low levels of infections by caryophyllids resulted in increased glycogen content in fish tissues but heavily parasitized fishes showed depleted glycogen contents. These may be attributed to the morphological and proliferative changes occurring in the worm-free ileum and large intestine (Cheema & Scofoeld, 1982).

5 Amino Acid Patterns of Trematodes

The amino acids found exclusively in proteocephalid tapeworms were cysteine and hydroxyproline (Banerjee, 1987). The composition of amino acids in trypanorhynchan, tetraphyllidean and pseudophyllidean cestodes in comparison with that of cyclophyllidean cestodes (Malhotra, 1990a, b) indicated the presence of cysteine and hydroxyproline in the latter group (Banerjee, 1987).

Table 16.3 Effect of polyparasitic infections on protein level in parasitized and non-parasitized organs of fishes of high altitude and Gangetic plains

	Gangetic plains (Prayagraj)					High altitude (Garhwal)	
Parameter	Tissues	*Channa punctatus*		*Sciaena coitor*		*Schizothorax richardsonii*	
		UI	I	UI	I	UI	I
Protein(mg/g)	Liver	126.07 ± 2.01	123.95 ± 13.2	–	–	–	–
		31.5[a]	39.89[a]	–	–	–	–
		(92.1–168.0)	(105.0–201.0)	–	–	–	–
	Intestine	56.67 ± 1.05	114.0 ± 4.07	27.9 ± 0.9	155.77 ± 1.14	156.76 ± 3.99	115.15 ± 2.11
		1.25[b]	21.19[b]	1.87[c]	41.94[c]	5.0[b]	3.65[b]
		(55.0–58.0)	(90.9–142.9)	(26.2–30.5)	(97.5–194.5)	(85.0–175.0)	(31.5–198.8)

UI non-infected, *I* infected
[a]$P > 0.50$, non-significant
[b]$P > 0.20$, non-significant
[c]$P < 0.05$, significant

On the contrary, the absence of cysteine, cystine and threonine in *Fasciola hepatica*; cysteine and cystine in *Clonorchis sinensis*; valine, cystine and cysteine in *Fasciolopsis indica*; cysteine in *Dicrocoelium dendriticum*; cysteine and cystine in *Diclidophora merlangi*; arginine, cysteine and cystine in *Opisthorchis pedicellata*; and arginine, aspartic acid, cysteine, cystine, hydroxyproline and phenylalanine further confirmed that the relative deprivation of harder parts on body of the aforesaid worms resulted in the absence of cysteine and cystine, as reported previously (Banerjee, 1987).

The concurrence of cysteine detection in the investigations quoted as above in the cyclophyllidean cestodes with exclusive occurrence of cysteine in proteocephalid tapeworms was emphatically proved by the application of the latest technique, energy-dispersive X-ray microanalysis (EDXMA), recently by Yadav and Malhotra (2023) and Yadav et al. (2022). These authors enumerated findings of Heckmann et al. (2007) emphasizing the association between cysteine and cystine as contributors of sulphur ion as well as the disulphide bonds of these two amino acids. This chemical composition was structurally the part of protein infrastructure of the mammalian hair and horns, as well as egg shells and hooks of acanthocephalans.

The published literature on amino acids of trematodes has been scarce. The amino acid patterns were investigated by Daugherty (1952), Kravica et al. (1964), and Czeeuga in *Fasciola hepatica*, Lynch and Bogtish (1962) in metacercarial cyst and adult *Posthodiplostomum minimum*, Ma (1963) in *Clonorchis sinensis*, Singhal (1963) in *Paramphistomum cervi*, Tandon (1968) in *F. indica*, Osikovaki and Bankov (1972) in *Dicrocoelium dendriticum*, Arme and Whyte (1975) in *Diclidophora merlangi*, Srivastava and Gupta (1976) in *Isoparorchis hypselobagri*, Gupta and Gupta (1977) in *Opisthorchis pedicellata*, and Gupta and Agarwal (1979) in *Gastrothylax crumenifer*. Recently, Gupta and Bahadur (1985) analysed the free amino acids of two rumen trematodes *G. crumenifer* and *Cotylophoron orientale*; Rajvanshi (1989) investigated the amino acids of *Parallelotestis indicus* from *Bubulcus ibis coromandus* in India.

The findings studied by Banerjee (1987) were in agreement with those of Daugherty (1952), Ma (1963), Tandon (1968), Osikovaki and Bankov (1972), Arme and Whyte (1975) and Gupta and Gupta (1977) in respect of the presence of arginine, aspartic acid, glutamic acid, glycine and leucine in *G. crumenifer* and *C. orientale*. The study was further in agreement with that of Rajvanshi (1989) regarding the presence of alanine, glutamic acid, glycine, isoleucine, leucine, proline, serine, tyrosine and valine in *P. indicus*. It differed in the absence of cysteine, cystine and threonine in *F. hepatica*; cysteine and cystine in *C. sinensis;* valine, cystine and cysteine in *F. indica*; cysteine in *D. dendriticum*; cysteine and cystine in *D. merlangi*; and arginine, cysteine and cystine in *O. pedicellata*, and arginine, aspartic acid, cysteine, cystine, hydroxyproline and phenylalanine further differ from those of Lynch and Bogtish (1962), Singhal (1963), Srivastava and Gupta (1976), and Gupta and Agarwal (1979) in respect of the absence of cysteine, cystine, glycine and proline in *P. minimum*; cysteine, glutamic acid and phenylalanine, serine, threonine and tyrosine in *T. hypselobagri*; and aspartic acid, cysteine, isoleucine, glutamic acid, serine, tyrosine and valine in *G. crumenifer*. The lower level of proline in *C. complanatum* also corroborated with similar observations reported in *F. hepatica* by Czeezuga and Gierasimow (1977), and lower levels of proline, serine, tyrosine and valine in *P. indicus* reported by Rajvanshi (1989).

6 Amino Acids of Tapeworms

The comparative composition of amino acid pools of Pseudophyllidea, Tetraphyllidea, Trypanorhyncha and Cyclophyllidea revealed that eight amino acids, viz. alanine, histidine, leucine, lysine, serine, threonine, tyrosine and valine, were common in all the four orders of cestoda (Malhotra & Chauhan, 1987a), but two of these amino acids, viz. serine and tyrosine, were not detected in whole-body hydrolysates of *S. teleostei*. In addition to these, 3,4-dihydroxyphenylalanine and DL-norleucine were reported from the pseudophyllidean tapeworm, *Bothriocephalus teleostei*, but only 3,4-dihydroxyphenylalanine could be detected in *S. teleostei*.

Among the two pseudophyllidean cestode, whose amino acid accounts are available to date, seven amino acids were common in both *B. teleostei* and *S. teleostei*. These were D-alanine, aspartic acid, cysteine, β-amino-n-butyric acid, 3,4-dihydroxyphenylalanine, leucine and valine. Four amino acids, viz. arginine, aspartic acid, β-amino-n-butyric acid and glycine, were exclusively found in *S. teleostei*, while the other five, viz. isoleucine, methionine, norleucine, serine and tyrosine, were exclusively present in *B. teleostei*. Simultaneously five amino acids, viz. glutamic acid, ornithine, phenylalanine, proline and taurine, that were reported in earlier studies from tetraphyllid, and trypanorhynchan *cestodes* (Simmons, 1969), and cyclophyllidean tapeworms (Goodchild & Dennis, 1966; Litchford, 1970; Von Brand, 1973; Malhotra, 1981) were not recorded from *S. teleostei* and from *B. teleostei*. The latter worm also did not have β- amino-n-butyric acid, arginine, glycine, isoleucine and methionine, which were detected in above groups of tapeworms by earlier workers. The presence of cysteine was noted in both the pseudophyllid worms in similarity to cyclophyllidean tapeworms. However, *B. teleostei* also contained tryptophan in conformity with cyclophyllidean cestodes. On the other hand, *B. teleostei* had aspartic acid, cysteine and hydroxyproline that were hitherto unreported from cyclophyllidean cestodes in earlier work, and *S. teleostei* contained only hydroxyproline that was unreported from cyclophyllidean worms. Histidine and alanine were relatively deficient in *S. teleostei*. A change in amino acid composition was noted with alteration in the duration of hydrolysis. Lysine appeared only in milder hydrolysate.

A comparison of the pool compositions of the specimens which are the subjects of present report with that of *Lacistorhynchus tenuis*, *Callibothrium verticillatum*, *Phyllobothrium foliatum*, *Rhodobothrium pulvinatum* and *Acanthobothrium* sp., which showed qualitative similarity in amino acid constituents among themselves, exhibited notable differences with the pseudophyllid tapeworms. The predominant amino acids of the latter five species of cestodes infesting elasmobranch fishes are unquestionably alanine, glutamic acid, glycine and taurine (Simmons, 1969), and while alanine was also abundant and others not recorded in teleostean cestodes, the latter was unusually rich in several other components, including leucine, lysine, serine, threonine, tryptophan and tyrosine (Malhotra, 1990a).

However, the trypanorhynchan and tetraphyllid cestode had more similar amino acid constituents to proteocephalid tapeworms, reported by Nanda et al. (1987), whose larvae were recorded by Gairola and Malhotra (1988) in fish from Allahabad, India, and had 16 common amino acids among them. These common amino acids were alanine, arginine, aspartic acid, glutamic acid, glycine, histidine, isoleucine, leucine, lysine, methionine, ornithine, phenylalanine, serine, threonine, tyrosine and valine. The amino acids exclusively found in trypanorhynchan and tetraphyllidean cestodes were β-amino-n-butyric acid, citrulline, ethanolamine, glutamine and taurine, while the amino acids found exclusively in proteocephalid tapeworms were cysteine and hydroxyproline.

The composition of amino acids in trypanorhynchan, tetraphyllidean and pseudophyllidean cestodes in comparison with those of cyclophyllidean cestodes (Malhotra, 1990a, b) indicated the presence of cysteine and hydroxyproline in the latter group. Of the remaining amino acids hitherto unreported from pseudophyllid cestodes (Malhotra & Chauhan, 1987a, b), the absence of arginine, isoleucine, methionine, ornithine, phenylalanine and threonine is in agreement with the study of Campbell (1963) on *Hymenolepis diminuta*. The reports of Fink, et al. (1956) and Litchford (1970) indicated β-amino-n-butyric acid to be the end product of thymine degradation; and alanine glutamic acid and ornithine to be the products of urea cycle, respectively—the metabolic processes whose details have not yet been completely known from the pseudophyllidean cestodes. The absence of cystine in *B. teleostei* (Malhotra & Chauhan, 1987a) is in conformity with similar reports of *Raillietina cesticillus* and *Taenia pisiformis* by Foster and Daugherty (1959) and Goodchild and Dennis (1966), respectively, and that of taurine by Litchford (1970) in *Cittotaenia perplexa*.

Hence, the biochemical differences between pseudophyllidean, proteocephalid, trypanorhynchan, tetraphyllidean and cyclophyllidean cestodes could clearly be useful in establishing differentiating chemotaxonomic interrelationship amongst parasitic organisms.

7 Amino Acids of Nematodes

Sixteen amino acids were reportedly common in the whole-body hydrolysates of both the roundworms analysed in the study, i.e. *E. himalayii* and *R. (F.) nayari*. These amino acids were alanine, arginine, aspartic acid, β-n-butyric acid, cysteine, 3,4-dihydroxyphenylalanine, glutamic acid, glycine, hydroxyproline, leucine, methionine, ornithine, phenylalanine, proline, serine and threonine. The exclusive amino acids detected only in *E. himalayii* were cysteine, histidine and isoleucine, while lysine, tyrosine and valine occurred only in *R. (F.) nayari* (Table 16.4).

Alanine, glycine, hydroxyproline, lysine and proline are known to be some of the important constituents of the parenchymatous tissue (Bell et al., 1972) that provide strength to the body of endohelminths. These amino acids have invariably been reported in a variety of roundworms (Malhotra & Rautela, 1986), and the tapeworms of Bothriocephalidae, Proteocephalidae, Hymenolepididae and Davaineidae Malhotra, 1982b; Bhalya et al., 1983; Malhotra & Chauhan, 1987a, b; Nanda et al., 1987) except that the absence of lysine in cysticercoid larvae of *H. himalayii* and in larvae of *Hydatigera balaniceps* (Malhotra & Chauhan, 1987b) has been noted. Since the body builds up the worms of genus *Eustrongylides*, in particular is very strong, the occurrence of alanine, proline and lysine in these roundworms could be great comparative functional significance. The transamination reactions involving interconversion from pyruvate to alanine as well as from α-ketoglutarate to glutamic acid were apparently of prominent occurrence in both the nematodes under study. These were evidently supported by the uniform appearance of alanine and glutamic acid in the chromatograms of *Eustrongylides* and *Rhabdochona*.

All the amino acids of urea cycle, arginine, aspartic acid, glutamic acid and ornithine except citrulline, appeared in both the nematodes. However, there is a controversy among the modern parasitologists (like Chappell, 1980) that the occurrence of urea cycle in worms, that contain only one or few of the enzymes of the cycle, could be doubtful. Thus, the presence of Aspartic acid and ornithine might be suggestive of their involvement in some alternate biosynthesis activity and not in the urea cycle.

Another amino acid, threonine, has the significance as a precursor for acetate and acetyl-CoA in *Trypanosoma* and *Crithidia* (Chappell, 1980). It has also been argued by Bell et al. (1972) that sulphur is ingested chiefly, during metabolism, in the form of sulphur-containing amino acids, viz. cysteine, cystine and methionine. Methionine may lose its methyl group in the process of transmethylation, leaving hemocysteine. This in its turn may be broken down to cystine by way of the intermediate compound cystathionine. The utility of sulphur-containing amino acids in repair and build-up of tissue proteins has also been illustrated in this work.

Table 16.4 Concentration distribution of amino acid [mean + S.E.] in 24-h and 36-h hydrolysates of different helminth parasites and their parasitized and non-parasitized tissues of host fishes. [A, acanthocephala; II, infected intestine; IL, infected liver; NII, non-infected intestine; NIL, non-infected liver; T, trematode]

Amino acids	*C. compla natum* 24 NIL	T	IL	36 h NIL	T	IL	*P. allahabadii* 24 h NIL	A
Alanine	0.3 ± 0.01	0.25 ± 0.01	0.06 ± 0.01	0.28 ± 0.01	–	0.11 ± 0.002	–	–
Arginine	–	2.9 ± 0.02	–	0.68 ± 0.006	2.2 ± 0.04	0.41 ± 0.002	–	–
Aspartic acid	1.12 ± 0.04	0.88 ± 0.03	0.94 ± 0.003	0.62 ± 0.01	0.82 ± 0.01	0.56 ± 0.02	0.81 ± 0.01	0.39 ± 0.01
Cysteine	1.42 ± 0.006	0.8 ± 0.02	0.04 ± 0.001	0.28 ± 0.001	0.9 ± 0.002	0.18 ± 0.003	–	–
Cystine	–	1.1 ± 0.04	–	–	–	–	–	–
3,4-dihydroxy phenylalanine	0.11 ± 0.006	–	–	0.1 ± 0.002	–	0.08 ± 0.002	0.24 ± 0.002	0.08 ± 0.001
Glutamic acid	0.16 ± 0.01	0.14 ± 0.01	0.04 ± 0.001	0.06 ± 0.01	0.05 ± 0.001	0.04 ± 0.001	–	–
Glycine	0.12 ± 0.03	0.36 ± 0.01	0.16 ± 0.003	0.09 ± 0.01	0.33 ± 0.01	0.09 ± 0.01	0.28 ± 0.001	0.11 ± 0.001
Histidine	–	–	–	–	–	–	0.22 ± 0.01	0.05 ± 0.001
Hydroxy proline	0.78 ± 0.01	1.0 ± 0.02	0.007 ± 0.001	–	–	–	0.28 ± 0.01	0.09 ± 0.02
Isoleucine	0.62 ± 0.01	0.5 ± 0.02	–	0.25 ± 0.01	–	–	–	–
Leucine	0.62 ± 0.01	0.41 ± 0.02	0.09 ± 0.002	0.24 ± 0.02	–	–	0.3 ± 0.001	0.12 ± 0.001
Lysine	–	–	–	–	–	–	–	–
Methionine	–	–	–	–	–	–	0.36 ± 0.01	0.15 ± 0.002
Ornithine	–	–	–	–	–	–	–	–
Phenylalanine	0.22 ± 0.004	0.27 ± 0.1	0.4 ± 0.005	–	–	–	–	–
Proline	0.29 ± 0.01	1.05 ± 0.01	0.08 ± 0.002	1.02 ± 0.01	0.11 ± 0.002	0.12 ± 0.01	0.11 ± 0.002	0.12 ± 0.01
Serine	–	2.9 ± 0.02	–	0.52 ± 0.01	2.4 ± 0.002	0.42 ± 0.05	0.62 ± 0.03	0.31 ± 0.002
Threonine	–	–	–	0.16 ± 0.01	–	0.09 ± 0.002	–	–
Tyrosine	0.41 ± 0.01	0.22 ± 0.01	0.19 ± 0.02	0.34 ± 0.01	0.15 ± 0.001	0.12 ± 0.004	0.21 ± 0.002	0.11 ± 0.001
Valine	0.31 ± 0.07	0.61 ± 0.02	–	0.11 ± 0.001	0.48 ± 0.02	0.08 ± 0.001	0.84 ± 0.01	0.06 ± 0.001

II	36 h NIL	A	II	*A. armatus* 24 h NIL	A	II	36 h NIL	A	II
–	–	–	–	–	–	–	–	–	–
–	–	–	–	0.71± 0.002	0.74 ± 0.014	0.67 ± 0.004	0.42 ± 0.01	0.54 ± 0.012	0.37 ± 0.004
0.52 ± 0.002	0.33 ± 0.001	0.28 ± 0.002	–	–	–	–	–	–	–
–	–	–	–	–	–	–	–	–	–
–	–	–	–	–	–	–	–	–	–
0.18 ± 0.002	0.21 ± 0.001	0.1 ± 0.001	0.08 ± 0.001	0.21 ± 0.001	0.98 ± 0.02	0.18 ± 0.007	0.02 ± 0.001	0.47 ± 0.01	0.12 ± 0.001
–	–	–	–	–	–	–	–	–	–
0.26 ± 0.003	0.12 ± 0.001	0.1 ± 0.001	0.08 ± 0.001	0.32 ± 0.003	0.28 ± 0.001	0.24 ± 0.001	0.21 ± 0.001	0.22 ± 0.001	0.12 ± 0.009
0.18 ± 0.001	0.08 ± 0.001	0.08 ± 0.001	0.06 ± 0.001	0.22 ± 0.001	0.23 ± 0.001	0.18 ± 0.001	0.15 ± 0.001	0.22 ± 0.001	0.09 ± 0.001
0.2 ± 0.001	0.46 ± 0.03	0.1 ± 0.001	0.07 ± 0.001	0.3 ± 0.002	0.42 ± 0.06	0.18 ± 0.001	0.43 ± 0.02	0.31 ± 0.002	0.0
–	–	–	–	–	–	–	–	–	–
0.26 ± 0.003	–	–	–	–	–	–	–	–	–
–	–	–	–	–	–	–	–	–	–
0.32 ± 0.002	–	–	–	–	–	–	–	–	–
–	–	–	–	0.50+ 0.01	0.32+ 0.002	0.49+ 0.001	0.26+ 0.001	0.28+ 0.001	0.1+ 0.001
–	–	–	–	–	–	–	–	–	–
0.08 ± 0.001	0.07 ± 0.001	–	–	–	–	–	–	–	–
0.58 ± 0.01	0.88 ± 0.001	0.32 ± 0.002	0.82 ± 0.02	0.62 ± 0.02	0.42 ± 0.004	0.49 ± 0.003	0.72 ± 0.01	0.5 ± 0.01	0.69 ± 0.01
–	–	–	–	0.5 ± 0.01	0.52 ± 0.001	0.4 ± 0.002	0.42 ± 0.002	0.22 ± 0.001	0.11 ± 0.001
0.2 ± 0.002	–	–	–	–	–	–	–	–	–
0.82±0.001	–	–	–	–	–	–	–	–	–

8 Amino Acids of Acanthocephala

Six amino acids, *viz.* 3,4-dihydroxyphenylalanine, glycine, histidine, hydroxyproline, proline and serine, were common between the two acanthocephalan worms, *A. armatus* and *P. allahabadii*, while 3 amino acids, i.e. arginine, ornithine and threonine, were exclusively detected in *A. armatus*, and 6 amino acids, viz. aspartic acid, glutamic acid, leucine, methionine, tyrosine and valine, were found exclusively in *P. allahabadii* of the two acanthocephala (Table 16.4).

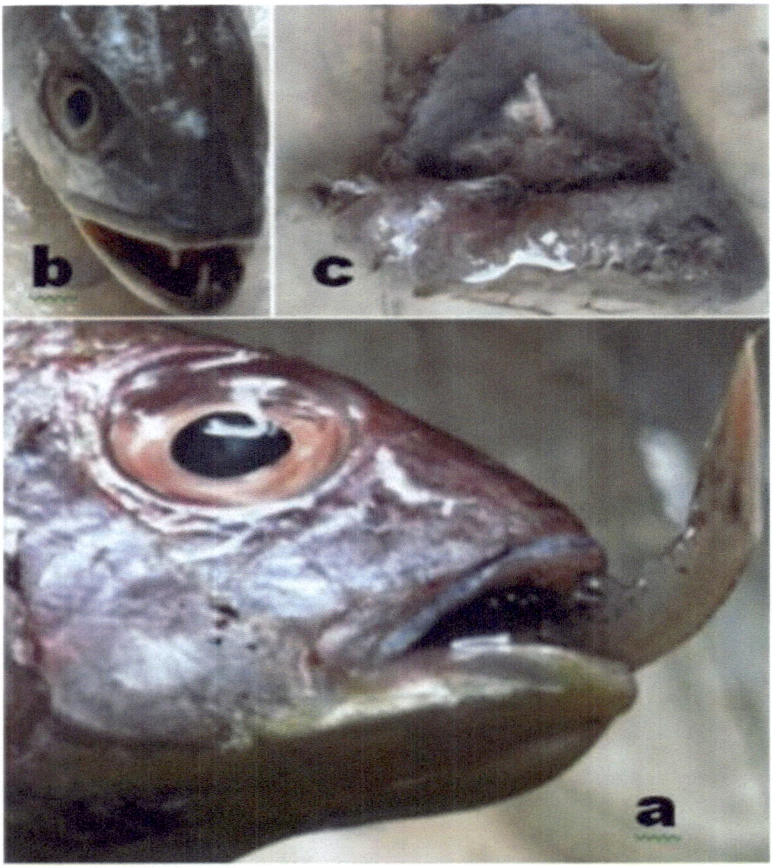

Fig. 16.1 *Lutjanus malabaricus*. (**a**) Signifying cannibalism to engulf smaller fish of the same genus, i.e. *Lutjanus malabaricus* from the Arabian Sea. (**b**) Anterior end (mouth) of the host fish, *Lutjanus malabaricus*. (**c**) Digested part of pyloric ceca of the engulfed smaller fish (the victim of cannibalism)

. The interconversion of certain amino acids, viz. from α-ketoglutarate to glutamic acid and pyruvate to alanine, is known to be of frequent occurrence in various endohelminths and other parasites (Nanda et al., 1987). The transmission reactions involving aminotransferases were attributed to be the most likely cause for the aforesaid interconversions within the parasite systems by Chappell (1980). The presence of former system is apparently signified by the occurrence of glutamic acid in *P. allahabadii*. Another observation on generalize occurrence of histidine in both the acanthocephalan worms (Table 16.4) was comparable with the findings of Saber and Wu (1985) who experimentally illustrated that histidine was rapidly taken up in vitro by adult *Schistosoma mansoni* and their metabolism occurred via decarboxylation, deamination and transamination reactions to produce histamine and glutamate as end products.

The detection of proline in both the acanthocephalan worms under study and of its oxidation products, *i.e.* glutamic acid in *P. allahabadii*, would suggest the efficient utilization of proline as an effective resource of energy under emergency conditions (Goodchild & Dennis, 1966) by these worms.

The occurrence of urea cycle in members of several tapeworm families, *viz.* Bothriocephalidae (Malhotra & Chauhan, 1987a), Proteocephalidae (Nanda et al., 1987), Hymenolepididae (Campbell, 1963; Smyth, 1969; Litchford, 1970) and Davaineidae (Malhotra, 1981), has been well demonstrated. The interrelationship of arginine and aspartic acid with the synthesis of urea was illustrated by Smyth (1969), Litchford (1970) and Von Brand (1973). But despite the presence of arginine and ornithine transcarbamylase in the cestodes, *Moniezia (Moniezia) expansa*, *Dipylidium caninum*, *T. pisiformis* and *Echinococcus granulosus*, the functional presence of ornithine cycle in these cestodes could not be experimentally demonstrated (Smyth, 1969). It is believed therefore that though aspartic acid was detected in the chromatograms of *P. allahabadii* and arginine and ornithine occurred in *A. armatus* (Table 16.4), the functional presence of urea cycle or ornithine cycle in these acanthocephalans was of doubtful significance until confirmed by further experimental work.

One of the interesting features of the study was the absence of methionine in *A. armatus* and occurrence of this amino acid in only milder (24 h) hydrolysate of *P. allahabadii* while it abstained from 36-h hydrolysate.

Instances of cannibalism (Fig. 16.1) in the host (*Lutjanus malabaricus*)-parasite (*Rotundocollarette capoori*) encountered during the current investigation also brought forth the phenomenon of multiple scale increase of accumulator effect as the trace metals would add up due to the already ingested trace elements in the body of pray smaller fish by the larger host fish that were initially parasitized by roundworms, *R. capoori*. The sulphur content was very high in the digested gill tissues (70.04–84.82 wt%) as compared to other partially digested organs, after reaching the stomach (44.05–64.66 wt%).

9 Conclusions

This chapter reviewed levels of parasitization by tapeworms, roundworms, trematodes and acanthocephalan parasitized by hill-stream fishes of Garhwal Himalayas, as well as the earlier surveys having been conducted by authors and co-workers on transmission pathways of these parasites in vertebrate and other related hosts, mainly from aquatic ecosystems in Garhwal Himalayas. The fulfilment of required energy input to make up its loss due to infection implications was attributable to the breakdown of cholesterol, showing thereby depletion of cholesterol in the infected fish. Glycogen depletion in the heavily parasitized fish than the converse being true in the lowly parasitized fish was concluded.

The amino acids found exclusively in proteocephalid tapeworms were cysteine and hydroxyproline (Banerjee, 1987). The composition of amino acids in trypanorhynchan, tetraphyllidean and pseudophyllidean cestodes in comparison with that of cyclophyllidean cestodes (Malhotra, 1990a, b) indicated the presence of cysteine and hydroxyproline in the latter group (Banerjee, 1987). On the contrary, the absence of cysteine, cystine and threonine in *Fasciola hepatica*; cysteine and cystine in *Clonorchis sinensis*; valine, cystine and cysteine in *Fasciolopsis indica*; cysteine in *Dicrocoelium dendriticum*; cysteine and cystine in *Diclidophora merlangi*; arginine, cysteine and cystine in *Opisthorchis pedicellata*; and arginine, aspartic acid, cysteine, cystine, hydroxyproline and phenylalanine further confirmed that the relative deprivation of harder parts on body of the aforesaid worms resulted in the absence of cysteine and cystine, as reported previously (Banerjee, 1987). The concurrence of cysteine detection in the investigations quoted as above in the cyclophyllidean cestodes with exclusive occurrence of cysteine in proteocephalid tapeworms was emphatically proved by the application of the latest technique, energy dispersive X-ray micro analysis (EDXMA) recently by Yadav et al. (2022). These authors enumerated findings of Heckmann (1996) and Heckmann et al. (2007) emphasizing association between cysteine and cystine as contributors of sulphur ion as well as the disulphide bonds of these two amino acids. This chemical composition was structurally the part of protein infrastructure of the mammalian hair and horns, as well as egg shells and hooks of acanthocephalans.

Conflict of Interest We do not have any conflict of interest.

References

Annon. (1979). Summer Institute in Pathology. Lecture notes. (ed. Varghese, T.J.) College of Fish, Mangalore.

Arme, C. (1968). Effects of the plerocercoid larva of a pseudophyllidean cestode, *L. intestinalis*, on the pituitary gland and gonads of its host. *The Biological Bulletin, 134*, 15–25.

Arme, C., & Owen, R. W. (1964). Massive infections of the three-spined stickleback with *Schistocephalus solidus* (Cestoda:Pseudophyllidae) and *Glugea anamals* (Sporozoa:Microsporidia). *Parasitology, 54*, 10–11.

Arme, C., & Whyte, A. (1975). Amino acids of *Dicliphora morlangi* (Monogenea). *Parasitology, 70*, 39–46.

Banerjee, S. (1987). *Population distribution and pathology of certain parasites in fishes at high altitude and gangetic plains*. Ph.D. thesis, Unpubl., Univ. Allahabad, pp. 551

Banerjee, S., & Malhotra, S. K. (1990). Post-parasitizational biochemical changes in five species of Indian hill-stream and freshwater fishes. *Geobios New Reports, 9*, 47–52.

Barker, I. K. (1973). A study of the pathogenesis of *Trichostrongylus colubriformis* infection in lambs with observations on the contribution of gastrointestinal plasma loss. *International Journal for Parasitology, 3*, 743–757.

Bell, G. H., Davidson, J. N., & Smith, D. E. (1972). *Text book of physiology and biochemistry*. The Eng. Lang. Book Soc. Church. Living, pp. 1160.

Bhalya, A., Seth, A., Malhotra Sandeep, K., & Capoor, V. N. (1983). Amino acids of *Amoebotaenia cuneata* (Cestoda:Dilepididea Wardle, McLeod and Radinovsky, 1974). *Journal of Helminthology, 57*, 9–10.

Campbell, J. W. (1963). Amino acids and nucleotides of the cestode *Hymenolepis diminuta*. *Comparative Biochemistry and Physiology Part A, 8*, 181–185.

Canedo, L., Garrido, F., & Gonzalez, M. (1976). In D. Van Bosseche (Ed.), *Comparative biochemistry of parasites*. Academic Press.

Chappell, L. H. (1980). *Physiology of parasites* (p. 229). Blackie & Sons Ltd..

Chauhan, R. S., Malhotra Sandeep, K., & Capoor, V. N. (1981a). The distribution and abundance of cestodes in eleven species of teleosts from Garhwal Himalayas with a note on host biology. *Himalayan Journal of Science, 1*, 15–30.

Chauhan, R. S., Malhotra, S. K., & Capoor, V. N. (1981b). An analysis of parasitization index and certain ecological parameters of cestode parasites infesting in hill-stream fishes of district Pauri-Garhwal, U.P (India). *Science & Environment, 3*(1), 47–56.

Cheema, K. J., & Scofoeld, A. M. (1982). Scanning electron microscopy of the intestines of rat infected with *Nippostrongylus brasiliensis*. *International Journal for Parasitology, 12*, 199–206.

Czeezuga, B., & Gierasimov, M. (1977). Differences in amino acid content in the protein of *Fasciola hepatica* L. individuals and of bovine liver. *Parasitology, 25*, 55–58.

Daugherty, J. W. (1952). Studies on the protein metabolism of certain helminth parasites. *The Journal of Parasitology, 38*, 32.

Dobson, S. I. (1974). Zooplankton competition and predation: An experimental test of the size-efficiency hypothesis. *Ecology, 55*, 606–613.

Dunsmore, J. D., & Dudzinski, M. L. (1968). Relationship of numbers of nematode parasites in wild rabbits, *Oryctolagus cuniculus* (L.), to host sex, age and season. *The Journal of Parasitology, 54*, 462–474.

Fink, K., McGaughey, C., Handerson, R. B., & Fink, R. M. (1956). Isotopic and enzymatic studies on thymine metabolites. *Federation Proceedings, 15*, 251.

Foster, W. B., & Daugherty, J. W. (1959). Establishment and distribution of *Raillietina cesticillus* in the fowl and comparative studies on amino acid metabolism of *R. cesticillus* and *Hymenolepis diminuta*. *Experimental Parasitology, 8*, 413–426.

Gairola, D. (1990). *Population biology and taxonomy of helminth parasites of certain economically important food fishes*. Unpubl. D.Phil. Thesis, Univ. Allahabad, pp. 600.

Gairola, D., & Malhotra, S. K. (1988). First report on *Proteocephalus* larvae from fish at Allahabad. *Geobios New Reports, 7*, 165–166.

Goodchild, C. G., & Dennis, E. S. (1966). Amino acids in seven species of cestodes. *The Journal of Parasitology, 52*, 60–62.

Gopalakrishnan, V., & Pal, R. N. (1964). Advancements in the study of parasites of migratory fishes. *Proceedings of the Zoological Society (Calcutta), 17*, 63–66.

Gupta, V., & Agarwal, S. K. (1979). Amino acid composition in *Gastrothylax crumenifer* (Trematoda). *Indian Journal of Helminthology, 29*, 140–143.

Gupta, P. C., & Bahadur, R. (1985). Free amino acid composition of rumen trematodes *Gastrothylax crumenifer* and *Cotylophoron orientale*. *Indian Journal of Parasitology, 9*, 155–157.
Gupta, S. P., & Gupta, R. C. (1977). Free amino acid composition of *Opisthorchis pedicellata* (trematoda) from the gall bladder of a freshwater fish, *Rita rita* (Ham.). *Zeitschrift für Parasitenkunde, 54*, 83–87.
Heckmann, R. A. (1996). Energy dispersive X-ray microanalysis in conjunction with election optics, a tool for analyzing aquatic animal diseases and deaths. *Microscopy and Analysis, 16*, 27–29.
Heckmann, R. A., Amin, O. M., & Standing, M. D. (2007). Chemical analysis of metals in acanthocephalans using energy dispersive X-ray analysis (EDXA) in conjunction with a scanning electron microscope 274 (SEM). *Comparative Parasitology, 74*(2), 388–391.
Jhingran, V. G. (1975). *Fish and fisheries of India* (2nd ed., p. 954). Hindustan Publ. Corpn.
Jones, W. O., & Symons, L. E. A. (1982). Protein synthesis in the whole body, liver, skeletal muscle and kidney cortex of lambs infected by the nematode *Trichostrongylus columbriformis*. *Indian Journal of Parasitology, 12*, 295–301.
Joshi, B. D. (1979). Biochemical changes in the liver and blood of a freshwater fish, *Rita rita*, infested with a trematode *Opisthorchis pedicellata. Folia Parasitologica, 26*, 143–144.
Kadav, M., & Agarwal, S. M. (1981). Parasitic effects on carbohydrate metabolism of *Clarias batachus* parasitized by caryophyllids. *Indian Journal of Helminthology, 33*, 153–155.
Kameshwari, M., Simha, S. S., & Rao, L. N. (1980). Seasonal variations of three trematodes of *Rana tigrina* and observations on the genus *Mehraorchis* Srivastava, 1934. *Helminthologia, 17*, 11–16.
Kennedy, C. R. (1968). Population biology of the cestode *Caryophyllaeus laticeps* (Pallas, 1781) in dace, *Leuciscus leuciscus* L., of the river Avon. *The Journal of Parasitology, 54*, 538–543.
Kennedy, C. R., & Lie, S. F. (1976). The distribution and pathogenicity of larvae of *Eustrongylides* (Nematoda) in brown trout *Salmo trutta* L. in Fernworthy reservoir Devon. *Journal of Fish Biology, 8*, 293–302.
Kosareva, A. F. N. A. (1961). Disturbance of the carbohydrate metabolism of cyprinidae in ligula and *Digramma interrupta* (cestoda) on fish. *Zoologicheskiy Zhurnal, 35*, 1629–1632.
Kravica, S., Maloseza, Z., Wagner, H., & Martinincic, T. (1964). The amino acid composition of the proteins and the distribution of amino acids in the lipids of eggs of *Fasciola hepatica* and *Ascaris suum. Veterinarski Arhiv, 34*, 163–166.
Levy, A. L., & Chung, D. (1953). Two-dimensional chromatography of amino acids on buffered papers. *Analytical Chemistry, 145*, 387–403.
Lincoln, R. F. (1981a). Sexual maturation in triploid male plaice (*Pleuronectes platessa*) and plaice x founder (*Platichthys flesus*) hybrids. *Journal of Fish Biology, 19*, 415–426.
Lincoln, R. F. (1981b). Sexual maturation in female triploid plaice, (*Pleuronectus platessa*) and plaice x founder, *Platichthys flesus*, hybrids. *Journal of Fish Biology, 19*, 499–507.
Litchford, R. G. (1970). Amino acids of *Hymenolepis microstoma* and host bile. *Comparative Biochemistry and Physiology, 32*, 61–67.
Lowry, O. H., Rosenbrough, N. J., Farr, A. L., & Randall, R. J. (1951). Protein measurement with Folin-phenol reagent. *Journal of Biological Chemistry, 193*, 265–275.
Lynch, D. L., & Bogtish, B. J. (1962). The chemical nature of metacercarical cyst. 2. Biochemical investigations on the cyst of *Posthodiplostomum minimum. The Journal of Parasitology, 48*, 241–243.
Ma, L. (1963). Trace elements and polyphenol oxidase in *Clonorchis sinensis. The Journal of Parasitology, 49*, 197–203.
Malhotra, S. K. (1981). Systems models for parasite pathways in ichthyoparasitology of the Garhwal Himalayan ecosystem. *Current Science, 50*(19), 874–875.
Malhotra, S. K. (1982a). Studies on amino acids in *Raillietina (Raillietina) saharanpurensis* (Malhotra and Capoor, 1981) with a note on biochemical variations in cyclophyllidean cestodes. *Comparative Physiology and Ecology, 7*, 207–210.
Malhotra, S. K. (1982b). Nematode-weight to body-weight relationship in *Tor tor* (Ham.) in the Himalayan riverine ecosystem. *Current Science, 51*(8), 423–424.

Malhotra, S. K. (1990a) *Ecological and Karyological Segregation of Helminths of Vertebrate Hosts in Garhwal Himalayan Ecosystems*. Final Tech. Rep. Dst Maj. Res. Proj., Dept. Sci. Tech., India, pp. 294.
Malhotra, S. K. (1990b). Ecology, pathology and biochemical correlates to parasitic diseases of Indian freshwater fishes. *Advances in the Biosciences*, 19–44.
Malhotra, S. K., & Chaubey, A. K. (1988). Parasites as potential pathogens affecting growth of Garhwal mahseer *Tor tor* (Ham.). In *Proc. Nat. Sem. Rec. Tr. Fish Biol* (pp. 102–106).
Malhotra, S. K., & Chauhan, R. S. (1987a). Chemotaxonomy of pseudophyllidean cestodes. 2. Amino acids of *Botheriocephalus teleostei* (Malhotra) from Indian hill-stream teleosts. *Boletín Chileno de Parasitología, 42*, 3–7.
Malhotra, S. K., & Chauhan, R. S. (1987b). Amino acids of *Hydatigera balaniceps* (Eucestoda: Taeniidea). *Indian Journal of Parasitology, 11*, 83.
Malhotra, S. K., & Rautela, A. S. (1986). Amino acids of a poultry nematode *Heterakis kotdwarensis* (Malhotra and Rautela) from *Gallus gallus*. In *Proceedings of the national symposium new dimension parasitology* (pp. 18–19).
Malhotra, S. K., Dixit, S., & Capoor, V. N. (1981). A contribution to the study of taxa differentiation in cestode taxonomy. *Proceedings of the Indian Academy of Sciences – Animal Science, 90*, 343–349.
Margalef, R. (1981). Stress in ecosystem: A future approach. In G. W. Barrett & R. Rosenberg (Eds.), *Stress effects on natural ecosystems* (pp. 281–289).
Mawdesley-Thomas, L. E. (1972). *Diseases of fish* (p. 380). Academic Press.
Nanda, S., Bhalya, A., Gairola, D., Malhotra Sandeep, K., & Capoor, V. N. (1987). Comparative analysis of amino acids of three species of *Gangesia* (Cestoda: Proteocephalata). *Journal of Helminthology, 61*, 233–239.
Oser, B. L. (1965). *Hawk's physiological chemistry* (p. 147). McGraw-Hill.
Osikovaki, E., & Bankov, I. (1972). Amino acid content of *Dicrocoelium dendriticum. Izv. Tsent. Khelmint. Lab., 15*, 151–158.
Rae, B. B. (1963). The incidence of larvae of *Porrocaecum decipiens* in the flesh of cod. *Marine Environmental Research, 2*, 28.
Rajavanshi, I. (1989). Chromatographic study of amino acid in a digenetic trematode-*Parallelotestis indicus. Indian Journal of Parasitology, 13*, 19–20.
Rasheed, S., & Simha, S. S. (1982). Cholesterol content in *Lytocestus indicus* and its host. *Proceedings/Indian Academy of Sciences, 3*, 46–47.
Read, C. P., Rothman, A. H., & Simmons, J. E., Jr. (1963). Studies on membrane transport, with special reference to parasite-host integration. *Annals of the New York Academy of Sciences, 113*(1), 154–205.
Reddy, M. V. (1988). Ecology of the threatened habitats. In V. P. Agarwal & L. D. Chaturvedi (Eds.), *Threatened habitats* (pp. 83–93). IndiaBioscience.
Saber, M. A., & Wu, G. (1985). In vitro catabolism of L-histidine by adult *Schistosoma mansoni. Journal of the Egyptian Society of Parasitology, 15*, 19–39.
Simmons, J. E. (1969). Composition of amino acid pools of some cestodes of elashmobranch fishes of the woods hole area. *Experimental Parasitology, 26*, 264–271.
Singhal, P. D. (1963). Free amino acids in diagenetic trematode, *Paramphistomum cervi. Current Science, 32*, 552–553.
Smyth, J. D. (1969). *The physiology of Cestodes* (p. 255). Oliver & Boyd.
Snieszko, S. F. (1974). The effects of environmental stress on outbreaks of infectious diseases of fishes. *Journal of Fish Biology, 6*, 197–208.
Snieszko, S. F. (1984). Therapy of bacterial fish diseases. *Transactions of the American Fisheries Society, 83*, 313–330.
Srivastava, M., & Gupta, S. P. (1976). Amino acid composition in *Isoparorchis hypselobagri. Z. Parasitenk., 49*, 183–185.
Stahl, E. (1969). *Thin layer chromatography* (2nd ed., p. 925). Springer-Verlag.

Symons, L. E. A. (1965). Kinetics of the epithelial cells and morphology of villi and crypts in the in the jejunum of the rat infected by the nematode *Nippostrongylus brasiliensis*. *Gastroenterology, 49*, 158–168.

Symons, L. E. A. (1976). Malabsorption. In E. J. L. Soulsby (Ed.), *Pathology of parasitic infections* (pp. 11–21). Academic Press.

Symons, L. E. A., & Jones, W. O. (1983). Intestinal protein synthesis in Guinea pigs infected with *Trichostrongylus colubriformis*. *International Journal for Parasitology, 13*, 309–312.

Taliaferro, W. H., & Sarles, M. P. (1939). The cellular reactions in the skin, lungs and intestine of normal and immune rats after infection with *Nippostrongylus muris*. *The Journal of Infectious Diseases, 64*, 157–192.

Thomson, J. F., & Morris, C. J. (1959). Quantitative determination of amino acids. *Analytical Chemistry, 31*, 1031.

Tondon, R. S. (1968). Amino acid composition of the Indian liver fluke, *Fasciola indica*. *Z. Parasitenk., 30*, 149–151.

Von Brand, T. (1973). *Biochemistry of parasites* (2nd ed., p. 429).

Wedemeyer, G. A., Meyer, F. P., & Smith, L. (1976). Environmental stress and fish diseases. In S. F. Snieszko & H. R. Axelford (Eds.), *Diseases of fishes* (p. 192). T.F.H. Publ. Ltd.

Yadav, A., & Malhotra Sandeep, K. (2023). Monitoring of surrogate zoonotic parasites of Anisakidae and *Echinobothrium deeghai* as bioindicators by application of energy dispersive X-ray microanalysis (EDXMA). *Journal of Parasitic Diseases, 47*, 793. https://doi.org/10.1007/s12639-023-01625-3

Yadav, A., Kapoor, N., Arif, A., & Malhotra, S. K. (2022). Energy dispersive X-ray microanalysis in conjunction with scanning electron micrography to establish nematodes as bioindicators in marine fish environment. *Journal of Parasitic Diseases, 46*(3), 664–671.

Chapter 17
Epigenetic Mechanisms in Fish: Insights into Adaptation, Resilience, and Conservation

Asra Imtiyaz Mattoo, Malik Wasim, Farooz Ahmad Bhat, and Darve Sabina Iqbal

1 Introduction

Epigenetics is described as "the study of phenomena and mechanisms that cause chromosome-bound, heritable changes to gene expression that are not dependent on changes to DNA sequence" (Deans & Maggert, 2015). The concept of "heritable" carries two distinct meanings: these changes can be transferred from a parent cell to its daughter cells during mitotic division or passed from parents to offspring through the formation of gametes during meiotic division. The primary mechanisms involved in epigenetic regulation of gene expression include DNA methylation, histone modifications, histone variants, and noncoding RNAs (ncRNAs) (Brock & Fisher, 2005). Notably, epigenetic clocks developed so far are based on analyses of DNA methylation (Fig. 17.1).

Epigenetics plays a vital role in fisheries by offering a deeper understanding of how environmental factors influence the growth, reproduction, and survival of fish species (Liu et al., 2022). Unlike genetic changes, epigenetic modifications are dynamic and reversible, allowing fish to adapt to fluctuating conditions such as temperature shifts, salinity changes, and pollution (Shen et al., 2023). This adaptability is particularly significant for managing fish populations in the face of climate change and habitat degradation. Epigenetic research can inform sustainable fisheries management by identifying stress-resilient species or stocks, improving breeding programs, and enhancing aquaculture productivity. Additionally, it provides insights into how human-induced environmental changes affect fish health and

A. I. Mattoo (✉) · F. A. Bhat · D. S. Iqbal
Division of Fisheries Resource Management, FoFy, SKUAST, Kashmir, Jammu and Kashmir, India

M. Wasim
Division of Fish Genetics and Biotechnology, FoFy, SKUAST, Kashmir, Jammu and Kashmir, India

I. Ahmed, I. Ahmad (eds.), *Aquaculture: Enhancing Food Security and Nutrition*, https://doi.org/10.1007/978-3-031-92858-1_17

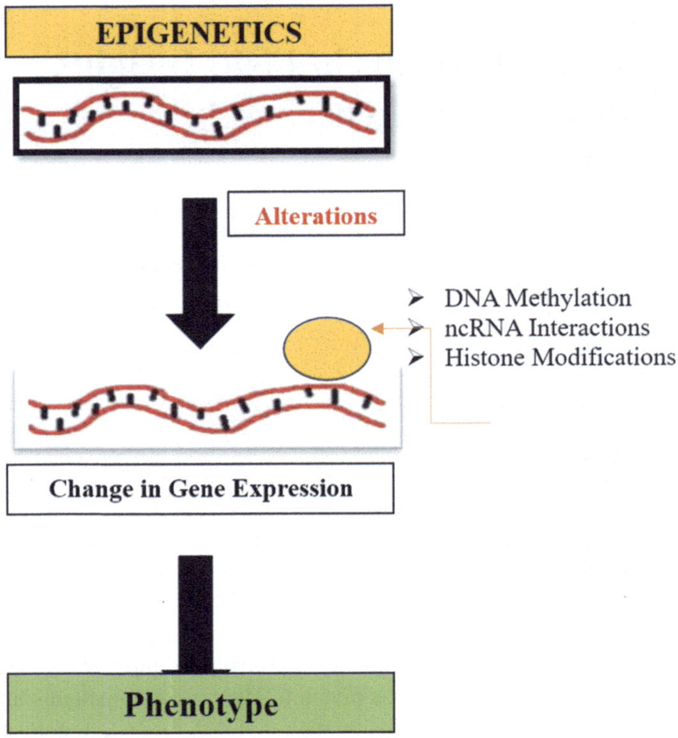

Fig. 17.1 Diagrammatic representation of the process of Epigenetics in Fish

ecosystem stability, guiding conservation efforts to preserve genetic diversity and ensure long-term viability of fisheries (Kim et al., 2024).

Aquaculture production is expanding rapidly worldwide. To meet the growing demand, advancements in technology are essential, particularly in areas like aquaculture feeds, breeding, and disease management. Exploring the role of epigenetic mechanisms in aquaculture practices offers significant potential. By understanding environment-induced epigenetic markers associated with traits like disease resistance, it may be possible to optimize breeding conditions. Epigenetics involves the study of inheritable changes in gene expression that occur without altering the underlying DNA sequence. These changes are driven by mechanisms that regulate gene activity. Over the past decade, it has become widely recognized that environmental factors can induce phenotypic changes in organisms through these epigenetic mechanisms. These processes can modulate gene expression during transcription, translation, or posttranslation stages, all without altering the DNA nucleotide sequence (Jaenisch & Bird, 2003).

Epigenetics in fish encompasses a range of mechanisms that influence growth, development, and adaptation to environmental changes. Key epigenetic processes, such as DNA methylation, histone modifications, and noncoding RNAs, are sensitive to nutritional inputs and environmental stressors, which can lead to lasting changes in gene expression and phenotypic traits (Abdelnour et al., 2024). For instance, nutritional programming can alter epigenetic regulation in broodstock and larvae, affecting gene accessibility and expression patterns that persist beyond the initial exposure (Epigenetics in Fish Nutritional Programming, 2023). Additionally, the epigenetic profiles established during gametogenesis play a crucial role in transmitting information to offspring, with environmental factors potentially modifying these profiles and influencing reproductive success (Epigenetics in Sexual Maturation and Gametes of Fish, 2023). Furthermore, fish exhibit remarkable phenotypic plasticity, allowing them to adapt to varying environmental conditions through epigenetic modifications, which can be inherited across generations (Abdelnour et al., 2024; Navarro-Martín et al., 2023). This intricate interplay between epigenetics and environmental factors highlights the potential for enhancing aquaculture practices through a deeper understanding of these mechanisms (Navarro-Martín et al., 2023). Epigenetics has profound implications for understanding biological adaptation, development, and environmental responses in various organisms, including fish. Fish are particularly significant in epigenetic studies due to their diverse habitats and the range of environmental stressors they face. This chapter explores the fundamental mechanisms of epigenetics in fish, its role in their adaptation to environmental challenges, and its implications for conservation and aquaculture.

2 Fundamental Epigenetic Mechanisms in Fish

To understand epigenetic mechanisms and how they interact and complement one another, it is essential to first grasp the basic structure and function of chromatin. In eukaryotic cells, DNA is tightly compacted into chromatin, with approximately 146 base pairs of DNA wrapped around histone octamers to form nucleosomes. Each nucleosome contains a histone octamer composed of two copies each of histones H2A, H2B, H3, and H4. The nucleosomes are connected by the linker histone H1 and are further folded into chromatin fibers, enabling the efficient packaging of genetic material within the nucleus (Conaway, 2012). Epigenetic mechanisms can modify the chromatin structure, ultimately determining specific patterns of gene expression. The epigenome represents the dynamic range of chromatin states, with various mechanisms marking regions associated with distinct genome functions (Siggens & Ekwall, 2014). This section highlights key measurable epigenetic marks, such as DNA methylation, histone methylation or acetylation, and chromatin remodeling.

2.1 DNA Methylation

DNA methylation involves the addition of a methyl group to the DNA molecule, typically at cytosine bases, leading to gene silencing. In fish, this process is vital for regulating gene expression, maintaining genome stability, and facilitating cellular differentiation (Smith et al., 2018). DNA methylation is a fundamental epigenetic mechanism in fish that plays a crucial role in their adaptation to environmental changes and nutritional programming. It involves the addition of methyl groups to DNA, which can alter gene expression without changing the DNA sequence, thereby influencing phenotypic traits such as growth, reproduction, and stress responses (Metzger, 2023; Navarro-Martín et al., 2023). Environmental stressors, including climate change and pollution, can induce irregular DNA methylation patterns, leading to transgenerational adaptations and potential impacts on reproductive competence (Abdelnour et al., 2024). Additionally, nutritional factors significantly affect DNA methylation, with specific feed compositions capable of permanently altering gene expression patterns in offspring through mechanisms like maternal nutrient allocation (Epigenetics in Fish Nutritional Programming, 2023). Furthermore, the quality of sperm DNA methylation is linked to offspring performance, highlighting the importance of understanding these epigenetic changes for aquaculture management and conservation efforts (Zhang et al., 2023). Overall, the interplay between DNA methylation and environmental factors underscores the adaptive capacities of fish populations in fluctuating ecosystems.

2.2 Histone Modifications

Histones are proteins around which DNA is wrapped. Chemical modifications to these proteins can alter the compactness of chromatin, influencing gene accessibility. Studies have shown that histone acetylation and methylation play key roles in the regulation of genes associated with immune responses and development in fish.

Fundamental epigenetic mechanisms in fish, particularly histone modifications, play a crucial role in regulating gene expression in response to environmental changes and nutritional factors. Histone modifications, such as methylation and acetylation, influence chromatin structure and accessibility, thereby affecting transcriptional activity and phenotypic traits in fish species. For instance, studies on the threespine stickleback have demonstrated that specific histone marks correlate with gene expression levels, highlighting their importance in identifying cis-regulatory elements that drive phenotypic variation (Okude et al., 2024). Additionally, nutritional epigenetics reveals that dietary components can induce lasting epigenetic changes, impacting gene expression and metabolic processes across generations (Epigenetics in Fish Nutritional Programming, 2023). Furthermore, comprehensive analyses of histone posttranslational modifications in species like Mozambique tilapia have shown that these modifications are tissue-specific and responsive to environmental stressors, underscoring their significance in fish physiology and adaptation (Mojica & Kultz, 2023). Overall, understanding these mechanisms is vital for advancing aquaculture practices and conserving fish populations in changing environments (Metzger, 2023; Navarro-Martín et al., 2023).

2.3 Noncoding RNA

Noncoding RNAs, such as microRNAs and long noncoding RNAs (lncRNAs), regulate gene expression posttranscriptionally. These molecules are involved in a variety of processes, including stress response and cell cycle control in fish (Lee & Brown, 2021). Fundamental epigenetic mechanisms in fish, particularly involving noncoding RNAs (ncRNAs), play a crucial role in regulating gene expression and adapting to environmental changes. Key mechanisms include DNA methylation, histone modifications, and the action of various ncRNAs, especially microRNAs (miRNAs), which are pivotal in posttranscriptional regulation and have been extensively studied in fish physiology and development (Metzger, 2023; Epigenetic Regulation of Gene Expression by Noncoding RNAs, 2023; Solis et al., 2022). These epigenetic modifications can be influenced by environmental factors, such as nutrition, which can lead to lasting changes in gene expression patterns through processes like nutritional programming (Epigenetics in Fish Nutritional Programming, 2023). Moreover, ncRNAs are integral to various biological processes, including development, reproduction, and immune responses, highlighting their importance in fish adaptation and resilience to climate-related stressors (Huang, 2023; Solis et al., 2022). Understanding these mechanisms not only enhances knowledge of fish biology but also informs conservation and aquaculture practices (Metzger, 2023).

3 Epigenetics and Environmental Adaptation

3.1 Response to Temperature Changes

Fish, like other ectotherms, are highly sensitive to temperature fluctuations. Epigenetic modifications such as DNA methylation have been linked to adaptive changes that help fish survive in varying thermal environments (Martinez et al., 2019). Epigenetics significantly influences how fish adapt to environmental changes, particularly temperature fluctuations. Research indicates that environmental stressors, such as rising temperatures, can induce epigenetic modifications, including DNA methylation and histone alterations, which regulate critical biological processes like reproduction, growth, and stress responses (Abdelnour et al., 2024; Shen et al., 2023; McCaw et al., 2020). For instance, a study on rainbow trout revealed that increased rearing temperatures during spermatogenesis led to differential methylation of numerous gene promoters, suggesting a mechanism for intergenerational inheritance of environmental adaptations (Brionne et al., 2022). Additionally, genome-wide analyses in marine fish like capelin demonstrated that life history variations are closely linked to epigenetic changes, highlighting the role of epigenetics in facilitating rapid adaptation to diverse habitats (Venney et al., 2023). Overall, these findings underscore the importance of epigenetic mechanisms in enabling fish to cope with climate change and environmental stressors.

3.2 Impact of Pollutants and Toxic Substances

Epigenetics significantly influences how fish adapt to environmental pollutants and toxic substances, with various studies highlighting the mechanisms involved. Environmental stressors, such as heavy metals and polycyclic aromatic hydrocarbons (PAHs), can induce epigenetic modifications that affect gene expression and phenotypic plasticity, leading to both immediate and transgenerational adaptations in fish populations (Abdelnour et al., 2024; Pierron et al., 2022). For instance, research on killifish has shown that chronic exposure to PAHs results in an epigenetic memory at the CYP1A gene, which reduces cancer risk but may compromise long-term health (Carrothers et al., 2024). Similarly, zebrafish exposed to cadmium demonstrated transgenerational epigenetic changes that were linked to genetic variations, allowing for rapid adaptation to metal pollution (Pierron et al., 2022). Overall, these findings underscore the critical role of epigenetics in mediating fish responses to pollutants, influencing their health, reproductive success, and survival in increasingly polluted environments (Satkar et al., 2024). Exposure to environmental pollutants can lead to significant epigenetic alterations in fish. For instance, heavy metals and endocrine disruptors can modify DNA methylation patterns and histone marks, resulting in altered gene expression that impacts growth, reproduction, and overall health.

3.3 Transgenerational Effects of Stress

Epigenetic marks can be passed from one generation to the next, enabling fish to "remember" past environmental stressors. This phenomenon can lead to improved resistance in progeny but may also contribute to maladaptive responses under changing conditions. Epigenetics significantly influences environmental adaptation and transgenerational effects of stress in fish, as evidenced by various studies. Environmental stressors, such as climate change, pollution, and hypoxia, can induce epigenetic modifications that affect gene expression and phenotypic plasticity, enabling fish to adapt to changing environments (Abdelnour et al., 2024). For instance, alterations in miRNA expression due to stress can lead to transgenerational epigenetic inheritance, impacting behavior and physiological traits across generations (Hanlin & Wang, 2024). Additionally, specific genes, like CYP1A in pollution-adapted killifish, exhibit epigenetic memory that provides protection against environmental toxins, although this may come with long-term costs, such as increased cancer risk (Carrothers et al., 2024). Overall, these findings underscore the critical role of epigenetics in facilitating rapid adaptation and survival of fish in fluctuating ecosystems (Bogan et al., 2024).

4 Implications for Conservation and Aquaculture

Conservation and aquaculture are two interconnected fields that play a critical role in addressing global challenges related to biodiversity loss, food security, and sustainable resource management. Understanding the ecological and genetic implications of these practices is essential for ensuring their long-term success. Conservation efforts aim to preserve biodiversity and ecosystem stability, while aquaculture seeks to meet the growing demand for seafood through controlled cultivation. However, these objectives often intersect, as aquaculture can impact wild populations and ecosystems, and conservation strategies can influence aquaculture practices. This interplay highlights the need for integrated approaches that balance ecological sustainability with economic and social needs.

4.1 Conservation Genetics

Understanding epigenetic responses in fish is critical for conservation efforts. Epigenetic adaptation can enhance the resilience of fish populations to climate change and habitat loss, suggesting potential strategies for conservation through genetic and epigenetic management. Understanding epigenetic responses in fish is vital for advancing conservation efforts, particularly in the face of climate change and environmental stressors. Epigenetic mechanisms, such as DNA methylation and histone modifications, play a key role in enabling phenotypic plasticity and transgenerational adaptation. These processes help fish adjust to rapidly changing conditions, such as temperature variations, hypoxia, and pollution (Metzger, 2023; Abdelnour et al., 2024). Such modifications influence essential biological functions, including reproduction and stress responses, which are critical for population resilience and maintaining biodiversity (Jeremias et al., 2018). Incorporating epigenetic insights into conservation strategies can enhance the identification of evolutionarily significant units (ESUs) and improve management practices. By identifying biomarkers associated with environmental stressors and physiological states, this approach supports better monitoring of fish health and guides adaptive management strategies. Ultimately, leveraging epigenetic data contributes to safeguarding aquatic ecosystems amid ongoing global changes (Metzger, 2023; Rey et al., 2020).

4.2 Aquaculture and Breeding

The aquaculture industry is now a major contributor to the global supply of food fish, with its growth heavily reliant on technological advancements. To address challenges related to increasing production demands and species diversity, improvements in key areas such as feeds, breeding, sex control, and disease management are essential. Understanding the epigenetic mechanisms linked to traits like enhanced stress and disease tolerance, as well as other economically valuable characteristics, offers significant potential for promoting more sustainable development within the industry. In aquaculture, manipulating epigenetic processes can improve traits such as growth rate, disease resistance, and stress tolerance. This practice, however, raises ethical concerns and necessitates rigorous research to ensure sustainable development. In aquaculture, the ability to regulate biological processes such as sexual differentiation, maturation, and reproduction is highly valuable and increasingly necessary. This capability enhances husbandry management and boosts productivity, ultimately leading to significant economic benefits. Epigenetics has been found to play a role in fish sex determination and differentiation, facilitating interactions between these processes and environmental factors. This may help explain some of the differences observed between sexes that cannot be fully attributed to genetic factors alone (Pfeifer et al., 2013; Budd et al., 2015).

Research has shown that changes in DNA methylation patterns caused by elevated temperatures can lead to the masculinization of genetically female fish, with these alterations potentially being passed down to subsequent generations. Chen et al. (2014) showed that in female half-smooth tongue soles (*Cynoglossus semilaevis*), exposure to elevated temperatures (28 °C) during the critical developmental stage results in 73% undergoing sex reversal to pseudo-males, compared to a 14% spontaneous sex reversal rate under normal conditions (22 °C). This phenomenon is linked to significant methylation of genes within the sex determination network during the reversal process, with these epigenetic changes being heritable by their offspring (Shao et al., 2014). Interestingly, sex-reversed fish can mate with normal females and produce viable offspring, which exhibit a high sex reversal rate (approximately 94%) even when reared under normal temperature conditions. While this phenomenon does not offer direct advantages for aquaculture—since female tongue soles grow faster and achieve larger final body sizes than males—it provides valuable insights into the molecular mechanisms and epigenetic patterns underlying environmentally induced sex reversal in fish.

5 Epigenetic Clocks for Age Estimation in Fishes: Implications for Fisheries Management and Conservation Biology

The distribution of age classes is a vital demographic factor for populations, making accurate age estimation essential for fisheries management and conservation biology. Traditionally, fish age estimation has relied on analyzing growth marks in hard structures like otoliths. However, this approach has limitations: it is lethal, time-intensive, sometimes inaccurate for certain species, and inapplicable to others. Consequently, there is a growing need for alternative methods. One promising avenue is DNA methylation, an epigenetic modification involving the addition of a methyl group to cytosine-guanine loci. Aging is linked to changes in DNA methylation patterns. While there is widespread, subtle genome hypomethylation with age, certain loci exhibit predictable, clock-like methylation changes. These patterns form the basis of chronological age estimators known as "epigenetic clocks." Over the past decade, epigenetic clocks have been developed for various species, including several fish species.

Aging is linked to changes in DNA methylation, which are influenced by intrinsic factors, extrinsic factors, and stochastic events. Generally, DNA methylation changes with age can be characterized by global genomic hypomethylation (Heyn et al., 2012), along with hypermethylation at specific CpG sites. Beyond this "epigenetic drift," some changes exhibit a predictable, clock-like pattern (De Paoli-Iseppi et al., 2019). Chronological age predictors, often referred to as cAge in some studies, are built using DNA methylation data from specific CpG loci and are termed "epigenetic clocks" (Zhang et al., 2019). These clocks incorporate age-related methylation changes (both hyper- and hypomethylation) across carefully chosen genomic loci, yielding an estimated or predicted age and, in some cases, an epigenetic age (eAge). The strong correlation between eAge and cAge underscores the accuracy of epigenetic clocks in predicting chronological age.

Despite their utility, there are still unresolved questions about what epigenetic clocks truly measure (Bell et al., 2019). While some suggest they reflect age-related epigenetic drift, the ability to measure eAge in both proliferating and non-proliferating tissues indicates that passive demethylation likely does not drive these clocks (Simpson & Chandra, 2021). Thus, while the exact nature of eAge remains unclear, it does not hinder the application of epigenetic clocks for age estimation across taxa, particularly in vertebrates. Furthermore, it has been proposed that epigenetic clocks may provide insights into both chronological and biological age (analogous to current measures like creatinine, fasting glucose, or telomere length) in varying proportions depending on the CpG loci used to construct the clock (Bell et al., 2019).

Although DNA methylation changes with age form a robust basis for epigenetic clock development, other epigenetic mechanisms, such as histone modifications and noncoding RNAs (ncRNAs), especially micro-RNAs (miRNAs), could also be considered (van Gelderen et al., 2022). Quantifying miRNAs is relatively straightforward (Lee & Shin, 2012), and fish miRNAs have been observed to vary across developmental stages, such as during flatfish metamorphosis (Campos et al., 2014). However, to date, no clock-like variation in miRNAs has been documented.

5.1 DNA methylation Changes with Age in Fishes

DNA methylation, an epigenetic modification involving the addition of methyl groups to cytosine bases in DNA, shows distinct age-related changes in fish. As fish age, global patterns of DNA methylation often shift, typically characterized by genome-wide hypomethylation alongside site-specific hypermethylation at particular CpG loci. These changes influence gene expression and are associated with key physiological and developmental processes. Age-related methylation changes in fish vary across tissues, genes, and species, reflecting their diverse environmental interactions and lifecycles.

Following the general pattern observed in vertebrates, it was discovered shortly after the publication of Horvath's human clock in 2013 that zebrafish (*Danio rerio*) exhibited age-related hypomethylation of CpGs. Later, DNA methylation changes in Chinook salmon (*Oncorhynchus tshawytscha*) were found to vary based on gene, tissue, and age (Venney et al., 2023). These findings suggested that epigenetic clocks could be successfully developed for fish species.

5.2 Epigenetics Provides Demographic Information for Wild Populations

Accurately determining the age of wild individuals without resorting to lethal sampling is crucial for effective management and represents an exciting avenue for applying epigenetic data. Epigenetic clocks estimate an individual's age by measuring methylation levels across specific loci that correlate with age in a given species (Bell et al., 2019). While only a limited number of studies ($n = 9$) have explicitly focused on assessing epigenetic marker changes over an individual's lifespan, all reported positive and often predictable relationships between age and methylation at certain loci (Ito et al., 2018; Tanabe et al., 2020). Although environmental factors such as UV exposure can influence an individual's epigenome, age-related and environmentally induced methylation changes typically occur at different loci (Grönniger et al., 2010; Polanowski et al., 2014).

A significant challenge is the need for species- or even population-specific epigenetic models to estimate age, as markers identified in one species are generally not transferable to others (Tanabe et al., 2020). However, loci identified in closely related species may serve as a starting point for marker development. This task is further complicated by the rarity of age-related regions, which are estimated to comprise only about 1% of the genome (Grönniger et al., 2010). Despite these challenges, established protocols for identifying relevant epigenetic markers in new species show great promise (Polanowski et al., 2014) and could revolutionize the study and management of wild populations.

5.3 Epigenetics in Fish in the Era of Climate Change

The study of epigenetics in fish has gained significant attention in the era of climate change, as it offers insights into how environmental stressors influence gene expression without altering the underlying DNA sequence. Epigenetic mechanisms, such as DNA methylation, histone modifications, and noncoding RNAs, enable fish to adapt to rapidly changing environments by regulating physiological and developmental processes. As climate change intensifies, altering water temperatures, oxygen levels, and pH, understanding these mechanisms becomes crucial for predicting fish populations' resilience and adaptability. Furthermore, epigenetics provides a valuable framework for addressing conservation challenges and optimizing aquaculture practices to sustain biodiversity and food security in the face of global environmental change.

Epigenetics plays a crucial role in how fish respond to climate change, acting as a bridge between environmental stressors and phenotypic expression. Fish, as a diverse group of vertebrates, face numerous challenges due to climate change, including temperature rise, acidification, and hypoxia, which can trigger epigenetic modifications such as DNA methylation, histone modifications, and changes in noncoding RNAs (Metzger, 2023). These modifications influence various biological processes, including sex determination, reproduction, stress response, and growth, thereby facilitating fish acclimation and adaptation to changing environments (Shen et al., 2023). For instance, in zebrafish, exposure to high temperatures during sex differentiation leads to male-skewed sex ratios, mediated by differentially expressed miRNAs, which serve as potential epigenetic markers of environmental changes (Solomos, 2022).

Moreover, epigenetic changes can be inherited across generations, allowing fish populations to adapt over time to persistent environmental changes (Skinner, 2022) (Abdelnour et al., 2024). This transgenerational inheritance is significant for understanding the long-term impacts of climate change on fish populations and their resilience (Skinner, 2023). However, challenges remain in establishing causality between specific epigenetic changes and phenotypic outcomes, as well as in identifying reliable epigenetic biomarkers for monitoring fish health and resilience (Metzger, 2023). Overall, advancing our understanding of epigenetic mechanisms in fish is essential for developing effective conservation strategies and ensuring the stability of aquatic ecosystems in the face of climate change (Abdelnour et al., 2024).

6 Conclusion and Future Perspectives

Epigenetic patterns and their mechanisms hold significant promise for advancing our understanding of ecological, evolutionary, and conservation biology. As global climate regimes continue to shift, many species will need to adapt rapidly, often relying on epigenetic responses to environmental changes. Gaining a deeper understanding of epigenetic mechanisms, including their evolutionary dynamics and rates, could help identify species at risk and inform potential management and intervention strategies. Investigating epigenetic processes in various species and populations through eco-epigenetic research offers valuable insights into the complex relationships between epigenetic variation, genetic adaptation, and population conservation.

Epigenetics in fish is a rapidly growing field that sheds light on how organisms adapt to environmental challenges. From DNA methylation and histone modification to the role of noncoding RNAs, the epigenetic landscape in fish underpins their adaptability and resilience. Further research is necessary to understand the long-term implications of these mechanisms, particularly in the context of global climate change and pollution. Such knowledge is vital for developing strategies for fish conservation and enhancing aquaculture sustainability. Remarkably, sex-reversed fish can successfully breed with normal females, producing viable offspring that exhibit a high rate of sex reversal (around 94%) even under standard temperature conditions. Although this does not offer immediate benefits for aquaculture—given that female tongue soles grow faster and reach larger sizes than males—it provides significant insights into the molecular mechanisms and epigenetic dynamics driving environmentally induced sex reversal in fish.

Gaining a deeper understanding of the epigenetic mechanisms that regulate key traits in aquatic farming is valuable for several reasons. Insights into stress and disease susceptibility or tolerance in economically important species can provide a significant competitive edge by helping reduce production costs and losses. In terms of sex determination, differentiation, morphological plasticity, and environmental adaptation, having control over these parameters would support stable breeding programs and enhance both production efficiency and product quality. Additionally, the potential for nutritional programming could reduce the industry's reliance on fish-based diets and fisheries. For these reasons, researching epigenetics in aquaculture will play a crucial role in advancing economically sustainable commercial aquaculture. Identifying suitable model species to address specific aquaculture challenges is essential for progress in species improvement. Furthermore, the continued development and enhancement of bioinformatics tools and methods for evaluating the epigenetic status of organisms will help bridge existing knowledge gaps, contributing to an innovative, twenty-first-century aquaculture industry. Also, a significant knowledge gap lies in understanding how environmental factors, particularly temperature and food availability, influence the ticking rate of piscine epigenetic clocks. Variations in these factors could alter the accuracy or consistency of age predictions, potentially complicating their application in dynamic ecosystems. Addressing this gap is essential, as improved age prediction using epigenetic clocks could revolutionize fisheries management by providing more precise data on population demographics. This advancement is especially critical in the context of globally overexploited fish stocks, where sustainable practices are urgently needed, and for conservation efforts aimed at accurately estimating age classes in endangered species, ultimately supporting their recovery and long-term survival.

In the era of climate change, epigenetics offers a transformative lens through which we can better understand and address the challenges facing biodiversity and ecosystem resilience. As rapid environmental shifts impose novel stressors on species, epigenetic mechanisms such as DNA methylation and histone modification provide critical insights into how organisms adapt or fail to adapt to these changes. Importantly, the potential for transgenerational inheritance of epigenetic modifications highlights how populations might respond dynamically to environmental pressures, even within a single generation. The integration of epigenetic data into conservation biology and resource management offers powerful tools for monitoring population health, predicting adaptive potential, and designing effective interventions. However, this emerging field also underscores the complexity of linking epigenetic changes to ecological outcomes, given the interplay of genetic, environmental, and stochastic factors. As climate change continues to reshape ecosystems at an unprecedented pace, leveraging the power of epigenetics will not only deepen our understanding of biological responses but also enhance our ability to implement proactive, science-based strategies for conserving the natural world.

References

Abdelnour, S. A., Naiel, M. A., Said, M. B., Alnajeebi, A. M., Nasr, F. A., Al-Doaiss, A. A., et al. (2024). Environmental epigenetics: Exploring phenotypic plasticity and transgenerational adaptation in fish. *Environmental Research, 252*, 118799.

Bell, C. G., Lowe, R., Adams, P. D., Baccarelli, A. A., Beck, S., Bell, J. T., et al. (2019). DNA methylation aging clocks: Challenges and recommendations. *Genome Biology, 20*, 1–24.

Bogan, S. N., Strader, M. E., & Hofmann, G. E. (2024). Transgenerational effects alter the fitness consequences and genetic architecture of phenotypic plasticity and its regulatory pathways. *bioRxiv*, 2024–10.

Brionne, A., Goupil, A. S., Kica, S., Lareyre, J. J., Labbé, C., & Laurent, A. (2022). Epigenetic landscape of heat stress intergenerational inheritance in a teleost fish. *bioRxiv*, 2022–10.

Brock, H. W., & Fisher, C. L. (2005). Maintenance of gene expression patterns. *Developmental Dynamics, 232*, 633–655.

Budd, A. M., Banh, Q. Q., Domingos, J. A., & Jerry, D. R. (2015). Sex control in fish: Approaches, challenges and opportunities for aquaculture. *Journal of Marine Science and Engineering, 3*(2), 329–355.

Campos, C., Sundaram, A. Y., Valente, L. M., Conceição, L. E., Engrola, S., & Fernandes, J. M. (2014). Thermal plasticity of the miRNA transcriptome during Senegalese sole development. *BMC Genomics, 15*, 525.

Carrothers, S., Trevisan, R., Jayasundara, N., Pelletier, N., Weeks, E., Meyer, J. N., … & Weinhouse, C. (2024). An epigenetic memory at the CYP1A gene in cancer-resistant, pollution-adapted killifish. *bioRxiv*.

Chen, S., Zhang, G., Shao, C., Huang, Q., Liu, G., Zhang, P., et al. (2014). Whole-genome sequence of a flatfish provides insights into ZW sex chromosome evolution and adaptation to a benthic lifestyle. *Nature Genetics, 46*(3), 253–260.

Conaway, J. W. (2012). Introduction to theme "chromatin, epigenetics, and transcription". *Annual Review of Biochemistry, 81*(1), 61–64.

De Paoli-Iseppi, R., Deagle, B. E., Polanowski, A. M., McMahon, C. R., Dickinson, J. L., Hindell, M. A., & Jarman, S. N. (2019). Age estimation in a long-lived seabird (Ardenna tenuirostris) using DNA methylation-based biomarkers. *Molecular Ecology Resources, 19*(2), 411–425.

Deans, C., & Maggert, K. A. (2015). What do you mean,"epigenetic"? *Genetics, 199*(4), 887–896.

Grönniger, E., Weber, B., Heil, O., Peters, N., Stäb, F., Wenck, H., et al. (2010). Aging and chronic sun exposure cause distinct epigenetic changes in human skin. *PLoS Genetics, 6*(5), e1000971.

Hanlin and Wang. (2024). Transgenerational epigenetic inheritance through miR-212/132 expression alterations induced by environmental stress. *Medscien, 1*(9).

Heyn, H., Li, N., Ferreira, H. J., Moran, S., Pisano, D. G., Gomez, A., et al. (2012). Distinct DNA methylomes of newborns and centenarians. *Proceedings of the National Academy of Sciences, 109*(26), 10522–10527.

Ito, H., Udono, T., Hirata, S., & Inoue-Murayama, M. (2018). Estimation of chimpanzee age based on DNA methylation. *Scientific Reports, 8*(1), 9998.

Jaenisch, R., & Bird, A. (2003). Epigenetic regulation of gene expression: How the genome integrates intrinsic and environmental signals. *Nature Genetics, 33*(3), 245–254.

Jeremias, G., Barbosa, J., Marques, S. M., Asselman, J., Gonçalves, F. J., & Pereira, J. L. (2018). Synthesizing the role of epigenetics in the response and adaptation of species to climate change in freshwater ecosystems. *Molecular Ecology, 27*(13), 2790–2806.

Kim, M. S., Kim, D. H., & Lee, J. S. (2024). A review of environmental epigenetics in aquatic invertebrates. *Marine Pollution Bulletin, 208*, 117011.

Lee, D., & Shin, C. (2012). MicroRNA–target interactions: New insights from genome-wide approaches. *Annals of the New York Academy of Sciences, 1271*(1), 118–128.

Liu, Z., Zhou, T., & Gao, D. (2022). Genetic and epigenetic regulation of growth, reproduction, disease resistance and stress responses in aquaculture. *Frontiers in Genetics, 13*, 994471.

McCaw, B. A., Stevenson, T. J., & Lancaster, L. T. (2020). Epigenetic responses to temperature and climate. *Integrative and Comparative Biology, 60*(6), 1469–1480.

Metzger, D. C. (2023). Epigenetic mechanisms in response to environmental change.

Mojica, E. A., & Kultz, D. (2023). A strategy to characterize the global landscape of histone post-translational modifications within tissues of nonmodel organisms. *Journal of Proteome Research.*

Navarro-Martín, L., Mennigen, J. A., & Asselman, J. (2023). Transcriptional epigenetic mechanisms in aquatic species. *Epigenetics in Aquaculture*, 45–64.

Okude, G., Yamasaki, Y. Y., Toyoda, A., Mori, S., & Kitano, J. (2024). Genome-wide analysis of histone modifications can contribute to the identification of candidate cis-regulatory regions in the threespine stickleback fish. *BMC Genomics, 25*(1), 685.

Pfeifer, G. P., Kadam, S., & Jin, S. G. (2013). 5-hydroxymethylcytosine and its potential roles in development and cancer. *Epigenetics & Chromatin, 6*, 1–9.

Pierron, F., Heroin, D., Daffe, G., Daramy, F., Barré, A., Bouchez, O., et al. (2022). Genetic and epigenetic interplay allows rapid transgenerational adaptation to metal pollution in zebrafish. *Environmental. Epigenetics, 8*(1), dvac022.

Polanowski, A. M., Robbins, J., Chandler, D., & Jarman, S. N. (2014). Epigenetic estimation of age in humpback whales. *Molecular Ecology Resources, 14*(5), 976–987.

Rey, O., Eizaguirre, C., Angers, B., Baltazar-Soares, M., Sagonas, K., Prunier, J. G., & Blanchet, S. (2020). Linking epigenetics and biological conservation: Towards a conservation epigenetics perspective. *Functional Ecology, 34*(2), 414–427.

Satkar, S. G., Kumar, A., Bhusare, S., Sahu, A., & Gautam, R. K. (2024). Impact of pollution and toxic stress on fish health: Mechanisms, consequences, and mitigation strategies. *Uttar Pradesh Journal of Zoology, 45*(6), 29–45.

Shao, C., Li, Q., Chen, S., Zhang, P., Lian, J., Hu, Q., et al. (2014). Epigenetic modification and inheritance in sexual reversal of fish. *Genome Research, 24*(4), 604–615.

Shen, Z. G., Yu, Y., & Wang, H. P. (2023). Environmental epigenetics in fish: Response to climate change stressors. In *Epigenetics in aquaculture* (pp. 127–147).

Siggens, L., & Ekwall, K. (2014). Epigenetics, chromatin and genome organization: Recent advances from the ENCODE project. *Journal of Internal Medicine, 276*(3), 201–214.

Simpson, D. J., & Chandra, T. (2021). Epigenetic age prediction. *Aging Cell, 20*(9), e13452.

Skinner, M. K. (2023). Environmental epigenetics and climate change. *Environmental Epigenetics, 9*(1), dvac028.

Smith, R. G., Hannon, E., De Jager, P. L., Chibnik, L., Lott, S. J., Condliffe, D., et al. (2018). Elevated DNA methylation across a 48-kb region spanning the HOXA gene cluster is associated with Alzheimer's disease neuropathology. *Alzheimer's & Dementia, 14*(12), 1580–1588.

Solis, F. J. T., Fernandes, J. M., Sarropoulou, E., & Monzon, I. F. (2022). Noncoding RNAs in fish physiology and development: miRNAs as a cornerstone in gene networks. In *Cellular and molecular approaches in fish biology* (pp. 105–159). Academic Press.

Tanabe, A., Shimizu, R., Osawa, Y., Suzuki, M., Ito, S., Goto, M., et al. (2020). Age estimation by DNA methylation in the Antarctic minke whale. *Fisheries Science, 86*, 35–41.

van Gelderen, T. A., Montfort, J., Álvarez-Dios, J. A., Thermes, V., Piferrer, F., Bobe, J., & Ribas, L. (2022). Deciphering sex-specific miRNAs as heat-recorders in zebrafish. *Scientific Reports, 12*(1), 18722.

Venney, C. J., Cayuela, H., Rougeux, C., Laporte, M., Mérot, C., Normandeau, E., et al. (2023). Genome-wide DNA methylation predicts environmentally driven life history variation in a marine fish. *Evolution, 77*(1), 186–198.

Zhang, Q., Vallerga, C. L., Walker, R. M., Lin, T., Henders, A. K., Montgomery, G. W., et al. (2019). Improved precision of epigenetic clock estimates across tissues and its implication for biological ageing. *Genome Medicine, 11*, 1–11.

Zhang, S., Cheng, Y., Vechtova, P., Boryshpolets, S., Shazada, N. E., Alavi, S. M. H., et al. (2023). Potential implications of sperm DNA methylation functional properties in aquaculture management. *Reviews in Aquaculture, 15*(2), 536–556.

Chapter 18
Aquaculture for Improving Productivity, Income, and Nutrition

Ishrat Mohd and Syed Talia Mushtaq

1 Overview of Aquaculture

The world's population is projected to exceed 9 billion by the year 2050, leading to a 70% surge in global demand for food, feed, and fiber. This growth will be accompanied by significant shifts in lifestyle and consumption patterns, primarily due to increasing urbanization. Aquaculture, with a long history, is a significant contributor of high-quality proteins to humanity (Nash, 2010; Gui et al., 2018). As the global population expands, the challenges of resource scarcity and environmental degradation are becoming increasingly pronounced. Land resources are needed to cater to the demands of societal progress. Consequently, the oceans are emerging as a new frontier for human survival and development, with fish evolving into a vital source of high-quality protein and a fundamental food staple for humanity (Baiden & Bissiri, 2007). Over the past few decades, aquaculture has emerged as the fastest-growing sector within agriculture. Since 2013, its production has surpassed wild fisheries (FAO, 2020). This formidable challenge presents alarming situations, particularly in developing countries where poverty and hunger persist (World Bank, 2014). In such a scenario, fisheries and aquaculture are expected to play a vital role in ensuring nutritional security (Subasinghe, 2014). Fish is a significant component of the human diet, providing roughly 20% of animal protein globally (Thorpe et al., 2006). On one hand, it serves as a healthy food option for the more affluent world due to its high polyunsaturated fatty acid (PUFA) content. On the other hand, small indigenous fishes contribute to the nutrition of populations on the lower end of the socioeconomic scale, offering proteins, oils, vitamins, minerals, and other benefits (Mohanty et al., 2010). Fisheries and aquaculture are crucial in enhancing food security and livelihoods worldwide. The aquaculture sector has gained remarkable momentum, registering an impressive annual growth rate of 7%, while fish products

I. Mohd (✉) · S. T. Mushtaq
Division of Fisheries Resource Management, FoFy, SKUAST-Kashmir, Kashmir, India

I. Ahmed, I. Ahmad (eds.), *Aquaculture: Enhancing Food Security and Nutrition*, https://doi.org/10.1007/978-3-031-92858-1_18

account for more than 37% of internationally traded food production by volume. The fisheries and aquaculture industry has emerged as one of the fastest-growing sectors globally (Tacon, 2023). It has been pivotal in advancing economic development, contributing significantly to food and nutritional security, bolstering national income, creating employment opportunities, and generating livelihoods (Kumar & Shivani, 2014). This industry serves as the primary source of animal protein for billions of people worldwide, with both capture fishery and aquaculture playing a vital role in supporting the livelihoods of over 10% of the global population (Anonymous, 2020). The 2022 State of Fisheries and Aquaculture (SOFIA) report showcased a remarkable pinnacle in global fisheries and aquaculture production, hitting 214 million tons in 2020. This surge was primarily fueled by significant growth in Asian aquaculture, encompassing 178 million tons of aquatic animals and 36 million tons of algae. Aquatic animal production notably exceeded averages from previous decades, skyrocketing by 30%, primarily attributed to a substantial increase in aquaculture output, soaring to 87.5 million tons. However, this rapid expansion in the sector necessitates pivotal, transformative changes for a more sustainable, inclusive, and equitable industry. The focus is on ensuring sustainable harvesting, safeguarding livelihoods, and preserving aquatic habitats and biodiversity. A significant portion of the 157 million tons of aquatic animal production was directly consumed by humans, substantially contributing to global animal protein consumption. In 2020, Asian nations took the lead in global fisheries and aquaculture production, accounting for a staggering 70% of aquatic animal production, led primarily by China. The growth trajectory of aquaculture continues to shape the future of aquatic foods, outstripping the growth of capture fisheries. Despite declines in capture fisheries due to the COVID-19 pandemic and natural fluctuations, aquaculture production surged and is forecasted to reach 106 million tons by 2030. This increasing demand is propelled by rising incomes, evolving post-harvest practices, urbanization, and dietary shifts focused on enhanced health and nutrition. These trends signify a substantial transformation in the landscape of global food production, with aquaculture taking center stage in meeting evolving dietary needs and demands. Aquaculture plays a vital role in enhancing productivity, generating income, and improving nutrition by providing a sustainable source of high-quality protein and supporting livelihoods in rural and urban areas. It promotes food security, economic growth, and environmental sustainability through efficient resource use and innovative farming practices (Fig. 18.1).

2 The Role of Aquaculture in Enhancing Productivity

Historically, aquaculture was geographically diverse and evolved in isolation, with distinct practices developing in different regions. The advent of advanced communication and transportation technologies has, however, facilitated the standardization of aquaculture practices, enabling species to be cultured in locations best suited to their biological and environmental requirements. The origins of aquaculture can

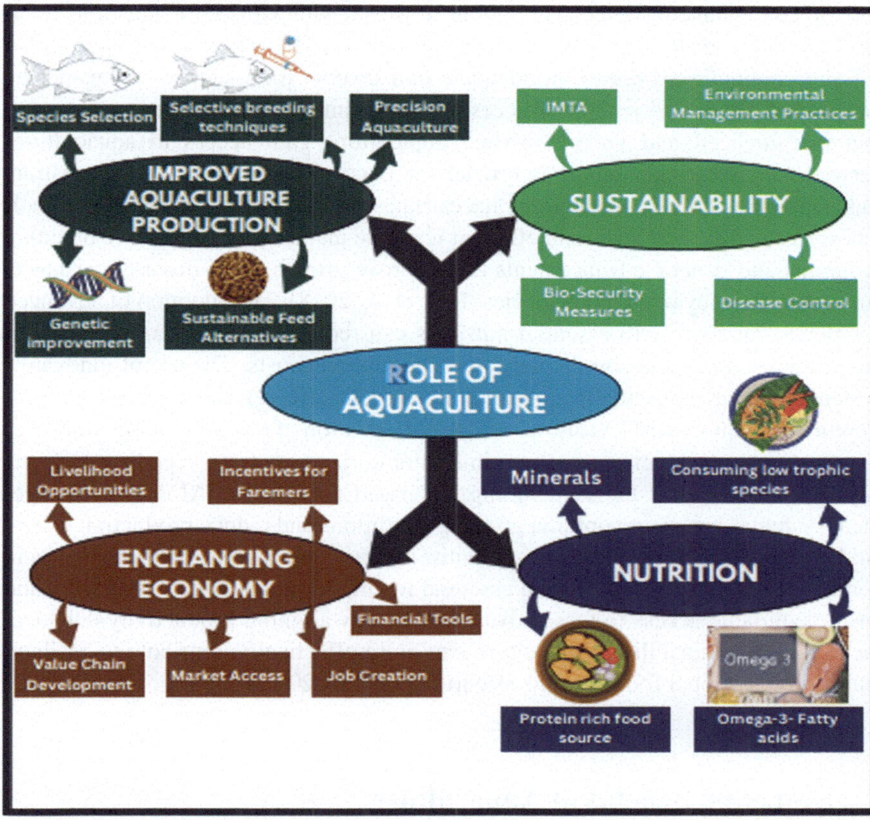

Fig. 18.1 Role of aquaculture in (1) improving aquaculture productivity, (2) enhancing economy, (3) providing nutrition, and (4) achieving sustainability

be traced back thousands of years. Evidence suggests that the ancient Egyptians were among the first to engage in fish farming as early as 2500 BC, as depicted in tomb engravings showing tilapia being harvested from artificial ponds. Similarly, the Romans practiced aquaculture in circular ponds with designated breeding areas. In China, the culture of Chinese carp was widespread by 2000 BC. In India, ancient texts from 300 BC refer to the use of fish as a strategic resource during warfare, implying the existence of fish culture in Indian reservoirs. By 1127 AD, historical records describe methods for fattening fish in ponds, and the cultivation of Gangetic carps in Bengal has been deeply rooted in tradition for centuries (Rogers, 2024). Aquaculture plays a pivotal role in addressing global challenges related to nutrition and food security. It is a vital source of high-quality protein, essential amino acids, omega-3 fatty acids, vitamins, and minerals, making it a cornerstone of a balanced diet. Moreover, aquaculture has significant potential to alleviate malnutrition and address micronutrient deficiencies, particularly in developing regions. The sector contributes to food security and improved nutrition through multiple channels: increasing the availability of affordable fish, generating employment and income, and directly enhancing fish consumption. However, achieving these nutritional benefits requires more than merely increasing production. Households must have physical and economic access to an adequate and diverse food supply, alongside the knowledge, time, and resources necessary to optimize food utilization. Ensuring that marginalized and undernourished populations gain access to aquaculture-derived resources and income is crucial for maximizing its impact on nutrition. Enhancing aquaculture productivity necessitates the integration of modern technologies, sustainable practices, and efficient resource management. Selective breeding programs and genetic advancements can improve growth rates, disease resistance, and feed efficiency in cultured species (Føre et al., 2018). The adoption of advanced aquafeeds enriched with essential nutrients can reduce dependence on wild fish-based feed sources and significantly lower production costs. The use of innovative systems such as recirculating aquaculture systems (RAS) and integrated multi-trophic aquaculture (IMTA) allows for the optimization of resources while minimizing environmental impacts. Real-time monitoring of water quality and the deployment of digital tools, including IoT-based sensors and AI-driven management systems, can ensure optimal growing conditions and reduce production losses. Additionally, capacity-building initiatives, farmer training programs, and community-driven approaches are essential to empower small-scale producers and ensure equitable access to these advancements. By aligning productivity enhancements with sustainability, aquaculture can more effectively contribute to resilient and equitable global food systems (Henriksson et al., 2021).

3 Economic Benefits of Aquaculture

Aquaculture is a significant economic driver, particularly in rural and coastal regions. The industry creates employment opportunities, fosters economic diversification, and plays a crucial role in poverty alleviation. Activities associated with aquaculture, such as fish breeding, processing, and distribution, provide livelihoods to millions of people globally. Farmed aquatic organisms are widely consumed worldwide as part of a nutritious diet and for their health benefits. Beyond these dietary advantages, aquaculture contributes substantially to economic growth. According to the Food and Agriculture Organization (FAO), this sector supports the creation of over 20 million jobs globally (Shang, 1985). Aquatic food is a vital component of global food security, offering a significant source of high-quality protein and healthy fats to people worldwide. Compared to land-based animal protein, seafood requires fewer resources to produce. As the climate continues to change and the global demand for protein grows, while there is an urgent need to reduce environmental impacts, wild-caught and farmed aquatic organisms will play a crucial role in meeting these challenges (Jones et al., 2022). Aquaculture offers many exciting career opportunities for individuals with diverse backgrounds and skill sets. In particular, it has proven to be a valuable tool for revitalizing working waterfronts and providing stable, well-paying jobs in rural communities. For young people facing an uncertain future due to climate change, aquaculture presents an opportunity to pursue "green careers" that contribute directly to developing sustainable solutions. The industry requires expertise in various fields, including seamanship, marine trades, engineering, product development, marketing, research, and more, making it a dynamic and multidisciplinary career path. While sustainable wild fisheries exist worldwide, sourcing fish from aquaculture operations significantly reduces pressure on wild fish stocks. This approach is fundamental in regions like Maine, where native wild Atlantic salmon are endangered. Producing fish in controlled aquaculture environments helps preserve biodiversity and ensures that fish are raised in conditions that prioritize their health, safety, and suitability for consumption.

Significant advancements in fish feed technology have further transformed the industry, with most feed ingredients now derived from plant-based sources rather than wild fish meal. These innovations have revolutionized aquaculture, substantially reducing its global ecological footprint. Additionally, species such as bivalves and seaweeds, which do not require feed, provide added environmental benefits by removing excess nutrients from the water, thereby improving water quality and promoting ecosystem health (Overton et al., 2024).

Aquaculture operations are vital in supporting the state's seafood economy by providing fresh, high-quality seafood to both local and domestic markets. With the United States currently importing over 85% of the seafood it consumes, increasing domestic aquaculture production helps reduce the trade deficit while lowering greenhouse gas emissions associated with seafood imports. The economic benefits of aquaculture are extensive. It generates jobs, ensures a stable and reliable food supply, and supports preserving wildlife and working waterfronts. Additionally, fish is a nutritious, climate-friendly source of protein that can contribute to healthier diets for people worldwide, addressing both nutritional and environmental challenges (Khan et al., 2011).

4 Nutritional Value of Aquatic Foods

As the global population grows, so does the demand for nutritious, climate-friendly food. Meeting this increasing demand sustainably is a significant challenge, but potential solutions lie in aquatic resources. Blue foods—from oceans, seas, rivers, and lakes—are among the most highly traded food products globally. These foods provide livelihoods for millions of people and nourish billions daily (Tigchelaar et al., 2022). With demand for blue foods expected to double by 2050, aquaculture will play an indispensable role in meeting this need. According to the UN Food and Agriculture Organization (FAO), aquaculture is projected to account for 106 million metric tons of global fish production by 2030, marking a 32% increase from 2020. Aquatic foods are an essential component of a healthy, sustainable diet. They are rich in high-quality protein, omega-3 fatty acids, vitamins, and minerals such as iodine, selenium, calcium, iron, and zinc—nutrients that are often scarce in other food sources. These foods are vital for improving public health outcomes, especially in developing regions with prevalent malnutrition and micronutrient deficiencies. For instance, iodine and long-chain omega-3 fatty acids are critical for a child's brain development, while omega-3s also protect against coronary heart disease. Recognized as one of the healthiest food options, aquatic foods are low in saturated fats and linked to a reduced risk of noncommunicable diseases. Aquaculture contributes significantly to global food security, as evidenced by the FAO, (2024). In 2022, global fish and seafood production reached 223 million tons, with aquaculture contributing 131 million tons for human consumption. Aquaculture now accounts for over 50% of the global per capita fish and seafood consumption, which stood at 20.5 kg in 2019, of which aquaculture contributed 11.2 kg. Consuming whole fish, rather than just fillets, provides critical vitamins and minerals, making it an affordable and nutritious option for low-income populations, including nutritionally vulnerable groups. Studies highlight the exceptional nutritional value and environmental benefits of aquatic foods. Research has shown that their production generates lower greenhouse gas emissions than terrestrial meat production, such as beef, pork, and chicken. Small pelagic fish, including anchovies and sardines, are nutritionally rich and have a lower climate impact. Non-fed farmed species, such as bivalves and seaweeds, have an even smaller carbon footprint and can positively affect the environment by improving water quality and removing excess nutrients. These qualities make aquaculture a sustainable and environmentally friendly alternative for meeting global nutritional needs while minimizing ecological impacts. By advancing sustainable practices, aquaculture can continue to serve as a key driver of global food security, public health, and environmental resilience (Crumlish & Norman, 2024).

5 Improving Aquaculture Productivity

To propel the Blue Revolution (Neel Kranti Mission) and modernize the fisheries industry into a world-class sector, it is crucial to integrate cutting-edge technologies, including the use of suitable species, advanced breeding techniques, blockchain, artificial intelligence (AI), the Internet of Things (IoT), and mobile applications. These technological advancements can potentially transform aquaculture operations by improving efficiency, transparency, and productivity. Among these, mobile apps and IoT technologies are rapidly expanding, offering new ways to monitor and manage aquaculture systems in real-time (Garlock et al., 2020). Advanced biotechnological tools, such as protein expression systems, microsatellites, gene mapping, genomics, DNA vaccines, proteomics, and transgenic technologies, present promising opportunities to enhance aquaculture practices. These technologies enable genetic manipulation, embryo manipulation, and the development of genetically engineered diagnostics and immunoprophylactic agents. Such tools are particularly valuable for selective breeding, hybridization, and improving cultured species' health, growth, nutrition, and overall productivity. The ability to manipulate genetics can also aid in cryopreservation and the conservation of genetic resources, which are essential for maintaining genetic diversity and ensuring the long-term sustainability of aquaculture. The application of genetic modification and biotechnology holds significant promise for improving the quality and quantity of fish in aquaculture. Enhancing selective breeding programs allows aquaculture to produce faster-growing, disease-resistant species with improved feed conversion rates. This can lead to higher productivity, lower operational costs, and more sustainable practices, as fewer resources are required to produce more fish (Vishwanatham et al., 2024). Moreover, biotechnology can be leveraged to develop fish strains with higher nutritional content, such as increased omega-3 fatty acids, which would further contribute to global food security by providing healthier food options. In addition to genetic improvements, integrating digital technologies like AI and IoT can greatly enhance aquaculture productivity. AI can optimize fish farming operations by predicting growth rates, disease outbreaks, and feeding requirements. At the same time, IoT sensors can monitor water quality, oxygen levels, and other critical environmental factors in real-time. These data-driven insights allow for more precise management of aquaculture systems, improving resource utilization, reducing waste, and increasing efficiency (Tamim et al., 2022). Blockchain technology can also ensure traceability and transparency in the supply chain, providing consumers with reliable information about the sustainability and safety of the seafood they consume. Combining biotechnology with digital innovations can ultimately unlock significant improvements in aquaculture productivity. By enhancing breeding programs, optimizing farming practices, and streamlining supply chains, these technologies can contribute to a more sustainable, efficient, and resilient aquaculture industry, ensuring it can meet the growing global demand for high-quality, nutritious seafood.

6 Best Practices for Sustainable Aquaculture

To address the key challenges faced by the aquaculture sector, its future will rely heavily on the ability to adapt to global pressures such as environmental degradation, water scarcity, population growth, food insecurity, and global hunger. Sustainable aquaculture has emerged as a promising solution to these challenges, offering a path to address current and future needs. The role of fisheries and aquaculture in ensuring food security and providing livelihoods for millions is undeniable. Aquaculture serves as a critical food source and an economic lifeline for many communities (Subasinghe et al., 2009). Scientists widely recognize aquaculture as an efficient producer of high-value proteins essential for human health. It requires fewer resources, such as space, food, energy, and water, than traditional agriculture. Moreover, modern sustainable aquaculture practices generate less waste and produce smaller carbon and nitrogen footprints than land-based food production. Therefore, promoting sustainable development within aquaculture is essential to overcoming the challenges it faces today and securing a stable, nutritious food supply for the future. Integrated Multi-Trophic Aquaculture (IMTA) is a sustainable approach that combines the farming of species from different trophic levels in a complementary system. For example, fish or shrimp are cultured alongside shellfish and seaweed, which utilize the nutrients and waste products generated by the primary species (Chopin et al., 2001). IMTA reduces environmental impact, enhances resource efficiency, and provides diversified income streams for farmers. The integration of species ensures ecological balance and improved water quality (Khanjani et al., 2022). Effective environmental management is essential to sustain aquaculture systems and preserve ecosystems. Regular assessment of parameters like pH, dissolved oxygen, ammonia, and temperature to ensure optimal conditions for cultured species. Effluent treatment systems prevent nutrient overload and contamination of surrounding water bodies (Novotny et al., 2011). Protecting critical habitats like mangroves, wetlands, and coastal areas serves as a natural buffer against environmental degradation. Using high-quality, sustainable feeds with optimal feed conversion ratios (FCR) further minimizes waste and nutrient pollution. Disease outbreaks pose significant risks to aquaculture productivity and sustainability. Comprehensive health management and biosecurity measures are vital to mitigate these risks. Regular health check-ups, vaccination programs, and the use of probiotics and immunostimulants strengthen the resilience of aquatic stocks (Murray, 2006). Quarantining new stock prevents the introduction of pathogens into existing systems, while maintaining strict hygiene standards for equipment, water sources, and handling procedures ensures a safer environment. Diagnostic tools and surveillance systems enable early identification and management of diseases (Hine et al., 2010). Additionally, employing biological controls and environmentally safe treatments effectively manages parasites and pathogens, contributing to long-term aquaculture viability and sustainability (Torrissen et al., 2013).

7 Economic Strategies for Enhancing Income Through Aquaculture

Access to markets and effective value chain development are critical for enhancing income in aquaculture. Developing strong market linkages and improving the distribution networks enable farmers to sell their products at fair prices, reducing market dependency and ensuring higher returns. Value chain development involves not only enhancing production capabilities but also improving post-harvest handling, processing, and packaging, which adds value to the product and opens up premium markets. Ensuring quality control and certifications (e.g., organic or sustainable labels) can further improve market access, especially in global markets where consumer preferences are shifting toward sustainable and traceable products (Trace, 2024). Moreover, supporting aquaculture cooperatives can help small-scale farmers share resources, reduce costs, and increase bargaining power in both input procurement and product marketing. Financial tools and incentives are vital for improving the financial viability of aquaculture operations. Access to affordable credit and insurance products can help aquaculture farmers invest in better technologies, infrastructure, and stock, which are essential for improving productivity and sustainability. Microfinancing models and targeted agricultural loans, along with low-interest rates, can increase financial inclusion for small-scale farmers. Additionally, subsidies or tax incentives for adopting sustainable practices such as the use of environmentally friendly feeds, energy-efficient systems, or waste management technologies can help reduce operational costs and increase profitability. Other financial incentives, such as grants for research and development, can support innovation in breeding techniques, feed development, and disease management .A supportive policy framework is crucial for the growth and sustainability of the aquaculture sector. Governments can play a significant role by developing and enforcing regulations that ensure food safety, environmental sustainability, and fair market practices. Policies that promote land and water rights, protect aquaculture from external threats (e.g., diseases and climate change), and encourage investment in infrastructure such as hatcheries, feed mills, and processing facilities are essential for long-term sector development. Furthermore, the establishment of clear and supportive land-use policies can reduce conflicts between aquaculture and other sectors, like agriculture or tourism, ensuring a balanced approach to resource use. Governments can also provide a framework for research and development to drive innovation, such as developing new breeds with higher growth rates or resilience to diseases. Public-private partnerships (PPPs) can facilitate knowledge transfer, technology adoption, and collaborative investments, ensuring that policies remain adaptable and relevant to changing industry needs (Investopedia, 2024).

8 Challenges Facing the Aquaculture Sector

Aquaculture has been associated with significant environmental challenges, particularly when improperly managed. Issues such as habitat destruction, nutrient pollution, and the spread of diseases to wild fish populations are among the key concerns (Phillips, 1990). For example, intensive aquaculture operations can lead to eutrophication in nearby water bodies due to excess feed and waste discharge, disrupting aquatic ecosystems and threatening biodiversity (Cavalli et al., 2019; Ngajilo & Jeebhay, 2019). The escape of farmed species, particularly genetically modified or non-native species, poses additional risks, as it can lead to competition with or genetic dilution of wild fish stocks (Naylor et al., 2005). Moreover, the reliance on fishmeal and fish oil in aquafeeds raises concerns about the sustainability of global fisheries used to supply these resources (Ghamkhar et al., 2021). Transitioning to alternative feed ingredients, such as plant-based proteins or insect meal, and adopting integrated multi-trophic aquaculture (IMTA) systems are considered critical steps to address these sustainability challenges (Periera et al., 2024). The aquaculture industry faces numerous regulatory and compliance hurdles that vary across regions. These include obtaining permits, navigating zoning restrictions, and adhering to environmental protection standards. Inconsistent or unclear regulations can create barriers for new entrants, especially small-scale farmers who lack the resources to comply with complex administrative requirements (Abate et al., 2016). Additionally, weak enforcement of existing regulations can lead to unsustainable practices, undermining the credibility of the industry and potentially causing long-term environmental and social harm. Harmonizing regulatory frameworks and promoting stakeholder engagement in policy formulation are vital to fostering an environment conducive to sustainable aquaculture growth. Climate change poses a growing threat to aquaculture by affecting the physical, chemical, and biological properties of aquatic ecosystems. Rising temperatures can alter the metabolic rates, growth, and reproduction of cultured species, potentially leading to reduced yields. Ocean acidification, caused by increased CO_2 absorption, affects shellfish calcification, making it harder for species like oysters and clams to form shells (Bressan et al., 2014). Furthermore, extreme weather events, such as storms and flooding, can damage aquaculture infrastructure and disrupt operations. To mitigate these impacts, aquaculture systems must adopt adaptive strategies, such as selecting heat-tolerant species, implementing resilient infrastructure designs, and integrating climate-smart practices into farming operations.

9 Future Directions for Aquaculture

The future of aquaculture lies in innovative approaches that ensure sustainability, resilience, and productivity. Genomics and biotechnology are at the forefront of these innovations, offering tools to enhance fish breeding, disease resistance, and growth performance. Advancements such as genome editing technologies (e.g., CRISPR-Cas9) and selective breeding programs allow for the development of robust species adapted to changing environments and disease pressures. Additionally, biotechnology innovations, including probiotics, immunostimulants, and biofloc systems, contribute to improved health management and feed efficiency, reducing environmental footprints. Alongside these technological advancements, the role of policy is critical in shaping sustainable aquaculture practices. Policies that promote responsible resource management, environmental safeguards, and incentives for adopting eco-friendly technologies ensure long-term sector growth while minimizing ecological impacts. Furthermore, community engagement and capacity-building initiatives are essential for inclusive development, particularly in rural and small-scale aquaculture sectors. Providing training, knowledge-sharing platforms, and access to technology empowers local communities to adopt best practices, improve livelihoods, and contribute to food security. A combination of innovation, supportive policies, and community-driven approaches will be pivotal in driving aquaculture toward a sustainable and resilient future.

10 Conclusion

Aquaculture has emerged as a vital solution to meet the growing global demand for seafood while supporting livelihoods and food security. However, the sector faces significant challenges, including environmental impacts, regulatory complexities, and the adverse effects of climate change on aquatic ecosystems. To address these concerns, strategies such as adopting sustainable practices like Integrated Multi-Trophic Aquaculture (IMTA), improving water quality management, and ensuring regulatory compliance are essential. Moreover, economic growth in aquaculture can be enhanced through market access development, financial tools, and supportive policies that encourage innovation and inclusivity. Advancements in genomics and biotechnology offer transformative opportunities to improve fish breeding, disease resistance, and resource efficiency. These innovations must be supported by robust policy frameworks that prioritize sustainability, alongside community engagement and capacity-building initiatives that empower small-scale farmers and stakeholders. By integrating technological innovation, effective policy measures, and community-driven approaches, aquaculture can achieve sustainable growth, ensuring ecological balance, economic viability, and resilience to future challenges.

References

Abate, T. G., Nielsen, R., & Tveterås, R. (2016). Stringency of environmental regulation and aquaculture growth: A cross-country analysis. *Aquaculture Economics & Management, 20*(2), 201–221.

Anonymous. (2020). The state of world fisheries and aquaculture 2020. In *Sustainability in action* (Vol. 244). FAO. https://doi.org/10.4060/ca9229en

Baiden, G., & Bissiri, Y. (2007). High bandwidth optical networking for underwater untethered telerobotic operation. In *OCEANS 2007* (pp. 1–9). IEEE.

Bressan, A., Čanić, S., Garavello, M., Herty, M., & Piccoli, B. (2014). Flows on networks: Recent results and perspectives. *EMS Surveys in Mathematical Sciences, 1*(1), 47–111.

Cavalli, L., Jeebhay, M. F., Marques, F., Mitchell, R., Neis, B., Ngajilo, D., & Watterson, A. (2019). Scoping global aquaculture occupational safety and health. *Journal of Agromedicine, 24*(4), 391–404.

Chopin, T., Buschmann, A. H., Halling, C., Troell, M., Kautsky, N., Neori, A., et al. (2001). Integrating seaweeds into marine aquaculture systems: A key toward sustainability. *Journal of Phycology, 37*(6), 975–986.

Crumlish, M., & Norman, R. (Eds.). (2024). *Aquatic food security*. 5m Books Ltd.

FAO. (2020). *The state of world fisheries and aquaculture 2020*. State of World Fisheries and Aquaculture.

FAO. (2024). *The state of world fisheries and aquaculture 2024—Blue transformation in action*. Rome. https://www.fao.org/publications/home/fao-flagship-publications/the-state-of-world-fisheries-and-aquaculture/en

Føre, M., Frank, K., Norton, T., Svendsen, E., Alfredsen, J. A., Dempster, T., et al. (2018). Precision fish farming: A new framework to improve production in aquaculture. *Biosystems Engineering, 173*, 176–193.

Garlock, T., Asche, F., Anderson, J., Bjørndal, T., Kumar, G., Lorenzen, K., et al. (2020). A global blue revolution: Aquaculture growth across regions, species, and countries. *Reviews in Fisheries Science & Aquaculture, 28*(1), 107–116.

Ghamkhar, R., Boxman, S. E., Main, K. L., Zhang, Q., Trotz, M. A., & Hicks, A. (2021). Life cycle assessment of aquaculture systems: Does burden shifting occur with an increase in production intensity? *Aquacultural Engineering, 92*, 102130.

Gui, J. F., Tang, Q., Li, Z., Liu, J., & De Silva, S. S. (2018). *Aquaculture in China: Success stories and modern trends*. Wiley.

Henriksson, P. J. G., Troell, M., Banks, L. K., Belton, B., Beveridge, M. C. M., Klinger, D. H., et al. (2021). Interventions for improving the productivity and environmental performance of global aquaculture for future food security. *One Earth, 4*(9), 1220–1232.

Hine, M., Adams, S., Arthur, J. R., Bartley, D., Bondad-Reantaso, M. G., Chávez, C., et al. (2010, September). Improving biosecurity: A necessity for aquaculture sustainability. In *Farming the waters for people and food. Proceedings of the global conference on aquaculture* (pp. 22–25).

Investopedia. (2024). https://www.investopedia.com/

Jones, A. R., Alleway, H. K., McAfee, D., Reis-Santos, P., Theuerkauf, S. J., & Jones, R. C. (2022). Climate-friendly seafood: The potential for emissions reduction and carbon capture in marine aquaculture. *Bio Science, 72*(2), 123–143.

Khan, M. A., Khan, S., & Miyan, K. (2011). Aquaculture as a food production system: A review. *Biology and medicine, 3*(2), 291–302.

Khanjani, M. H., Zahedi, S., & Mohammadi, A. (2022). Integrated multitrophic aquaculture (IMTA) as an environmentally friendly system for sustainable aquaculture: Functionality, species, and application of biofloc technology (BFT). *Environmental Science and Pollution Research, 29*(45), 67513–67531.

Kumar, S. T., & Shivani, P. (2014). Marine fisheries; its current status, sustainable management and socio-economic status of the marine fishers of Odisha, through Indian marine policy: A case study. *Research Journal of Animal, Veterinary and Fishery Sciences, 2*(7), 10–19.

Mohanty, B. P., Behera, B. K., & Sharma, A.P. (2010). *Nutritional significance of small indigenous fishes in human health.*

Murray, A. G. (2006). A model of the emergence of infectious pancreatic necrosis virus in Scottish salmon farms 1996–2003. *Ecological Modelling, 199*(1), 64–72.

Nash, C. (2010). *The history of aquaculture*. Wiley.

Naylor, R., Hindar, K., Fleming, I. A., Goldburg, R., Williams, S., Volpe, J., & Mangel, M. (2005). Fugitive salmon: Assessing the risks of escaped fish from net-pen aquaculture. *Bioscience, 55*(5), 427–437.

Ngajilo, D., & Jeebhay, M. F. (2019). Occupational injuries and diseases in aquaculture–a review of literature. *Aquaculture, 507*, 40–55.

Novotny, E. V., Bechle, M., Millet, D. B., & Marshall, J. D. (2011). National satellite-based land use regression: NO2 in the United States. *Epidemiology, 22*(1), S81.

Overton, K., Dempster, T., Swearer, S. E., Morris, R. L., & Barrett, L. T. (2024). Achieving conservation and restoration outcomes through ecologically beneficial aquaculture. *Conservation Biology, 38*(1), e14065.

Pereira, R., Yarish, C., & Critchley, A. T. (2024). Seaweed aquaculture for human foods in land based and IMTA systems. In *Applications of seaweeds in food and nutrition* (pp. 77–99). Elsevier.

Phillips, M. J. (1990, August). Environmental aspects of seaweed culture. In *Regional workshop on the culture and utilization of seaweeds* (Vol. 2).

Rogers, A. J. (2024). Aquaculture in the ancient world: Ecosystem engineering, domesticated landscapes, and the first blue revolution. *Journal of Archaeological Research, 32*(3), 427–491.

Shang, Y. C. (1985). Aquaculture economics: An overview. *GeoJournal, 10*, 299–305.

Subasinghe, R. (2014). *Fishing for development, background paper for session 3: Green growth in fisheries and aquaculture*. https://www.oecd.org/tad/events/Fishing-for-development-2014-Session-3-GG-in-aquaculture.pdf.

Subasinghe, R., Soto, D., & Jia, J. (2009). Global aquaculture and its role in sustainable development. *Reviews in Aquaculture, 1*(1), 2–9.

Tacon, A. G. (2023). Contribution of fish and seafood to global food and feed supply: An analysis of the FAO food balance sheet for 2019. *Reviews in Fisheries Science & Aquaculture, 31*(2), 274–283.

Tamim, A. T., Begum, H., Shachcho, S. A., Khan, M. M., Yeboah-Akowuah, B., Masud, M., & Al-Amri, J. F. (2022). Development of IoT based fish monitoring system for aquaculture. *Intelligent Automation & Soft Computing, 32*(1).

Thorpe, A., Reid, C., Van Anrooy, R., Brugere, C., & Becker, D. (2006). Poverty reduction strategy papers and the fisheries sector: An opportunity forgone? *Journal of International Development, 18*(4), 489–517.

Tigchelaar, M., Leape, J., Micheli, F., Allison, E. H., Basurto, X., Bennett, A., et al. (2022). The vital roles of blue foods in the global food system. *Global Food Security, 33*, 100637.

Torrissen, O., Jones, S., Asche, F., Guttormsen, A., Skilbrei, O. T., Nilsen, F., et al. (2013). Salmon lice–impact on wild salmonids and salmon aquaculture. *Journal of Fish Diseases, 36*(3), 171–194.

Trace, X. (2024). https://tracextech.com/

Vishwanatham, K., Kumar, B. S., & Kumar, D. R. (2024). Application of biotechnology in aquaculture. *Journal of Advanced Zoology, 45*(3).

World Bank. (2014). Fish to 2030: Prospects for fisheries and aquaculture.

Chapter 19
Impact of Gut Microbiota on Fish Nutrition and Health

Sudeshna Mandal, Ravi Kumar Goswami, Dhitri Borah, Sristi Sharma, Bronson Kumar Khangembam, and Samar Pal Singh

1 Introduction

The microorganisms found in the intestine that affect all biochemical processes and activate the host's immune system are known as the gut microbiota (Gómez & Balcázar, 2008; Hanning & Diaz-Sanchez, 2015). Gaining a deeper understanding of the circumstances surrounding this taxonomic and metagenomic study aids in appreciating the importance of this approach. The term "microbiota"or "microbiome" describes a collection of microbial genomes found in the environment. Often used interchangeably, the term "microbiota"refers to the many microorganisms found in the environment that are connected to the concentration of microorganisms associated with habitat. An ecosystem's chemical and physical characteristics include a habitat that supports each unique niche space; in other words, it establishes the microbial interaction along the habitat's chemical and physical

S. Mandal
Department of Medicine, Indiana University, School of Medicine, Indianapolis, IN, United States

R. K. Goswami (✉)
Department of Zoology, Hindu College, University of Delhi, Delhi, India

D. Borah
Department of Zoology, Biswanath College, Gauhati University, Biswanath Chariali, Assam, India

S. Sharma
Department of Zoology, University of Delhi, Delhi, India

B. K. Khangembam
Department of Zoology, Tripura University (A Central University), Suryamaninagar, Tripura, India

S. P. Singh
Department of Zoology, Ramjas College, University of Delhi, Delhi, India

I. Ahmed, I. Ahmad (eds.), *Aquaculture: Enhancing Food Security and Nutrition*, https://doi.org/10.1007/978-3-031-92858-1_19

dimensions (Shade & Handelsman, 2012; Venkatesh et al., 2014). From a molecular genetics perspective, metagenomics is thought to be a method for examining the connections among those components in an ecosystem.

Whiteside et al. (2015) have described concepts like microbiota and microbiome, where the former is associated with the dominant environment and the latter is defined as the connection between biotic and abiotic elements that influence the microbes' habitat and symbiotic environment. Using the 16S rRNA sequencing approach, metagenomic analysis encompasses the genetic capacity of this microbial taxonomic group in its natural habitat at both the genomic and plasmid levels (Whiteside et al., 2015).

When it comes to acquiring nutritional advantages from the topic that confers the recipient's metabolic and immunological roles, the intestinal microbiome is the makeup of the intestinal lumen microorganisms connected with the microbial community. An effective resource that can satisfy the need for premium animal feed that is necessary for human consumption is fish. As a result, it is acknowledged as a very efficient source for supplying the food requirements of our expanding world population. However, creating cost-effective and sustainable feed production techniques is a major obstacle to guaranteeing sustainable and reasonably priced fish production. Fish meal and oil prices have uncontrollably increased as a result of the ongoing global climate catastrophe and changes in the price of fossil fuels, which have raised freight costs. These ingredients are essential when making fish feed (Omosowone & Ogunrinde, 2018).

Even while fish meal (FM) is still generally considered the best source of protein for aquaculture feeds, its use is limited by its expensive cost and the fact that it competes with fish that are meant for direct human consumption (Tacon & Metian, 2008). Furthermore, it is getting more and more unsustainable to employ traditional fishing methods to obtain fish for FM production (Turchini et al., 2019). Much research focuses on lowering dependency on wild fish stocks for FM production and using more efficient and sustainable feed components in fish feeds to promote sustainable aquaculture techniques (Welch et al., 2010). Consequently, the effectiveness of protein substitutes for fish meal in aquafeeds has been the subject of several investigations (Yiğit et al., 2012; Bulut et al., 2014; Yıldırım et al., 2014). It is crucial to remember that there are other measures available for evaluating feed modification than performance and quality. Additionally, it is crucial that changes in feed composition do not negatively impact the fish's health and well-being (Fan et al., 2023; Xie et al., 2021). This emphasizes how important vegetable protein sources are as necessary choices for supplying the feed needs of the quickly growing aquaculture sector (Hassaan et al., 2018). Thus, even though fishmeal and fish oil are natural resources, it is clear that they are not sustainable raw materials that may be used to expand fish farming. The use of forage plants in fish feeds is anticipated to be vital to aquaculture's future due to the possible production issues with fish meal and oil. A major contributor to meet the growing demand for premium animal protein is the aquaculture sector. High death rates and a poor feed conversion ratio inhibit the growth of fish, which lowers fish productivity.

The development of resistant bacterial infections in people is linked to the use of antibiotics in aquaculture (Allameh et al., 2016). Concerns about consumer safety and the rise in antibiotic resistance were the main factors that spurred researchers to look for other control methods (Pandiyan et al., 2013). In place of antibiotic treatments, probiotics as dietary additions offer an alternate method of infection management (Bandyopadhyay et al., 2015; Wu et al., 2015; Kuebutornye et al., 2019).

Gut microbiota is linked to the host's metabolism, immunity, and gut development (Pérez et al., 2010; Nicholson et al., 2012; O'Shea et al., 2012; Butel, 2014; Koch et al., 2018). Poor metabolism, stunted growth, stress, and the start of disease are all consequences of an unbalanced fish gut microbiota. The makeup of gut microbial communities determines the overall effect of gut microbiota on fish physiology (Vigneri, 2014). The main factor influencing gut flora is nutrition (Miyake et al., 2015). Target regulation of bacterial communities for improved fish productivity and health depends on the comprehensive understanding of the bacterial community provided by next-generation sequencing (Ou et al., 2021). Prebiotics, antibiotics, and probiotics can all be used to alter the gut microbiota of fish in order to study its function (Butt & Volkoff, 2019). New food additives, particularly probiotics, prebiotics, and synbiotics as feed additives, regulate growth, health, and physiology. They are attracting the interest of researchers as additions for fish feed because of their effects on the gut flora (Miyake et al., 2015; Kuebutornye et al., 2019; Butt & Volkoff, 2019). Fishes from different geographical and taxonomic groups have comparable gut core microbiota, which suggests that they serve as regulators of metabolism and physiology (Talwar et al., 2018).

The goal of the chapter is to provide a comprehensive overview of the gut microbial system, including its functional impacts and aquaculture-related influences. In the wide range of metagenomics, prebiotics, and probiotics, the gut microbial diversity is unique to each species, and different physiological factors like temperature, pH, nutrient availability, and environmental conditions influence the gut microbial system and immune system.

2 Gut Microbiota and Growth Performance

Individual growth variations within a fish stock provide serious obstacles to the effectiveness of commercial farming, impacting both the performance of farmed animals and the total output of aquaculture. It does, however, also present a chance for selective breeding. In order to reduce size heterogeneity, common industry methods group cultivated species with different growth rates into more homogenous groups during their production cycle (Conte, 2004), using growth rate responses per generation as an aquaculture breeding technique (Olesen et al., 2003). The gut is home to a microbiome that is closely related to host development, immunity, health, and energy metabolism (Brahe et al., 2015; Zheng et al., 2017; Patil et al., 2020). In order to understand the complex interaction between the gut microbiota and other phenotypic features, such as obesity and leanness, the metabolic activity and capacity of gut microorganisms have been thoroughly investigated (Ley et al., 2006; Turnbaugh et al., 2006; Brahe et al., 2015).

Although research on the relationship between genetic background, gut microbiota, and individual performance differences has mostly concentrated on mammals, including humans, there are still issues with planning studies with large enough sample sizes and successfully accounting for host diet and environmental variations. Fish have significant benefits in this regard since they have many siblings available for well-monitored studies. Because of variations in food, culture duration, sample size, and developmental stage, it might be difficult to compare the gut microbiota different between research results. But for aquaculture development, it is essential to comprehend how the gut microbiota affects performance characteristics like growth (Olafsen, 2001).

3 Growth Related to Gut Microbiome in Fish

Genetics and other host-related variables are blamed for variations in individual development (De-Santis & Jerry, 2007; Piazzon et al., 2020); feeding habits and digestive processes come next (Tamayo et al., 2011). Through a variety of selection pressures, such as host–microbiota interactions, ontogenetic development alters the gut microbiota (Wong et al., 2015). Since most research in this field has focused on the impact of food and ecological variables on farmed species, little is known about the diversity and structure of the gut microbiome as well as the ontogenetic process in fish with varying development rates. According to Forberg et al. and Trinh et al., genetic variants in fish may be related to growth rate and contribute to alterations in the larval gut microbiota (Forberg et al., 2016; Trinh et al., 2017), indicating a relationship between a fish's size and gut microbiome. Early research used a culture-based methodology to show relationships between fish development and the gut microbiota in dominant and subordinate individuals (Ringø et al., 2008). According to Sun et al., grouper larvae that grew more quickly had higher abundances of *Bacillus pumilus*, *Bacillus clausii*, and *Psychrobacter* sp. and lower abundances of opportunistic *Vibrio* spp. (Sun et al., 2009).

The three taxa showed antagonistic effects against Vibrio isolates in in vitro tests, which may account for the reduction of opportunistic infections and improved fish development performance (Sun et al., 2009). The investigation of aquatic species' gut microbiomes has gained more attention since the development of high-throughput sequencing (HTS). This emphasis encompasses important facets of aquaculture operations, such as probiotic products, disease incidence, and nutritional replacement (Knipe et al., 2021).

Research on aquatic animals shows that there are more differences between host species than within host species in the gut microbial makeup (Fietz et al., 2018), promoting the idea that aquatic creatures exhibit phylosymbiosis (Tang et al., 2021). This discovery emphasizes how important ecological and evolutionary forces are in regulating how the gut microbial population is created (Zhang & Li, 2018; Fietz et al., 2018; Tang et al., 2021), suggesting that different methods are used by the hosts to choose and filter microbial populations. The composition and functional profiles of the gut microbiota may differ with different growth performances, as shown in transgenic fast-growth and conventionally raised fish (Li et al., 2013). Research on genetically modified fish with increased development in the PROGENSA selection program has shown that genotype influences certain microbial taxa in the gut mucosa (Torrecillas et al., 2023). Fish that grew quickly demonstrated better use of various foods, a more robust and adaptable microbiota with wider metabolic responses, and suppression of proinflammatory pathway expression (Torrecillas et al., 2023). Interestingly, in sea bass that had been genetically modified for growth, a decreased abundance of *Paracoccus* was linked to lower levels of the proinflammatory cytokine interleukin-1β (IL-1β) (Torrecillas et al., 2023). *Bacilli*, *Clostridium*, *Gemella*, *Reyranella*, *Lactobacillus*, and *Bifidobacterium* were among the beneficial gut taxa that proliferated more in fast-growing genetically modified gilthead sea bream (Naya-Català et al., 2022a, b). Fast-growing families' gut microbiome demonstrated increased metabolic adaptability in response to dietary modifications without sacrificing fitness or growth results (Naya-Català et al., 2022a, b).

However, because of increased genetic selection pressure per generation, the diet impact may be obscured in genetically modified fish (Naya-Català et al., 2022a, b). When given plant-based meals, fast-growing families can modify their intestines to maximize nutritional absorption (Perera et al., 2019), proving that the dynamics of gut microbial populations are changed by host selection pressure and dietary influences (Yih et al., 2019). Combining microbiome and metabolome analysis further showed that the gut microbial composition may influence carp growth in rice-fish coculture farming systems without feed supplements through distinct metabolic pathways that generate particular metabolites, such as by reducing unsaturated fatty acid biosynthesis and bile secretion and impairing spermidine biosynthesis by producing protease inhibitors (Nie et al., 2023). This discovery offers a potential way to speed up our comprehension of how the gut microbiota contributes to the development of features that are valuable to the market. Fast-growing grass carp exposed to two extruded diets consistently displayed an enrichment of bacteria that break down cellulose and amylose, along with bacteria linked to health, including *Erysipelatoclostridium*, *Leptotrichia*, and *Anaerorhabdus furcosa*. On the other hand, opportunistic pathogenic bacteria, including *Ralstonia*, *Brevundimonas*, and *Pandoraea,* may be more prevalent in slow-growing grass carp. Despite the fact that several studies have shown a reduction of gut microbial diversity as a finding of gut dysbiosis, both circumstances consistently imply that it may not be a sign of intestinal disruption in fast-growing fish. Recent research on aquatic animals indicates that the development rate is unaffected by the variety of gut microbes (Lin et al., 2019; Piazzon et al., 2020; Nie et al., 2023). A possible reason for the limited physiological function and immunological resistance shown in some slow-growing fish is that their guts are co-inhabited by more opportunistic pathogenic species. Only two studies have examined gut microbial populations linked to growth rate in non-aquaculture animals; most recent research has concentrated on aquaculture species

(Forberg et al., 2016; Zhang et al., 2021). Just one research has examined the gut microbial populations in fish at various stages and growth rates (Zeng et al., 2019). This study found that whereas slow-growing people had larger amounts of *Plesiomonas* in their guts, fast-growing juveniles had higher levels of Bacillus and Lactococcus. Adult fish that grew more slowly had greater amounts of *Enterococcus*, whereas fish that grew more quickly had higher levels of *Lactococcus*. Fish trophic level and ecological niche have also been shown to have a substantial impact on the gut microbiota (Zhang et al., 2021). According to reports, the guts of herbivorous fishes are more abundant with cellulose-degrading bacteria like *Clostridium*, *Citrobacter*, and *Leptotrichia*, but the guts of carnivorous fishes are dominated by protease-producing bacteria like *Halomonas* and *Cetobacterium* (Liu et al., 2016). However, there is bias in the species chosen for research on fish-eating patterns with respect to examining the gut microbiota communities in both slow-growing and fast-growing fish, as carnivorous species have been the subject of the majority of research. The growing market demand for these farmed species and the increased economic significance given to carnivorous species in recent years might be the causes of this development.

Different fish species had different gut microbiomes, and there were inconsistent differences in the gut microbiota's alpha and beta diversity between fish with quick growth and those with sluggish growth across dietary patterns. While herbivorous species revealed notable differences in gut microbial communities between fast-growing and slow-growing individuals, certain omnivore and carnivorous species demonstrated gut microbiota independence from individual growth. Interestingly, since only one herbivorous species has been investigated to date, this conclusion may be biased by nature. Furthermore, it is important to take into account these research's small sample sizes. The gut microbiota of fish, particularly carnivorous species, may be impacted by the establishment of dominance hierarchies through social behaviors including eating, aggression, and even cannibalism (Ringo et al., 2008), and finally decide on performance or growth rate.

Currently, less than 30 species or species groupings account for 90% of aquatic animal products, with herbivorous and omnivorous fish like tilapia and carp making major contributions. The gut microbiome of the most widely cultivated species with varying growth characteristics, as well as the relationships between the gut microbiome and fish trophic level and growth performance, require more research.

4 Gut Microbiota's Physiological Role

4.1 Feeding, Digestion, and Metabolism

Research conducted on mammals demonstrates that the GIT's microbes control appetite/ingestion, digestion, and metabolism (Duca et al., 2012; Fetissov, 2017; Read & Holmes, 2017). For instance, even when given more calories, germ-free animals with no gut flora are thinner than typical control mice (Duca et al., 2012). Additionally, these mice exhibit reduced levels of chemicals that control hunger, including ghrelin and leptin (Duca et al., 2012), suggesting that hunger and metabolism are regulated by the intestinal flora. Secretions from microorganisms, including particular metabolites such as indoles, propionate, butyrate, acetate, and short-chain fatty acids (SCFAs) (Zhang & Davies, 2016), impact metabolism and digestive functions. Additionally, the microbiota and GIT neurotransmitters interact [e.g., serotonin (Yano et al., 2015) & dopamine and norepinephrine, two catecholamines (Asano et al., 2012)]. They thus impact how they affect eating behavior, gastrointestinal (GI) motility, function, and hormone release (Strandwitz, 2018; Yang & Chiu, 2017). On the other hand, serotonin and catecholamines that are produced by enteric neurons can change the production of cytokines and bacterial compounds as well as affect the gut microbiota (Mittal et al., 2017). Some of these metabolites can act on enterocytes and regulate their intestinal barrier function (Venkatesh et al., 2014), absorptive capacity (e.g., monosaccharide absorption), and nutrient uptake and storage [e.g., altered enzymatic activity in the gut and fat storage (Franchini et al., 2014)], thus influencing metabolism such as cholesterol metabolism and adipogenesis (Agustí et al., 2018). Additionally, the generation of gut peptides that regulate gut motility and enzyme secretion can be impacted by metabolites from the gut microbiota that alter enterocytes' secretory activity (Borre et al., 2014; Cani & Knauf, 2016). For instance, it has been demonstrated that SCFA receptors interact with enteroendocrine L cells that carry the gut hormone peptide YY (PYY) to affect the production of PPY in the rat colon and could have additional effects on metabolism. Additionally, these microbiological substances may affect dietary habits (Duca et al., 2012; Breton et al., 2016) either directly by entering the bloodstream and making their way to the brain or indirectly by either triggering vagal terminals or influencing the release of gut peptides that regulate appetite (such as CCK, ghrelin, and gastrin), which in turn influence the release of central neuropeptides that regulate appetite (such as neuropeptide Y, NPY; proopiomelanocortin, POMC) (Cussotto et al., 2018; Kim & Serre, 2018) (Fig. 19.1).

It is currently unknown what precise processes govern the "microbiota-gut-brain axis," or communication between the gut microbiota and the brain, and how modifications to the gut microbiota may affect the brain's neuropeptide systems (Cani & Knauf, 2016). Although there is currently relatively little research on the impact of microbiota on eating and metabolism in fish, those that have been done suggest that there are some parallels with mammals in this area. A few research studies that link feeding rates and microbiome alterations have looked at the impact of bacteria on food intake. However, when several studies and additives are employed, the results are inconsistent and challenging to compare. For instance, zebrafish fed *Lactobacillus rhamnosus* exhibit less hunger than fish kept as controls (Falcinelli et al., 2016; Falcinelli et al., 2018). However, compared to fish fed a control diet, carp administered a fructo-oligosaccharide (FOS) supplement show changes in microbiota composition (higher levels of lactic acid bacteria and heterotrophic aerobic bacteria), but no changes in eating rates (Hoseinifar et al., 2014). In several fish species, the

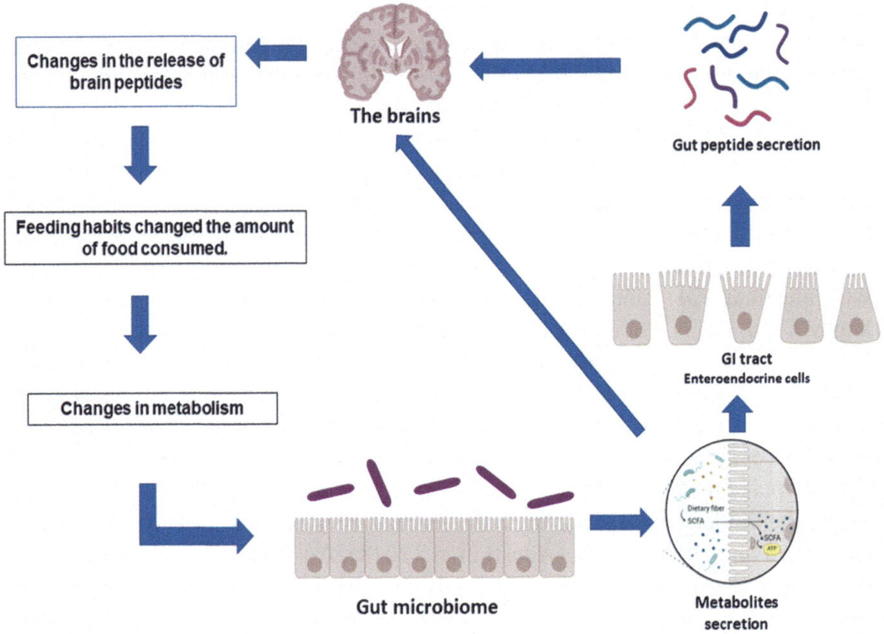

Fig. 19.1 An outline of the food and digesting axis involving the stomach, microbiome, and brain. When the gut lumen contains substrates, the gut microbiota responds by releasing metabolites. These metabolites either locally activate the gastrointestinal tract's (GIT) enteroendocrine cells or go to the brain. The gut peptides released by the activated enteroendocrine cells operate locally in the GIT, influencing feeding centers in the brain, changing the release of neuropeptides, and affecting eating behavior and energy balance. (Adapted from Butt & Volkoff, 2019)

possible impact of the gut microbiota on metabolism has been investigated. Numerous routes for the production and metabolism of lipids, carbohydrates, and amino acids alter in grass carp (*Ctenopharyngodon idella*) when the microbiota's makeup shifts (Ni et al., 2014). Fatty acid epithelium absorption in zebrafish is facilitated by microbial colonization of the gut (Semova et al., 2012). When compared to germ-free fish without microbiota, fish with intact microbiota exhibit higher fat buildup in the intestinal epithelium as well as higher expression of genes linked to lipid metabolism (Sheng et al., 2018). Furthermore, as compared to fish fed control diets, Japanese flounder (*Paralichthys olivaceus*) fed a diet supplemented with *Bacillus clausii* exhibit improved weight gain, feed efficiency, and growth performance (Ye et al., 2011). According to the scientists, this may be explained by better nutritional digestion and greater food consumption (Ye et al., 2011). All of these information points to the microbiome having a significant impact on fish metabolism.

4.2 *Additional Roles in Energy Homeostasis and Stress Reaction*

Numerous hormones mediate the stress response, which is the outcome of two-way communication between the brain and peripheral organs (Sandrini et al., 2015). Numerous environmental conditions, such as poor water quality, excessive particulate matter, inappropriate photoperiod, oxygen levels, and temperature, as well as high population density, poor food or malnutrition, and handling and transportation, can stress fish (Cantas et al., 2012). In order to prime the body to handle stress, the hypothalamic-pituitary-adrenal (HPA) axis releases corticotrophin-releasing hormone (CRH), which in turn triggers the anterior pituitary to release adrenocorticotropic hormone (ACTH), which in turn triggers the release of adrenal glucocorticoids (Sudo, 2014).

Similar to mammals, fish's microbiome influences the HPA axis, stress response, and behavior, particularly anxiety-like and locomotor activities, which may have an impact on eating habits and energy balance (Sudo, 2014; Garcia-Reyero, 2018). For instance, improving the microbiota in zebrafish with pro- and prebiotics diminishes anxiety-like behavior and the stress response by reducing cortisol levels and CRH expression (Davis et al., 2016; Forsatkar et al., 2017). Because stress hormones have been demonstrated to restrict eating, disruption of the gut microbiota may thus reduce the fish's capacity to forage for food and reduce feeding e.g., rainbow trout *Oncorhynchus mykiss (Ortega et al.,* 2013); goldfish *Carassius auratus (de Pedro et al.,* 1998). On the other hand, stress can modify the intestinal mucosa's structure and cause changes in the intestinal mucus, which can impact nutrient absorption and the gut immune system (also known as gut-associated lymphoid tissue, or GALT), which can result in opportunistic pathogen infections (Sekirov & Finlay, 2009). Acute stressors like netting cause fish to slough off mucus more often and eliminate autochthonous microorganisms, which protect the fish from possible infections (Vatsos, 2016). All things considered, stress causes changes in the gut microbiota, may change the immune system, and raises the risk of infection and pathogen colonization/invasion (Sandrini et al., 2015). These changes may cause other fish to eat less often [e.g., goldfish (Volkoff & Peter, 2004); chinook salmon (*Oncorhynchus tshawytscha*) (Bernier, 2010; Pirhonen et al., 2000)]. However, various fish species may handle stress in different ways, which might result in variable species-specific impacts on gut microbiota (Vatsos, 2016).

4.3 *Reproduction*

Since reproduction requires a lot of energy and can only be carried out properly when there is enough energy storage available, it is intimately tied to energy homeostasis (Carnevali et al., 2017). According to studies, the host's gonad development and subsequent reproductive success may be influenced by the gut bacteria.

Lactobacillus rhamnosus, for instance, changes the gut microbiota and speeds up zebrafish larval development by enhancing growth and sex differentiation when given consistently from birth until sexual maturity (Avella et al., 2012; Carnevali et al., 2013). The probability of successful reproduction is increased in adult female zebrafish treated with *L. rhamnosus* because they exhibit higher numbers of vitellogenic follicles, higher gonadal indexes (GSI), more ovulated eggs, and higher expression levels of reproductive hormones (leptin, GnRH3, and kisspeptinss) than control fish (Carnevali et al., 2013). Similarly, adding probiotics to the diet improves spawning females' GSI, fecundity, and fry production as well as the length and weight of the fry in ornamental livebearer fish species (Ghosh et al., 2007) and goldfish (Mehdinejad et al., 2019). The control of eating, food absorption, and energy balance are probably among the processes governing the activities of gut microbiota in host reproduction; however, these mechanisms are still being studied.

4.4 Development

The existence and makeup of the microbiota have a significant impact on the development of the tissues and organs that comprise an animal, particularly the neurological and digestive systems (Stephens et al., 2016), which are both essential for proper food absorption and ingestion. The gut microbiota composition of fish during development exhibits distinct patterns that demonstrate the connection between the host's development and the gut microbiota of the fish (Yan et al., 2016). Bacterial communities of grass carp (*Ctenopharyngodon idella*) differ between the eggs and the larvae, with Bacteroidetes and Proteobacteria predominating in the larvae and eggs, respectively. As the fish grows from egg to larvae, bacterial diversity rises (Wang et al., 2015). Grass carp, Chinese perch (*Siniperca chuatsi*), and southern catfish (*Silurus meridionalis*) have all shown an increase in variety from the larval to the adult stage, which further suggests that the levels of gut microbiota variance rise with fish growth (Yan et al., 2016). The presence of microbiota in zebrafish stimulates the proliferation of epithelial cells in the developing gut, demonstrating the direct involvement of gut microbiota in GIT development (Cheesman et al., 2011). Additionally, the introduction of bacteria can correct the inadequate development and decreased function of the germ-free zebrafish's GIT (Lescak & Milligan-Myhre, 2017). There is evidence that the microbiota has a role in neurological development as well and is necessary for the zebrafish's early neurobehavioral development to proceed normally. After receiving antibiotics, fish with disturbed microbiota exhibit aberrant locomotor activity (Phelps et al., 2017). This might influence foraging and eating habits. However, the process governing this relationship is still unknown.

4.5 Immune Responses

Pathogens may impair intestine and brain function, as well as hinder development and nutrition (Ribet & Cossart, 2015). The concept of "colonization resistance" suggests that the microbiota shields the host from environmental pathogen colonization and growth (Lawley & Walker, 2013; Kim et al., 2017). It has been proposed that commensal bacterial species manufacture and release antimicrobial peptides in addition to competing with pathogens for niche space, albeit the mechanisms behind this resistance remain unclear (Kim et al., 2017). Thus, infections and GALT activation may result from any disturbance of the intestinal equilibrium mucosa (Lazado & Caipang, 2014). Since the gut microbiota is crucial to the growth and maturation of the GALT, the related commensal microbiota of the mucosal immune system contributes significantly to the immunity and metabolism of host fish (Adel et al., 2017; Wang et al., 2018). For instance, giving beneficial bacteria (probiotics) to gilthead seabream (*Sparus aurata*) (Tapia-Paniagua et al., 2011) and rainbow trout (Adel et al., 2017) improves their gut microbiota and immune response.

5 Gut Microbiota Manipulation and Applications

With more study being done in this area, it is becoming clearer how the gut microbiota of fish affects host physiology. Numerous experimental techniques, such as gnotobiotic, antibiotic, probiotic and prebiotic, and symbiotic research, have been employed to manipulate these communities in order to evaluate the function of the fish gut microbiota (Table 19.1).

5.1 Gnotobiotic Fish

Animals having a known microbiota makeup are referred to as gnotobiotics. These include axenic (or germ-free) animals as well as axenic animals that have had an inoculation with recognized microbes. Gnotobiotic fish studies give researchers control over a variety of factors that influence the microbiota's growth and the examination of host reactions to certain gut microbes. The intricate processes required for the cultivation and upkeep of gnotobiotes are one of the study's drawbacks.

Table 19.1 Fish gut microbiota through research manipulation techniques

Modification of the microbiota	Goal	Possible application	Benefits
Gnotobiotic	Create a host that is germ-free (has no gut microbiota) or has a known microbiota composition (predefined microbiota)	Analyze how the host's physiology is affected by either a predefined microbiota or the impacts of certain gut microbes and/or a germ-free microbiota	Multiple variable control and analysis of the host's reactions to certain microbiota modifications
Antibiotic	Targeted gut microbiota microorganisms and/or bacterial pathogens should be inhibited or eliminated	Prevention and treatment of bacterial diseases	Treatment and prevention of bacterial diseases
Probiotic	Create a healthy gut microbiome of bacteria	Use in aquaculture, enhance fish health (growth and digestion), control health, and prevent disease	Improvement of the host's immunity, resistance to infections, and general well-being
Prebiotic	Encourage the development of probiotic microorganisms	Use in aquaculture, enhance fish health (growth and digestion), and prevent illness	Improvement of the host's immunity, resistance to infections, and general well-being
Symbiotic	Put probiotic and prebiotic microorganisms in place	Use in aquaculture, enhance fish health (immune system, growth, and digestion), maintain health, and prevent disease	Enhancement in the host's development and feed consumption, as well as an improvement in immunological response (better than probiotics alone)

Adapted from Butt and Volkoff (2019)

5.2 Antibiotics

One category of environmental variables influencing the gut microbiota is antibiotics, sometimes known as antibacterials. They can be found naturally in aquatic environments or as pollutants released as metabolites [such as oxytetracycline (OTC) and sulfamethoxazole (SMX)] through treated human or animal urine or feces (Zhou et al., 2018). In comparison to untreated fish, mosquitofish (*Gambusia affinis*) subjected to antibiotics exhibit reduced community diversity and taxonomic composition from both skin and gut microbiomes (Carlson et al., 2017). and the intestinal flora of zebrafish is disrupted with exposure to OTC (Zhou et al., 2018). Since antibiotics can eradicate or stop the growth of particular bacteria, they can be used to control the microbiota.

Antibiotic treatment can result in notable alterations in the microbial composition but does not totally eradicate gut microbiota groups. Antibiotics are often used in aquaculture to prevent and cure illness. Antibiotics, however, have the potential to alter microbial ecosystems and raise the risk of illness (Carnevali et al., 2017; Hassan et al., 2020). Additionally, according to Carnevali et al. (2017), antibiotics have the ability to bioaccumulate in animal tissues and cause the emergence of drug-resistant bacteria that may be transferred up the food chain (Banerjee & Ray, 2017; Fu et al., 2017). Strict rules have been put in place in the aquaculture sector as a result of increased knowledge of the drawbacks of antibiotics, and other techniques are being created and evaluated (Defoirdt et al., 2011). By generating antimicrobial peptides and competing with harmful bacteria for nutrition, probiotics can prevent the colonization of dangerous pathogens in fish, making their treatment a feasible alternative to antibiotics (Gatesoupe, 1999). Thus, probiotics may enhance fish health and counteract the harmful effects of antibiotics. For instance, the effective colonization of two probiotic organisms (*Phaeobacter inhibens* and *Bacillus pumilus*) in black molly *Poecilia sphenops* reverses the detrimental effects of drugs and lowers death rates (Schmidt et al., 2017).

5.3 Probiotics, Prebiotics, and Symbiotics

Prebiotics are indigestible dietary elements that specifically promote the formation of probiotics, whereas probiotics are living or dead components of microbial cells that help the host's health by encouraging good intestinal bacterial species (Ghanbari et al., 2015). It is thought that adding probiotics and/or prebiotics to fish diets may improve gut microbiota, which will then modify metabolism and energy expenditure in ways that are advantageous to the host (Llewellyn et al., 2014; Haygood & Jha, 2018). In addition to improving overall health and indirectly supporting eating and development, the administration of these supplements strengthens the host's immune system and increases its resilience to infections (Topic Popovic et al., 2017). These findings demonstrate how probiotics and prebiotics may improve fish health by influencing the gut microbiome of fish. Probiotics belonging to the *Lactobacillus*, *Lactococcus*, *Leuconostoc*, *Enterococcus*, *Carnobacterium*, *Shewanella*, *Bacillus*, *Aeromonas*, *Vibrio*, *Enterobacter*, *Pseudomonas*, *Clostridium*, and *Saccharomyces* genera are frequently utilized in aquaculture (Carnevali et al., 2017). Because the several strains included in multispecies probiotics boost the likelihood of survival in the gut, they may be more effective than single-strain probiotics [as seen in rainbow trout (Ramos et al., 2013)]. Probiotics boost the quantity of good microorganisms in the stomach. For instance, probiotic-fed fish exhibit increased bacterial diversity in rainbow trout (Ramos et al., 2013) and *Malaysian mahseer* (Asaduzzaman et al., 2018), as well as a greater abundance of core gut bacteria in zebrafish (Siriyappagouder et al., 2018). Probiotics often increase feed efficiency and growth in fish (such as Japanese flounder, carp, tilapia, and red seabream); however, there are few exceptions. This is probably due to increased nutrient absorption and perhaps feeding. Changes in intestinal shape, such as larger absorptive surface areas and denser microvilli in the gut (as in tilapia, zebrafish, and Malaysian mahseer), lead to an increase in absorption. Changes in growth-related gene expression may act as a mediator between the effects on growth. For instance, after taking probiotics, the pituitary of Malaysian mahseer and the liver of yellow perch (*Perca flavescens*) both exhibit higher levels of growth hormone (GH) expression, and the liver of both species as well as the body of European seabass exhibit increased expression of insulin-growth factor (IGF-1). In general, probiotics' effects on eating have not been well studied and are yet unknown. Fish given probiotics (*Bacillus subtilis*) and control diets have comparable feeding ratios, and probiotics cause zebrafish (*Lactobacillus rhamnosus*) to lose their appetite while pacu (*Bacillus subtilis*) enhance their food intake. The differences in the types and dosages of probiotics utilized are probably the cause of the disparities in the outcomes.

These changes in eating may be caused in part by modification of appetite-related gene expression. The probiotic *Lactobacillus rhamnosus* reduces brain NPY expression and increases adipose tissue leptin expression in larval zebrafish, while *Lactobacillus acidophilus*-fed goldfish exhibit down-regulated intestinal ghrelin expression in comparison to fish fed control diets.

The potential of prebiotics for aquaculture has not gotten as much attention as that of probiotics. All things considered, research indicates that prebiotics improve the host's ability to develop, the function of digestive enzymes, and its tolerance to illness and stress. Inulin, fructooligosaccharides (FOS), short-chain fructooligosaccharides (scFOS), and mannanoligosaccharides (MOS) are common prebiotics used in fish. In certain fish species, such as rainbow trout, pacu (*Piaractus mesopotamicus*), Nile tilapia, and Caspian roach (*Rutilus rutilus*), prebiotics have been demonstrated to improve growth performance and feed consumption; however, this is not always the case. For instance, feeding inulin to common carp fry (*Cyprinus carpio*) or β-glucan to pacu had little effect on their growth performance. Changes in intestinal structure (i.e., increases in intestinal villi height and digestive enzyme activity) that occur after probiotics are administered may be partially responsible for the growth increase observed in certain studies. Fish with stronger immune systems (like rainbow trout and Caspian roach) and better stress responses (like zebrafish) most likely also do better in terms of growth.

Synbiotics are mixtures of probiotics and prebiotics that are also given. According to studies, synbiotics enhance the GIT's probiotic health-promoting bacteria's implantation, survival, and metabolism (Cerezuela et al., 2011). It has been demonstrated that synbiotics improve the host's development and feed utilization, which might be due to improved digestive processes, the provision of energy and nutrients, or both (Ringø & Song, 2016). For instance, Nile tilapia (*Oreochromis niloticus*) exhibit improved growth performance after taking probiotics (Bacillus licheniformis) and prebiotics (yeast extract), which is followed by an increase in feed intake and feed utilization (Hassaan et al., 2014). Better immunological responses are also produced by dietary combinations of probiotics and prebiotics than by probiotics alone [e.g., rainbow trout] (Rodriguez-Estrada et al., 2013).

6 Conclusions and Future Aspects

Several studies provide compelling evidence that the gut microbiota of fish has an impact on the host fish's general health, including its immune system, digestion, stress response, reproduction, and general physiology. The effects of the microbiota on energy balance have been the subject of very little research to date, and the results of these investigations vary greatly, making comparisons challenging. First, many more species must be studied before any conclusions can be drawn, considering the potential impact of genetics and the environment as well as the small number of species studied. Second, the gut microbiota of fish has been studied using a range of techniques, and the outcomes may differ based on the experimental setup. This emphasizes the necessity of creating suitable, consistent techniques to characterize fish microbiota. It is difficult to investigate how the gut microbiota may affect fish energy balance because there are multiple mechanisms of action at play, including endocrine and local pathways, various physiological systems (such as stress and immunity), molecules (such as hormones and metabolites), and interactions between all of these systems (such as gut-microbiota-brain axis communication). Additionally, every microbe in the microbiome may behave differently. Furthermore, fish are more exposed to ongoing environmental changes that may have an impact on the microbiota than terrestrial species. Aquaculture offers a lot of promise for using fish gut microbiota manipulation to enhance growth. Despite their promise, probiotics and prebiotics face a number of obstacles in the future, such as determining the best dosages and modalities of therapy (oral or in water), characterizing the modes of action of distinct probiotic organisms, and maintaining quality control and regulation. The gut microbiota of other vertebrate species, including humans, can also be better understood using the fish model. Due to their tiny size, ease of maintenance in experimental settings, and quick growth and generation durations, zebrafish and three-spined sticklebacks have found extensive employment. Furthermore, their genomes are easily accessible and have genetic resemblances to humans in both structure and function. As a result, even if there has been progress, there is still much to be determined about gut microbiota using fish models. However, the research to date has provided valuable insights into the mechanisms through which these communities can control the fish host. It has also helped to improve aquaculture practices, better understand the relationships between host microbes and other vertebrates, including humans, and develop potential treatments for pathological conditions.

References

Adel, M., Lazado, C. C., Safari, R., Yeganeh, S., & Zorriehzahra, M. J. (2017). Aqualase®, a yeast-based in-feed probiotic, modulates intestinal microbiota, immunity and growth of rainbow trout *Oncorhynchus mykiss*. *Aquaculture Research, 48*(4), 1815–1826.

Agustí, A., García-Pardo, M. P., López-Almela, I., Campillo, I., Maes, M., Romaní-Pérez, M., & Sanz, Y. (2018). Interplay between the gut-brain axis, obesity and cognitive function. *Frontiers in Neuroscience, 12*, 155.

Allameh, S. K., Yusoff, F. M., Ringø, E., Daud, H. M., Saad, C. R., & Ideris, A. (2016). Effects of dietary mono-and multiprobiotic strains on growth performance, gut bacteria and body composition of J avanese carp (*Puntius gonionotus*, Bleeker 1850). *Aquaculture Nutrition, 22*(2), 367–373.
Asaduzzaman, M., Sofia, E., Shakil, A., Haque, N. F., Khan, M. N. A., Ikeda, D., et al. (2018). Host gut-derived probiotic bacteria promote hypertrophic muscle progression and upregulate growth-related gene expression of slow-growing Malaysian Mahseer *Tor tambroides*. *Aquaculture Reports, 9*, 37–45.
Asano, Y., Hiramoto, T., Nishino, R., Aiba, Y., Kimura, T., Yoshihara, K., et al. (2012). Critical role of gut microbiota in the production of biologically active, free catecholamines in the gut lumen of mice. *American Journal of Physiology Gastrointestinal and Liver Physiology, 303*, G1288–G1295.
Avella, M. A., Place, A., Du, S. J., Williams, E., Silvi, S., Zohar, Y., & Carnevali, O. (2012). Lactobacillus rhamnosus accelerates zebrafish backbone calcification and gonadal differentiation through effects on the GnRH and IGF systems. *PLOS One, 7*, e45572.
Bandyopadhyay, P., Mishra, S., Sarkar, B., Swain, S. K., Pal, A., Tripathy, P. P., & Ojha, S. K. (2015). Dietary Saccharomyces cerevisiae boosts growth and immunity of IMC Labeo rohita (Ham.) juveniles. *Indian Journal of Microbiology, 55*, 81–87.
Banerjee, G., & Ray, A. K. (2017). The advancement of probiotics research and its application in fish farming industries. *Research in Veterinary Science, 115*, 66–77.
Bernier, N. J. (2010). Food intake regulation and disorders. In *Fish diseases and disorders* (Vol. 2, pp. 238–266). CABI.
Borre, Y. E., Moloney, R. D., Clarke, G., Dinan, T. G., & Cryan, J. F. (2014). The impact of microbiota on brain and behavior: Mechanisms & therapeutic potential. In *Microbial endocrinology: The microbiota-gut-brain axis in health and disease* (pp. 373–403).
Brahe, L. K., Le Chatelier, E., Prifti, E., Pons, N., Kennedy, S., Hansen, T., et al. (2015). Specific gut microbiota features and metabolic markers in postmenopausal women with obesity. *Nutrition & Diabetes, 5*(6), e159–e159.
Breton, J., Tennoune, N., Lucas, N., Francois, M., Legrand, R., Jacquemot, J., et al. (2016). Gut commensal *E. coli* proteins activate host satiety pathways following nutrient-induced bacterial growth. *Cell Metabolism, 23*(2), 324–334.
Bulut, M., Yiğit, M., Ergün, S., Kesbiç, O. S., Acar, Ü., Karga, M., & Güroy, D. (2014). Incorporation of corn gluten meal as a replacement for fish meal in the diets of two banded seabream (*Diplodus vulgaris*) juveniles. *International Journal of AgriScience, 4*(1), 60–65.
Butel, M. J. (2014). Probiotics, gut microbiota and health. *Médecine et Maladies Infectieuses, 44*(1), 1–8.
Butt, R. L., & Volkoff, H. (2019). Gut microbiota and energy homeostasis in fish. *Frontiers in Endocrinology, 10*, 9.
Cani, P. D., & Knauf, C. (2016). How gut microbes talk to organs: The role of endocrine and nervous routes. *Molecular Metabolism, 5*(9), 743–752.
Cantas, L., Sørby, J. R. T., Aleström, P., & Sørum, H. (2012). Culturable gut microbiota diversity in zebrafish. *Zebrafish, 9*(1), 26–37.
Carlson, J. M., Leonard, A. B., Hyde, E. R., Petrosino, J. F., & Primm, T. P. (2017). Microbiome disruption and recovery in the fish Gambusia affinis following exposure to broad-spectrum antibiotic. *Infection and Drug Resistance, 10*, 143–154.
Carnevali, O., Avella, M. A., & Gioacchini, G. (2013). Effects of probiotic administration on zebrafish development and reproduction. *General and Comparative Endocrinology, 188*, 297–302.
Carnevali, O., Maradonna, F., & Gioacchini, G. (2017). Integrated control of fish metabolism, wellbeing and reproduction: the role of probiotic. *Aquaculture, 472*, 144–155.
Cerezuela, R., Meseguer, J., & Esteban, M. A. (2011). Current knowledge in synbiotic use for fish aquaculture: A review. *Journal of Aquaculture Research & Development S, 1*, 1–7.
Cheesman, S. E., Neal, J. T., Mittge, E., Seredick, B. M., & Guillemin, K. (2011). Epithelial cell proliferation in the developing zebrafish intestine is regulated by the Wnt pathway and

microbial signaling via Myd88. *Proceedings of the National Academy of Sciences, 108*(supplement_1), 4570–4577.

Conte, F. S. (2004). Stress and the welfare of cultured fish. *Applied Animal Behaviour Science, 86*(3–4), 205–223.

Cussotto, S., Sandhu, K. V., Dinan, T. G., & Cryan, J. F. (2018). The neuroendocrinology of the microbiota-gut-brain axis: A behavioural perspective. *Frontiers in Neuroendocrinology, 51*, 80–101.

Davis, D. J., Bryda, E. C., Gillespie, C. H., & Ericsson, A. C. (2016). Microbial modulation of behavior and stress responses in zebrafish larvae. *Behavioural Brain Research, 311*, 219–227.

Defoirdt, T., Sorgeloos, P., & Bossier, P. (2011). Alternatives to antibiotics for the control of bacterial disease in aquaculture. *Current Opinion in Microbiology, 14*(3), 251–258.

de Pedro, N., Pinillos, M. L., Valenciano, A. I., Alonso-Bedate, M., & Delgado, M. J. (1998). Inhibitory effect of serotonin on feeding behavior in goldfish: involvement of CRF. *Peptides, 19(3), 505–511.*

De-Santis, C., & Jerry, D. R. (2007). Candidate growth genes in finfish—Where should we be looking? *Aquaculture, 272*(1–4), 22–38.

Duca, F. A., Swartz, T. D., Sakar, Y., & Covasa, M. (2012). Increased oral detection, but decreased intestinal signaling for fats in mice lacking gut microbiota. *PLoS One, 7*(6), e39748.

Falcinelli, S., Rodiles, A., Unniappan, S., Picchietti, S., Gioacchini, G., Merrifield, D. L., & Carnevali, O. (2016). Probiotic treatment reduces appetite and glucose level in the zebrafish model. *Scientific Reports, 6*(1), 18061.

Falcinelli, S., Rodiles, A., Hatef, A., Picchietti, S., Cossignani, L., Merrifield, D. L., et al. (2018). Influence of probiotics administration on gut microbiota core: A review on the effects on appetite control, glucose, and lipid metabolism. *Journal of Clinical Gastroenterology, 52*, S50–S56.

Fan, J., Jiang, T., & He, D. (2023). Advances in the implications of the gut microbiota on the treatment efficacy of diseasemodifying anti-rheumatic drugs in rheumatoid arthritis. *Frontiers in Immunology, 14, 1189036.*

Fetissov, S. O. (2017). Role of the gut microbiota in host appetite control: Bacterial growth to animal feeding behaviour. *Nature Reviews Endocrinology, 13*(1), 11–25.

Fietz, K., Rye Hintze, C. O., Skovrind, M., Kjærgaard Nielsen, T., Limborg, M. T., Krag, M. A., et al. (2018). Mind the gut: Genomic insights to population divergence and gut microbial composition of two marine keystone species. *Microbiome, 6*, 1–16.

Forberg, T., Sjulstad, E. B., Bakke, I., Olsen, Y., Hagiwara, A., Sakakura, Y., & Vadstein, O. (2016). Correlation between microbiota and growth in Mangrove Killifish (*Kryptolebias marmoratus*) and Atlantic cod (*Gadus morhua*). *Scientific Reports, 6*(1), 21192.

Forsatkar, M. N., Nematollahi, M. A., Rafiee, G., Farahmand, H., & Lawrence, C. (2017). Effects of the prebiotic mannan-oligosaccharide on the stress response of feed deprived zebrafish (*Danio rerio*). *Physiology & Behavior, 180*, 70–77.

Franchini, P., Fruciano, C., Frickey, T., Jones, J. C., & Meyer, A. (2014). The gut microbial community of Midas cichlid fish in repeatedly evolved limnetic-benthic species pairs. *PLoS One, 9*(4), e95027.

Fu, J., Yang, D., Jin, M., Liu, W., Zhao, X., Li, C., et al. (2017). Aquatic animals promote antibiotic resistance gene dissemination in water via conjugation: Role of different regions within the zebra fish intestinal tract, and impact on fish intestinal microbiota. *Molecular Ecology, 26*(19), 5318–5333.

Garcia-Reyero, N. (2018). The clandestine organs of the endocrine system. *General and Comparative Endocrinology, 257*, 264–271.

Gatesoupe, F. J. (1999). The use of probiotics in aquaculture. *Aquaculture, 180*(1–2), 147–165.

Ghanbari, M., Kneifel, W., & Domig, K. J. (2015). A new view of the fish gut microbiome: Advances from next-generation sequencing. *Aquaculture, 448*, 464–475.

Ghosh, S., Sinha, A., & Sahu, C. (2007). Effect of probiotic on reproductive performance in female livebearing ornamental fish. *Aquaculture Research, 38*(5), 518–526.

Gómez, G. D., & Balcázar, J. L. (2008). A review on the interactions between gut microbiota and innate immunity of fish. *FEMS Immunology & Medical Microbiology, 52*(2), 145–154.

Hanning, I., & Diaz-Sanchez, S. (2015). The functionality of the gastrointestinal microbiome in non-human animals. *Microbiome, 3*, 1–11.

Hassaan, M. S., Soltan, M. A., & Ghonemy, M. M. R. (2014). Effect of synbiotics between B*acillus licheniformis* and yeast extract on growth, hematological and biochemical indices of the Nile tilapia (*Oreochromis niloticus*). *The Egyptian Journal of Aquatic Research, 40*(2), 199–208.

Hassaan, M. S., Soltan, M. A., Mohammady, E. Y., Elashry, M. A., El-Haroun, E. R., & Davies, S. J. (2018). Growth and physiological responses of Nile tilapia, *Oreochromis niloticus* fed dietary fermented sunflower meal inoculated with *Saccharomyces cerevisiae* and *Bacillus subtilis*. *Aquaculture, 495*, 592–601.

Hassan, S., Abdel-Rahman, M., Mansour, E. S., & Monir, W. (2020). Prevalence and antibiotic susceptibility of bacterial pathogens implicating the mortality of cultured Nile tilapia, *Oreochromis niloticus*. *Egyptian Journal for Aquaculture, 10*(1), 23–43.

Haygood, A. M., & Jha, R. (2018). Strategies to modulate the intestinal microbiota of Tilapia (*Oreochromis* sp.) in aquaculture: A review. *Reviews in Aquaculture, 10*(2), 320–333.

Hoseinifar, S. H., Soleimani, N., & Ringø, E. (2014). Effects of dietary fructo-oligosaccharide supplementation on the growth performance, haemato-immunological parameters, gut microbiota and stress resistance of common carp (*Cyprinus carpio*) fry. *British Journal of Nutrition, 112*(8), 1296–1302.

Kim, S., Covington, A., & Pamer, E. G. (2017). The intestinal microbiota: Antibiotics, colonization resistance, and enteric pathogens. *Immunological Reviews, 279*(1), 90–105.

Knipe, H., Temperton, B., Lange, A., Bass, D., & Tyler, C. R. (2021). Probiotics and competitive exclusion of pathogens in shrimp aquaculture. *Reviews in Aquaculture, 13*(1), 324–352.

Koch, B. E., Yang, S., Lamers, G., Stougaard, J., & Spaink, H. P. (2018). Intestinal microbiome adjusts the innate immune setpoint during colonization through negative regulation of MyD88. *Nature Communications, 9*(1), 4099.

Kuebutornye, F. K., Abarike, E. D., & Lu, Y. (2019). A review on the application of Bacillus as probiotics in aquaculture. *Fish & Shellfish Immunology, 87*, 820–828.

Lawley, T. D., & Walker, A. W. (2013). Intestinal colonization resistance. *Immunology, 138*(1), 1–11.

Lazado, C. C., & Caipang, C. M. A. (2014). Mucosal immunity and probiotics in fish. *Fish & Shellfish Immunology, 39*(1), 78–89.

Lescak, E. A., & Milligan-Myhre, K. C. (2017). Teleosts as model organisms to understand host-microbe interactions. *Journal of Bacteriology, 199*(15), 10–1128.

Ley, R. E., Turnbaugh, P. J., Klein, S., & Gordon, J. I. (2006). Human gut microbes associated with obesity. *Nature, 444*(7122), 1022–1023.

Li, X., Yan, Q., Xie, S., Hu, W., Yu, Y., et al. (2013). Gut microbiota contributes to the growth of fast-growing transgenic common carp (*Cyprinus carpio* L.). *PLoS One, 8*(5), e64577.

Lin, M., Zeng, C. X., Jia, X. Q., Zhai, S. W., Li, Z. Q., & Ma, Y. (2019). The composition and structure of the intestinal microflora of Anguilla marmorata at different growth rates: A deep sequencing study. *Journal of Applied Microbiology, 126*(5), 1340–1352.

Liu, H., Guo, X., Gooneratne, R., Lai, R., Zeng, C., Zhan, F., & Wang, W. (2016). The gut microbiome and degradation enzyme activity of wild freshwater fishes influenced by their trophic levels. *Scientific Reports, 6*(1), 24340.

Llewellyn, M. S., Boutin, S., Hoseinifar, S. H., & Derome, N. (2014). Teleost microbiomes: the state of the art in their characterization, manipulation and importance in aquaculture and fisheries. *Frontiers in Microbiology, 5*, 207.

Mehdinejad, N., Imanpour, M. R., & Jafari, V. (2019). Combined or individual effects of dietary probiotic, Pediococcus acidilactici and nucleotide on reproductive performance in goldfish (*Carassius auratus*). *Probiotics and Antimicrobial Proteins, 11*, 233–238.

Mittal, R., Debs, L. H., Patel, A. P., Nguyen, D., Patel, K., O'Connor, G., et al. (2017). Neurotransmitters: The critical modulators regulating gut–brain axis. *Journal of Cellular Physiology, 232*(9), 2359–2372.

Miyake, S., Ngugi, D. K., & Stingl, U. (2015). Diet strongly influences the gut microbiota of surgeonfishes. *Molecular Ecology, 24*(3), 656–672.

Naya-Català, F., Piazzon, M. C., Calduch-Giner, J. A., Sitjà-Bobadilla, A., & Pérez-Sánchez, J. (2022a). Diet and host genetics drive the bacterial and fungal intestinal metatranscriptome of gilthead sea bream. *Frontiers in Microbiology, 13*, 883738.

Naya-Català, F., Piazzon, M. C., Torrecillas, S., Toxqui-Rodríguez, S., Calduch-Giner, J. À., Fontanillas, R., et al. (2022b). Genetics and nutrition drive the gut microbiota succession and host-transcriptome interactions through the gilthead sea bream (*Sparus aurata*) production cycle. *Biology, 11*(12), 1744.

Ni, J., Yan, Q., Yu, Y., & Zhang, T. (2014). Factors influencing the grass carp gut microbiome and its effect on metabolism. *FEMS Microbiology Ecology, 87*(3), 704–714.

Nicholson, J. K., Holmes, E., Kinross, J., Burcelin, R., Gibson, G., Jia, W., & Pettersson, S. (2012). Host-gut microbiota metabolic interactions. *Science, 336*(6086), 1262–1267.

Nie, Z., Xu, X., Shao, N., He, J., Li, P., Xu, P., et al. (2023). Integrative analysis of microbiome and metabolome reveals the linkage between gut microbiota and carp growth. *Environmental Research, 220*, 115133.

Olafsen, J. A. (2001). Interactions between fish larvae and bacteria in marine aquaculture. *Aquaculture, 200*(1–2), 223–247.

Olesen, I., Gjedrem, T., Bentsen, H. B., Gjerde, B., & Rye, M. (2003). Breeding programs for sustainable aquaculture. *Journal of Applied Aquaculture, 13*(3–4), 179–204.

Omosowone, O. O., & Ogunrinde, A. M. (2018). Effect of partial or total supplementation of soybean meal with fluted pumpkin (*Telfairia occidentalis*) seed meal in the diet of hybrid catfish (Heteroclarias) fingerlings. *Food Science & Nutrition, 6*(5), 1190–1195. https://doi.org/10.1002/fsn3.659

Ortega, V. A., Lovejoy, D. A., & Bernier, N. J. (2013). Appetite-suppressing effects and interactions of centrally administered corticotropin-releasing factor, urotensin I and serotonin in rainbow trout (*Oncorhynchus mykiss*). *Frontiers in Neuroscience, 7*, 196.

O'Shea, E. F., Cotter, P. D., Stanton, C., Ross, R. P., & Hill, C. (2012). Production of bioactive substances by intestinal bacteria as a basis for explaining probiotic mechanisms: Bacteriocins and conjugated linoleic acid. *International Journal of Food Microbiology, 152*(3), 189–205.

Ou, W., Yu, G., Zhang, Y., & Mai, K. (2021). Recent progress in the understanding of the gut microbiota of marine fishes. *Marine Life Science & Technology, 3*, 434–448.

Pandiyan, P., Balaraman, D., Thirunavukkarasu, R., George, E. G. J., Subaramaniyan, K., Manikkam, S., & Sadayappan, B. (2013). Probiotics in aquaculture. *Drug Invention Today, 5*(1), 55–59.

Patil, Y., Gooneratne, R., & Ju, X. H. (2020). Interactions between host and gut microbiota in domestic pigs: A review. *Gut Microbes, 11*(3), 310–334.

Perera, E., Simó-Mirabet, P., Shin, H. S., Rosell-Moll, E., Naya-Catalá, F., de las Heras, V., et al. (2019). Selection for growth is associated in gilthead sea bream (*Sparus aurata*) with diet flexibility, changes in growth patterns and higher intestine plasticity. *Aquaculture, 507*, 349–360.

Pérez, T., Balcázar, J. L., Ruiz-Zarzuela, I., Halaihel, N., Vendrell, D., de Blas, I., & Múzquiz, J. L. (2010). Host–microbiota interactions within the fish intestinal ecosystem. *Mucosal Immunology, 3*(4), 355–360.

Phelps, D., Brinkman, N. E., Keely, S. P., Anneken, E. M., Catron, T. R., Betancourt, D., et al. (2017). Microbial colonization is required for normal neurobehavioral development in zebrafish. *Scientific Reports, 7*(1), 11244.

Piazzon, M. C., Naya-Català, F., Perera, E., Palenzuela, O., Sitjà-Bobadilla, A., & Pérez-Sánchez, J. (2020). Genetic selection for growth drives differences in intestinal microbiota composition and parasite disease resistance in gilthead sea bream. *Microbiome, 8*, 1–17.

Pirhonen, J., Schreck, C. B., & Gannam, A. (2000). Appetite of Chinook salmon (*Oncorhynchus tshawytscha*) naturally infected with bacterial kidney disease. *Aquaculture, 189*(1–2), 1–10.

Ramos, M. A., Weber, B., Gonçalves, J. F., Santos, G. A., Rema, P., & Ozório, R. O. A. (2013). Dietary probiotic supplementation modulated gut microbiota and improved growth of juve-

nile rainbow trout (*Oncorhynchus mykiss*). *Comparative Biochemistry and Physiology Part A: Molecular & Integrative Physiology, 166*(2), 302–307.
Read, M. N., & Holmes, A. J. (2017). Towards an integrative understanding of diet–host–gut microbiome interactions. *Frontiers in Immunology, 8, 538.*
Ribet, D., & Cossart, P. (2015). How bacterial pathogens colonize their hosts and invade deeper tissues. *Microbes and Infection, 17*(3), 173–183.
Ringø, E., & Song, S. K. (2016). Application of dietary supplements (synbiotics and probiotics in combination with plant products and β-glucans) in aquaculture. *Aquaculture Nutrition, 22*(1), 4–24.
Ringø, E., Sperstad, S., Kraugerud, O. F., & Krogdahl, Å. (2008). Use of 16S rRNA gene sequencing analysis to characterize culturable intestinal bacteria in Atlantic salmon (*Salmo salar) fed diets with cellulose or non-starch polysaccharides from soy. Aquaculture Research, 39*(10), 1087–1100.
Rodriguez-Estrada, U., Satoh, S., Haga, Y., Fushimi, H., & Sweetman, J. (2013). Effects of inactivated *Enterococcus faecalis* and mannan oligosaccharide and their combination on growth, immunity, and disease protection in rainbow trout. *North American Journal of Aquaculture, 75*(3), 416–428.
Sandrini, S., Aldriwesh, M., Alruways, M., & Freestone, P. (2015). Microbial endocrinology: Host–bacteria communication within the gut microbiome. *Journal of Endocrinology, 225*(2), R21–R34.
Schmidt, V., Gomez-Chiarri, M., Roy, C., Smith, K., & Amaral-Zettler, L. (2017). Subtle microbiome manipulation using probiotics reduces antibiotic-associated mortality in fish. *Msystems, 2*(6), 10–1128.
Sekirov, I., & Finlay, B. B. (2009). The role of the intestinal microbiota in enteric infection. *The Journal of Physiology, 587*(17), 4159–4167.
Semova, I., Carten, J. D., Stombaugh, J., Mackey, L. C., Knight, R., Farber, S. A., & Rawls, J. F. (2012). Microbiota regulate intestinal absorption and metabolism of fatty acids in the zebrafish. *Cell Host & Microbe, 12*(3), 277–288.
Shade, A., & Handelsman, J. (2012). Beyond the Venn diagram: the hunt for a core microbiome. *Environmental Microbiology, 14*(1), 4–12.
Sheng, Y., Ren, H., Limbu, S. M., Sun, Y., Qiao, F., Zhai, W., et al. (2018). The presence or absence of intestinal microbiota affects lipid deposition and related genes expression in zebrafish (*Danio rerio*). *Frontiers in Microbiology, 9*, 1124.
Siriyappagouder, P., Galindo-Villegas, J., Lokesh, J., Mulero, V., Fernandes, J. M., & Kiron, V. (2018). Exposure to yeast shapes the intestinal bacterial community assembly in zebrafish larvae. *Frontiers in Microbiology, 9*, 1868.
Stephens, W. Z., Burns, A. R., Stagaman, K., Wong, S., Rawls, J. F., Guillemin, K., & Bohannan, B. J. (2016). The composition of the zebrafish intestinal microbial community varies across development. *The ISME Journal, 10*(3), 644–654.
Strandwitz, P. (2018). Neurotransmitter modulation by the gut microbiota. *Brain Research, 1693*, 128–133.
Sudo, N. (2014). Microbiome, HPA axis and production of endocrine hormones in the gut. In *Microbial endocrinology: the microbiota-gut-brain axis in health and disease* (pp. 177–194). New York, Springer.
Sun, Y., Yang, H., Ling, Z., Chang, J., & Ye, J. (2009). Gut microbiota of fast and slow growing grouper *Epinephelus coioides*. *African Journal of Microbiology Research, 3*(11), 713–720.
Tacon, A. G., & Metian, M. (2008). Global overview on the use of fish meal and fish oil in industrially compounded aquafeeds: Trends and future prospects. *Aquaculture, 285*(1–4), 146–158.
Talwar, C., Nagar, S., Lal, R., & Negi, R. K. (2018). Fish gut microbiome: Current approaches and future perspectives. *Indian Journal of Microbiology, 58*, 397–414.
Tamayo, D., Ibarrola, I., Urrutia, M., & Navarro, E. (2011). The physiological basis for inter-individual growth variability in the spat of clams (*Ruditapes philippinarum*). *Aquaculture, 321*, 113–120. https://doi.org/10.1016/j.aquaculture.2011.08.024

Tang, Y., Ma, K., Cheung, M. K., et al. (2021). Gut microbiota in decapod shrimps: Evidence of Phylosymbiosis. *Microbial Ecology, 82*, 1–14. https://doi.org/10.1007/s00248-021-01720-z

Tapia-Paniagua, S. T., Reyes-Becerril, M., Ascencio-Valle, F., Esteban, M. Á., Clavijo, E., Balebona, M. C., & Moriñigo, M. A. (2011). Modulation of the intestinal microbiota and immune system of farmed *Sparus aurata* by the administration of the yeast *Debaryomyces hansenii* L2 in conjunction with inulin. *Journal of Aquaculture Research & Development, 1*(12), 10–4172.

Topic Popovic, N., Strunjak-Perovic, I., Sauerborn-Klobucar, R., Barisic, J., Jadan, M., Kazazic, S., et al. (2017). The effects of diet supplemented with Lactobacillus rhamnosus on tissue parameters of rainbow trout, *Oncorhynchus mykiss* (Walbaum). *Aquaculture Research, 48*(5), 2388–2401.

Torrecillas, S., Rimoldi, S., Montero, D., Serradell, A., Acosta, F., Fontanillas, R., et al. (2023). Genotype x nutrition interactions in European sea bass (*Dicentrarchus labrax*): Effects on gut health and intestinal microbiota. *Aquaculture, 574*, 739639.

Trinh, L. T., Bakke, I., & Vadstein, O. (2017). Correlations of age and growth rate with microbiota composition in Atlantic cod (*Gadus morhua*) larvae. *Scientific Reports, 7*(1), 8611.

Turchini, G. M., Trushenski, J. T., & Glencross, B. D. (2019). Thoughts for the future of aquaculture nutrition: Realigning perspectives to reflect contemporary issues related to judicious use of marine resources in aquafeeds. *North American Journal of Aquaculture, 81*(1), 13–39.

Turnbaugh, P. J., Ley, R. E., Mahowald, M. A., Magrini, V., Mardis, E. R., & Gordon, J. I. (2006). An obesity-associated gut microbiome with increased capacity for energy harvest. *Nature, 444*(7122), 1027–1031.

Vatsos, I. N. (2016). Standardizing the microbiota of fish used in research. *Laboratory animals, 51(4), 353–364.*

Venkatesh, M., Mukherjee, S., Wang, H., Li, H., Sun, K., Benechet, A. P., et al. (2014). Symbiotic bacterial metabolites regulate gastrointestinal barrier function via the xenobiotic sensor PXR and Toll-like receptor 4. *Immunity, 41*(2), 296–310.

Vigneri, S. (2014). The brain-gut axis: From pathophysiology to possible future strategies of treatment. *Brain Disorders & Therapy, 3*(5), 10–4172.

Volkoff, H., & Peter, R. E. (2004). Effects of lipopolysaccharide treatment on feeding of goldfish: Role of appetite-regulating peptides. *Brain Research, 998*(2), 139–147.

Wang, W., Wu, S., Zheng, Y., Cheng, Y., Li, W., Zou, H., & Wang, G. (2015). Characterization of the bacterial community associated with early-developmental stages of grass carp (*Ctenopharyngodon idella*). *Aquaculture Research, 46*(11), 2728–2735.

Wang, A. R., Ran, C., Ringø, E., & Zhou, Z. G. (2018). Progress in fish gastrointestinal microbiota research. *Reviews in Aquaculture, 10*(3), 626–640.

Welch, A., Hoenig, R., Stieglitz, J., Benetti, D., Tacon, A., Sims, N., & O'Hanlon, B. (2010). From fishing to the sustainable farming of carnivorous marine finfish. *Reviews in Fisheries Science, 18*(3), 235–247.

Whiteside, S. A., Razvi, H., Dave, S., Reid, G., & Burton, J. P. (2015). The microbiome of the urinary tract—A role beyond infection. *Nature Reviews Urology, 12*(2), 81–90.

Wong, S., Stephens, W. Z., Burns, A. R., Stagaman, K., David, L. A., Bohannan, B. J., et al. (2015). Ontogenetic differences in dietary fat influence microbiota assembly in the zebrafish gut. *MBio, 6*(5), 10–1128.

Wu, Z. Q., Jiang, C., Ling, F., & Wang, G. X. (2015). Effects of dietary supplementation of intestinal autochthonous bacteria on the innate immunity and disease resistance of grass carp (*Ctenopharyngodon idellus*). *Aquaculture, 438*, 105–114.

Xie, M., Xie, Y., Li, Y., Zhou, W., Zhang, Z., Yang, Y., et al. (2021). The effects of fish meal replacement with ultra-micro ground mixed plant proteins (uPP) in practical diet on growth, gut and liver health of common carp (*Cyprinus carpio*). *Aquaculture Reports, 19*, 100558.

Yan, Q., Li, J., Yu, Y., Wang, J., He, Z., Van Nostrand, J. D., et al. (2016). Environmental filtering decreases with fish development for the assembly of gut microbiota. *Environmental Microbiology, 18*(12), 4739–4754.

Yang, N. J., & Chiu, I. M. (2017). Bacterial signaling to the nervous system through toxins and metabolites. *Journal of Molecular Biology, 429*(5), 587–605.

Yano, J. M., Yu, K., Donaldson, G. P., Shastri, G. G., Ann, P., Ma, L., et al. (2015). Indigenous bacteria from the gut microbiota regulate host serotonin biosynthesis. *Cell, 161*(2), 264–276.

Ye, J. D., Wang, K., Li, F. D., & Sun, Y. Z. (2011). Single or combined effects of fructo-and mannan oligosaccharide supplements and Bacillus clausii on the growth, feed utilization, body composition, digestive enzyme activity, innate immune response and lipid metabolism of the Japanese flounder *Paralichthys olivaceus*. *Aquaculture Nutrition, 17*(4), e902–e911.

Yiğit, M., Bulut, M., Ergün, S., Güroy, D., Karga, M., Kesbiç, O. S., et al. (2012). Utilization of corn gluten meal as a protein source in diets for gilthead sea bream (*Sparus aurata* L.) juveniles. *Journal of FisheriesSciences. com, 6*(1), 63–73.

Yıldırım, Ö., Acar, Ü., Türker, A., Sunar, M. C., & Kesbic, O. S. (2014). Effects of replacing fish meal with peanut meal (*Arachis hypogaea*) on growth, feed utilization and body composition of Mozambique Tilapia Fries (*Oreochromis mossambicus*). *Pakistan Journal of Zoology, 46*(2), 377.

Zeng, Q., Li, D., He, Y., Li, Y., Yang, Z., Zhao, X., ... & Dai, W. (2019). Discrepant gut microbiota markers for the classification of obesity-related metabolic abnormalities. *Scientific Reports, 9(1), 13424.*

Zhang, L. S., & Davies, S. S. (2016). Microbial metabolism of dietary components to bioactive metabolites: Opportunities for new therapeutic interventions. *Genome Medicine, 8*, 1–18.

Zhang, Z., & Li, D. (2018). Thermal processing of food reduces gut microbiota diversity of the host and triggers adaptation of the microbiota: Evidence from two vertebrates. *Microbiome, 6*, 1–14.

Zhang, Y., Wen, B., David, M. A., Gao, J. Z., & Chen, Z. Z. (2021). Comparative analysis of intestinal microbiota of discus fish (*Symphysodon haraldi*) with different growth rates. *Aquaculture, 540*, 736740.

Zheng, Y., Yu, M., Liu, J., Qiao, Y., Wang, L., Li, Z., et al. (2017). Bacterial community associated with healthy and diseased Pacific white shrimp (*Litopenaeus vannamei*) larvae and rearing water across different growth stages. *Frontiers in Microbiology, 8*, 1362.

Zhou, L., Limbu, S. M., Shen, M., Zhai, W., Qiao, F., He, A., et al. (2018). Environmental concentrations of antibiotics impair zebrafish gut health. *Environmental Pollution, 235*, 245–254.

Chapter 20
Nutritional Strategies for Improving Reproductive Performance in Fishes

Pujadebi Bera, Hiranmoy Dhara, Sukalpa Mandal, Surya Kanta Sau, and Srinibas Das

1 Introduction

Enhancing reproductive performance in fish through effective nutritional strategies is essential for sustainable aquaculture and the production of superior offspring. Providing a well-balanced diet rich in key nutrients, including proteins, lipids, vitamins, and minerals, is crucial for improving gamete quality and gonadal development. Vital elements such as omega-3 fatty acids, antioxidants like vitamins C and E, and trace minerals like zinc and selenium offer significant benefits. Additionally, customized broodstock diets, functional feed supplements, and probiotics help optimize reproductive success by promoting hormone production, strengthening immune function, and improving nutrient utilization. By tailoring these nutritional interventions (Fig. 20.1) to specific species and environmental conditions, aquaculture operations can achieve enhanced reproductive performance and greater overall efficiency.

2 Balanced Diet

A balanced diet plays a crucial role in optimizing fish reproductive performance. Adequate levels of protein, lipids, carbohydrates, vitamins, and minerals are essential for gamete development, fertilization, embryonic development, larval growth,

P. Bera (✉) · H. Dhara · S. K. Sau
Department of Aquaculture, Faculty of Fishery Sciences, West Bengal University of Animal and Fishery Sciences, Kolkata, West Bengal, India

S. Mandal · S. Das
Department of Fish Nutrition, Faculty of Fishery Sciences, West Bengal University of Animal and Fishery Sciences, Kolkata, West Bengal, India

I. Ahmed, I. Ahmad (eds.), *Aquaculture: Enhancing Food Security and Nutrition*, https://doi.org/10.1007/978-3-031-92858-1_20

Fig. 20.1 Nutritional strategies in fish reproduction

and parental care. Protein provides the building blocks for tissue growth and repair, while lipids serve as an energy source and are crucial for membrane structure. Carbohydrates provide energy for metabolic processes. Vitamins and minerals are essential for various physiological functions, including hormone production, immune function, and antioxidant defense (Halver & Hardy, 2002). A well-balanced diet thus ensures that fish have the necessary nutrients to support their reproductive needs, resulting in increased fecundity, improved egg quality, enhanced larval survival, and overall reproductive success.

2.1 Effect of Dietary Protein

Protein is an essential nutrient for fish, playing a critical role in reproductive performance. Within the broad framework of "Fish Nutrition," protein is considered a primary and indispensable component. It serves as the sole nitrogen source for synthesizing amino acids and building proteins in fish. Although the term "protein requirement" is used for consistency with existing literature, it is worth noting that body protein remains in a state of dynamic equilibrium, and requirements vary significantly depending on the species and life stage. Insufficient protein intake leads to growth retardation as protein is redirected from less vital tissues to sustain critical functions. Protein needs differ not only between species but also within the same species based on factors such as age, life stage, physiological activity, sex, and environmental conditions (Ahmed & Ahmad, 2020). To achieve optimal growth and health in fish, all essential amino acids and proteins must be provided in adequate amounts, tailored to the specific requirements of each species.

2.1.1 Optimal Protein Levels

The optimal protein level in fish diets depends on different factors, as discussed earlier, with the quality of the protein source being a crucial consideration. High-quality sources like fish meal or insect meal are often preferred due to their superior amino acid profiles. A balanced amino acid composition is essential for efficient protein utilization, as fish cannot synthesize essential amino acids and must obtain them through their diet (Craig et al., 2017). Research has shown similarities in protein and amino acid requirements among major carps. For instance, the growth and weight gain of Catla and Rohu fingerlings were comparable when fed diets containing 30–40% protein. Studies also revealed that the essential amino acid requirements of Catla fry closely resemble those of common carp, with minor differences in individual amino acids. Additionally, incorporating nonprotein energy sources like carbohydrates and lipids can reduce protein catabolism, allowing for lower dietary protein levels without compromising growth. For example, *C. mrigala* fry demonstrated similar growth and feed conversion ratios (FCR) when fed diets with 30% protein and 40% dextrin, highlighting a protein-sparing effect. Extensive research has also been conducted on protein requirements in catfish species, with channel catfish (*Ictalurus punctatus*) being a primary focus due to its widespread cultivation. Studies have investigated protein needs for maintenance and growth in small channel catfish fed purified diets, providing valuable insights for optimizing their nutrition.

2.1.2 Protein Requirement for Reproduction

Reproductive performance and egg quality in fish are influenced by various factors, including nutrition, stress, egg over-ripening, genetics, broodstock age, water quality, and environmental conditions. Protein, a fundamental building block for tissues such as gonads, plays a pivotal role in reproductive health (Ogino & Chen, 1973). Adequate protein intake supports (Fig. 20.2) the development of healthy ovaries and testes, facilitates the synthesis of protein-based hormones like gonadotropins and sex steroids, and enhances gonadal development and fecundity, particularly in synchronized spawners like *P. hypophthalmus*. In females, dietary protein directly impacts ovarian development, spawning rates, gametogenesis, and egg quality, while in males, it contributes to motile and fertile sperm production, improving sperm count and fertilization success. Protein also supports embryonic development by providing essential amino acids for cell division and differentiation, as well as larval growth post-hatching.

During reproductive maturation, the energy demands shift: fat is initially utilized, but as fat reserves diminish, muscle protein becomes a key energy source. Protein is vital for yolk formation in eggs, secondary sexual character development, and morphological changes, as cartilage, a primary component of these characters, cannot be synthesized from fat. In migratory fish, like salmon, protein may also serve as an energy reserve during long-distance migrations. The transfer of muscle

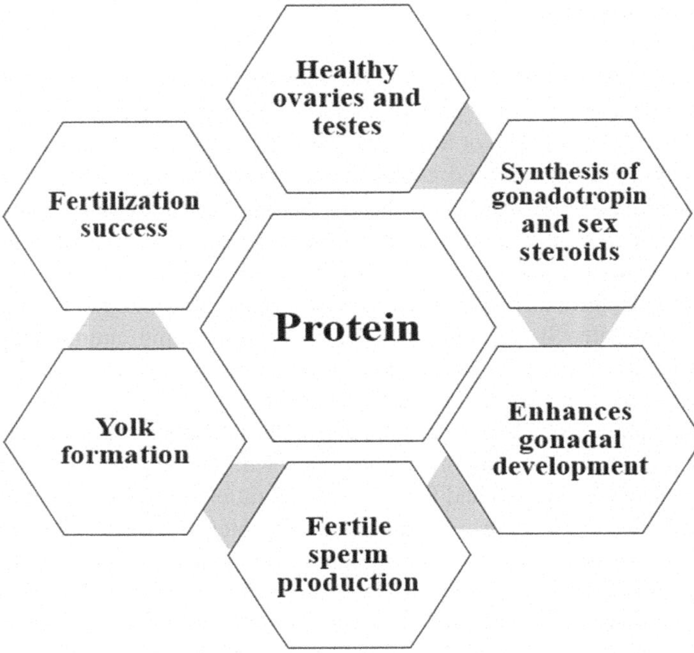

Fig. 20.2 Requirement of protein in fish reproduction

protein to ovaries in females and to secondary sexual structures in males underscores the critical role of protein in reproductive development (Bowen, 1987). The gonadosomatic index, calculated as the gonad weight relative to body weight, provides an indirect measure of energy allocation to reproduction, reflecting the intricate balance of nutritional and physiological demands during this life stage (Symons, 1970).

2.2 *Effect of Dietary Lipids and Fatty Acids*

Dietary lipids, composed of fats and oils (triglycerides), play a critical role in fish as energy sources in metabolic processes, stress resistance, immunity development, hormone precursors, and essential fatty acids (EFAs). These lipids and their fatty acid (FA) components provide the necessary energy for gonadogenesis and embryogenesis, serving as integral elements of cell structures and organelles during organ development (Nandi et al., 2007). The specific fatty acid requirements during these processes depend on the individual levels and proportions of dietary FAs, which significantly influence reproductive success. Adjusting lipid levels in broodstock diets has been shown to enhance reproductive performance in various fish species. For instance, studies on white seabream (*Diplodus sargus*) and rainbow trout

(*Oncorhynchus mykiss*) demonstrate that dietary lipids influence lipid movement to gonads and the FA composition in reproductive tissues, directly impacting gonadogenesis. In Chinese sturgeon (*Acipenser sinensis*), the depletion of muscle lipid reserves during ovarian development and the ability to reach advanced ovarian stages on high-lipid diets highlight the combined role of exogenous food and endogenous lipid sources (liver, muscle, and visceral fat) in reproduction. Including long-chain polyunsaturated fatty acids (LCPUFAs) in fish feed improves egg quality, fertility, and fertilization success. Lipids are essential for both female egg development and male reproductive behaviors, such as enhanced swimming, parental care, and courtship. Additionally, they act as precursors for steroid hormones and prostaglandins (PGs) and as transporters for fat-soluble vitamins in the intestine. Research in freshwater fish species, such as Nile tilapia (*Oreochromis niloticus*), striped catfish (*Pangasianodon hypophthalmus*), and common carp (*Cyprinus carpio*), has underscored the significance of lipids in processes like gametogenesis, fecundity, fertilization, egg morphology, hatching success, larval quality, and survival, solidifying their critical role in reproductive nutrition (Zakeri et al., 2011; Mir et al., 2020).

2.2.1 Gametogenesis

Gamete formation in fish involves processes such as generation, motility, multiplication, specialization, and ovulation or spermiation, which are consistent across species. In wild fish, starvation and insufficient fat reserves often lead to skipped spawning, highlighting the critical role of proper nutrition in effective gametogenesis. Including dietary lipids in broodstock diets in appropriate quantities positively influences reproductive performance and seed production. For example, inclusion of fish oil at levels of 10%, 14%, and 18% in Chinese sturgeon broodstock diets promoted puberty onset and ovarian development. Fish fed diets with 10% and 14% lipid achieved stage II ovarian development, with small white eggs measuring 0.3–0.9 mm in diameter, whereas those fed 18% lipid progressed to stage III, producing yellow oocytes (2.0–2.6 mm) and grey oocytes (2.6–3.7 mm). Insufficient dietary lipid intake was linked to stagnation in ovarian development from stage II to stage III due to inadequate lipid storage. Similar benefits of dietary lipid inclusion have been observed in rabbitfish (*Siganus guttatus*), where proper broodstock nutrition and balanced lipid sources promoted early puberty and gamete formation in both sexes. By optimizing broodstock feeding practices and incorporating sufficient lipid levels, early oocyte maturation and improved seed quality can be achieved, ensuring better reproductive outcomes in fish farming.

2.2.2 Fecundity

Fecundity, defined as the total egg production by fish either as the number of eggs per spawn or per unit of body weight, is a key indicator of reproductive performance and is heavily influenced by nutrient intake. Nutritional deficiencies in broodstock diets have been shown to negatively impact both the quantity and quality of eggs produced. Studies have demonstrated that increasing lipid levels in broodstock diets can enhance fecundity and the gonadosomatic index (GSI). For instance, raising lipid levels from 100 g/kg to 180 g/kg in snakehead murrel (*Channa striata*) diets significantly improved these parameters. Similarly, increasing lipid levels from 60 g/kg to 120 g/kg in striped catfish (*Pangasianodon hypophthalmus*) diets (Kabir et al., 2012) and including 110 g/kg lipids in common carp (*Cyprinus carpio*) diets led to higher GSI and fecundity. The observed increase in fecundity with higher dietary lipid levels is attributed to the corresponding rise in essential fatty acid (EFA) content, which plays a crucial role in enhancing reproductive performance in fish.

2.2.3 Fertilization

Fertilization serves as an important parameter to accurately estimate the quality of an egg. An increase in lipid levels from 60 g/kg to 180 g/kg in the broodstock diet of striped catfish, *P. hypophthalmus*, resulted in a significantly higher fertilization rate. A better fertilization rate in the eggs of common carp, *C. carpio*, is reported to be at the optimal crude lipid level of 14%.

2.2.4 Egg Morphology and Hatching Rate

The quality of fish eggs is typically assessed based on fertilization rates and the successful development of viable offspring. Several factors influence egg quality, including the size, age, nutrition, genetics, origin, and environmental conditions of the female fish. Lipids are particularly important in broodstock nutrition, significantly impacting both egg quality and larval development. Different dietary lipid sources have been found to have notable effects on egg quality, with a direct correlation between the fatty acid composition of broodstock diets and the eggs produced. For example, a study on European eel broodstock demonstrated that diets rich in essential fatty acids (EFAs) resulted in a higher number of floating eggs, increased total lipid content in the eggs, and improved hatching success (Zakeri et al., 2009; Zakeri et al., 2011).

2.2.5 Larval Quality and Survival

The quality and survival of larvae are closely linked to the provision of adequate dietary lipids. Fish larvae, upon hatching, often have underdeveloped mouthparts and intestinal tracts, preventing them from immediately beginning exogenous feeding. Instead, they rely on endogenous energy reserves from yolk during embryogenesis and the early stages of larval growth before initiating external feeding. Studies have shown that European eel broodstock fed diets enriched with essential fatty acids (EFAs), including varying ratios of arachidonic acid (ARA), docosahexaenoic acid (DHA), and eicosapentaenoic acid (EPA), resulted in higher total lipid content in the eggs and improved larval survival. Similarly, increasing the levels of n-3 long-chain polyunsaturated fatty acids (LCPUFAs) in the diet of European perch (*Perca fluviatilis*) broodstock significantly enhanced larval weight, survival rate, and osmotic shock tolerance. Likewise, common carp broodstock fed a diet containing 11% lipids showed the lowest mortality rate in larvae (Sink & Lochmann, 2008) (Fig. 20.3).

2.3 Effect of Dietary Carbohydrate

Carbohydrates play an important role in fish reproduction by serving as a readily available energy source, often complementing the functions of proteins and lipids. Their contributions are evident in various reproductive processes, such as energy provision for behaviors like migration, courtship, and territorial defense. The distribution of carbohydrate reserves varies across tissues during reproductive phases: the brain and pituitary show the highest carbohydrate content during the prespawning phase, while the ovaries peak during spawning. In contrast, carbohydrate levels in the liver and muscles are highest in the preparatory phase and lowest during spawning. This reflects a dynamic interplay of metabolic changes among the nervous, endocrine, somatic, and reproductive systems (Wilson, 1994).

For instance, in rainbow trout (*Oncorhynchus mykiss*), dietary carbohydrate utilization increases during gonadal maturation, with significant metabolic adjustments observed in the liver and gonads at the onset of ovarian recrudescence (Kvale et al., 2007). Carbohydrates are typically stored as glycogen in the muscles and

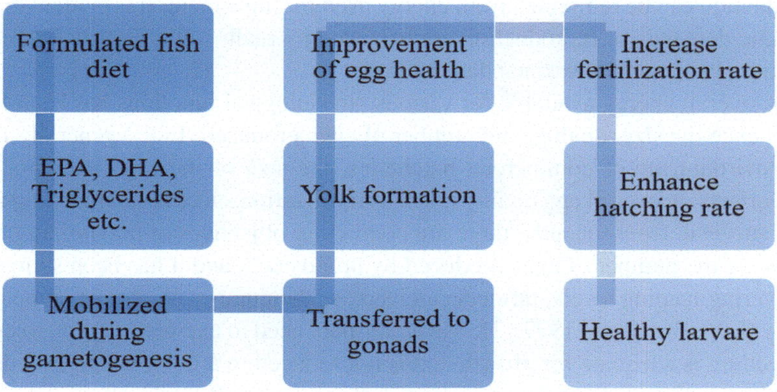

Fig. 20.3 Action of lipids and fatty acids during fish reproduction

liver, which is mobilized rapidly during high-energy-demand periods. In species such as Plaice (*Pleuronectus platessa*) and Arctic capelin (*Mallotus villosus*), marked depletion of glycogen and body lipids occurs during gonadal development, highlighting their role as critical energy reserves. Additionally, carbohydrates are vital for developing embryos, supplying energy for cell division, growth, and differentiation. They also support the rapid growth of larvae, particularly during the transition from endogenous to exogenous feeding. In *Trichogaster pectoralis*, however, carbohydrates are present in negligible amounts in female gonads, underscoring species-specific differences in carbohydrate metabolism during reproduction. A balanced diet with adequate carbohydrates is therefore essential for maintaining energy demands, supporting vitellogenesis, and ensuring optimal reproductive outcomes in fish.

2.4 Energy Partitioning

Animals allocate energy among the different physiological functions necessary for maintenance, growth, and reproduction. Consequently, energy cannot be used for more than one function; for instance, energy allocated to growth cannot simultaneously be used for reproduction. Typically, individual maintenance needs are prioritized, and any surplus energy is then distributed between growth and reproduction. The way energy is divided between growth and reproduction varies not only across different species but also among various strains within a single species, making further generalizations challenging. Generally, the energy intake is divided among protein (for growth), fat (for energy storage), and reproduction (for gonadal development) (Brody, 1945). Fishes that do not use energy for reproductive activities tend to accumulate excess energy as fat instead of enhancing their somatic growth or body protein. Increased energy allocated to reproduction comes at the cost of somatic growth. Research on energy allocation has shown that energy intended for reproduction can be redirected from energy used for movement. It is evident that the energetic demands of reproduction are balanced by reallocating energy from other activities (Pandian & Vivekanandan, 1985).

The overall energy available for various physiological functions has been found to influence the size, quality, and number of eggs produced. In the generally favorable environments of commercial hatcheries, the size of the eggs does not have direct effects on overall egg quality (such as fertilization success and hatching rates) or fry survival. Consequently, there are conflicting objectives in maximizing either the size or the number of eggs produced by broodstock, and it has been shown that by lowering feeding levels, farmers can increase the number of eggs of comparable quality (Li & Brocksen, 1977). However, farmers need to exercise caution to ensure that feeding is adequate for growth and is not reduced to a point where broodstock individuals can only meet their basic maintenance needs (Fig. 20.4).

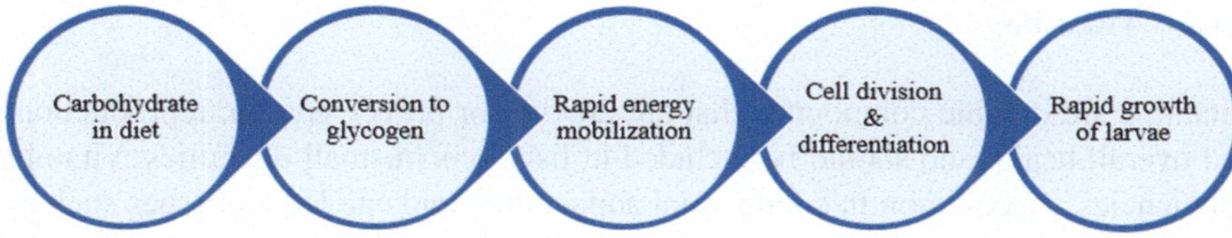

Fig. 20.4 Effect of carbohydrate in fish reproduction

3 Functional Additives Influencing Reproductive Performances

3.1 Nutritional Supplementation

3.1.1 Amino Acids

Amino acids are the fundamental building blocks of proteins that are essential for fish at all stages of development. They serve as energy substrates and are involved in various metabolic pathways. About 50% of the amino acids ingested by fish are stored in muscle tissue. However, since fish cannot synthesize all the amino acids they need, these must be supplied through their diet. Recognizing that amino acid requirements differ by species and growth stages, the National Research Council (NRC) has provided recommendations for the essential amino acids that are crucial for fish. Broodstock diets should provide essential amino acids either through protein or other supplements (Grisdale-Helland et al., 2013). The amino acid composition of broodstock diets has been studied in species like *S. dumerili*, *S. dorsalis*, and *Sparus aurata*. While essential amino acids are crucial for fish, certain nonessential amino acids, such as taurine, also play a significant role in reproductive performance, as seen in *S. quinqueradiata.* However, there is limited documentation on the inclusion of taurine in broodstock diets. Further research is needed to determine the optimal dosage and to confirm its effectiveness across different fish species. The impact of taurine supplementation on spawning in marine fish was studied in species from the *Seriola* genus. In *S. quinqueradiata*, the absence of taurine in the broodstock diet led to reduced spawning success, while supplementation at 1.0% resulted in higher hatching success and improved spawning quality (Costas et al., 2008). Similarly, in *S. dorsalis* broodstock, taurine supplementation at 2.67% significantly enhanced reproductive performance compared to those without taurine. However, in *S. dumerili*, histidine supplementation at 0.5–1.5% proved more effective than taurine in improving reproductive performance. Still, taurine supplementation at levels ranging from 0.3 to 1.1% resulted in better fecundity compared to broodstock fed a higher protein diet. Nevertheless, it is crucial to continue expanding knowledge on amino acid supplementation, as there is relatively limited information available regarding their potential impact on the reproductive performance of marine fish.

3.1.2 Vitamins

Vitamins are organic compounds that are crucial for proper growth, reproduction, and overall health and should be included in fish diets in small quantities. Vitamin deficiencies are common in commercial aquaculture and can lead to issues such as abnormal swimming, skin diseases, bone deformities, edema, eye and gill problems, hemorrhage, liver disease, stunted growth, and other remarkable symptoms (Andia et al., 2001). The most important vitamins for fish diets include fat-soluble vitamins A, D, E, and K, as well as water-soluble vitamins C and the B vitamins. During fertilization, certain nutrients are especially important for fish reproduction, such as vitamin E (alpha-tocopherol), vitamin C (ascorbic acid), and carotenoids. In salmon, vitamin C is vital for reproduction, steroidogenesis, and yolk sac production. Both vitamins C and E act as antioxidants, helping to protect sperm during spermatogenesis and fertilization, reducing the risk of lipid peroxidation, which negatively impacts sperm motility. A deficiency in vitamin C has been shown to reduce sperm motility during egg laying. Supplementing the diet of *Sparus aurata* with 22–125 mg/kg of vitamin E (alpha-tocopherol) significantly reduced abnormal egg incidence and improved fertility and vitality. In salmon, the optimal vitamin E requirement for successful spawning is 250 mg/kg. On the other hand, low levels of vitamin E (<22 mg/kg) have been linked to lower reproductive and breeding success. Vitamin C in the broodstock diet is critical for collagen synthesis during embryonic development and enhances embryonic survival. Additionally, vitamin A is essential for egg laying, larval development, gonadal maturation, and embryo development (Mejri et al., 2017) (Table 20.1).

3.1.3 Minerals

Aquatic organisms have a unique physiological mechanism for absorbing minerals both from their diet and the surrounding aquatic environment. However, research on the role of minerals in fish nutrition has progressed slowly. Ten minerals—calcium, phosphorus, magnesium, potassium, copper, iron, zinc, manganese, selenium, and iodine—are essential for fish health and development. Macro-minerals such as calcium, magnesium, phosphorus, potassium, and sodium are crucial for the formation of bones and other hard tissues like fins, scales, and exoskeletons. While fish can often obtain these minerals from the environment, deficiencies in their diet can lead to various physiological issues. For instance, a lack of phosphorus, magnesium, and zinc can cause bone deformities, while copper deficiency can impair growth and lead to cataract formation. Although there is limited information on the mineral requirements of marine broodstock, it is clear that understanding the mineral needs of juvenile fish is an important step in formulating optimal broodstock diets. Additionally, while commercial multivitamin complexes are commonly included in diets, further exploration of mineral requirements is essential for improving broodstock nutrition.

Table 20.1 A summary of the effects of vitamins on reproductive performance of different farmed fish species

Species	Dietary vitamin levels		Description	References
	Treatment	Optimum		
Kissing gourami, (*Helostoma temminckii*) (broodstock)	Vit E 300–700 mg kg^{-1}	300–700 mg kg^{-1}	Increased gonadosomatic index (GSI), hepatosomatic index (HSI), and higher fecundity and larger egg dia	Fitriliyani et al. (2022)
P. Hypophthalmus (broodstock)	Vit E 28.08, 146.65, 189.65 mg kg^{-1}	189.65 mg kg^{-1}	Improvement in gonadosomatic index, fecundity, egg diameter, hatching rate, reduction in the number of abnormal larvae, and increased larvae number	Yulfiperius et al. (2003)
Japanese eel, *Anguilla japonica* (Broodstock)	Vit C 32, 206,423, 840, 1686 mg kg^{-1}	410.8 and 911.8 mg kg^{-1}	Increased GSI, initiates the proliferation of spermatogonia, changes in testes color, and increased collagen level	Shahkar et al. (2015)
Rainbow trout, *Oncorhynchus mykiss* (Broodstock)	Vit A 0, 20, 700 or 0, 20, 200 IU/g	Vit A 0, 20, 700 or 0, 20, 200 IU/g	Females fed diet 200 showed better fecundity and increased growth. Females fed 700 showed a decline in survival from the eyed stage	Fontagn ́e-Dicharry et al. (2010)

3.2 Non-nutrient Supplementation

3.2.1 Carotenoids

Carotenoids in the diet play a significant role in animal health by acting as antioxidants, helping to neutralize free radicals produced during normal cellular functions and stress. The antioxidant effects of carotenoids vary depending on their structure and the type of oxidant involved. For example, carotenoids are known to be strong inhibitors of simple oxygen, while astaxanthin, a specific carotenoid, has been shown to slow oxidation rates. Additionally, carotenoids contribute to hatching success, larval viability, and egg pigmentation. They can enhance the immune system of fish and improve larval survival, as they are precursors to vitamin A. Carotenoids are quickly absorbed by the ovary and larvae, often within 48 hours, and influence egg pigmentation. Astaxanthin, in particular, is a carotenoid that significantly impacts egg quality. In a study with *S. quinqueradiata*, adding 10% krill meal to their diet resulted in eggs with a strong yellowish color and higher levels of zeaxanthin and lutein. However, total egg production, hatching rate, fertilization success, and larval survival were greater in broodstock fed a diet without krill meal compared to those fed diets containing 20% or 30% krill meal. A possible contributing factor could be that the diet without krill included 20% more fish meal. Similarly, broodstock that received dry pellets with 30 ppm of astaxanthin showed improved reproductive performance. Additionally, egg production and survival rates were generally higher when different carotenoid sources, such as dry pellets with squid meal and paprika, were used, compared to those fed pellets with astaxanthin or paprika alone. The inclusion of 73.7 mg/kg astaxanthin in the diet of *Gadus morhua* broodstock improved egg production and quality, as well as increased the number of viable and fertilized eggs, with noticeable enhancement after 15 days of egg production, thus boosting reproductive performance. Broodstock fed diets with higher water content and higher astaxanthin (100 ppm/kg with 30% water) produced more eggs, had lower mortality, and experienced higher fertilization rates compared to those given lower astaxanthin and water levels (50 ppm/kg with 30% water or 50 ppm with 10% water). The addition of β-carotene, canthaxanthin, or astaxanthin to the basal diet of *Pagrus major* also increased the percentage of fertilized eggs, although the hatching rate did not improve. However, a reduction in lipid droplet anomalies was observed. Furthermore, the addition of carotenoids from animal sources, like 10% krill meal, improved both egg and larval quality, and adding 10% squid to the diet resulted in improved egg numbers.

3.2.2 Acidifiers

Acidifiers, a group of compounds consisting of weak acids and their salts, primarily function to improve gut health by lowering the pH level in the digestive tract, can enhance nutrient absorption as well as be beneficial for bacterial growth and potentially optimize reproductive performance by ensuring proper nutrient delivery for developing eggs and sperm. Most acidifiers are organic carboxylic acids varying in carbon number from 1 to 6, along with their salts of calcium, magnesium, potassium, or sodium. Because of their low pH, these substances have bacteriostatic and bactericidal effects when added to food at weight-based concentrations of 1–3%. Additionally, they may limit the growth of gram-negative bacteria and lower the pH of the intestines and stomach. Numerous species have demonstrated that citric and formic acids lower the pH of the diet and gastrointestinal tract, which usually increases the bioavailability of dietary minerals, particularly phosphorus. It has been noted that the effects of other acids, such as potassium diformate, sodium formate, or sodium butyrate, on lowering the pH of the GI tract and affecting the digestive process are less consistent. Another acid that has been demonstrated in multiple studies to promote the growth of the Arctic char is sodium lactate. The rise in activity of digestive enzymes such as lipase, trypsin, and pepsin seemed to aid in enhancing performances; levels of leucine-aminopeptidase and phosphatases also increased with the addition of organic acids to the diet. Therefore, further research involving various acidifiers might be essential to optimize their inclusion in the diets of different aquatic species, considering the observed positive effects.

3.2.3 Beta Glucans

All naturally occurring polysaccharides that include glucose as one of their components and are linked by beta bonds are referred to as beta-glucans (BGs). These BG molecules can be found in a variety of organisms, including plants, yeasts, fungi, algae, protozoa, and bacteria. The interaction and recognition of BG molecules by pattern recognition receptors (PRRs) from fish leukocytes and other cells are influenced by their structural conformation, which encompasses branching levels, polymer charge, molecular weight, and solubility. When PAMPs engage with leukocyte PRRs, it initiates a downstream intracellular signaling cascade. This triggers transcription factors (NFAT and NF-kB), leading to the production of inflammatory cytokines and a respiratory burst. The extraction processes and methods used to prepare BGs as supplements for animal feed may involve multiple steps, depending on the source of the BG and the intended molecular configuration. For instance, several patented extraction methods for BGs could utilize enzymatic incubation, hot water and detergent washes, and acid and alkaline treatments, along with extractions facilitated by ultrasound and microwave technologies. A study by Abu-Elala et al. (2021) demonstrated that dietary fermented extracts derived from *Saccharomyces cerevisiae* (including nucleotides, β-glucans, and MOS) serve as a strategic measure to enhance tilapia production by improving the productivity and health of broodstock, as well as seed survival and performance. The bioencapsulation of BGs with live feed like rotifers could help reduce the harmful effects of acid digestion on the stability of BG molecules in stomachless fish such as common carp. Further research is recommended to explore the effects of BGs on gut microbiota using advanced techniques such as next-generation sequencing, which could provide a clearer understanding of how dietary BGs influence the intestinal bacterial community.

3.2.4 Probiotics

To improve the nutritional status of their broodstock, most fish farms use fresh organisms like squid, tubifex, daphnia, mussels, krill, cuttlefish, and small crustaceans. However, these organisms do not provide the necessary nutrient levels required by the broodstock and can increase the risk of pathogen transmission to both the parents and offspring (Sharifuzzaman & Austin, 2017). To mitigate these issues, probiotics are added to food or water to prevent infections and enhance reproductive performance, gamete quality, hatching rates, and fertilization success (Nayak, 2010). Despite the potential benefits, few studies have explored the effects of probiotics on fish reproduction (Rahman et al., 2018). One study involving the supplementation of the commercial probiotic Primalac in the artificial feeds of swordtail, *Xiphophorus helleri*, for 26 weeks found that the probiotic significantly improved gonadosomatic indices, fecundity, and larval quality of female broodstock. In a study on zebrafish, Carnevali et al. (2013) examined the effects of *Lactobacillus rhamnosus* on reproductive performance, using zebrafish as a biomedical model (Chorong et al., 2019). Their findings indicated that long-term administration of *L. rhamnosus* positively affected the physiology of the reproductive system. Sexual differentiation of gonocytes was observed at 6 weeks postfertilization, and gonadotropin-releasing hormone was detectable at the larval stage. Additionally, broodstock fed a diet supplemented with *L. rhamnosus* showed improved ovulation. However, PCR-DGGE analysis of the ovaries revealed that *L. rhamnosus* was absent, and bacterial rRNA was not detectable in the gonads, suggesting that the probiotic did not directly affect the ovary but instead operated systemically. Gioacchini et al. (2012) concluded that their results clearly demonstrated the positive effects of probiotic administration on the reproductive performance of both commercial and ornamental fish species. The significance of their findings highlighted the relationship between gut microbiota and fish reproduction (Miccoli et al., 2015).

3.2.5 Prebiotics

Prebiotics are characterized as nondigestible components of food that positively influence the host by specifically encouraging the growth of and/or enhancing the activity of certain beneficial bacteria in the gut, thereby promoting a healthier intestinal balance. The beneficial bacteria that are typically enhanced by prebiotics belong to the genera *Lactobacillus* and *Bifidobacterium*, which help reduce the presence of harmful bacteria. Prebiotics can include substances such as mannanoligosaccharides, lactose, galactoglucomannan, oligofructose, and inulin. A significant number of these prebiotic substances are carbohydrates, primarily consisting of short-chain oligosaccharides with 3–10 sugar units, sourced from various plants or the cell wall components of yeast. Other commercially available products that demonstrate prebiotic properties include GroBiotic-A, a yeast-based mixture of partially autolyzed brewer's yeast, components from dairy ingredients, and dried fermentation products. Generally, the different prebiotic compounds remain unaffected by food processing and require minimal regulatory approval, making their use in diets far easier than what is necessary for pharmaceuticals or chemical therapies (Mehrabi et al., 2018).

3.2.6 Nucleotides

The use of nucleotide supplementation in fish diets produced in aquaculture has gained significant global interest over the last 20 years. Nucleotides, which are the building blocks of nucleic acids, play a critical role in aquaculture. It is recognized in aquaculture that the organisms being cultured may experience stress related to physiological changes such as maturation and reproduction. These stressful environments may necessitate that the cultured organisms receive extra nucleotides for effective cell growth. The importance of ribonucleotides becomes more evident in these situations. Recent years have seen considerable focus on commercial nucleotide products like AccelerAid and NuPro. Furthermore, nucleotide supplements were incorporated into the base diet, replacing some of the cellulose. Individual supplementation with adenosine monophosphate (AMP), uridine monophosphate (UMP), cytidine monophosphate (CMP), guanosine monophosphate (GMP), and inosine monophosphate (IMP) resulted in significantly improved performance in fish when compared with those on the standard diet.

3.2.7 Phytochemicals

A diverse array of chemical substances obtained from plants, referred to as phytochemicals, has been studied as supplements for various aquatic species. These substances originate from a broad range of plants, including herbs, spices, seaweeds, and traditional Chinese medicinal practices. Many of these components have been sourced from plants historically recognized for their therapeutic properties and are typically extracted using solvents like alcohols, then administered either through injections or more conveniently incorporated into the diet. Researchers have explored phytochemicals as sustainable and beneficial additives aimed at enhancing reproductive traits and boosting the immune system in farmed fish species while also reducing environmental and public health hazards. Phytochemicals present a promising strategy for managing reproductive performance and increasing productivity in intensive aquaculture practices. Given the rising demand for eco-friendly approaches, phytochemicals are effective in diminishing oxidative stress and improving disease resistance among various farmed fish species. These substances exhibit numerous bioactive compounds with distinct physiological effects, particularly concerning the reproductive capabilities of aquatic animals. However, the precise mechanisms through which these additives impact aquatic reproductive traits remain inadequately documented. Overall, phytochemicals can serve as a viable substitute for synthetic additives and possess the capacity to foster sustainability in diverse aquaculture sectors worldwide, depending on the availability and cost-effectiveness of medicinal plants in specific regions. Well-known phytochemical compounds such as polyphenols, flavonoids, isoflavones, anthocyanins, phytoestrogens, terpenoids, carotenoids, limonoids, phytosterols, and glucosinolates offer a multitude of benefits, including antibacterial, antiviral, antioxidant, and anti-inflammatory properties. These compounds are frequently delivered orally to aquatic animals. Importantly, plant extracts that contain bioactive substances like saponins and flavonoids are crucial for promoting sexual inversion and addressing fertility issues. Nevertheless, the concentration of phenolic compounds can vary based on the genetics of the herb (including genus, species, cultivar, and genotype), methods of plant extraction, environmental factors, and the maturation status of the aquatic organisms. Most research has indicated that phytochemicals significantly influence fish reproduction through their diverse bioactive components. They can also facilitate sexual maturation by interacting with hormone receptors, encourage gonadal development by boosting gonadotropin production, enhance sperm quality due to their antioxidant effects, modulate hormone concentrations, and support reproductive behaviors.

4 Feeding Management

The timing and frequency of feeding significantly impact fish breeding performance. Optimal feeding schedules can enhance growth, improve feed utilization, and promote reproductive health.

4.1 Timing of Feeding

The timing of feeding significantly influences fish reproductive performance. Feeding fish during their natural feeding periods, typically in the morning and evening, aligns with their biological rhythms, optimizing digestion and nutrient absorption. During the pre-spawning period, increasing the frequency and quantity of feed can stimulate gonadal development and enhance egg quality. Conversely, post-spawning, a balanced diet is crucial for rapid recovery and subsequent spawning cycles. Frequent, small meals can improve digestion, reduce feed wastage, and optimize nutrient utilization. However, overfeeding can lead to water quality deterioration and reduced feed efficiency. The optimal feeding frequency varies based on factors such as fish size, water temperature, and reproductive stage. By carefully considering the timing and frequency of feeding, aquaculture producers can enhance fish growth, reproductive performance, and overall health.

4.2 Frequency of Feeding

The frequency of feeding significantly impacts fish reproductive performance. Frequent, small meals optimize digestion, reduce feed wastage, and ensure consistent nutrient absorption. This is particularly important during the pre-spawning period when increased nutrient intake stimulates gonadal development and enhances egg quality. Additionally, frequent feeding can improve the overall health and condition of broodstock, leading to better reproductive outcomes. However, it is crucial to avoid overfeeding, as excessive feed can degrade water quality and stress fish, negatively impacting their reproductive performance. The optimal feeding frequency varies depending on fish species, size, water temperature (Kausar & Salim, 2006), and reproductive stage. By carefully adjusting the frequency of feeding, aquaculture producers can maximize fish growth, reproductive output, and overall productivity.

By carefully considering the timing and frequency of feeding, aquaculture producers can optimize fish growth, reproductive performance, and overall health.

5 Molecular Approach

Understanding the internal processes that control particular biological and physiological functions and transformations is essential for manipulating and improving aquaculture. The utilization of molecular techniques like genomics, proteomics, transcriptomics, and metabolomics is increasingly recognized in various areas of aquaculture, including disease management, immune responses, stress reactions, and growth development. This section will concentrate on relevant molecular techniques and their potential significance in the context of broodstock nutrition.

5.1 Genotyping

Advanced genomic technologies like next-generation sequencing (NGS) and microarrays allow researchers to explore genetic variations such as single nucleotide polymorphisms (SNPs) and significant structural alterations in DNA. The access to abundant genomic data further facilitates the creation of genetic linkage maps and the execution of genome-wide association studies (GWAS), which enable researchers to connect DNA variants to phenotypes of interest. Findings from polymorphic markers, linkage maps, and GWAS play a crucial role in broodstock breeding programs, including marker-assisted selection (MAS), genome selection, and genome editing. Initially, researchers faced challenges in identifying any Amplified Fragment Length Polymorphism (AFLP) markers specific to sex across different species due to insufficient genetic information. Through the use of NGS, Vo et al. (2018) discovered 11,009 SNPs in a sample of 400 wild and farmed *P. hypophthalmus*, suggesting that two specific SNPs (SNP5, SNP9) unique to the population in Vietnam could be utilized for commercial traceability of *P. hypophthalmus*. At present, the complete genomes of various species are also accessible. The availability of whole genomes is crucial as a reference for GWAS studies aimed at pinpointing SNPs linked to sexual maturation and gonadal development. For instance, a GWAS study involving Atlantic salmon from the Cermaq Canada broodstock program was performed to uncover potential SNPs related to growth and sexual maturation. The identification of specific markers tied to growth as well as early and late sexual maturation, coupled with the determination of corresponding candidate genes, highlighted possible pathways that may be associated with these developments.

5.2 Gene Expression and Transcriptome Analysis

Considerable progress has been made in the field of nutritional genomics, particularly in research concerning lipid metabolism. A recent review focused on the roles and current understanding of elongase and desaturase genes in aquatic organisms. To date, only a few full-length genes associated with fatty acid biosynthesis pathways, such as putative Δ6 fatty acyl desaturases (fads), have been identified. Genes related to growth have been characterized, including fads2, which regulates the Δ6 desaturase enzyme, and growth hormone 1 (GH1), responsible for bone and tissue development.

With the rapid advancements in next-generation sequencing (NGS), researchers are investigating the transcriptomes of various species under specific conditions to identify any differentially expressed genes of interest. According to the whole-body transcriptome analysis of rainbow trout alevins from broodstock, the maternal dietary history did not significantly influence the progeny's gene response prior to their first feeding (Lazzarotto et al., 2016), which raises the potential for using plant-based diets for fish species. In species such as Atlantic salmon and rainbow trout, where early developmental nutritional programming (the introduction of plant-based diets) has been effectively implemented, it has been shown that nutritional history notably impacts gene expression in crucial pathways, including pyruvate metabolism, glycolysis, and fatty acid metabolism (Panserat et al., 2013). This highlights the potential molecular mechanisms behind nutritional programming as evidenced by the liver transcriptome profiles of Atlantic salmon that were challenged with a plant-based diet. To enhance the broodstock quality of gilthead seabream, selecting broodstock with elevated fads2 expression, combined with nutritional programming (replacing fish oil with rapeseed oil), led to offspring exhibiting increased fads2 expression and better growth outcomes.

5.3 Genome Editing

A significant amount of progress has been made in the field of nutritional genomics, particularly regarding lipid metabolism studies. A recent review covering elongase and desaturase in aquatic species outlines the present state of these genes and their roles. So far, there have only been a few full-length genes identified that are associated with fatty acid biosynthesis pathways, such as the putative Δ6 fatty acyl desaturases (fads). Genes linked to growth, such as fads2, which governs the Δ6 desaturase enzyme, and growth hormone 1 (GH1), responsible for the development of bones and tissues, have been characterized. With the emergence and progress of complete genome sequencing, genome editing techniques are becoming a crucial method for exploring biological processes and improving or modifying specific traits. Current genome editing methods encompass zinc finger nucleases (ZFN), transcription activator-like effector nucleases (TALEN), and clustered regularly interspaced short palindromic repeats (CRISPR). ZFN is a synthetic endonuclease that can be customized to target and cut a particular genomic sequence, which can lead to gene knockout or modification according to the subsequent cellular repair mechanisms. In the initial phases of DNA recombination and genome sequencing, the application of ZFN in fish mainly concentrated on model organisms like zebrafish. To address challenges encountered with ZFN, TALEN was developed, followed by the CRISPR/Cas9 system. CRISPR/Cas9 has gained widespread application in fish aquaculture to improve traits like disease resistance, growth rates, and fertility. Prominent species that have been subjected to CRISPR/Cas9 techniques in the aquaculture industry include salmon species, Nile tilapia, zebrafish, and medaka. From a nutritional standpoint, CRISPR/Cas9 has illustrated the role of elovl2 in the in vivo production of PUFAs, particularly docosahexaenoic acid, in Atlantic salmon. Moreover, CRISPR/Cas9 can be utilized specifically for elucidating the contributions and roles of essential genes in regulating germ line development and egg viability. For instance, Cas9 mRNA and sgRNA targeting the tyrosinase gene in the large yellow croaker (*Larimichthys crocea*) were effectively microinjected into fertilized eggs, emphasizing the promising potential of CRISPR/Cas9 in fish genome editing, particularly during embryonic development (Li et al., 2023). Nonetheless, the application of CRISPR/Cas9 in fish breeding requires extensive fundamental research, as CRISPR/Cas9 technologies are specific to particular species and genes, necessitating comprehensive knowledge for precise targeting.

6 Conclusion

A well-designed nutritional strategy is essential for optimizing fish reproductive performance. By providing a balanced diet that meets the specific nutritional requirements of each fish species, aquaculture producers can enhance gamete quality, improve fertilization rates, increase larval survival, and accelerate growth rates. Key factors to consider include protein quality and quantity, lipid composition, carbohydrate levels, vitamin and mineral supplementation, and feeding frequency and timing. Additionally, monitoring water quality, reducing stress, and implementing appropriate disease prevention measures are crucial for overall fish health and reproductive success. By carefully managing these factors, aquaculture producers can achieve sustainable and profitable fish production.

References

Abu-Elala, N. M., et al. (2021). Analysis of the productivity, immunity, and health performance of Nile tilapia (*Oreochromis niloticus*) Broodstock-fed dietary fermented extracts sourced from *Saccharomyces cerevisiae* (Hilyses): A field trial. *Animals, 11*, 815. https://doi.org/10.3390/ani11030815

Ahmed, I., & Ahmad, I. (2020). Effect of dietary protein levels on growth performance, hematological profile and biochemical composition of fingerlings rainbow trout, *Oncorhynchus mykiss* reared in Indian himalayan region. *Aquaculture Reports, 16*, 100268.

Andia, S. N., Cattopadhyay, D. N. C., Ermab, J. P. V., Arkarb, S. K. S., & Mukhopadhyaya, P. K. M. (2001). Effect of dietary supplementation of fatty acids and vitamins on the breeding performance of the carp *Catla catla*. *Reproduction Nutrition Development, 41*(4), 365–375.

Bowen, S. H. (1987). Dietary protein requirements of fishes – A reassessment. *Canadian Journal of Fisheries and Aquatic Sciences., 44*(11), 1995–2001.

Brody, S. (1945). *Bioenergetics and growth*. Hafner Publishing Company.

Chorong, L., Jee, E. H., Ji, E. K., Sung, H. K., Jae, W. K., Jong, S. E., et al. (2019). Effect of dietary supplementation of *bacillus* spp. on growth performance, and resistance of Pacific white shrimp (*Litopenaeus vannamei*) to acute Hepatopancreatic necrosis disease. *Israeli Journal of Aquaculture*.

Costas, B., Aragao, C., Mancera, J. M., Dinis, M. T., & Conceiçao, L. E. (2008). High stocking density induces crowding stress and affects amino acid metabolism in Senegalese sole *Solea senegalensis* (Kaup 1858) juveniles. *Aquaculture Research, 39*(1), 1–9.

Craig, S., Helfrich, L. A., Kuhn, D., & Schwarz, M. H. (2017). *Understanding fish nutrition, feeds, and feeding*. The Ohio State University.

Fitriliyani, I., Siswanto, S., & Luchas, L. (2022). Optimizing the gonadal performance of broodstock *Helostoma temminckii* with the addition of glutathione and vitamin E as enrichment ingredients in feed. *Journal Natural, 22*(1). https://doi.org/10.24815/jn.v22i1.23426

Fontagne-Dicharry, S., et al. (2010). Effects of dietary vitamin A on broodstock performance, egg quality, early growth, and retinoid nuclear receptor expression in rainbow trout (*Oncorhynchus mykiss*). *Aquaculture, 303*, 40–49.

Gioacchini, G., Giorgini, E., Merrifield, D.L., Hardiman, G., Borini, A., Vaccari, L., & Carnevali, O. (2012). Probiotics can induce follicle maturational competence: the *Danio rerio* case. *Biology of Reproduction, 86*(3), 65. https://doi.org/10.1095/biolreprod.111.094243

Grisdale-Helland, B., Lemme, A., & Helland, S. J. (2013). Threonine requirement for maintenance and efficiency of utilization for threonine accretion in Atlantic salmon smolts determined using increasing ration levels. *Aquaculture, 372*, 158–166.

Halver, J. E., & Hardy, R. W. (2002). *Fish nutrition*. Academic Press.

Kabir, M. A., Ghaedi, A., & Hashim, R. (2012). Ovarian development and sexual maturation of female striped catfish, *Pangasianodon hypophthalmus* (Sauvage, 1878) reared in captivity. *Asian Fisheries Science., 25*(3), 232–244.

Kausar, R., & Salim, M. (2006). Effect of water temperature on the growth performance and feed conversion ratio of *Labeo rohita*. *Pakistan Veterinary Journal, 26*(3), 105–108.

Kvale, A., Nordgreen, A., Tonheim, S. K., & Hamre, K. (2007). The problem of meeting dietary protein requirements in intensive aquaculture of marine fish larvae, with emphasis on Atlantic halibut (*Hippoglossus hippoglossus* L.). *Aquaculture Nutrition, 13*(3), 170–185.

Lazzarotto, V., Corraze, G., Larroquet, L., Mazurais, D., et al. (2016). Does broodstock nutritional history affect the response of progeny to different firstfeeding diets? A whole-body transcriptomic study of rainbow trout alevins. *The British Journal of Nutrition, 115*, 2079–2092.

Li, H. W., & Brocksen, R. W. (1977). Approaches to the analysis of energetic costs of intraspecific competition for space by rainbow trout (*Salmo gairdneri*). *Journal of Fish Biology, 11*(4), 329–341.

Li, Q., Shao, G., Ding, Y., Xu, L., Shao, J., Ao, J., & Chen, X. (2023). Effective CRISPR/Cas9-based genome editing in large yellow croaker (*Larimichthys crocea*). *Aquaculture and Fisheries, 8*, 26–32. https://doi.org/10.12016/j.aaf.2021.04.008

Mehrabi, F., Khalesi, M., & Kaivan, H. (2018). Effects of pre- and probiotics on growth, survival, body composition, and hematology of common carp (*Cyprinus carpio* L.) fry from the Caspian Sea. *Turkish Journal of Fisheries and Aquatic Sciences, 18*, 597–602. https://doi.org/10.4194/1303-2712-v18_4_11

Mejri, S., Tremblay, R., Vandenberg, G., Moren, M., Khemis, I. B., & Audet, C. (2017). Differences in nutrient content of eggs and larvae as indicators for improvement of broodstock nutrition in walleye (*Sander vitreus*) production. *Canadian Journal of Zoology, 95*(5), 299–310.

Miccoli, A., Gioacchini, G., Maradonna, F., Benato, F., Skobo, T., & Carnevali, O. (2015). Beneficial bacteria affect *Danio rerio* development by the modulation of maternal factors involved in autophagic, apoptotic and dorsalizing processes. *Cellular Physiolology Biochemistry, 35*(5), 1706–1718. https://doi.org/10.1159/000373983

Mir, I. N., Srivastava, P. P., Bhat, I. A., Jaffar, Y. D., Sushila, N., Sardar, P., et al. (2020). Optimal dietary lipid and protein level for growth and survival of catfish *Clarias magur* larvae. *Aquaculture, 520*, 734678.

Nandi, B. S., Routray, P., Gupta, S. D., et al. (2007). Reproductive performance of carp, *Catla catla* (ham.), reared on a formulated diet with PUFA supplementation. *Journal of Applied Ichthyology, 23*(6), 684–691.

Nayak, S. K. (2010). Probiotics and immunity: A fish perspective. *Fish Shellfish Immunology, 29*(1), 2–14. https://doi.org/10.1016/j.fsi.2010.02.017

Ogino, C., & Chen, M. S. (1973). Protein nutrition in fish. 5. Relation between biological value of dietary proteins and their utilization in carp. *Bulletin of the Japanese Society of Scientific Fisheries, 39*(9), 955–959.

Pandian, T. J., & Vivekanandan, E. (1985). Energetics of feeding and digestion. In P. Tykr & P. Calow (Eds.), *Fish energetics – New perspectives* (pp. 99–124). Johns Hopkins University Press.

Panserat, S., Kaushik, S., & Medale, F. (2013). Rainbow trout as a model for nutrition and nutrient metabolism studies. In *Trout: From physiology to conservation* (pp. 131–153). Nova Science Publishers.

Rahman, M. L., Akhter, S., Mallik, M. K. M., & Rashid, I. (2018). Probiotic enrich dietary effect on the reproduction of butter catfish, *Ompok pabda* (Hamilton, 1872). *International Journal of Current Research in Life Sciences, 7*(2), 866–873.

Shahkar, E., Yun, H., Kim, D. J., Kim, S. K., Lee, B. I., & Bai, S. C. (2015). Effects of dietary vitamin C levels on tissue ascorbic acid concentration, hematology, nonspecific immune response, and gonad histology in broodstock Japanese eel, *Anguilla japonica*. *Aquaculture, 438*, 115–121.

Sharifuzzaman, S. M., & Austin, B. (2017). Probiotics for disease control in aquaculture shellfish. In B. Austin & A. Newaj-Fyzul (Eds.), *Diagnosis and control of diseases of fish and shellfish* (pp. 189–222). Wiley & Sons, Ltd.

Sink, T. D., & Lochmann, R. T. (2008). Effects of dietary lipid source and concentration on channel catfish (*Ictalurus punctatus*) egg biochemical composition, egg and fry production, and egg and fry quality. *Aquaculture, 283*(1–4), 68–76.

Symons, P. E. (1970). The possible role of social and territorial behaviour of Atlantic salmon parr in the production of smolts. In *Fisheries research Board of Canada, Biological Station*.

Vo, T. T. B., Nguyen, H. D., Nguyen, T. T. H., & Tran, H. T. T. (2018). Identification and analysis of snps in population of Vietnamese catfish (*Pangasianodon hypophthalmus*), using next generation sequencing and snp validation. *MOJ Current Research & Reviews, 1*, 12–19.

Wilson, R. P. (1994). Utilization of dietary carbohydrate by fish. *Aquaculture, 124*(1–4), 67–80.

Yulfiperius, et al. (2003). Effect of vitamin E levels in feed on quality of catfish (*Pangasius hypophthalmus*) eggs. *Indonesian Journal of Ichthyology, 3*, 11–18. https://doi.org/10.24815/jn.v22i1.23426

Zakeri, M., Marammazi, J. G., Kochanian, P., Savari, A., Yavari, V., & Haghi, M. (2009). Effects of protein and lipid concentrations in broodstock diets on growth, spawning performance and egg quality of yellowfin sea bream (*Acanthopagrus latus*). *Aquaculture, 295*(1–2), 99–105.

Zakeri, M., Kochanian, P., Marammazi, J. G., Yavari, V., Savari, A., & Haghi, M. (2011). Effects of dietary n-3 HUFA concentrations on spawning performance and fatty acids composition of broodstock, eggs and larvae in yellowfin sea bream, *Acanthopagrus latus*. *Aquaculture, 310*(3–4), 388–394.

Chapter 21
Role of Epigenetics in Fisheries and Aquaculture

Javeria Fatima and Yasir Hasan Siddique

1 Introduction

Fisheries, as a vital component of global food security and economic sustainability, are influenced by a range of biological and environmental factors (Jablonka & Lamb, 2014). Human activities significantly impact aquatic ecosystems by introducing various environmental stressors, including climate change, nutrient shifts, and chemical or biological pollution (Hawes et al., 2018; Halpern et al., 2019; Mushtaq et al., 2020). Over the past decade, growing interest in epigenetics has led to a shift in understanding inheritance mechanisms. The term "epigenetics," derived from the Greek prefix *epi,* meaning "over" or "in addition to," and genetics, refers to heritable changes that go beyond the traditional genetic framework (Roy et al., 2022). Research indicates that traits like growth, disease resistance, feed conversion ratio (FCR), and stress tolerance in aquaculture species are influenced by both genetic and epigenetic variations (Moore et al., 2013; Zeng & Chen, 2019).

Epigenetics is a process that leads to inheritable changes in gene activity without altering the fundamental DNA sequence. These alterations arise from the epigenetic mechanisms that regulate gene activity (Jablonka & Lamb, 2002). Over the past few years, it has been commonly identified that environmental factors can trigger phenotypic changes in the organism through epigenetic processes. These processes regulate gene expressions at various stages, including transcription, translation, and posttranslational, without altering the nucleotide sequence (Jaenisch & Bird, 2003). Epigenetic mechanisms, such as DNA methylation, histone modifications, and the activity of noncoding RNAs, regulate gene expression by altering the structure and accessibility of chromatin (Weaver et al., 2004). These changes, which primarily take place during embryonic development, are believed to support the

J. Fatima · Y. H. Siddique (✉)
Laboratory of Alternative Animal Models, Section of Genetics, Department of Zoology, Faculty of Life Sciences, Aligarh Muslim University, Aligarh, Uttar Pradesh, India

I. Ahmed, I. Ahmad (eds.), *Aquaculture: Enhancing Food Security and Nutrition*, https://doi.org/10.1007/978-3-031-92858-1_21

differentiation of cells into specific types and can be inherited through mitosis and/or meiosis (Daneshfozouna et al., 2015).

Epigenetic marks can develop over a lifetime, often remaining relatively stable, though some may be reversible (Bishop & Ferguson, 2015). Environmentally induced epigenetic changes can influence an organism's phenotype, with effects that may be either harmful (Mirbahai et al., 2011) or beneficial (Babu et al., 2012). The dynamic balance of the epigenome enables rapid adjustments to external stimuli, facilitating gradual changes in gene expression and phenotype variation (Turner, 2009; Skinner et al., 2014). However, in germline and stem cells, certain epigenetic patterns can be stably maintained across generations, supporting long-term adaptation (Youngson & Whitelaw, 2008; Gapp et al., 2014). In cancer, studies suggest that epigenetic changes, such as alterations in DNA methylation patterns, can precede genetic mutations, highlighting the interplay between epigenetic and genetic factors in cancer progression (Pogribny, 2010; Sharma et al., 2010). Conversely, epigenetic modifications can also confer adaptive advantages. For instance, in polyploid organisms, they provide flexible mechanisms for neopolyploids to quickly adapt to significant genetic changes, supporting their survival and reproduction (Li et al., 2011).

Significant advancements have been achieved in aquaculture genomics and genetics research over the past decade. While epigenomic studies in fish species are still in their early stage, researchers are increasingly exploring the role of epigenetic regulation in key traits. This chapter provides an overview of progress in epigenetic research for fisheries and aquaculture species, highlighting potential applications and addressing existing challenges.

2 Epigenetic Modifications

Epigenetics explores heritable changes in gene expression that do not involve alterations to the DNA sequence. Environmental factors can trigger phenotypic changes in animals through epigenetic mechanisms such as DNA methylation, noncoding RNA activity, histone modifications, and chromatin remodeling. These mechanisms regulate gene activity at the molecular level without modifying the nucleotide sequence and play a critical role in controlling gene expression (Rasal et al., 2024). In eukaryotic cells, DNA is compacted into chromatin, where approximately 146 base pairs of DNA are wrapped around histone octamers (comprising two copies each of histones H2A, H2B, H3, and H4), forming nucleosomes. These nucleosomes, linked by the histone H1, are folded into chromatin fibers to efficiently compact genetic material within the nucleus (Conaway, 2012).

Epigenetic mechanisms can modify this chromatin structure, resulting in distinct gene expression patterns (Siggens & Ekwall, 2014). The epigenome encompasses various chromatin states marked by specific mechanisms that regulate genome function. Key epigenetic mechanisms include DNA methylation, histone modifications, and noncoding RNA activity, which individually or collectively modulate gene expression by altering chromatin locally (Granada et al., 2018). These mechanisms enable or suppress the production of specific proteins and mediate phenotypic responses to environmental cues, underscoring their vital role in gene regulation and adaptability (Roy et al., 2022).

2.1 DNA Methylation

DNA methylation is a key epigenetic mechanism involving the addition of a methyl group to the cytosine base at CpG dinucleotides, forming 5-methylcytosine (5mC). This modification, primarily catalyzed by DNA methyltransferases (DNMTs), regulates gene expression by repressing transcription through blocking transcription factor binding or attracting repressor proteins (Jin et al., 2011; Moore et al., 2013). DNMT1 maintains existing methylation patterns during DNA replication, while DNMT3A and DNMT3B establish new methylation marks (Jablonka & Raz, 2009). Methylation patterns are tissue-specific and influence diverse biological processes, including genomic imprinting, X-chromosome inactivation, and transposable element silencing, contributing to genomic stability and development (Maloisel & Rossignol, 1998; Csankovszki et al., 2001).

DNA methylation typically suppresses transcription when located in gene promoters but may enhance expression when present in gene bodies (Jones, 1999). Its regulatory role depends on genomic context and environmental factors, such as nutrition, toxins, and light exposure, which can dynamically alter methylation patterns (Weaver et al., 2004; Dolinoy et al., 2006; Azzi et al., 2014). Transgenerational inheritance of DNA methylation is rare in mammals due to extensive reprogramming during embryonic development but occurs more frequently in plants, where methylation marks are less reset between generations (Daxinger & Whitelaw, 2012; Hauser et al., 2011). Recent studies also suggest that DNA demethylation, involving intermediates like 5-hydroxymethylcytosine (5hmC), plays a functional role in gene regulation and environmental adaptation (Tahiliani et al., 2009; Branco et al., 2012).

2.2 Histone Modification and Chromatin Remodeling

Chromatin is composed of DNA and histone proteins that enable DNA compaction within the nucleus. Each nucleosome consists of an octamer of core histones (two H3-H4 homodimers and two H2A-H2B heterodimers) around which ~147 base pairs of DNA are wrapped. The histone tails, flexible N-terminal extensions rich in basic amino acids, undergo various posttranslational modifications (PTMs), influencing chromatin structure and gene regulation (Jenuwein & Allis, 2001; Berger, 2007). Chromatin exists in two primary states: euchromatin and heterochromatin. Euchromatin, associated with active gene expression, has a relaxed structure, while heterochromatin is tightly packed, resulting in transcriptional silencing (Khalaf &

Gräff, 2016; Araki & Mimura, 2017). These states are modulated by PTMs, including acetylation, methylation, phosphorylation, ubiquitylation, and others, which are mediated by specific enzymes like histone acetyltransferases (HATs), histone deacetylases (HDACs), histone methyltransferases (HMTs), and histone demethylases (Greer et al., 2014). For instance, histone acetylation typically promotes transcription by loosening chromatin, whereas deacetylation results in tighter chromatin and transcriptional repression. Methylation can have dual roles, depending on the modified residue (Annunziato & Hansen, 2018; Verdone et al., 2006).

Epigenetic modifications, including those of histones, play pivotal roles in gene regulation, development, and cellular responses to environmental stimuli. These modifications can be heritable through mitosis and meiosis and are critical during gametogenesis and early embryogenesis, where epigenetic reprogramming resets the parental epigenome to establish totipotency (Morgan et al., 2005; Henikoff & Smith, 2015). Aberrant reprogramming or exposure to harmful environmental stimuli can lead to heritable vulnerabilities in offspring. Histone variants also influence chromatin dynamics and gene expression. While some, like macroH2A, are conserved across species, others are lineage-specific and facilitate unique regulatory mechanisms. Variants and PTMs, often detected using chromatin immunoprecipitation (ChIP) or mass spectrometry, are crucial for processes such as embryogenesis, tissue regeneration, and stress responses (Talbert & Henikoff, 2010; Vastenhouw & Schier, 2012).

In aquatic species, histone modifications and variants have been shown to mediate physiological responses to environmental factors. For instance, temperature influences histone methylation in oysters, and diet modulates histone acetylation in fish (Marandel et al., 2016; Fellous et al., 2015). Unlike mammals, bivalves retain histones in sperm, which are replaced by protamine-like proteins in certain species, highlighting evolutionary diversity in chromatin organization (Eirín-López & Ausio, 2009). Overall, chromatin remodeling and histone modifications are integral to genomic regulation, developmental programming, and adaptive responses, acting as a nexus between environmental cues and gene expression (Cheung et al., 2000; Lawrence et al., 2016).

2.3 Noncoding RNAs

Noncoding RNAs (ncRNAs) are transcribed from most eukaryotic genomes and do not encode proteins. They play significant regulatory roles in gene expression, particularly through epigenetic mechanisms, by modulating DNA methylation and histone modifications (Moazed, 2009; Bizuayehu & Babiak, 2014). Two primary classes of ncRNAs are recognized: long ncRNAs (lncRNAs), typically >200 nucleotides, and small ncRNAs, which are <200 nucleotides and include microRNAs (miRNAs), small interfering RNAs (siRNAs), and PIWI-interacting RNAs (piRNAs) (Castel & Martienssen, 2013; Wang & Chang, 2011). NcRNAs perform various roles, including regulating transposon activity, mediating X-chromosome inactivation, and influencing chromatin remodeling (Lee, 2009). For instance, the lncRNA XIST silences one X chromosome in female cells by forming a complex with chromatin (Ng et al., 2007). Similarly, HOTAIR, another lncRNA, directs histone modifications to transcriptionally silence specific chromosomal regions (Rinn et al., 2007). Small ncRNAs, such as piRNAs, are crucial in silencing repetitive elements, as demonstrated in zebrafish studies (Houwing et al., 2007).

NcRNAs are also implicated in genome stability, environmental plasticity, and embryonic development. They interact with other epigenetic mechanisms, including histone modifications and DNA methylation, to regulate gene activity (Peschansky & Wahlestedt, 2014). MiRNAs, in particular, have emerged as key posttranscriptional regulators involved in immune functions, transgenerational inheritance, and adaptive responses to environmental stress (Klosin et al., 2017; Wang & Zhu, 2017). Recent research highlights the importance of RNA methylation as an additional epigenetic control mechanism, affecting RNA stability and translation (Yue et al., 2015; Romano et al., 2018). In aquaculture species like Atlantic salmon and rainbow trout, miRNAs have been studied for their roles in immune function and development, while piRNAs and lncRNAs show associations with silencing transposable elements and disease (Paneru et al., 2016; Yu et al., 2016). NcRNAs, particularly miRNAs, have also been linked to transgenerational inheritance in species such as *C. elegans* and mammals, where environmental cues like starvation or temperature influence their expression and epigenetic memory (Rechavi et al., 2014; Klosin et al., 2017).

3 Role of Epigenetics in Sustainable Aquaculture Development

The aquaculture sector, a critical source of global food fish, increasingly leverages technological advancements to meet rising production demands and enhance species management. Among these innovations, epigenetics offers significant potential for improving traits like stress tolerance, disease resistance, and growth, thereby promoting sustainability and economic viability (Macleod et al., 1999; Hanson & Skinner, 2016). Epigenetics offers transformative potential in aquaculture, addressing critical challenges like disease control, stress management, and nutritional efficiency while promoting sustainable practices. Integrating epigenetic insights into breeding programs and production strategies can enhance resilience, productivity, and environmental sustainability, ensuring the industry's long-term viability (Abdelrahman et al., 2017; Canário, 2019). Table 21.1 shows different epigenetic studies carried out on various aquatic species.

Table 21.1 Epigenetic studies carried out on different aquatic species

S. no.	Species	Exposure condition	Epigenetic mark	Phenotypic outcome	Result	References
1.	*Dicentrarchus labrax* (European sea bass)	Exposure to higher temperature during the thermosensitive period	Differential methylation of gonadal cyp19a and b-actin promoters	Masculinization of females	Elevated methylation levels in the promoter region of the gonadal aromatase (*cyp19a*) gene	Navarro-Martín et al. (2011))
2.	*Solea senegalensis* (Senegalese sole)	Larvae undergoing metamorphosis reared at different temperatures	Cytosine methylation of the myog gene promoter	Increased thermal plasticity of myogenesis of larvae reared at higher temperatures	Enhanced methylation of the myog gene promoter in skeletal muscle at lower temperatures, accompanied by elevated expression levels of dnmt1 and dnmt3s	Campos et al. (2013)
3.	*Gadus morhua* (Atlantic cod)	Exposure to two different photoperiod regimes during early juvenile stages	DNA methyltransferases (dnmts) expression in fast muscle tissue	Organisms reared under continuous light during the early juvenile stages presented larger sizes compared to fish reared at simulated natural photoperiod	Elevated expression of dnmt1 and dnmt3a in the fast muscle of Atlantic cod exposed to continuous illumination compared to those reared under a natural photoperiod	Giannetto et al. (2013)
4.	*Salmo trutta* (Brown trout)	Exposure of freshwater fish to seawater conditions, after salt-enriched diet intake	Cytosine methylation in CpG sites of samples from gill tissue	Freshwater fish physiologically adapted to seawater	Transient and reversible changes in global DNA methylation induced by salt exposure	Morán et al. (2013))
5.	*Cynoglossus semilaevis* (Half-smooth tongue sole)	Exposure to higher temperature during the sensitive developmental period in early life	DNA methylation of samples from gonadal tissue, methylation status of dmrt1	Masculinization of females	Significant methylation of genes associated with the sex determination network during sexual reversal, with global patterns inherited by the offspring	Shao et al. (2014)

6.	*Artemia* (Brine shrimp)	Nonlethal heat shock during early life stages	Global DNA methylation, histone H3, and H4 total acetylation	Tolerance to lethal heat stress and resistance against the pathogenic *Vibrio campbellii*, phenotype inherited by three subsequent unstressed generations	Changes in global DNA methylation levels are accompanied by increased overall acetylation of histones H3 and H4	Norouzitallab et al. (2014)
7.	*Anguillla anguilla* (European eel)	Chronic exposure of immature organisms to cadmium	Global and site-specific CpG methylation in liver tissue	–	Elevated global CpG methylation levels, along with hypermethylation of genes involved in intracellular trafficking and phospholipid biosynthesis	Pierron et al. (2014)
8.	*Danio rerio* (Zebrafish)	Exposure to copper and heat stress during early embryogenesis	Global DNA methylation; stress-related gene expression	High mortality, delayed embryo hatching, and reduced hatching rate	No change in global cytosine methylation, but copper exposure altered mRNA expression of *dnmt3* genes, and increased mt2 and hsp70 mRNA expression, with elevated hsp70 mRNA expression observed at higher temperatures	Dorts et al. (2016)
9.	*Oryzias melastigma* (Medaka fish)	Hypoxia exposure of sexually mature (F_0) fish	Global DNA methylation in sperm cells (F_0–F_2)	Transgenerational reproductive impairments such as aberrant sperm motility, decreased testis weight, percentage of spermatids, and percentage of fertilization success	Global DNA hypermethylation was observed in the sperm of F2 fish, with hypomethylation in the promoter regions of ehmt2 and ptk2b and hypermethylation in the exonic regions of foxp2	Wang et al. (2016)
10.	*Daphnia magna* (Waterflea)	Hypoxia exposure of neonates (F_0) during their whole life	Expression of histone-related genes	Possible adverse effects on embryonic development and carryover effects on larval development	Dysregulation of histone-related genes, including histones H2B, H3, H4, and histone deacetylase 4 (HDAC4), in F1 embryos	Lai et al. (2016)

(continued)

Table 21.1 (continued)

S. no.	Species	Exposure condition	Epigenetic mark	Phenotypic outcome	Result	References
11.	*Gasterosteus aculeatus* (Three-spine stickleback)	Different environmental salinities	Genome-wide DNA methylation patterns between three-spine stickleback males and females and between stickleback reared at different environmental salinities	–	Increased methylation of the X chromosome relative to the Y chromosome	Metzger and Schulte (2018)
12.	*Culter alburnus*	–	State of CpG methylation in the dmrt1 promoter of *C. alburnus* in the ovaries and testes	–	*Dmrt1* gene is hypo-methylated and highly expressed in males	Jia et al. (2019)
13.	*Crassostrea gigas*	High temperature (35 °C) exposure for 30 minutes	Expression of histone-related genes	–	One HSP70, with a binding site in its promoter region, was confirmed to be regulated by HSF1 under ocean warming stress	Liu et al. (2020)
14.	*Danio rerio* (Zebrafish)	High temperature (33 °C) exposure of fish from 20 to 30 days postfertilization (dpf)	DNA methylation levels in sox9b and esr1 promoter	Masculinization induced due to high temperature	Elevated temperature-induced hypermethylation of the promoters of sox9b and esr1	Han et al. (2021)
15.	*Crassostrea gigas*	–	Promoter methylation in the diacylglycerol kinase delta gene (*Dgkd*)	Dgkd: An epigenetic marker to distinguish male *C. gigas*	Methylomes are associated with infertility of triploid oysters	Sun et al. (2022)

16.	Ictalurus punctaus (Channel catfish)	Prior to sexual differentiation at 9, 12, and 16 days postfertilization	–	Epigenetic control of sex determination in channel catfish	Sex-specific epigenetically marked locus was discovered in the sex determination region (SDR)	Yang et al. (2022)
17.	*Pseudobagrus ussuriensis* (Ussuri catfish)	Sampling of female and male at the age of 2	–	Amh, gsdf, and hsd11b2 in brain tissues and slco3a1, socs2, and trim47 in gonad tissues	Distinct methylation patterns between males and females	Li et al. (2022)
18.	*Apostichopus japonicus*	High-temperature exposure	H3K9ac is an extensive epigenetic modulator	–	H3K9ac in promoters is negatively correlated with *NECPA1* gene expression in response to thermal stress	Xu et al. (2022)
19.	*Channa maculata*	30, 60, 90, 120, 150, and 180 days postfertilization	CpG methylation levels of *cyp19a1a and cyp19a1b*	–	Epigenetic modifications are essential for regulating the sexually dimorphic expression of cyp19a1a, as well as for ovarian differentiation and oogenesis	Huang et al. (2023)

3.1 Epigenetic Inheritance in Aquatic Species

Epigenetic inheritance is more feasible in aquatic species than in mammals due to less extensive embryonic reprogramming. Environmental factors during early developmental stages can establish epigenetic states that influence morphology, behavior, and stress resilience, reducing reliance on antibiotics and improving productivity (González-Recio, 2012; Burggren & Blank, 2009).

3.2 Enhancing Disease and Stress Resistance

Epigenetic modifications, such as DNA methylation and histone acetylation, play a crucial role in enhancing stress and disease resistance (Liu et al., 2022). For example, heat-stressed brine shrimp (Artemia) exhibited increased thermotolerance and pathogen resistance, with these traits passed across generations (Norouzitallab et al., 2014). Similarly, poly(I:C)-treated oysters (*Crassostrea gigas*) produced offspring with enhanced antiviral defenses, showcasing inheritable epigenetic reprogramming (Green et al., 2016). In Atlantic salmon (*Salmo salar*), early-life stress influenced the methylome and transcriptome, enhancing immune responses and disease resistance (Moghadam et al., 2017; Uren Webster et al., 2018).

3.3 Nutrigenomics and Early-Life Nutrition

Nutritional programming through epigenetic pathways significantly affects growth and adaptation in aquaculture. Salt-enriched diets in brown trout (*Salmo trutta*) improved seawater survival by modulating DNA methylation during smoltification (Morán et al., 2013). Similarly, plant-based diets induced epigenetic changes in rainbow trout (*Oncorhynchus mykiss*) and gilthead sea bream (*Sparus aurata*), enabling adaptation to alternative feeds (Izquierdo et al., 2015; Geurden et al., 2013).

3.4 Temperature-Driven Sex Control

Epigenetic mechanisms mediate sex determination in species like European sea bass (*Dicentrarchus labrax*), where elevated temperatures during early development increased DNA methylation in the cyp19a gene, masculinizing genetic females (Navarro-Martín et al., 2011). In half-smooth tongue sole (*Cynoglossus semilaevis*), similar temperature-induced sex reversal was associated with inheritable DNA methylation patterns (Shao et al., 2014).

3.5 Broodstock Conditioning and Epigenetic Inheritance

Broodstock exposed to environmental stressors can pass adaptive advantages to offspring through transgenerational epigenetic inheritance. For instance, stressed brine shrimp produced more resilient offspring, while Manila clams exhibited enhanced growth and survival rates under stress, highlighting epigenetics role in ecological fitness (Norouzitallab et al., 2014; Zhao et al., 2017).

3.6 Integrating Epigenetics into Selective Breeding

Epigenetic selection can complement genetic breeding programs, enhancing aquaculture outcomes. Research on Pacific oysters (*Crassostrea gigas*) demonstrated links between epigenetic variation and environmental factors, while hatchery-reared salmon exhibited epigenetic changes potentially impacting fitness in natural habitats (Jiang et al., 2013; Le Luyer et al., 2017).

4 Potential Aquaculture Applications of Epigenetics

Environmental manipulation is a promising application of epigenetics in aquaculture, especially during sensitive life stages such as larval development and broodstock conditioning. Environmental factors can influence epigenetic mechanisms, which may drive phenotypic changes beneficial for aquaculture, such as growth, disease resistance, and sex determination. This concept, known as developmental programming, suggests that early-life environmental conditions can affect later phenotypic traits (Gluckman et al., 2008; Jonsson & Jonsson, 2014). In fish, exposure to high temperatures during early development has been shown to alter sex determination through epigenetic changes, such as DNA methylation of sex-related genes (Navarro-Martín et al., 2011; Shao et al., 2014).

In shellfish, early environmental conditions, such as exposure to ocean acidification, have been linked to adaptive changes in later-life traits. For example, Olympia oyster larvae exposed to acidification showed changes in juvenile traits that persisted into adulthood (Hettinger et al., 2012). Additionally, studies have demonstrated that environmental stressors, such as immune challenges or changes in temperature, can induce transgenerational plasticity, where offspring inherit beneficial traits acquired by their parents. Nutritional epigenetics is another area of interest, with studies showing that early-life nutrition can influence growth and feed efficiency in species like rainbow trout (Geurden et al., 2013). Epigenetic modifications, such as DNA methylation and histone modifications, have been linked to these effects (Balasubramanian et al., 2016; Panserat et al., 2017). Similarly, dietary interventions like vitamin supplementation or plant-based feeds can induce lasting epigenetic changes that affect growth and survival (Morán et al., 2013; Campos et al., 2013). Broodstock conditioning, which influences offspring through epigenetic inheritance, has also gained attention. For instance, Manila clams and Sydney rock oysters exposed to low pH during gonadal maturation produced faster-growing and more robust offspring (Zhao et al., 2017; Parker et al., 2012). These findings suggest that epigenetic mechanisms can mediate transgenerational responses to environmental conditions, contributing to aquaculture sustainability.

The potential for epigenetic selection in aquaculture is also being explored. Epigenetic markers could be integrated into breeding programs to improve traits like disease resistance or growth. Epigenetic changes can complement genetic selection, offering a more significant approach to breeding (Ong-Abdullah et al., 2015; Gavery & Roberts, 2017). However, epigenetic variation can also complicate genetic selection, as environmental stressors may induce changes in the genome, such as the activation of transposable elements (Dowen et al., 2012; Yu et al., 2013).

5 Future Perspectives

To advance the integration of epigenetics into aquaculture, future research should prioritize understanding the interaction between genome sequences and epigenetic modifications in shaping phenotypic traits under varying environmental conditions. Overcoming the challenges of limited genomic and epigenomic data for commercially important species, as well as the resource-intensive nature of studying species with long life cycles, is essential. Initial efforts could focus on species with shorter life cycles to generate cost-effective and timely insights. Expanding studies on histone modifications, noncoding RNAs, and epigenome editing technologies will be critical for identifying causal relationships between epigenetic changes and desirable traits. Additionally, the development of advanced bioinformatics tools will facilitate the mapping of key epigenetic markers, enabling their application in selective breeding and other genomic studies. With continued advancements in genomic sequencing, computational methods, and editing technologies, epigenetics has the potential to revolutionize aquaculture by enabling precise and sustainable practices that enhance the industry's competitiveness and environmental adaptability.

6 Conclusion

Epigenetics offers a transformative approach to improving productivity and sustainability in aquaculture. By elucidating the mechanisms that regulate gene expression and phenotype development, epigenetic research can enhance breeding programs, increase production efficiency, and support sustainable practices. Despite its potential, the field is still in its early phase, with significant knowledge gaps, particularly in understanding the role of epigenetic mechanisms in stress resistance, disease tolerance, nutritional programming, and environmental adaptation. Addressing these gaps will be crucial for translating epigenetic insights into actionable strategies for the aquaculture industry.

Acknowledgments The fellowship received from the University Grants Commission, New Delhi to Javeria Fatima (NTA Ref. No.: 211610179057) is thankfully acknowledged. We are also thankful to the Chairperson, Department of Zoology, AMU for providing laboratory facilities.

References

Abdelrahman, H., et al. (2017). Aquaculture genomics, genetics and breeding in the United States: Current status, challenges, and priorities for future research. *BMC Genomics, 18*, 1–23.

Annunziato, A. T., & Hansen, J. C. (2018). Role of histone acetylation in the assembly and modulation of chromatin structures. *Gene Expression, 9*(1–2), 37.

Araki, Y., & Mimura, T. (2017). The histone modification code in the pathogenesis of autoimmune diseases. *Mediators of Inflammation, 2017*(1), 2608605.

Azzi, A., Dallmann, R., Casserly, A., Rehrauer, H., Patrignani, A., Maier, B., Kramer, A., & Brown, S. A. (2014). Circadian behavior is light-reprogrammed by plastic DNA methylation. *Nature Neuroscience, 17*(3), 377–382.

Babu, K., Zhang, J., Moloney, S., Pleasants, T., McLean, C. A., Phua, S. H., & Sheppard, A. M. (2012). Epigenetic regulation of ABCG2 gene is associated with susceptibility to xenobiotic exposure. *Journal of Proteomics, 75*(12), 3410–3418.

Balasubramanian, M. N., Panserat, S., Dupont-Nivet, M., Quillet, E., Montfort, J., Le Cam, A., Medale, F., Kaushik, S. J., & Geurden, I. (2016). Molecular pathways associated with the nutritional programming of plant-based diet acceptance in rainbow trout following an early feeding exposure. *BMC Genomics, 17*, 1–20.

Berger, S. L. (2007). The complex language of chromatin regulation during transcription. *Nature, 447*(7143), 407–412.

Bishop, K. S., & Ferguson, L. R. (2015). The interaction between epigenetics, nutrition and the development of cancer. *Nutrients, 7*(2), 922–947.

Bizuayehu, T. T., & Babiak, I. (2014). MicroRNA in teleost fish. *Genome Biology and Evolution, 6*(8), 1911–1937.

Branco, M. R., Ficz, G., & Reik, W. (2012). Uncovering the role of 5-hydroxymethylcytosine in the epigenome. *Nature Reviews Genetics, 13*(1), 7–13.

Burggren, W., & Blank, T. (2009). Physiological study of larval fishes: Challenges and opportunities. *Scientia Marina, 73*(S1), 99–110.

Campos, C., Valente, L., Conceição, L., Engrola, S., & Fernandes, J. (2013). Temperature affects methylation of the myogenin putative promoter, its expression and muscle cellularity in Senegalese sole larvae. *Epigenetics, 8*(4), 389–397.

Canário, A. V. (2019). Aquaculture genomics. *Aquaculture and Fisheries, 4*(1), 1–2.

Castel, S. E., & Martienssen, R. A. (2013). RNA interference in the nucleus: Roles for small RNAs in transcription, epigenetics and beyond. *Nature Reviews Genetics, 14*(2), 100–112.

Cheung, P., Allis, C. D., & Sassone-Corsi, P. (2000). Signaling to chromatin through histone modifications. *Cell, 103*(2), 263–271.

Conaway, J. W. (2012). Introduction to theme "chromatin, epigenetics, and transcription". *Annual Review of Biochemistry, 81*(1), 61–64.

Csankovszki, G., Nagy, A., & Jaenisch, R. (2001). Synergism of Xist RNA, DNA methylation, and histone hypoacetylation in maintaining X chromosome inactivation. *The Journal of Cell Biology, 153*(4), 773–784.

Daneshfozouna, H., Panjvinib, F., Ghorbanic, F., & Farahmandd, H. (2015). A review of epigenetic imprints in aquatic animals. *Fisheries and Aquaculture Journal, 6*(1), 1.
Daxinger, L., & Whitelaw, E. (2012). Understanding transgenerational epigenetic inheritance via the gametes in mammals. *Nature Reviews Genetics, 13*(3), 153–162.
Dolinoy, D. C., Weidman, J. R., Waterland, R. A., & Jirtle, R. L. (2006). Maternal genistein alters coat color and protects Avy mouse offspring from obesity by modifying the fetal epigenome. *Environmental Health Perspectives, 114*(4), 567–572.
Dorts, J., Falisse, E., Schoofs, E., Flamion, E., Kestemont, P., & Silvestre, F. (2016). DNA methyltransferases and stress-related genes expression in zebrafish larvae after exposure to heat and copper during reprogramming of DNA methylation. *Scientific Reports, 6*(1), 34254.
Dowen, R. H., Pelizzola, M., Schmitz, R. J., Lister, R., Dowen, J. M., Nery, J. R., Dixon, J. E., & Ecker, J. R. (2012). Widespread dynamic DNA methylation in response to biotic stress. *Proceedings of the National Academy of Sciences, 109*(32), E2183–E2191.
Eirín-López, J. M., & Ausio, J. (2009). Origin and evolution of chromosomal sperm proteins. *BioEssays, 31*(10), 1062–1070.
Fellous, A., Favrel, P., & Riviere, G. (2015). Temperature influences histone methylation and mRNA expression of the Jmj-C histone-demethylase orthologues during the early development of the oyster *Crassostrea gigas*. *Marine Genomics, 19*, 23–30.
Gapp, K., von Ziegler, L., Tweedie-Cullen, R. Y., & Mansuy, I. M. (2014). Early life epigenetic programming and transmission of stress-induced traits in mammals: How and when can environmental factors influence traits and their transgenerational inheritance? *BioEssays, 36*(5), 491–502.
Gavery, M. R., & Roberts, S. B. (2017). Epigenetic considerations in aquaculture. *PeerJ, 5*, e4147.
Geurden, I., Borchert, P., Balasubramanian, M. N., Schrama, J. W., Dupont-Nivet, M., Quillet, E., Kaushik, S. J., Panserat, S., & Médale, F. (2013). The positive impact of the early-feeding of a plant-based diet on its future acceptance and utilisation in rainbow trout. *PLoS One, 8*(12), e83162.
Giannetto, A., Nagasawa, K., Fasulo, S., & Fernandes, J. M. (2013). Influence of photoperiod on expression of DNA (cytosine-5) methyltransferases in Atlantic cod. *Gene, 519*(2), 222–230.
Gluckman, P. D., Hanson, M. A., Cooper, C., & Thornburg, K. L. (2008). Effect of in utero and early-life conditions on adult health and disease. *New England Journal of Medicine, 359*(1), 61–73.
González-Recio, O. (2012). Epigenetics: A new challenge in the post-genomic era of livestock. *Frontiers in Genetics, 2*, 106.
Granada, L., Lemos, M. F., Cabral, H. N., Bossier, P., & Novais, S. C. (2018). Epigenetics in aquaculture–the last frontier. *Reviews in Aquaculture, 10*(4), 994–1013.
Green, T. J., Helbig, K., Speck, P., & Raftos, D. A. (2016). Primed for success: Oyster parents treated with poly (I: C) produce offspring with enhanced protection against Ostreid herpesvirus type I infection. *Molecular Immunology, 78*, 113–120.
Greer, E. L., Beese-Sims, S. E., Brookes, E., Spadafora, R., Zhu, Y., Rothbart, S. B., Aristizábal-Corrales, D., Chen, S., Badeaux, A. I., Jin, Q., & Wang, W. (2014). A histone methylation network regulates transgenerational epigenetic memory in *C. elegans*. *Cell Reports, 7*(1), 113–126.
Halpern, B. S., Frazier, M., Afflerbach, J., Lowndes, J. S., Micheli, F., O'Hara, C., Scarborough, C., & Selkoe, K. A. (2019). Recent pace of change in human impact on the world's ocean. *Scientific Reports, 9*(1), 11609.
Han, J., Hu, Y., Qi, Y., Yuan, C., Naeem, S., & Huang, D. (2021). High temperature induced masculinization of zebrafish by down-regulation of sox9b and esr1 via DNA methylation. *Journal of Environmental Sciences, 107*, 160–170.
Hanson, M. A., & Skinner, M. K. (2016). Developmental origins of epigenetic transgenerational inheritance. *Environmental Epigenetics, 2*(1), dvw002.
Hauser, M. T., Aufsatz, W., Jonak, C., & Luschnig, C. (2011). Transgenerational epigenetic inheritance in plants. *Biochimica et Biophysica Acta (BBA)-Gene Regulatory Mechanisms, 1809*(8), 459–468.

Hawes, N. A., Tremblay, L. A., Pochon, X., Dunphy, B., Fidler, A. E., & Smith, K. F. (2018). Effects of temperature and salinity stress on DNA methylation in a highly invasive marine invertebrate, the colonial ascidian *Didemnum vexillum*. *PeerJ, 6*, e5003.

Henikoff, S., & Smith, M. M. (2015). Histone variants and epigenetics. *Cold Spring Harbor Perspectives in Biology, 7*(1), a019364.

Hettinger, A., Sanford, E., Hill, T. M., Russell, A. D., Sato, K. N., Hoey, J., Forsch, M., Page, H. N., & Gaylord, B. (2012). Persistent carry-over effects of planktonic exposure to ocean acidification in the Olympia oyster. *Ecology, 93*(12), 2758–2768.

Houwing, S., Kamminga, L. M., Berezikov, E., Cronembold, D., Girard, A., van den Elst, H., Filippov, D. V., Blaser, H., Raz, E., Moens, C. B., & Plasterk, R. H. (2007). A role for Piwi and piRNAs in germ cell maintenance and transposon silencing in zebrafish. *Cell, 129*(1), 69–82.

Huang, S., Wu, Y., Chen, K., Zhang, X., Zhao, J., Luo, Q., Liu, H., Wang, F., Li, K., Fei, S., & Zhang, X. (2023). Gene expression and epigenetic modification of aromatase during sex reversal and gonadal development in blotched snakehead (*Channa maculata*). *Fishes, 8*(3), 129.

Izquierdo, M. S., Turkmen, S. E. R. H. A. T., Montero, D., Zamorano, M. J., Afonso, J. M., Karalazos, V., & Fernández-Palacios, H. (2015). Nutritional programming through broodstock diets to improve utilization of very low fishmeal and fish oil diets in gilthead sea bream. *Aquaculture, 449*, 18–26.

Jablonka, E., & Lamb, M. J. (2002). The changing concept of epigenetics. *Annals of the New York Academy of Sciences, 981*(1), 82–96.

Jablonka, E., & Lamb, M. J. (2014). *Evolution in four dimensions, revised edition: Genetic, epigenetic, behavioral, and symbolic variation in the history of life*. MIT press.

Jablonka, E., & Raz, G. (2009). Transgenerational epigenetic inheritance: Prevalence, mechanisms, and implications for the study of heredity and evolution. *The Quarterly Review of Biology, 84*(2), 131–176.

Jaenisch, R., & Bird, A. (2003). Epigenetic regulation of gene expression: How the genome integrates intrinsic and environmental signals. *Nature Genetics, 33*(3), 245–254.

Jenuwein, T., & Allis, C. D. (2001). Translating the histone code. *Science, 293*(5532), 1074–1080.

Jia, Y., Zheng, J., Chi, M., Liu, S., Jiang, W., Cheng, S., Gu, Z., & Chen, L. (2019). Molecular identification of dmrt1 and its promoter CpG methylation in correlation with gene expression during gonad development in Culter alburnus. *Fish Physiology and Biochemistry, 45*, 245–252.

Jiang, Q., Li, Q., Yu, H., & Kong, L. F. (2013). Genetic and epigenetic variation in mass selection populations of Pacific oyster *Crassostrea gigas*. *Genes & Genomics, 35*, 641–647.

Jin, B., Li, Y., & Robertson, K. D. (2011). DNA methylation: Superior or subordinate in the epigenetic hierarchy? *Genes and cancer, 2*(6), 607–617.

Jones, P. A. (1999). The DNA methylation paradox. *Trends in Genetics, 15*(1), 34–37.

Jonsson, B., & Jonsson, N. (2014). Early environment influences later performance in fishes. *Journal of Fish Biology, 85*(2), 151–188.

Khalaf, O., & Gräff, J. (2016). Structural, synaptic, and epigenetic dynamics of enduring memories. *Neural Plasticity, 2016*(1), 3425908.

Klosin, A., Casas, E., Hidalgo-Carcedo, C., Vavouri, T., & Lehner, B. (2017). Transgenerational transmission of environmental information in C. Elegans. *Science, 356*(6335), 320–323.

Lai, K. P., Li, J. W., Chan, C. Y. S., Chan, T. F., Yuen, K. W. Y., & Chiu, J. M. Y. (2016). Transcriptomic alterations in Daphnia magna embryos from mothers exposed to hypoxia. *Aquatic Toxicology, 177*, 454–463.

Lawrence, M., Daujat, S., & Schneider, R. (2016). Lateral thinking: How histone modifications regulate gene expression. *Trends in Genetics, 32*(1), 42–56.

Le Luyer, J., Laporte, M., Beacham, T. D., Kaukinen, K. H., Withler, R. E., Leong, J. S., Rondeau, E. B., Koop, B. F., & Bernatchez, L. (2017). Parallel epigenetic modifications induced by hatchery rearing in a Pacific salmon. *Proceedings of the National Academy of Sciences, 114*(49), 12964–12969.

Lee, J. T. (2009). Lessons from X-chromosome inactivation: Long ncRNA as guides and tethers to the epigenome. *Genes and Development, 23*(16), 1831–1842.

Li, Z., Lu, X., Gao, Y., Liu, S., Tao, M., Xiao, H., Qiao, Y., Zhang, Y., & Luo, J. (2011). Polyploidization and epigenetics. *Chinese Science Bulletin, 56*, 245–252.

Li, P., Chen, J., Zhu, C., Pan, Z., Li, Q., Wei, H., Wang, G., Cheng, W., Fu, B., & Sun, Y. (2022). DNA methylation difference between female and male ussuri catfish (*Pseudobagrus ussuriensis*) in brain and gonad tissues. *Life, 12*(6), 874.

Liu, Y., Zhu, Q., Li, L., Wang, W., & Zhang, G. (2020). Identification of HSF1 target genes involved in thermal stress in the Pacific oyster *Crassostrea gigas* by ChIP-seq. *Marine Biotechnology, 22*, 167–179.

Liu, Z., Zhou, T., & Gao, D. (2022). Genetic and epigenetic regulation of growth, reproduction, disease resistance and stress responses in aquaculture. *Frontiers in Genetics, 13*, 994471.

Macleod, D., Clark, V. H., & Bird, A. (1999). Absence of genome-wide changes in DNA methylation during development of the zebrafish. *Nature Genetics, 23*(2), 139–140.

Maloisel, L., & Rossignol, J. L. (1998). Suppression of crossing-over by DNA methylation in Ascobolus. *Genes and Development, 12*(9), 1381–1389.

Marandel, L., Lepais, O., Arbenoits, E., Véron, V., Dias, K., Zion, M., & Panserat, S. (2016). Remodelling of the hepatic epigenetic landscape of glucose-intolerant rainbow trout (*Oncorhynchus mykiss*) by nutritional status and dietary carbohydrates. *Scientific Reports, 6*(1), 32187.

Metzger, D. C., & Schulte, P. M. (2018). The DNA methylation landscape of stickleback reveals patterns of sex chromosome evolution and effects of environmental salinity. *Genome Biology and Evolution, 10*(3), 775–785.

Mirbahai, L., Yin, G., Bignell, J. P., Li, N., Williams, T. D., & Chipman, J. K. (2011). DNA methylation in liver tumorigenesis in fish from the environment. *Epigenetics, 6*(11), 1319–1333.

Moazed, D. (2009). Small RNAs in transcriptional gene silencing and genome defence. *Nature, 457*(7228), 413–420.

Moghadam, H. K., et al. (2017). Impacts of early life stress on the methylome and transcriptome of Atlantic salmon. *Scientific Reports, 7*(1), 5023.

Moore, L. D., Le, T., & Fan, G. (2013). DNA methylation and its basic function. *Neuropsychopharmacology, 38*(1), 23–38.

Morán, P., Marco-Rius, F., Megías, M., Covelo-Soto, L., & Pérez-Figueroa, A. (2013). Environmental induced methylation changes associated with seawater adaptation in brown trout. *Aquaculture, 392*, 77–83.

Morgan, H. D., Santos, F., Green, K., Dean, W., & Reik, W. (2005). Epigenetic reprogramming in mammals. *Human Molecular Genetics, 14*(suppl_1), R47–R58.

Mushtaq, N., Singh, D. V., Bhat, R. A., Dervash, M. A., & Hameed, O. B. (2020). Freshwater contamination: Sources and hazards to aquatic biota. In *Fresh water pollution dynamics and remediation* (pp. 27–50).

Navarro-Martín, L., Vinas, J., Ribas, L., Díaz, N., Gutierrez, A., Di Croce, L., & Piferrer, F. (2011). DNA methylation of the gonadal aromatase (cyp19a) promoter is involved in temperature-dependent sex ratio shifts in the European Sea bass. *PLoS Genetics, 7*(12), e1002447.

Ng, K., Pullirsch, D., Leeb, M., & Wutz, A. (2007). Xist and the order of silencing. *EMBO Reports, 8*(1), 34–39.

Norouzitallab, P., Baruah, S. K., Vandegehuchte, M., Van Stappen, G., Catania, F., Vanden Bussche, J., Vanhaecke, L., Sorgeloos, P., & Bossier, P. (2014). Environmental heat stress induces epigenetic transgenerational inheritance of robustness in parthenogenetic Artemia model. *FASEB Journal, 28*(8), 3552–3563.

Ong-Abdullah, M., Ordway, J. M., Jiang, N., Ooi, S. E., Kok, S. Y., Sarpan, N., Azimi, N., Hashim, A. T., Ishak, Z., Rosli, S. K., & Malike, F. A. (2015). Loss of Karma transposon methylation underlies the mantled somaclonal variant of oil palm. *Nature, 525*(7570), 533–537.

Paneru, B., Al-Tobasei, R., Palti, Y., Wiens, G. D., & Salem, M. (2016). Differential expression of long non-coding RNAs in three genetic lines of rainbow trout in response to infection with *Flavobacterium psychrophilum*. *Scientific Reports, 6*(1), 36032.

Panserat, S., Marandel, L., Geurden, I., Veron, V., Dias, K., Plagnes-Juan, E., Pegourié, G., Arbenoits, E., Santigosa, E., Weber, G., & Trichet, V. V. (2017). Muscle catabolic capacities and global hepatic epigenome are modified in juvenile rainbow trout fed different vitamin levels at first feeding. *Aquaculture, 468*, 515–523.

Parker, L. M., Ross, P. M., O'Connor, W. A., Borysko, L., Raftos, D. A., & Pörtner, H. O. (2012). Adult exposure influences offspring response to ocean acidification in oysters. *Global Change Biology, 18*(1), 82–92.

Peschansky, V. J., & Wahlestedt, C. (2014). Non-coding RNAs as direct and indirect modulators of epigenetic regulation. *Epigenetics, 9*(1), 3–12.

Pierron, F., Baillon, L., Sow, M., Gotreau, S., & Gonzalez, P. (2014). Effect of low-dose cadmium exposure on DNA methylation in the endangered European eel. *Environmental Science & Technology, 48*(1), 797–803.

Pogribny, I. P. (2010). Epigenetic events in tumorigenesis: Putting the pieces together. *Experimental Oncology, 32*(3), 132–136.

Rasal, K. D., Asgolkar, P., Shinde, S., Dey, D., & Sundaray, J. K. (2024). Role of epigenetics in fisheries and aquaculture. In *Current trends in fisheries biotechnology* (pp. 27–44). Singapore.

Rechavi, O., et al. (2014). Starvation-induced transgenerational inheritance of small RNAs in *C. elegans*. *Cell, 158*(2), 277–287.

Rinn, J. L., Kertesz, M., Wang, J. K., Squazzo, S. L., Xu, X., Brugmann, S. A., Goodnough, L. H., Helms, J. A., Farnham, P. J., Segal, E., & Chang, H. Y. (2007). Functional demarcation of active and silent chromatin domains in human HOX loci by noncoding RNAs. *Cell, 129*(7), 1311–1323.

Romano, G., Veneziano, D., Nigita, G., & Nana-Sinkam, S. P. (2018). RNA methylation in ncRNA: Classes, detection, and molecular associations. *Frontiers in Genetics, 9*, 243.

Roy, S., Kumar, V., Behera, B. K., & Das, B. K. (2022). Epigenetics: Perspectives and potential in aquaculture. In *Advances in fisheries biotechnology* (pp. 133–150). Singapore.

Shao, C., Li, Q., Chen, S., Zhang, P., Lian, J., Hu, Q., Sun, B., Jin, L., Liu, S., Wang, Z., & Zhao, H. (2014). Epigenetic modification and inheritance in sexual reversal of fish. *Genome Research, 24*(4), 604–615.

Sharma, S., Kelly, T. K., & Jones, P. A. (2010). Epigenetics in cancer. *Carcinogenesis, 31*(1), 27–36.

Siggens, L., & Ekwall, K. (2014). Epigenetics, chromatin and genome organization: Recent advances from the ENCODE project. *Journal of Internal Medicine, 276*(3), 201–214.

Skinner, M. K., Gurerrero-Bosagna, C., Haque, M. M., Nilsson, E. E., Koop, J. A., Knutie, S. A., & Clayton, D. H. (2014). Epigenetics and the evolution of Darwin's finches. *Genome Biology and Evolution, 6*(8), 1972–1989.

Sun, D., Li, Q., & Yu, H. (2022). DNA methylation differences between male and female gonads of the oyster reveal the role of epigenetics in sex determination. *Gene, 820*, 146260.

Tahiliani, M., Koh, K. P., Shen, Y., Pastor, W. A., Bandukwala, H., Brudno, Y., Agarwal, S., Iyer, L. M., Liu, D. R., Aravind, L., & Rao, A. (2009). Conversion of 5-methylcytosine to 5-hydroxymethylcytosine in mammalian DNA by MLL partner TET1. *Science, 324*(5929), 930–935.

Talbert, P. B., & Henikoff, S. (2010). Histone variants – Ancient wrap artists of the epigenome. *Nature Reviews Molecular Cell Biology, 11*(4), 264–275.

Turner, B. M. (2009). Epigenetic responses to environmental change and their evolutionary implications. *Philosophical Transactions of the Royal Society B: Biological Sciences, 364*(1534), 3403–3418.

Uren Webster, T. M., Rodriguez-Barreto, D., Martin, S. A., Van Oosterhout, C., Orozco-terWengel, P., Cable, J., Hamilton, A., Garcia de Leaniz, C., & Consuegra, S. (2018). Contrasting effects of acute and chronic stress on the transcriptome, epigenome, and immune response of Atlantic salmon. *Epigenetics, 13*(12), 1191–1207.

Vastenhouw, N. L., & Schier, A. F. (2012). Bivalent histone modifications in early embryogenesis. *Current Opinion in Cell Biology, 24*(3), 374–386.

Verdone, L., Agricola, E., Caserta, M., & Di Mauro, E. (2006). Histone acetylation in gene regulation. *Briefings in Functional Genomics, 5*(3), 209–221.

Wang, K. C., & Chang, H. Y. (2011). Molecular mechanisms of long noncoding RNAs. *Molecular Cell, 43*(6), 904–914.

Wang, Z., & Zhu, F. (2017). MicroRNA-100 is involved in shrimp immune response to white spot syndrome virus (WSSV) and vibrio alginolyticus infection. *Scientific Reports, 7*(1), 42334.

Wang, S. Y., Lau, K., Lai, K. P., Zhang, J. W., Tse, A. C. K., Li, J. W., Tong, Y., Chan, T. F., Wong, C. K. C., Chiu, J. M. Y., & Au, D. W. T. (2016). Hypoxia causes transgenerational impairments in reproduction of fish. *Nature Communications, 7*(1), 12114.

Weaver, I. C., Cervoni, N., Champagne, F. A., D'Alessio, A. C., Sharma, S., Seckl, J. R., Dymov, S., Szyf, M., & Meaney, M. J. (2004). Epigenetic programming by maternal behavior. *Nature Neuroscience, 7*(8), 847–854.

Xu, D., Fang, H., Liu, J., Chen, Y., Gu, Y., Sun, G., & Xia, B. (2022). ChIP-seq assay revealed histone modification H3K9ac involved in heat shock response of the sea cucumber *Apostichopus japonicus*. *Science of the Total Environment, 820*, 153168.

Yang, Y., Zhou, T., Liu, Y., Tian, C., Bao, L., Wang, W., Zhang, Y., Liu, S., Shi, H., Tan, S., & Gao, D. (2022). Identification of an epigenetically marked locus within the sex determination region of channel catfish. *International Journal of Molecular Sciences, 23*(10), 5471.

Youngson, N. A., & Whitelaw, E. (2008). Transgenerational epigenetic effects. *Annual Review of Genomics and Human Genetics, 9*(1), 233–257.

Yu, A., Lepère, G., Jay, F., Wang, J., Bapaume, L., Wang, Y., Abraham, A. L., Penterman, J., Fischer, R. L., Voinnet, O., & Navarro, L. (2013). Dynamics and biological relevance of DNA demethylation in Arabidopsis antibacterial defense. *Proceedings of the National Academy of Sciences, 110*(6), 2389–2394.

Yu, H., Zhao, X., & Li, Q. (2016). Genome-wide identification and characterization of long intergenic noncoding RNAs and their potential association with larval development in the Pacific oyster. *Scientific Reports, 6*(1), 20796.

Yue, Y., Liu, J., & He, C. (2015). RNA N6-methyladenosine methylation in post-transcriptional gene expression regulation. *Genes & Development, 29*(13), 1343–1355.

Zeng, Y., & Chen, T. (2019). DNA methylation reprogramming during mammalian development. *Genes (Basel), 10*, 257.

Zhao, L., Schöne, B. R., Mertz-Kraus, R., & Yang, F. (2017). Sodium provides unique insights into transgenerational effects of ocean acidification on bivalve shell formation. *Science of the Total Environment, 577*, 360–366.

Chapter 22
Emerging Trend of Nanotechnology: Applications, Mode of Action, and Challenges

Kousar Jan, Imtiaz Ahmed, and Ishtiyaq Ahmad

1 Introduction to Nanoparticles

Aquaculture is regarded as one of the most significant food production systems when it comes to food security and economic impact. As such, the industry's continued growth is essential to the plan for ensuring the nutritional safety of the world's population. Over half of the aquatic products used for food are produced by aquaculture, considered a vital food industry having the fastest growth and development (Hussain et al., 2019). Global fish production in 2018 was predicted to be 180 million tons, with a first sale value of \$401 billion, and out of this amount, aquaculture production accounted for 82 million tons, or \$250 billion. On the other hand, the average annual growth rate of edible fish consumption per capita increased from 9.0 kg in 1961 to 20.5 kg in 2018 (FAO, 2020). Accordingly, aquaculture has the potential to be one of the industries that mostly helps to achieve the 2030 sustainable development goals (Ahmed et al., 2023). This industry employs around 20.5 million people worldwide, including the support of traditional fishing families, and holds a significant proportion of the population particularly in developing nations (FAO, 2020). However, a number of concerns that are detrimental to this sector for achieving global food security include disease prevalence, inadequate nutrition, chemical pollution, and environmental degradation (Shah & Mraz, 2020; Rather et al., 2024). To improve its productivity, efficiency, and sustainability, many nanotechnology-based approaches are being used nowadays (Fajardo et al., 2022). As a requirement to affecting our ability to feed a population of almost 7 billion

K. Jan · I. Ahmed (✉)
Fish Nutrition Research Laboratory, Department of Zoology, University of Kashmir, Srinagar, Jammu and Kashmir, India

I. Ahmad
Department of Zoology, University of Kashmir, Srinagar, Jammu and Kashmir, India

I. Ahmed, I. Ahmad (eds.), *Aquaculture: Enhancing Food Security and Nutrition*, https://doi.org/10.1007/978-3-031-92858-1_22

people, nanotechnology has immense potential in delivering creative enhancements to aquaculture systems in order to lower prices, boost efficiency, and lessen our impact on the environment. China has been at the forefront of the rapid development and application of nanotechnology in the aquaculture sector. Applications are considered successful only if they meet the requirements of high quality, low cost, environmental sustainability, and low risk to human health (Chen & Yada, 2011). Nowadays, the market is home to over a thousand items that include nanoparticles, demonstrating the multibillion-dollar and quickly growing nature of the nanotechnology business. More than 300 nanofood items have been introduced to global markets in the last ten years (Ramsden, 2018). The nanotechnology sector is expected to generate at least $3 trillion in economic effect by 2020 and employ roughly 6 million people (He et al., 2019). According to a report by Technavio (2019), the global market for nanotechnology applied to the food sector is expected to develop at a rate of more than 24% per year between 2019 and 2023, with a projected value of $112.48 billion. The primary driver of this market surge is the growing use of nanotechnology in nutraceuticals. Based on the information that is currently available, the United States appears to be the leading country in the world in the nanofood sector, followed by China, Japan, and the European Union (EU) (Thiruvengadam et al., 2018). With a 4-year plan of investment of $3.7 billion funneled through its National Nanotechnology Initiative (NNI), the United States leads this group as the world's largest funding source for nanotechnology research (Thiruvengadam et al., 2018; Meghani et al., 2020). In this industry, some of the leading names today are AQUANOVA (Germany), BASF (Germany), NanoPack (Belgium), and PEN (USA).

1.1 Etymology

The prefix "nano" is derived from the Greek word "nanos," meaning "a dwarf." In 1947, at the 14th conference of the International Union of Pure and Applied Chemistry (IUPAC), the prefix "nano" was officially adopted to describe the one-billionth part (10^{-9}) of a unit 1 (Joudeh & Linke, 2022). In scientific literature, the prefix "nano" has been adopted as a popular label in many fields of modern science to describe small entities and processes. These terms include, but are not limited to, nanoscience, nanotechnology, nanorobots, nanomagnets, nanoelectronics, nanoencapsulation, etc. (Buzea et al., 2007). In all of these cases, it is used to describe "very small" entities or processes, most often at the actual nanometer scale.

1.2 Definition and Characteristics of Nanoparticles

Nanoscience is a branch of science that comprises the study of properties of matter at nanoscale and particularly focuses on the unique, size-dependent properties of solid-state materials (Mulvaney, 2015; Joudeh & Linke, 2022). This branch mainly comprises the synthesis, engineering, and utilization of materials whose size ranges from 1 to 100 nanometers (nm), known as nanomaterials (Hasan, 2015). It is also defined as the "understanding and control of matter at the nanoscale, at dimensions between approximately 1 and 100 nm, where unique phenomena enable novel applications" by the United States National Nanotechnology Initiative (US-NNI) (Fajardo et al., 2022). This scale corresponds to the size of individual molecules and tiny atomic groupings. Owing to the small size and high surface area-to-volume ratio, nanoparticles are suitable for a variety of applications, and their increased surface area also increases their reactivity (Campelo et al., 2009). Compared to their larger counterparts, they have distinct and frequently improved physical, chemical, and biological properties due to their wide size range and high surface area-to-volume ratio. These characteristics may include improved strength or conductivity, changed optical qualities, and higher reactivity. The higher percentage of atoms or molecules at the nanoparticles surface is thought to be the cause of their enhanced reactivity (Sirelkhatim et al., 2015). Quantum effects intensify at the nanoscale, which can alter a material's optical, electrical, and magnetic characteristics and set nanoparticles apart from bigger particles of the same composition (Asha & Narain, 2020). Applications such as magnetic resonance imaging (MRI) and magnetic targeted medication delivery (MTMD) may benefit from certain improved magnetic characteristics of nanoparticles, especially those containing magnetic materials. Aside from this, the interaction of light at the nanoscale with the material can give nanoparticles distinctive optical features, including color shifts and fluorescence (Patra et al., 2012). When it comes to applications like thermal management and nanocomposite materials, nanoparticles may have different thermal conductivity than bulk materials. For their manufacture, a variety of techniques, including chemical reduction, sol-gel procedures, physical vapor deposition, and biological methods, can be employed; each has an impact on the properties of the final nanoparticles. Because of these physiochemical characteristics, nanomaterials are widely used in the healthcare, water treatment, and food preservation industries (Ogunkalu, 2019).

The synthesis of nanomaterials can be done in two ways: bottom-up (atom by atom) or top-down (converting big materials to nanoscale). These processes can be aided by the properties of nanomaterials, which include physical, optical, chemical, magnetic, and electric activity (Roy et al., 2012). A wide range of materials can be used to create nanomaterials, including metals (gold, silver), metal oxides (titanium dioxide, zinc oxide), polymers, carbon-based materials (carbon nanotubes, graphene), and lipids. Composites such as quantum dots, nano-ceramics, and nanoshells can also be created using these materials (Stone et al., 2010). Various shapes and types of nanoparticles are employed, including dendrimers (Torchilin, 2006), nanospheres (Donbrow, 1991), nanotubes (Reilly, 2007), and others (Wu et al., 2015). These days, complicated three-dimensional structures and the presence of multiple distinct chemical compounds are also being seen in nanomaterials (Handy, 2012). Several industries, including wastewater treatment (Zahra et al., 2020), environmental monitoring (Rassaei et al., 2011), antimicrobial agents (Islam et al., 2022), and food additives (Chen et al., 2023), along with biotechnology, biomedicine and medicines, drug delivery, and imaging are among the many sectors in which these particles find extensive use.

1.3 Importance of Nanoparticles in Aquaculture

Research on the application of nanoparticles in aquaculture is still in progress, but it has the potential to benefit the sector in a number of ways. The use of nanoparticles in aquaculture offers the potential to support more sustainable practices by increasing the overall production, decreasing the environmental effect of aquaculture operations, and better-utilizing resources (Nasr-Eldahan et al., 2021). Aquaculture feeds can benefit from the addition of nanoparticles to improve nutrient uptake and utilization. Nanoparticles can enhance the feed efficiency ratio, which will promote the growth and well-being of aquatic species by making the nutrients more soluble (Chris et al., 2018). By encapsulating nutrients at the nanoscale, one can increase their bioavailability and make sure that the aquatic organisms, including fish, receive the necessary elements more effectively. This may be especially crucial for animals that have distinct food needs. Fish diseases can be controlled and prevented with the use of nanoparticles, particularly those having antibacterial qualities such as silver nanoparticles (Khursheed et al., 2023; Khan et al., 2024a, b). On inhibiting the growth of dangerous bacteria, viruses, or parasites, the antimicrobial qualities may help to lower the frequency of infections. As a means for the controlled release of antimicrobial drugs or other protective chemicals, nanoparticles may be able to support biosecurity efforts in aquaculture (Okeke et al., 2022). This could potentially mitigate the transmission of diseases among aquaculture establishments. In addition, nanoparticles can be used in aquaculture as vaccination carriers. By ensuring a controlled and targeted release of vaccines, this strategy may improve vaccination programs efficacy and may strengthen the immune responses in fish as well (Vinay et al., 2018). Besides, toxins and other pollutants in water may bind to certain types of nanoparticles. Hence, by adding these nanoparticles to aquaculture systems, water can be made more hygienic, and detoxified by virtue of this, the surrounding environment of aquatic life can be improved (Yaqoob et al., 2020). Aquaculture systems that monitor the environment can use sensors and diagnostic tools equipped with nanoparticles. In order to enable prompt treatments, these sensors may be able to identify changes in pathogen presence, nutrient levels, and water quality.

2 Types of Nanoparticles Used in Aquaculture Feed

Aquaculture is using nanoparticles more and more owing to their special qualities, which include tiny size, large surface area, and the capacity to improve nutrient bioavailability. The following categories of nanoparticles are frequently found in the aquaculture feed industry:

2.1 Metallic Nanoparticles

Metallic nanoparticles are nanoscale particles composed of metal atoms. Several metals, such as gold, silver, copper, platinum, palladium, iron, and others, can be included in these nanoparticles (Roszczenko et al., 2022). Each metal gives the nanoparticles unique characteristics. The well-known localized surface plasmon resonance (LSPR) effect gives noble metal nanoparticles, especially gold and silver, unique optical features (Loiseau et al., 2019). Because of this phenomenon, they are useful in sensing, imaging, and medical diagnostic applications, which increases light absorption and scattering. Metallic nanoparticles are created by a variety of techniques, including chemical reduction, sol-gel procedures, physical vapor deposition, and biological techniques (Khan et al., 2023). The size, form, and characteristics of the nanoparticles are influenced by the process used. Certain sizes and geometries, such as spheres, rods, wires, or other forms, can be manufactured into metallic nanoparticles. The size and shape of nanoparticles are considered important factors influencing their optical, electrical, and catalytic capabilities (Khan et al., 2019). The heightened surface area and reactivity of metallic nanoparticles make them efficient catalysts. The reduction of pollutants, and hydrogenation and oxidation processes are just a few of the chemical reactions where they can be used to catalyze. Owing to their special qualities, such as their small size, high surface area, and particular biological activities, these nanoparticles are also being added to aquaculture feed on a growing basis (Moges et al., 2020). The nutritive value of feed, disease resistance, and feed consumption efficiency can all be improved by the use of these nanoparticles. Some of the common nanoparticle industries are listed below:

2.1.1 Silver Nanoparticles

Silver atoms combine to form tiny particles known as silver nanoparticles. Their small size, high surface area-to-volume ratio, and capacity to absorb and scatter visible near-infrared light give them those distinctive physical and chemical features (Altammar, 2023; Khan et al., 2024b). Silver nanoparticles may also have additional antimicrobial properties not exhibited by their ionic silver nature due to their small size and high surface-to-volume ratios, which result in physical and chemical differences in their properties compared to their bulk counterparts (Shenashen et al., 2014). Additionally, the manufacturing process, chemical reduction being the most common process, can produce silver nanoparticles in a variety of sizes and shapes. Silver nanoparticles play a vital role in aquaculture due to their special qualities and possible uses. Strong antibacterial capabilities are a well-known characteristic of silver nanoparticles. It has the ability to effectively combat a variety of aquatic diseases, such as fungus, viruses, and bacteria, and can therefore be helpful in aquaculture settings for the prevention and treatment of diseases. Besides, silver nanoparticles can be employed in water treatment procedures to preserve water quality as well

(Fiorati et al., 2020). They can lessen the need for dangerous pesticides and aid in the management of microbial growth. A further possible use is adding silver nanoparticles to fish meal that can accelerate growth rates and strengthen the immune systems of aquatic animals (Popoola et al., 2023). However, to fully comprehend the ramifications of this application for the health of consumers and fish, more research needs to be done. Although silver nanoparticles have many advantages, however, there are some worries regarding their environmental impacts and safety (Pulit-Prociak & Banach, 2016). These nanoparticles run the risk of building up in aquatic life and maybe making their way into the human food chain. Furthermore, it is unclear how employing nanoparticles in aquaculture may affect the environment in the long run. Many nations have restrictions governing the use of silver nanoparticles in aquaculture as a result of these worries (Ogunfowora et al., 2021). Understanding their effects on aquatic ecosystems, evaluating their safety for the environment and human health, and creating standards for their safe and efficient usage are the main goals of ongoing studies of various researchers.

2.1.2 Copper Nanoparticles

It is a copper-based nanoparticle having a size range of 1–100 nm (Khan et al., 2019). The characteristics of copper nanoparticles are mostly determined by their size and form. The antibacterial efficaciousness and increased reactivity of small-sized nanoparticles are common. There are several ways to create copper nanoparticles, such as chemical reduction, green synthesis with plant extracts, physical approaches, and electrochemical techniques (Crisan et al., 2021). Nanoparticle stability, size, and form are all influenced by the synthesis and method used. The physical and chemical properties of nanoparticles can be influenced by their form, which can be spherical, rod-shaped, or have other geometries (Sabourian et al., 2020). A variety of chemical processes can effectively be catalyzed by copper nanoparticles. In the realm of heterogeneous catalysis, they are employed for the processes of oxidation and reduction because of their high surface area and reactivity. It also has antibacterial properties, just like silver nanoparticles, that make them helpful in water treatment, textiles, and medical equipment (Mahmoodi et al., 2018). Owing to their special qualities that set them apart from bulk copper, these nanoparticles are useful for a wide range of applications in numerous industries. To give surfaces antibacterial qualities, surfaces are coated with copper nanoparticles. This application is especially pertinent in businesses and healthcare settings where it is critical to stop the growth of fungus and bacteria. Numerous microorganisms, including fungi and bacteria, have been demonstrated to be inhibited by them. The purpose of antimicrobial apparel and wound dressings using copper nanoparticles is to prevent the growth of bacteria (Kim et al., 2023). The application of copper nanoparticles in heat transfer fluids is being researched. They can be added to fluids to improve heat transfer efficiency in a variety of applications, including electronics cooling, because of their high thermal conductivity (Zehri et al., 2019). Potential uses for copper nanoparticles in water treatment include disinfection and can support efforts to purify water by assisting in the removal of dangerous germs. Numerous biomedical uses, such as medication delivery, imaging, and therapeutic treatments, are being investigated for copper nanoparticles that make them useful in nanomedicine research owing to these special qualities (Elumalai et al., 2024).

2.2 *Organic Nanoparticles*

Nanoparticles composed of polymers, proteins, carbohydrates, lipids, polymers, and other organic compounds that are used to create organic nanoparticles. The most well-known members of this family are protein complexes like ferritin, liposomes, dendrimers, and micelles (Joudeh & Linke, 2022). These nanoparticles are generally nontoxic and biodegradable, and in certain instances may have a hollow center. Heat and light, as well as other forms of electromagnetic radiation, can affect organic nanoparticles (Ealia & Saravanakumar, 2017). Furthermore, they are frequently created by non-covalent intermolecular interactions, which increase their labile character and provide an exit path for the body (Ng & Zheng, 2015). The possible range of use of organic nanoparticles is determined by a number of factors, including composition, surface shape, stability, carrying capacity, etc. Nowadays, targeted drug delivery and cancer therapy are the two main uses of organic nanoparticles in the biomedical industry (Gujrati et al., 2014). With the help of the special qualities connected to the nanoscale, these nanoparticles can be designed for a range of uses. Applications for organic nanoparticles can be found in medicine, medication delivery, electronics, materials research, and imaging (Mitragotri & Stayton, 2014). Among the many varieties of organic nanoparticles are polymeric ones. These can be made of biodegradable polymers like chitosan, polyethylene glycol (PEG), or poly lactic-co-glycolic acid (PLGA). Due to their biocompatibility and customizable features, polymeric nanoparticles are frequently employed in medicine administration.

2.2.1 Lipid Nanoparticles

Lipid nanoparticles are usually spherical and have a diameter of 10–1000 nm. These are lipid- or lipid-like material-based organic nanoparticles. According to Khan et al. (2019), lipid nanoparticles consist of a matrix made up of soluble lipophilic molecules and a solid lipid core.

Types of lipid-based nanoparticles:

- *Liposomes*: Liposomes are spherical vesicles with a lipid bilayer structure. They can encapsulate hydrophilic drugs within their aqueous core and hydrophobic drugs within the lipid bilayers (Cagdas et al., 2014).
- *Lipid nanocarriers*: This category includes various lipid-based nanoparticles such as solid lipid nanoparticles (SLNs) and nanostructured lipid carriers (NLCs). SLNs have a solid lipid core, while NLCs have a combination of solid and liquid lipids (Shirodkar et al., 2019).

Lipid-based nanoparticles are known for their excellent biocompatibility. This characteristic is essential for drug delivery applications, as the nanoparticles must have a positive interaction with biological systems in order to deliver medications to the intended locations (Kumar, 2019). These nanoparticles have the ability to encapsulate medications, enhancing their solubility and shielding them from deterioration. Additionally, lipid nanoparticles can improve therapeutic efficacy by enabling continuous and regulated drug release (Scioli Montoto et al., 2020). Lipid-based nanoparticles special structure enables them to incorporate both hydrophilic and hydrophobic medications. It is possible to integrate hydrophobic medications into the lipid core and encapsulate hydrophilic drugs within the watery compartments. Targeted medication delivery can be achieved by functionalizing or altering the surface of lipid-based nanoparticles (Priya et al., 2022). The precision of medication administration can be increased by ligands or antibodies on the surface of the nanoparticles that can identify certain receptors on target cells. Lipid-based nanoparticles can help with controlled release and improve the stability of medications that are encapsulated (Lim et al., 2012). Lipid matrixes have the ability to both shield delicate medications from deterioration and offer a medium for long-term sustained drug release. Liposomes have the ability to encapsulate various agents, such as fluorescent dyes for optical imaging or contrast compounds for magnetic resonance imaging (Niesman et al., 1994; Mulder et al., 2006). In addition, the delivery of vaccines using lipid nanoparticles has also been investigated and can effectively transport antigens, stimulating the immune system. Particularly, liposomes have been studied in relation to vaccine development. Since lipid-based nanoparticles may encapsulate and shield nucleic acids (such as DNA or RNA), they are therefore also utilized in gene delivery systems (Li & Szoka, 2007). Applications in genetic medicine and gene therapy require this. Lipid-based nanoparticles can have their surfaces modified to accommodate targeted ligands, polymers, or other functional moieties, thereby customizing the nanoparticles for particular uses (Priya et al., 2022). Moreover, lipid-based nanoparticles are also a good option for oral drug delivery, since they can improve the absorption of medications that are poorly soluble in water and remain stable in the gastrointestinal tract (Kumar, 2019). Biomedical applications can benefit from the reduced toxicity of lipid-based nanoparticles. Nevertheless, thorough toxicity tests are necessary to guarantee these nanoparticles safety. The promise of lipid-based nanoparticles to improve medication delivery, medical imaging, and other biomedical applications keeps researchers interested in this field, and they are useful in addressing issues related to drug solubility, stability, and targeted delivery owing to their adaptability.

2.2.2 Polymeric Nanoparticles

Polymeric nanoparticles are huge molecules consisting of repeating subunits that are assembled into nanoscale particles. Active ingredients may be surface-adsorbed onto the polymeric core of polymeric nanoparticles ranging in size from 1 to 1000 nm, or they may be confined inside the polymeric body (Gouthami et al., 2023). The phrase "polymer nanoparticle" is frequently used in the literature to refer to these nanoparticles, which are frequently organic. For the most part, they resemble nanospheres or nanocapsules (Khan et al., 2019; Zielinska et al., 2020). The advantages of polymeric nanoparticles are their biocompatibility, their adjustable characteristics, and their capacity to encapsulate a broad variety of compounds.

Polymer types:

- *Biodegradable polymers*: These polymers can break down into nontoxic byproducts, making them suitable for drug delivery. Examples include poly-lactic-co-glycolic acid (PLGA), poly-lactic acid (PLA), and poly-caprolactone (PCL).
- *Natural polymers*: They are biocompatible and can be used for drug delivery and tissue engineering. Polymers such as chitosan, alginate, and gelatin are derived from natural sources.
- *Synthetic polymers*: Synthetic polymers like polyethylene glycol (PEG) and polyvinyl alcohol (PVA) are designed with specific properties for drug delivery and other applications.

Polymeric nanoparticles are extensively used for drug delivery. The ability to encapsulate pharmaceuticals results in improved solubility, controlled release, and protection against degradation (Kamaly et al., 2016). The release of medications can be facilitated by biodegradable polymers. Hydrophobic as well as hydrophilic medicines can be encapsulated in polymeric nanoparticles (Vrignaud et al., 2011). Targeted drug administration, increased stability, and extended bloodstream circulation are all possible with surface modifications of polymeric nanoparticles (Ahmed et al., 2021). For particular interactions with target cells, surface modification may entail the attachment of ligands, antibodies, or other molecules. In medical imaging approaches, polymeric nanoparticles can act as carriers for imaging agents. They can, for instance, contain fluorescent dyes for optical imaging or contrast agents for MRI (Elsabahy et al., 2015). Gene delivery makes use of polymeric nanoparticles, which enable the secure and effective transfer of nucleic acids such as DNA and RNA. Polymers have the ability to shield genetic material from deterioration and promote cellular absorption. These combine therapeutic and diagnostic roles in theranostic applications (Lim et al., 2015). Since they carry both imaging and medicinal chemicals, they can be used for simultaneous treatment and response monitoring. High biocompatibility polymeric nanoparticles make them appropriate for a range of biological applications (Elsabahy & Wooley, 2012). In particular, biodegradable polymers are appealing for reducing long-term toxicity issues. The controlled release of molecules enclosed is made possible by polymeric nanoparticles, and upon modifications to the nanoparticle surface, suitable polymer choices and encapsulating methods can all help to achieve this controlled release. By offering scaffolds for cell growth and regeneration, polymeric nanoparticles contribute to tissue engineering (Smith et al., 2009). To assist in the creation of tissue, they can be integrated into three-dimensional structures. Certain polymeric nanoparticles are made to react to particular stimuli, including variations in temperature, pH, or enzyme activity. Furthermore, in reaction to external stimuli, responsive nanoparticles can release their cargo in a regulated manner.

3 Applications of Nanoparticles in Aquaculture Feed

The use of nanoparticles in aquaculture feed is an area of research with potential applications to enhance the health, growth, and sustainability of aquatic organisms. Here are some impending applications of nanoparticles in aquaculture feed:

3.1 Improved Nutrient Absorption

Nanoparticles can be used to encapsulate and provide vital nutrients in aquaculture feed. Owing to their small size and increased surface area, nanoparticles may be able to better absorb nutrients, which would promote aquatic species' growth and well-being (Fajardo et al., 2022). The application of nanoparticles to enhance nutrient absorption and boost the overall efficiency of nutrient utilization by aquatic species has been studied in aquaculture. Nanoparticles have a significantly higher surface area than larger particles (Khan et al., 2019). Higher nutrient absorption is encouraged during digestion by greater contact between this bigger surface area and the gut walls. Vitamins, minerals, and amino acids are among the nutrients that can be encapsulated in nanoparticles to prevent deterioration (Pateiro et al., 2021). This protection ensures that more of the nutrients are absorbed by the targeted tissues while enhancing their bioavailability. Under controlled circumstances, nutrients contained in nanoparticles can be released, and by exposing nutrients to the digestive system over an extended period of time, this controlled release technique maximizes absorption over time. Functionalized nanoparticles that are designed to deliver nutrients to certain tissues or cells of aquatic organisms can be produced (Yusuf et al., 2023). By concentrating on the appropriate areas, nutrients are supplied where they are most needed, thus boosting overall efficacy. Some nanoparticles, particularly those with antioxidant properties, can stop sensitive nutrients from oxidizing, which is necessary to preserve the stability and efficacy of nutrients in the feed as well as during digestion (Samrot et al., 2022). Lipophilic materials can be included in nanoparticles such as nanoemulsions or liposomes, which play an important role in ensuring omega-3 fatty acids and fat-soluble vitamins are absorbed effectively (Ashaolu, 2021). Besides this, the unique properties and small size of nanoparticles can also increase the digestion of feed. Such an increase in digestibility helps fish to absorb and use nutrients more effectively (Gelaye, 2024). Biocompatible nanoparticles are less likely to harm the digestive system, especially if they are made of lipids or naturally occurring polymers. Biocompatibility with feed formulations based on nanoparticles enhances the absorption of aquatic species (Fajardo et al., 2022). Moreover, nanoparticles can assist in minimizing nutrient loss by lowering the quantity of nutrients that spill into the adjacent water during digestion. The reduction in nutrient loss allows for a more efficient use of feed resources. Since nanoparticles are better at absorbing and using nutrients, they have the potential to improve the growth rates, weight gain, and overall performance of fish species in aquaculture systems (Ahmed et al., 2024). The application of nanoparticles in aquaculture feed is an emerging area of research that is quickly evolving; however, the factors, including the size, makeup, and surface characteristics of the particles, need to be carefully considered. It is also critical to thoroughly investigate any potential environmental and safety concerns associated with the use of nanoparticles in aquaculture in order to ensure moral and sustainable practices.

3.2 Role of Nanoparticles in Disease Prevention and Treatment

Nanoparticles have shown potential in disease prevention and treatment in aquaculture by leveraging their unique properties to combat microbial infections and enhance the overall health of aquatic organisms, e.g., silver nanoparticles have strong antimicrobial properties. These nanoparticles can be added to aquaculture systems to reduce disease risk by restricting the development and spread of bacteria, viruses, and other pathogens (Marquez et al., 2018). The constant presence of antibacterial substances in the aquatic environment is ensured by the controlled release mechanism that provides protection against infections (Wang et al., 2020). Using functionalized nanoparticles, antimicrobial medications can target the infected regions of the fish. This targeted approach boosts therapeutic efficacy while being helpful in reducing the nonspecific effects on beneficial bacteria. Nanoparticles have the ability to disturb biofilms, a typical source of dangerous bacteria in aquaculture systems (De Silva & Heo, 2023), which can help to reduce the chance of disease outbreaks. Besides, nanoparticles can serve as vaccine carriers in aquaculture. It boosts the immune systems of aquatic animals, including fish, and increases their resilience to disease by enabling the focused and controlled administration of immunizations (Nasr-Eldahan et al., 2021). It is also possible to encapsulate drugs in nanoparticles to stop them from degrading. This increased stability contributes to the more effective treatment of diseases by guaranteeing that therapeutic medications maintain their potency during delivery and storage. The use of nanoparticles in the delivery of antimicrobial medications may also help to stop the development of antibiotic resistance in aquatic infections (Makabenta et al., 2021). Furthermore, nanoparticles lessen the needless use of antibiotics, contributing to more environmentally friendly methods of disease control. Antiviral nanoparticles can be made to combat viral infections in aquaculture, as these nanoparticles have the potential to reduce the incidence of viral diseases by blocking viruses from proliferating or infiltrating the body (Pradhan et al., 2021). Nanoparticles have widely been used in diagnostic procedures that can detect diseases in aquaculture early on. Nanoscale sensors and probes can swiftly take action and take care of the problem when pathogens or abnormal situations are found (Wang et al., 2013). Certain types of nanoparticles can change the immune responses of the fish. Improved immune modulation may help fish or shrimp fight off infections and increase their general resilience to diseases. Thorough research and risk evaluations are necessary prior to introducing nanoparticles into aquaculture systems. Taking the severity and impact of nanoparticles into account, it is suggested that careful evaluation of variables such as nanoparticle toxicity, environmental impact, and other unintended repercussions is necessary before implementing these technologies in aquaculture for disease prevention and treatment, which will ensure the acceptable and sustainable use of nanoparticles in aquaculture (Ogunfowora et al., 2021).

3.3 Enhancing Growth and Performance

Nanoparticles may be used in aquaculture to boost the development and production of aquatic organisms. The unique properties of nanoparticles can facilitate improved nutrient absorption, disease resistance, and general physiological functions due to the increased surface area of nanoparticles, which allows them to interact with the intestinal wall and digesting enzymes (Bunglavan et al., 2014). Growth and overall performance of the fish are improved due to the improvement in the absorption of nutrients. Nanoparticles may directly be added to the aquaculture feed for the improvement of feed efficiency, as the former enhance nutrient delivery and help to reduce feed conversion ratios by virtue of an increase in nutrient absorption (Abd El-Naby et al., 2019). Being nanoencapsulated, nutrients such as vitamins, minerals, and amino acids are preserved and protected from degeneration. Consequently, a more consistent and dependable supply of essential nutrients may be available to support growth. Since some nanoparticles reduce stress, they may be able to assist aquatic animals in coping with environmental problems (Fajardo et al., 2022). Reduced stress has a positive impact on overall performance and growth rates of the organisms. Vitamins or other antioxidants encased in nanoparticles are the best examples of antioxidant-containing nanoparticles that can help in the protection of cells from the harmful effects of free radicals (Losada-Barreiro et al., 2022). This protection may lead to the development and improvement of the overall health of aquaculture species. For aquatic life to grow and perform at its peak, the water must be clear and well-maintained. The use of certain nanoparticles can improve water quality in aquaculture systems by removing the contaminants (Tayel et al., 2019). In addition, nanoparticles have the potential to enhance overall health and resistance to disease by altering the immune response of the organism (Liu et al., 2017). To prevent illnesses that could limit the growth of aquaculture, a strong immune system is necessary. Functionalized nanoparticles that intentionally supply nutrition to certain tissues or organs can be produced (Nickols-Richardson & Piehowski, 2008). Nutrients are ensured to reach the areas that require them the most for growth and development by employing a customized approach. Nanoparticles in biocompatible forms can be found in aquafeed. Aquatic animals have a strong capability to absorb and use biocompatible nanoparticles more readily, since they are less likely to induce adverse reactions. By altering the characteristics and composition of the nanoparticles in aquaculture feed, it can be useful to maximize the weight gain, growth performance, and other crucial aspects of the aquaculture industry. It has been observed in the past that young carp and sturgeon fed iron nanoparticles through feed grow at a higher rate of 30% and 24%, respectively, as reported by the Russian Academy of Sciences (ETC, 2003). On the other side, it was observed that the use of silver nanoparticle-coated water filters may shield rainbow trout fish from fungal infections. Juvenile trout have been observed to suffer when silver nanoparticles are directly added to water to cure fungal infections (Khursheed et al., 2023). According to studies, increasing the amount of selenium in the basal diet, specifically, selenomethionine and nano-Se, can improve the antioxidant status, final weight, and relative growth rate in *Carassius auratus gibelio* (Zhou et al., 2009). Furthermore, nano-Se appears to be more successful at raising the selenium content of muscle than organic selenomethionine. Zhou et al. (2009) reported that when these nutraceuticals were administered at the nanoscale, the fish that were being tested showed increased development and efficiency. Nanosensors for disease detection in fish farming systems, porous nanostructures for the delivery of veterinary products in fish food, and nanotechnological applications in aquaculture systems can all be used to control fish health (El-Chaghaby & Rashad, 2023). The application of nanoparticles in aquaculture may encourage more environmentally friendly methods by improving resource efficiency, lowering waste, and reducing the negative environmental effects of traditional aquaculture processes (Fajardo et al., 2022). Additionally, studies utilizing the previously described nanomaterials have also shown that nutraceuticals administered at the nanoscale proved improvement in fish growth (Aklakur et al., 2016).

4 Mechanisms of Action of Nanoparticles in Aquaculture Feed

Depending upon the unique characteristics of the nanoparticles utilized there can be a wide range of different mechanisms of action for them in aquaculture feed. It is noteworthy that the particular mechanisms may differ based on the kind, dimensions, and surface properties of the employed nanoparticles and their intended use in aquaculture feed. Furthermore, extensive research is required to guarantee the security and effectiveness of nanoparticle use in aquaculture. In order to protect aquatic life and reduce possible negative effects on the environment, ethical and sustainable techniques must be used. The following are some important ways that nanoparticles in aquaculture may work:

4.1 Cellular Uptake

The cellular uptake of nanoparticles in aquaculture is an area of active research, and the understanding of these mechanisms is continually evolving. The choice of nanoparticles and their surface modifications plays a critical role in determining their interactions with aquatic organisms at the cellular level (Kim et al., 2013). However, comprehensive studies are needed to assess the efficiency, safety, and potential impacts of nanoparticle cellular uptake in the context of aquaculture practices. The physiological circumstances in the aquatic environment, the kind of cells involved, and the characteristics of the nanoparticles themselves can all together influence the intricate processes of cellular uptake of nanoparticles in aquaculture (Rather et al., 2018). It has been suggested that the cellular absorption of nanoparticles is known to occur through a number of mechanisms:

- *Endocytosis*: Endocytosis is a common mechanism for the cellular uptake of nanoparticles. It involves the formation of vesicles around the nanoparticles, which are then internalized by the cell (Adjei et al., 2014). There are different types of endocytosis, including clathrin-mediated endocytosis, caveolae-mediated endocytosis, and macropinocytosis.
- *Clathrin-mediated endocytosis*: Clathrin-coated pits on the cell membrane recognize and bind to specific receptors on the nanoparticle surface. The coated pit invaginates, forming a vesicle that contains the nanoparticle. This vesicle then undergoes internalization into the cell (Rennick et al., 2021).
- *Caveolae-mediated endocytosis*: Caveolae are small invaginations of the cell membrane enriched in caveolin proteins. Nanoparticles can be internalized through caveolae-mediated endocytosis, which involves the formation of vesicles containing the nanoparticles (Foroozandeh & Aziz, 2018).
- *Macropinocytosis*: It is a nonspecific form of endocytosis, where cells engulf large volumes of fluid along with any suspended material, including nanoparticles. This process involves the formation of large vesicles known as macropinosomes (de Almeida et al., 2021).
- *Phagocytosis*: Phagocytosis is a process primarily associated with immune cells such as macrophages and neutrophils. These cells can engulf and internalize nanoparticles by forming phagosomes (Gustafson et al., 2015). This process is particularly relevant in the context of nanoparticle interactions with the immune system.
- *Direct penetration*: Some nanoparticles can directly penetrate the cell membrane through mechanisms such as passive diffusion or membrane disruption. This is more common for small and lipophilic nanoparticles that can cross cellular membranes (Barua & Mitragotri, 2014).
- *Active targeting*: Functionalized nanoparticles with ligands or targeting moieties can actively bind to specific receptors on the cell surface (Friedman et al., 2013). This targeted interaction enhances the specificity of cellular uptake, ensuring that the nanoparticles are taken up by specific cells or tissues.
- *Surface charge and size*: The surface charge and size of nanoparticles influence their interaction with cell membranes. Positively charged nanoparticles may interact more favorably with negatively charged cell membranes, while smaller nanoparticles may have increased cellular uptake due to their ability to penetrate more easily (Forest & Pourchez, 2017).

- *Cellular internalization pathways*: Nanoparticles may exploit specific cellular internalization pathways based on their physicochemical properties (Duan & Li, 2013). For example, certain nanoparticles may preferentially utilize lipid raft-mediated pathways or receptor-mediated pathways for uptake.
- *Physiological conditions*: The physiological conditions in the aquatic environment including pH, temperature, and salinity can influence the cellular uptake of nanoparticles (Yung et al., 2017). Therefore, understanding these factors is crucial for predicting and controlling nanoparticle behavior in aquaculture systems.

4.2 Immunomodulation

The immune response in aquaculture can be modulated by nanoparticles, which in turn affects how active the immune system is in aquatic creatures. The intricate processes behind nanoparticle-mediated immunomodulation can differ depending on the size, content, and surface features of the particles. Interactions between immune cells and nanoparticles can affect the synthesis of cytokines, signaling molecules which are essential in controlling the immunological responses of the organism (Liu et al., 2022). Immune cell activation, proliferation, and differentiation are all impacted by cytokines. Macrophages are important immune cells that help to find and get rid of infections. Both macrophages and nanoparticles may interact to change the activation state of the latter, which might have an effect on immune mediator release and phagocytosis efficiency (Qie et al., 2016). When it comes to presenting antigens and starting immune responses, dendritic cells are essential. Dendritic cells and nanoparticles can interact with each other, affecting the cells capacity to deliver antigens (Jia et al., 2018). Moreover, complement proteins and nanoparticles may interact, affecting complement activation and the ensuing immunological responses (Chen et al., 2017); however, pathogen clearance and inflammation may be impacted by this interaction. Antigens can be transported by nanoparticles, which make it easier for immune cells to present them. This may improve the immune system's ability to identify particular infections and trigger an adaptive immunological reaction (Look et al., 2010). The ratio of T-helper 1 (Th1) to T-helper 2 (Th2) immune responses may be influenced by nanoparticles, and the sort of immune response that is elicited can be affected by altering this balance. Th1 responses are often linked to cellular immunity, whereas Th2 responses are linked to humoral immunity (Constant & Bottomly, 1997). Immune cells may produce immunomodulatory chemicals such as reactive oxygen species or nitric oxide in response to nanoparticles. These compounds have the ability to influence the immune system as a whole and support antibacterial activity. Moreover, functionalized nanoparticles can interact directly with immunological receptors if they have particular ligands or chemicals on their surface (Mu et al., 2014). Immune pathway modification or activation may arise from this specific interaction. Some nanoparticles are made to stimulate the immune system by activating immune cells and causing an increased inflammatory response (Mitarotonda et al., 2019). The immune system's capacity to fight infections may be strengthened by using these nanoparticles. On the other hand, nanoparticles can be designed to suppress the immune system and reduce overreactions. This strategy might help to lessen the autoimmune reactions or stop inflammatory reactions (Anfray et al., 2020). It is noteworthy that although research on the immunomodulatory potential of nanoparticles in aquaculture is going on, however, the precise effects may vary depending on the environment. Therefore, a comprehensive examination is necessary to ensure the safety of immunomodulation generated by nanoparticles.

4.3 Antimicrobial Properties

Several nanoparticles have been used in aquaculture because of having good antibacterial qualities as well as their nanoscale interactions with microbes (Marquez et al., 2018). The targeted microorganisms and the type of nanoparticles utilized can affect the mechanisms of action. The cell membranes of bacteria, fungi, and other microbes can readily interact with nanoparticles. Cell death results from the integrity of the cell membrane being compromised, which causes leakage of the contents of the cell. Reactive oxygen species (ROS) can be produced by some nanoparticles when they come into contact with microorganisms, particularly those that include metals like copper and silver (Abdal Dayem et al., 2017). Moreover, oxidative stress brought on by ROS damages proteins, lipids, and DNA and ultimately results in cell death. The enzymatic activity that is essential to microbe viability can be disrupted by nanoparticles. The ability of microbes to survive is compromised by this interference, which throws off vital metabolic processes (Wang et al., 2017). Microbial DNA has a capability to interact with nanoparticles, causing disruption of transcription and replication. This disruption of genetic material has the potential to kill cells, thus preventing microbes from growing. Metal ions can be released into the surrounding environment by metal-based nanoparticles, including silver nanoparticles (Sharma et al., 2019). Through their disruption of microbial development and interference with biological activities, these metal ions demonstrate their antibacterial characteristics. Increased antibacterial effects could result from positively charged nanoparticles interacting more favorably with negatively charged microbial cell membranes (Haffner & Malmsten, 2017). The development and integrity of microbial biofilms can easily be stopped by the involvement of nanoparticles. Microorganisms create protective layers called biofilms, and when these layers are disrupted, the microbes become more vulnerable to antimicrobial treatments (Jain et al., 2007). Synergistic effects between nanoparticles and traditional antibiotics have been observed in the past. In addition to lowering the risk of developing antibiotic resistance, this combination can improve antimicrobial efficacy overall. It is possible to engineer functionalized nanoparticles to target particular microbes with precision. The specificity of antibacterial action is increased by surface modifications using ligands or compounds that bind to microbiological components. When exposed to light, certain nanoparticles, especially those possessing photothermal characteristics, have the ability to produce heat (Zharov & Lapotko, 2005). Particularly in confined spaces, photothermal effects can be used to kill or harm microorganisms only. Antimicrobial drugs can be carefully encapsulated and released via nanoparticles (Zhang et al., 2010). The antibacterial substance's efficacy is enhanced by the controlled release, which guarantees its continuous presence. Since nanoparticles use several methods of action at once, they may prevent the emergence of microbial resistance (Pelgrift & Friedman, 2013). The probability of bacteria responding to the antimicrobial actions is decreased by this multimodal strategy. Although aquaculture may benefit from the antibacterial qualities of nanoparticles, it is important to take into account aspects like toxicity, environmental impact, and resistance development (Sarkar et al., 2022). Therefore, it is recommended that before the use of nanoparticles for antimicrobial purposes in an ethical and sustainable manner, extensive study and risk evaluations should be carried out.

5 Challenges and Considerations in Nanoparticle Application

Numerous issues and concerns arise when using nanoparticles in a variety of industries, including aquaculture. To guarantee the ethical and sustainable usage of nanoparticles, these issues must properly be taken care of. For example, when compared to bigger particles of the same material, nanoparticles can have distinct toxicological characteristics. Nanoparticles toxicity may be influenced by their size, makeup, and surface features (Zoroddu et al., 2014). The possible negative consequences of nanoparticles on aquatic environments and species must carefully be considered. It has also been observed that there are worries regarding the effects of nanoparticles on the environment when they are released into aquatic settings. Ecosystems may be disrupted, and nontarget creatures may be impacted by nanoparticles that linger in sediments and water bodies (Banu et al., 2021). Understanding the fate and transport of nanoparticles in aquatic environments requires research. Aquatic organisms may absorb nanoparticles, which could result in bioaccumulation in their tissues (Ates et al., 2013). Extensive studies should be conducted to determine the long-term effects of nanoparticle bioaccumulation on aquatic animals and their translocation throughout the food chain. The regulatory environment pertaining to the use of nanoparticles in aquaculture and other sectors is still developing. It is crucial to follow current laws and create new policies for the responsible usage of nanoparticles. When employing nanoparticles in aquaculture, ethical issues such as environmental ethics and animal welfare must be taken into account (Nasr-Eldahan et al., 2021). In aquaculture systems, the stability and agglomeration of nanoparticles can affect their efficacy and possible environmental impact (Kah Sem et al., 2023). It is essential to comprehend the elements influencing the stability of nanoparticles, such as the chemistry of water and interactions with other substances. Recently, several concerns have been raised about the possibility of unintentional nanoparticle releases from manufacturing, handling, or disposal units (Amoabediny et al., 2009). Strong waste management procedures and containment strategies ought to be implemented in order to reduce the possibility of the release of accidental nanoparticles in the environment. Human health may also be at risk from occupational exposure to nanoparticles during the production and application procedures. The nanoparticle industry requires the implementation of appropriate safety protocols and protective gear in order to mitigate any health risks for its workforce (Oksel et al., 2016). One constraining element may be the expense of creating nanoparticles, particularly ones with certain functions or surface changes. It is imperative that researchers and practitioners of the industry investigate economical techniques for the production and utilization of nanoparticles. Society's adoption of nanoparticle uses can be influenced by public opinion, which includes worries about safety and environmental effects (Babatunde et al., 2019). Establishing public trust requires transparent and unambiguous communication about the advantages and disadvantages of using nanoparticles in various fields (Boholm & Larsson, 2019). In aquatic environments, nontarget creatures may interact with nanoparticles and experience possible behavioral and physiological effects. Evaluating the total effect of nanoparticle applications requires an understanding of these interactions and their ecological ramifications (Bordin et al., 2024). Although, it is unclear how long-term exposure to nanoparticles would affect aquatic ecosystems. Therefore, it is imperative that questions of cumulative effects, long-term exposure, and the possibility of nanoparticle persistence in the environment should be explored (Sajid et al., 2015).

5.1 Environmental Concerns

Environmental issues and opportunities are associated with the use of nanoparticles in the aquaculture industry. A comprehensive and interdisciplinary strategy involving scientists, aquaculture practitioners, regulators, and stakeholders is needed to address these issues and challenges. For the aquaculture industry to minimize environmental hazards and ensure the sustainability of nanoparticle applications, responsible usage of nanoparticles, risk assessments and continuous monitoring are essential. The following are some environmental issues, difficulties, and factors to take into account when using nanoparticles in aquaculture:

- *Potential ecotoxicity*: The release of nanoparticles into aquaculture systems may pose risks to aquatic organisms. The ecotoxicity of nanoparticles, especially those containing metals or other reactive materials needs thorough assessment to prevent any harm to fish, shrimp, and other aquatic species (Khosravi-Katuli et al., 2017).
- *Bioaccumulation in aquatic organisms*: Nanoparticles once released into the water can be taken up by aquatic organisms and may accumulate in their tissues (Ma & Lin, 2013). This accumulation can have long-term consequences on the health of aquatic species and potential impacts on higher trophic levels.

- *Impact on microbial communities*: Aquaculture systems rely on complex microbial communities for obtaining nutrients and water quality management. The introduction of nanoparticles may disrupt these microbial communities, affecting the overall ecosystem balance within aquaculture facilities (Baker et al., 2014).
- *Water quality concerns*: The use of nanoparticles in aquaculture feed or water treatment may influence water quality parameters (Ogunfowora et al., 2021). Changes in water chemistry, including pH and nutrient levels could have cascading effects on both aquatic organisms and the overall health of the aquatic ecosystem.
- *Potential for nanoparticle runoff*: Runoff from aquaculture facilities may introduce nanoparticles into surrounding ecosystems, leading to unintended environmental exposure (Tayel et al., 2019). This runoff may also affect nearby water bodies and terrestrial environments, raising concerns about nanoparticle dispersion.
- *Effects on nontarget species*: Nanoparticles may have unintended effects on nontarget species in aquaculture environments, including plankton, invertebrates, and other organisms (Abbas, 2021). Understanding the broader ecological impacts is thus crucial for the effective use of nanoparticles in aquaculture.
- *Fate and transport in water systems*: Nanoparticles can exhibit unique behaviors in aquatic environments, influencing their fate and transport. Understanding how nanoparticles disperse, sediment, or remain suspended in water is essential for predicting their overall environmental impact (Cross et al., 2015).
- *Nanoparticle stability in aquaculture systems*: The stability of nanoparticles in aquaculture systems, including their tendency to aggregate or undergo transformations, can affect their effectiveness and potential environmental impact (Shah & Mraz, 2020). Factors such as water chemistry and interactions with organic matter should be considered.
- *Development of antibiotic resistance*: Nanoparticles with antimicrobial properties may contribute to develop strong antibiotic resistance in aquaculture (Okeke et al., 2022). This is a concern, as it could compromise the effectiveness of conventional antibiotics in treating bacterial infections in both aquatic organisms and potentially humans.
- *Risk of nanoparticle release during production*: There is a risk of nanoparticle release during the production, handling, and disposal of aquaculture products containing nanoparticles (Khosravi-Katuli et al., 2017). Appropriate measures should be in place to minimize accidental releases and ensure proper waste management practices.
- *Regulatory compliance*: The regulatory landscape for nanoparticle use in aquaculture is evolving. Compliance with existing regulations and the development of new guidelines for safe nanoparticle use in aquaculture is critical for environmental protection (Fajardo et al., 2022).
- *Public perception and acceptance*: Finally, public perception of nanoparticle use in aquaculture can influence acceptance and support. Engaging with the public, addressing concerns, and providing transparent information about the environmental impacts of nanoparticle applications are important for building trust (Murphy, 2010).

- *Research gaps and monitoring*: There are still research gaps in understanding the long-term environmental effects of nanoparticle application in aquaculture (Shah & Mraz, 2020). Ongoing monitoring and research are necessary to assess the ecological consequences and refine management practices.

5.2 Health and Safety Implications

The use of nanoparticles in the aquaculture industry has the potential to enhance a number of aquaculture processes. A dedication to safeguarding the health and safety of employees, customers, and the environment is essential when using nanoparticles in aquaculture. Responsible nanoparticle applications in the aquaculture industry need careful consideration of risk assessments and adherence to safety procedures. The consequences of using nanoparticles in this situation for health and safety must, however, be carefully taken into account. Below are mentioned a few important things to take into consideration:

- *Occupational exposure*: Workers involved in the production, handling, and application of nanoparticle-based products in aquaculture may be at risk of occupational exposure. Inhalation, dermal contact, or ingestion of nanoparticles during these activities should be minimized, and appropriate safety measures, including personal protective equipment (PPE), should be implemented (Amoabediny et al., 2009).
- *Toxicity and health effects*: The toxicity of nanoparticles can vary based on their size, composition, and surface properties (Khan et al., 2019). Research is needed to assess the potential health effects associated with exposure to specific nanoparticles used in aquaculture, both for workers and consumers.
- *Bioavailability and bioaccumulation*: Understanding the bioavailability and potential accumulation of nanoparticles in aquatic organisms is essential. If nanoparticles accumulate in edible parts of fish, there is a need to evaluate the impact on human health through the consumption of nanoparticle-contaminated products (Parsai & Kumar, 2020).
- *Skin and respiratory sensitization*: Some nanoparticles may cause skin and respiratory sensitization. Workers involved in handling nanoparticles in aquaculture should be aware of the potential sensitization risks and preventive measures, such as using appropriate skin protection and respiratory equipment (Seaton et al., 2010).
- *Risk of ingestion*: Workers or consumers may inadvertently ingest nanoparticles present in aquaculture products. Understanding the potential for nanoparticle ingestion and its health implications is crucial for assessing overall safety (McCracken et al., 2016).
- *Safe handling and storage*: Proper protocols for the safe handling, storage, and disposal of nanoparticle-containing materials in aquaculture should be established (Amoabediny et al., 2009). This includes providing guidelines on minimizing spillage, waste management, and decontamination procedures.

- *Potential for nanoparticle release*: The potential release of nanoparticles into the environment either during production or application should be minimized (Ju-Nam & Lead, 2008). Robust containment measures and preventive strategies must be in place to avoid unintended exposure to nanoparticles.
- *Regulatory compliance*: Compliance with existing health and safety regulations and standards is essential. Regulations specific to nanomaterials and their applications in aquaculture should be considered and followed to ensure the protection of human health (Fajardo et al., 2022).
- *Consumer safety*: If nanoparticle residues are found in aquaculture products intended for human consumption, thorough risk assessments should be conducted to evaluate the safety of these products for consumers (Karimi et al., 2018).
- *Monitoring and surveillance*: Regular monitoring and surveillance programs should be conducted to assess the health and safety aspects of nanoparticle applications in aquaculture (Hardy et al., 2012). This includes monitoring the health of workers, consumers, and the broader ecosystem.
- *Emergency response plans*: Emergency response plans should be developed to address accidental spills, releases, or exposures (Cheremisinoff & Davletshin, 2010). Clear procedures for handling emergency situations and providing medical assistance should be in place.
- *Education and training*: Adequate education and training programs should be conducted for workers involved in nanoparticle applications in aquaculture. This includes training on safe handling practices, risk awareness, and emergency response procedures (Ramesh et al., 2023).
- *Public communication*: Transparent communication with the public regarding the use of nanoparticles in aquaculture, associated health and safety measures, and ongoing monitoring efforts is important for building trust and addressing public concerns (WHO, 2010).

5.3 Regulatory Aspects

The regulatory landscape for nanoparticle applications in aquaculture is evolving, and regulations may vary across different regions. Regulatory bodies aim to ensure the safety of nanoparticle use in aquaculture while considering environmental impact, human health, and sustainable practices. It is important for all stakeholders in the aquaculture sector to stay informed about relevant regulations and actively engage with regulatory bodies to ensure compliance and responsible nanoparticle use. Collaboration between researchers, industry, and regulatory agencies is crucial for the development of effective and balanced regulatory frameworks for nanoparticle applications in aquaculture. Here are some key aspects which should be taken into consideration for the regulation of nanoparticle applications in the aquaculture sector:

- *Risk assessment and approval process*: Regulatory agencies typically require a comprehensive risk assessment of nanoparticle applications in aquaculture. This involves evaluating potential environmental impacts, toxicity to aquatic organisms, and human health risks (Sarkar et al., 2022). Approval processes may include the submission of detailed data, studies, and safety assessments.
- *Environmental protection regulations*: Regulations governing the environmental release of nanoparticles in aquaculture focus on preventing adverse effects on ecosystems. These regulations may specify permissible concentrations, monitoring requirements, and measures to minimize nanoparticle release into aquatic environments (Khosravi-Katuli et al., 2017).
- *Human health and safety regulations*: Regulatory agencies address human health and safety concerns associated with nanoparticle exposure in aquaculture (Faunce & Watal, 2010). This includes setting permissible exposure limits for workers and consumers, as well as guidelines for safe handling, storage, and disposal of nanoparticle-containing materials (Handy & Shaw, 2007).
- *Labeling and reporting requirements*: Regulations may require clear labeling of aquaculture products containing nanoparticles to inform consumers. Reporting requirements may include the submission of data on nanoparticle formulations, concentrations, and potential risks during production and application (Allan et al., 2021).
- *Product registration*: Certain regions may require the registration or notification of aquaculture products containing nanoparticles before they can be introduced to the market. This process often involves demonstrating compliance with safety and efficacy standards (Dasgupta et al., 2015).
- *International standards*: International organizations and standard-setting bodies may establish standard guidelines for the use of nanoparticle applications in aquaculture (WHO, 2010, 2013). Compliance with international standards helps to ensure consistency and facilitates global trade of aquaculture products.
- *Good manufacturing practices (GMP)*: Adherence to GMP is crucial for ensuring the quality and safety of nanoparticle-containing aquaculture products. GMP guidelines provide a framework for the production, handling, and storage of these products.
- *Data requirements for regulatory submissions*: Regulatory submissions for nanoparticle applications in aquaculture often require comprehensive data on nanoparticle characterization, toxicity studies, environmental fate, and potential impacts (Blinova et al., 2020). This data is essential for regulators to make certain informed decisions.
- *Monitoring and compliance checks*: Regulatory agencies may conduct monitoring and compliance checks to ensure that aquaculture facilities using nanoparticles are following approved protocols (Okeke et al., 2022).
- *Emerging regulatory frameworks*: Some regulatory bodies are actively developing specific frameworks to address nanotechnology applications, including those in aquaculture. These frameworks aim to keep pace with technological advancements while ensuring safety and sustainability (Allan et al., 2021).

- *Dialogue with stakeholders*: Regulatory agencies often engage in a dialogue with industry stakeholders, researchers, and public. This communication helps gather input, address concerns, and adapt regulations based on emerging scientific knowledge and societal expectations (Vogel et al., 2012).
- *Research and development guidelines*: Regulatory agencies may provide guidelines for the research and development of nanoparticle applications in aquaculture (Allan et al., 2021). These guidelines help researchers design studies that meet regulatory standards and contribute to the overall safety assessment.

6 Future Directions and Innovations of Nanoparticle Application in the Aquaculture Sector

The need for sustainable and effective aquaculture methods, continued research, and technological improvements is anticipated to drive the future orientations and innovations of nanoparticle uses in the aquaculture sector. Further investigation could result in the creation of sophisticated medication delivery systems that make use of nanoparticles. These technologies have the potential to enhance aquaculture health and disease control by more focused administration of nutrients, vaccinations, or medicinal substances to aquatic species. The creation of responsive smart nanoparticles, such as temperature or pH sensitivity, might make it possible to release compounds under control in response to particular environmental circumstances. The accuracy and effectiveness of aquaculture applications using nanoparticles may be improved by this invention. Probiotics, enzymes, or sensitive feed additives can all be encapsulated and protected using nanoencapsulation technology. By enhancing the stability and bioavailability of these chemicals in aquaculture diets, this method can help aquatic animals thrive and better utilize nutrients. The development of sensors for the real-time monitoring of disease outbreaks, stress levels in aquaculture systems, and water quality factors may be useful for the effective use of nanoparticles in the field. These sensors might offer timely data for preventative management techniques. Aquaculture water treatment could make use of sophisticated nanoparticles with effective adsorption or catalytic capabilities. This improves the quality of the water by getting rid of contaminants, heavy metals, and harmful algae. Aquafeed compositions with nanostructures may be created to improve the ability of aquatic species to successfully absorb and use nutrients. This could lead to lower environmental impact, higher feed conversion rates, and better overall growth performance. A role for nanoparticles could be seen in the creation of quick and accurate diagnostic instruments for aquatic disease detection. Assays based on nanoparticles may be able to detect diseases in aquaculture systems early, allowing for prompt intervention and disease management. The transmission of genetic material, such as DNA or RNA, to aquatic creatures by nanoparticles is a potential research area. This may pave the way for genetic engineering, the augmentation of disease resistance, or the creation of genetically enhanced aquaculture species. It may be possible to construct bioresponsive nanoparticles that respond to particular physiological stimuli in aquatic creatures. In reaction to changes in the organism's health or surroundings, these nanoparticles may release therapeutic medicines or bioactive substances. Applications for nanoparticles could include aquatic habitat management. They might be applied, for instance, to the targeted delivery of substances to suppress invasive species, support advantageous microbial ecosystems, or regulate algal blooms. Concerns regarding the permanence of nanoparticles in the environment can be resolved by developing biodegradable nanoparticles.

However, applications of nanoparticles in aquaculture may have a shorter-term negative environmental impact if biodegradable formulations are used. Moreover, the applications for nanoparticles can be included in data-driven precision aquaculture systems to maximize production parameters, which might entail the targeted use of nanoparticles for analytics and real-time monitoring. In addition, the application of nanoparticles to reduce stress in aquaculture species may be the subject of future innovations. This could entail creating methods to increase the resilience of aquatic species or stress-reducing drugs based on nanoparticles. Cooperation among researchers, industry players, and regulatory agencies is expected to be beneficial for future developments in the applications of nanoparticles in aquaculture. Collaborations can help to turn research discoveries into useful and long-lasting applications. It is significant to note that the responsible and sustainable integration of nanoparticle applications in aquaculture will depend on a thorough understanding of the potential health and environmental impacts as well as adherence to ethical and regulatory considerations as these innovations advance. The future of nanoparticle technologies in the aquaculture industry will be significantly shaped by ongoing research and interdisciplinary collaboration.

7 Conclusion

The aquaculture industry has great potential to improve several elements of aquaculture practices, including feed efficiency, disease control, and water quality control, through the use of nanoparticle applications. Nanoparticles offer the potential for enhanced nutrient utilization, disease control, and water quality in aquaculture, with advanced formulations possibly enhancing sustainability and production. Antimicrobial nanoparticles in particular can be very helpful in controlling and avoiding infections in aquaculture. Targeted delivery methods and controlled release systems can improve the efficacy of medicinal substances. Systems for nanoencapsulation and nanodelivery can increase the bioavailability of the nutrients in aquaculture diets which could improve aquatic organisms overall nutritional results, growth performance, and feed efficiency. Aquaculture water treatment can benefit from the use of nanoparticles, which help to eliminate contaminants, heavy metals, and dangerous microbes. Water quality characteristics could be monitored in real time with the help of sensors based on nanoparticles. However, there are several obstacles associated with the use of nanoparticles in aquaculture, such as potential toxicity, regulatory complexity, and environmental issues. To solve these issues and guarantee the safe application of nanoparticles, responsible procedures and continuous research are crucial. For aquaculture operations to evolve responsibly and sustainably, it is imperative to weigh the possible hazards against the benefits of applying nanoparticles. Ethical considerations, continuous monitoring, and thorough risk assessments all play a part in the proper usage of nanoparticles. In conclusion, even though aquaculture uses for nanoparticles have great potential, it is critical to practice caution while utilizing them because of the possible negative effects on the environment, public health, and regulatory compliance. The sustainable expansion of the aquaculture industry depends on the full utilization of nanoparticle technologies, which will require sustained research, inventiveness, and ethical conduct.

References

Abbas, W. T. (2021). Advantages and prospective challenges of nanotechnology applications in fish cultures: A comparative review. *Environmental Science and Pollution Research, 28*(7), 7669–7690. https://doi.org/10.1007/s11356-020-12166-0

Abd El-Naby, F. S., Naiel, M. A., Al-Sagheer, A. A., & Negm, S. S. (2019). Dietary chitosan nanoparticles enhance the growth, production performance, and immunity in *Oreochromis niloticus*. *Aquaculture, 501*, 82–89. https://doi.org/10.1016/j.aquaculture.2018.11.014

Abdal Dayem, A., Hossain, M. K., Lee, S. B., Kim, K., Saha, S. K., Yang, G. M., Choi, H. Y., & Cho, S. G. (2017). The role of reactive oxygen species (ROS) in the biological activities of metallic nanoparticles. *International Journal of Molecular Sciences, 18*(1), 120–140. https://doi.org/10.3390/ijms18010120

Adjei, I. M., Sharma, B., & Labhasetwar, V. (2014). Nanoparticles: Cellular uptake and cytotoxicity. In *Nanomaterial: Impacts on Cell Biology and Medicine* (pp. 73–91).

Ahmed, A., Sarwar, S., Hu, Y., Munir, M. U., Nisar, M. F., Ikram, F., Asif, A., Rahman, S., Chaudhry, A. A., & Rehman, I. U. (2021). Surface-modified polymeric nanoparticles for drug delivery to cancer cells. *Expert Opinion on Drug Delivery, 18*(1), 1–24. https://doi.org/10.1080/17425247.2020.1822321

Ahmed, I., Khan, Y. M., Lateef, A., Jan, K., Majeed, A., & Shah, M. A. (2023). Effect of fish meal replacement by Azolla meal on growth performance, hemato-biochemical and serum parameters in the diet of scale carp, *Cyprinus carpio* var. *communis*. *Journal of the World Aquaculture Society, 54*(5), 1301–1316. https://doi.org/10.1111/jwas.12995

Ahmed, M. T., Ali, M. S., Ahamed, T., Suraiya, S., & Haq, M. (2024). Exploring the aspects of the application of nanotechnology system in aquaculture: A systematic review. *Aquaculture International, 1-30*. https://doi.org/10.1007/s10499-023-01370-7

Aklakur, M., Asharf Rather, M., & Kumar, N. (2016). Nanodelivery: An emerging avenue for nutraceuticals and drug delivery. *Critical Reviews in Food Science and Nutrition, 56*(14), 2352–2361. https://doi.org/10.1080/10408398.2013.839543

Allan, J., Belz, S., Hoeveler, A., Hugas, M., Okuda, H., Patri, A., Rauscher, H., Silva, P., Slikker, W., Sokull-Kluettgen, B., Tong, W., & Anklam, E. (2021). Regulatory landscape of nanotechnology and nanoplastics from a global perspective. *Regulatory Toxicology and Pharmacology, 122*, 104885. https://doi.org/10.1016/j.yrtph.2021.104885

Altammar, K. A. (2023). A review on nanoparticles: Characteristics, synthesis, applications, and challenges. *Frontiers in Microbiology, 14*, 1155622. https://doi.org/10.3389/fmicb.2023.1155622

Amoabediny, G. H., Naderi, A., Malakootikhah, J., Koohi, M. K., Mortazavi, S. A., Naderi, M., & Rashedi, H. (2009). Guidelines for safe handling, use and disposal of nanoparticles. *Journal of Physics: Conference Series, 170*(1), 012037. https://doi.org/10.1088/1742-6596/170/1/012037

Anfray, C., Mainini, F., & Andon, F. T. (2020). Nanoparticles for immunotherapy. *Frontiers of Nanoscience, 16*, 265–306. https://doi.org/10.1016/B978-0-08-102828-5.00011-5

Asha, A. B., & Narain, R. (2020). Nanomaterials properties. In *Polymer Science and Nanotechnology* (pp. 343–359). Elsevier. https://doi.org/10.1016/B978-0-12-816806-6.00015-7

Ashaolu, T. J. (2021). Nanoemulsions for health, food, and cosmetics: A review. *Environmental Chemistry Letters, 19*(4), 3381–3395.

Ates, M., Demir, V., Adiguzel, R., & Arslan, Z. (2013). Bioaccumulation, subacute toxicity, and tissue distribution of engineered titanium dioxide nanoparticles in goldfish (*Carassius auratus*). *Journal of Nanomaterials, 2013*, 1–6. https://doi.org/10.1155/2013/460518

Babatunde, D. E., Denwigwe, I. H., Babatunde, O. M., Gbadamosi, S. L., Babalola, I. P., & Agboola, O. (2019). Environmental and societal impact of nanotechnology. *IEEE Access, 8*, 4640–4667. https://doi.org/10.1109/ACCESS.2019.2961513

Baker, T. J., Tyler, C. R., & Galloway, T. S. (2014). Impacts of metal and metal oxide nanoparticles on marine organisms. *Environmental Pollution, 186*, 257–271. https://doi.org/10.1016/j.envpol.2013.11.014

Banu, A. N., Kudesia, N., Raut, A. M., Pakrudheen, I., & Wahengbam, J. (2021). Toxicity, bioaccumulation, and transformation of silver nanoparticles in aqua biota: A review. *Environmental Chemistry Letters, 19*(6), 4275–4296.

Barua, S., & Mitragotri, S. (2014). Challenges associated with penetration of nanoparticles across cell and tissue barriers: A review of current status and future prospects. *Nano Today, 9*(2), 223–243. https://doi.org/10.1016/j.nantod.2014.04.008

Blinova, I., Muna, M., Heinlaan, M., Lukjanova, A., & Kahru, A. (2020). Potential hazard of lanthanides and lanthanide-based nanoparticles to aquatic ecosystems: Data gaps, challenges and future research needs derived from bibliometric analysis. *Nanomaterials, 10*(2), 328–346. https://doi.org/10.3390/nano10020328

Boholm, A., & Larsson, S. (2019). What is the problem? A literature review on challenges facing the communication of nanotechnology to the public. *Journal of Nanoparticle Research, 21*, 86–106. https://doi.org/10.1007/s11051-019-4524-3

Bordin, E. R., Ramsdorf, W. A., Domingos, L. M. L., de Souza Miranda, L. P., Mattoso Filho, N. P., & Cestari, M. M. (2024). Ecotoxicological effects of zinc oxide nanoparticles (ZnO-NPs) on aquatic organisms: Current research and emerging trends. *Journal of Environmental Management, 349*, 119396. https://doi.org/10.1016/j.jenvman.2023.119396

Bunglavan, S. J., Garg, A. K., Dass, R. S., & Shrivastava, S. (2014). Use of nanoparticles as feed additives to improve digestion and absorption in livestock. *Livestock Research International, 2*(3), 36–47.

Buzea, C., Pacheco, I. I., & Robbie, K. (2007). Nanomaterials and nanoparticles: Sources and toxicity. *Biointerphases, 2*(4), MR17–MR71. https://doi.org/10.1116/1.2815690

Cagdas, M., Sezer, A. D., & Bucak, S. (2014). Liposomes as potential drug carrier systems for drug delivery. *Application of Nanotechnology in Drug Delivery, 1*, 1–50. https://doi.org/10.5772/58459

Campelo, J. M., Luna, D., Luque, R., Marinas, J. M., & Romero, A. A. (2009). Sustainable preparation of supported metal nanoparticles and their applications in catalysis. *ChemSusChem: Chemistry, Sustainability, Energy & Materials, 2*(1), 18–45. https://doi.org/10.1002/cssc.200800227

Chen, H., & Yada, R. (2011). Nanotechnologies in agriculture: New tools for sustainable development. *Trends in Food Science & Technology, 22*(11), 585–594. https://doi.org/10.1016/j.tifs.2011.09.004

Chen, F., Wang, G., Griffin, J. I., Brenneman, B., Banda, N. K., Holers, V. M., Backos, D. S., Wu, L., Moghimi, S. M., & Simberg, D. (2017). Complement proteins bind to nanoparticle protein corona and undergo dynamic exchange in vivo. *Nature Nanotechnology, 12*(4), 387–393.

Chen, M., Liu, Y., Bao, X., Yue, Y., Tong, B., Yang, X., Yu, H., Yang, Y., Liu, Y., & Yu, Y. (2023). Potential of Chinese yam (*Dioscorea polystachya* Turczaninow) by-product as a feed additive in largemouth bass (*Micropterus salmoides*): Turning waste into valuable resources. *Aquaculture Nutrition, 2023*(1), 9983499. https://doi.org/10.1155/2023/9983499

Cheremisinoff, N. P., & Davletshin, A. (2010). *Emergency response management of offshore oil spills: Guidelines for emergency responders* (p. 531). Wiley.

Chris, U. O., Singh, N. B., & Agarwal, A. (2018). Nanoparticles as feed supplement on growth behaviour of cultured catfish (*Clarias gariepinus*) fingerlings. *Materials Today: Proceedings, 5*(3), 9076–9081. https://doi.org/10.1016/j.matpr.2017.10.023

Constant, S. L., & Bottomly, K. (1997). Induction of Th1 and Th2 $CD4^+$ T cell responses: The alternative approaches. *Annual Review of Immunology, 15*(1), 297–322. https://doi.org/10.1146/annurev.immunol.15.1.297

Crisan, M. C., Teodora, M., & Lucian, M. (2021). Copper nanoparticles: Synthesis and characterization, physiology, toxicity and antimicrobial applications. *Applied Sciences, 12*(1), 141–158. https://doi.org/10.3390/app12010141

Cross, R. K., Tyler, C., & Galloway, T. S. (2015). Transformations that affect fate, form and bioavailability of inorganic nanoparticles in aquatic sediments. *Environmental Chemistry, 12*(6), 627–642. https://doi.org/10.1071/EN14273

Dasgupta, N., Ranjan, S., Mundekkad, D., Ramalingam, C., Shanker, R., & Kumar, A. (2015). Nanotechnology in agro-food: From field to plate. *Food Research International, 69*, 381–400. https://doi.org/10.1016/j.foodres.2015.01.005

de Almeida, M. S., Susnik, E., Drasler, B., Taladriz-Blanco, P., Petri-Fink, A., & Rothen-Rutishauser, B. (2021). Understanding nanoparticle endocytosis to improve targeting strategies in nanomedicine. *Chemical Society Reviews, 50*(9), 5397–5434. https://doi.org/10.1039/D0CS01127D

De Silva, L. A. D. S., & Heo, G. J. (2023). Biofilm formation of pathogenic bacteria isolated from aquatic animals. *Archives of Microbiology, 205*(1), 36–45. https://doi.org/10.1007/s00203-022-03332-8

Donbrow, M. (1991). Microcapsules and nanoparticles in medicine and pharmacy. CRC press.

Duan, X., & Li, Y. (2013). Physicochemical characteristics of nanoparticles affect circulation, biodistribution, cellular internalization, and trafficking. *Small, 9*(9–10), 1521–1532. https://doi.org/10.1002/smll.201201390

Ealia, S. A. M., & Saravanakumar, M. P. (2017). A review on the classification, characterisation, synthesis of nanoparticles and their application. *IOP Conference Series: Materials Science and Engineering, 263*(3), 032019. https://doi.org/10.1088/1757-899X/263/3/032019

El-Chaghaby, G. A., & Rashad, S. (2023). Nanosensors in agriculture: Applications, prospects, and challenges. In *Handbook of Nanosensors: Materials and Technological Applications* (pp. 1–29).

Elsabahy, M., & Wooley, K. L. (2012). Design of polymeric nanoparticles for biomedical delivery applications. *Chemical Society Reviews, 41*(7), 2545–2561.

Elsabahy, M., Heo, G. S., Lim, S. M., Sun, G., & Wooley, K. L. (2015). Polymeric nanostructures for imaging and therapy. *Chemical Reviews, 115*(19), 10967–11011. https://doi.org/10.1021/acs.chemrev.5b00135

Elumalai, K., Srinivasan, S., & Shanmugam, A. (2024). Review of the efficacy of nanoparticle-based drug delivery systems for cancer treatment. *Biomedical Technology, 5*, 109–122. https://doi.org/10.1016/j.bmt.2023.09.001

ETC (Action Group on Erosion, Technology and Concentration). (2003). *Down on the Farm: The Impact of Nanoscale Technologies on Food and Agriculture*, p. 64. http://www.etcgroup.org

Fajardo, C., Martinez-Rodriguez, G., Blasco, J., Mancera, J. M., Thomas, B., & De Donato, M. (2022). Nanotechnology in aquaculture: Applications, perspectives and regulatory challenges. *Aquaculture and Fisheries, 7*(2), 185–200. https://doi.org/10.1016/j.aaf.2021.12.006

Faunce, T., & Watal, A. (2010). Nanosilver and global public health: International regulatory issues. *Nanomedicine, 5*(4), 617–632. https://doi.org/10.2217/nnm.10.33

FAO. (2020). *State of world fisheries and aquaculture 2020 (Spanish).* Food & Agriculture Organization.

Fiorati, A., Bellingeri, A., Punta, C., Corsi, I., & Venditti, I. (2020). Silver nanoparticles for water pollution monitoring and treatments: Ecosafety challenge and cellulose-based hybrids solution. *Polymers, 12*(8), 1635–1657. https://doi.org/10.3390/polym12081635

Forest, V., & Pourchez, J. (2017). Preferential binding of positive nanoparticles on cell membranes is due to electrostatic interactions: A too simplistic explanation that does not take into account the nanoparticle protein corona. *Materials Science and Engineering: C, 70*, 889–896. https://doi.org/10.1016/j.msec.2016.09.016

Foroozandeh, P., & Aziz, A. A. (2018). Insight into cellular uptake and intracellular trafficking of nanoparticles. *Nanoscale Research Letters, 13*, 339–450.

Friedman, A., Claypool, S., & Liu, R. (2013). The smart targeting of nanoparticles. *Current Pharmaceutical Design, 19*(35), 6315–6329.

Gelaye, Y. (2024). Application of nanotechnology in animal nutrition: Bibliographic review. *Cogent Food & Agriculture, 10*(1), 2290308. https://doi.org/10.1080/23311932.2023.2290308

Gouthami, K., Lakshminarayana, L., Faniband, B., Veeraraghavan, V., Bilal, M., Bhargava, R. N., Ferreira, L. F. R., Rahdar, A., Kakkameli, S., & Mulla, S. I. (2023). Introduction to polymeric nanomaterials. In *Smart Polymer Nanocomposites* (pp. 3–25). Elsevier. https://doi.org/10.1016/B978-0-323-91611-0.00008-6

Gujrati, M., Malamas, A., Shin, T., Jin, E., Sun, Y., & Lu, Z. R. (2014). Multifunctional cationic lipid-based nanoparticles facilitate endosomal escape and reduction-triggered cytosolic siRNA release. *Molecular Pharmaceutics, 11*(8), 2734–2744. https://doi.org/10.1021/mp400787s

Gustafson, H. H., Holt-Casper, D., Grainger, D. W., & Ghandehari, H. (2015). Nanoparticle uptake: The phagocyte problem. *Nano Today, 10*(4), 487–510. https://doi.org/10.1016/j.nantod.2015.06.006

Haffner, S. M., & Malmsten, M. (2017). Membrane interactions and antimicrobial effects of inorganic nanoparticles. *Advances in Colloid and Interface Science, 248*, 105–128. https://doi.org/10.1016/j.cis.2017.07.029

Handy, R. D. (2012). FSBI briefing paper: Nanotechnology in fisheries and aquaculture. *Fisheries Society of the British Isles, 1*, 1–29.

Handy, R. D., & Shaw, B. J. (2007). Toxic effects of nanoparticles and nanomaterials: Implications for public health, risk assessment and the public perception of nanotechnology. *Health, Risk & Society, 9*(2), 125–144. https://doi.org/10.1080/13698570701306807

Hardy, C. L., LeMasurier, J. S., Belz, G. T., Scalzo-Inguanti, K., Yao, J., Xiang, S. D., & Plebanski, M. (2012). Inert 50-nm polystyrene nanoparticles that modify pulmonary dendritic cell function and inhibit allergic airway inflammation. *Journal of Immunology, 188*(3), 1431–1441. https://doi.org/10.4049/jimmunol.1100156

Hasan, S. (2015). A review on nanoparticles: Their synthesis and types. *Research Journal of Recent Sciences, 4*(ISC-2014), 1–3.

He, X., Deng, H., & Hwang, H. M. (2019). The current application of nanotechnology in food and agriculture. *Journal of Food and Drug Analysis, 27*(1), 1–21. https://doi.org/10.1016/j.jfda.2018.12.002

Hussain, S. M., Ahmad, N., Rasul, A., Shahzad, M. M., Latif, M., Arsalan, U. H., & M. Z., Umair, M., & Shafqat, H. H. (2019). Efficacy of Nano-Cr particles supplemented sunflower meal based diets on growth performance, digestibility and hematology of *Catla catla* fingerlings. *Pakistan Journal of Zoology, 51*(5), 1943–1952. https://doi.org/10.17582/journal.pjz/2019.51.5.1943.1952.

Islam, F., Shohag, S., Uddin, M. J., Islam, M. R., Nafady, M. H., Akter, A., Mitra, S., Roy, A., Emran, T. B., & Cavalu, S. (2022). Exploring the journey of zinc oxide nanoparticles (ZnO-NPs) toward biomedical applications. *Materials, 15*(6), 2160. https://doi.org/10.3390/ma15062160

Jain, A., Gupta, Y., Agrawal, R., Jain, S. K., & Khare, P. (2007). Biofilms – A microbial life perspective: A critical review. *Critical Reviews™ in Therapeutic Drug Carrier Systems, 24*(5), 393–443. https://doi.org/10.1615/CritRevTherDrugCarrierSyst.v24.i5.10

Jia, J., Zhang, Y., Xin, Y., Jiang, C., Yan, B., & Zhai, S. (2018). Interactions between nanoparticles and dendritic cells: From the perspective of cancer immunotherapy. *Frontiers in Oncology, 8*, 404–415. https://doi.org/10.3389/fonc.2018.00404

Joudeh, N., & Linke, D. (2022). Nanoparticle classification, physicochemical properties, characterization, and applications: A comprehensive review for biologists. *Journal of Nanobiotechnology, 20*(1), 262–290.

Ju-Nam, Y., & Lead, J. R. (2008). Manufactured nanoparticles: An overview of their chemistry, interactions and potential environmental implications. *Science of the Total Environment, 400*(1–3), 396–414. https://doi.org/10.1016/j.scitotenv.2008.06.042

Kah Sem, N. A. D., Abd Gani, S., Chong, C. M., Natrah, I., & Shamsi, S. (2023). Management and mitigation of *Vibriosis* in aquaculture: Nanoparticles as promising alternatives. *International Journal of Molecular Sciences, 24*(16), 12542–12562. https://doi.org/10.3390/ijms241612542

Kamaly, N., Yameen, B., Wu, J., & Farokhzad, O. C. (2016). Degradable controlled-release polymers and polymeric nanoparticles: Mechanisms of controlling drug release. *Chemical Reviews, 116*(4), 2602–2663. https://doi.org/10.1021/acs.chemrev.5b00346

Karimi, M., Sadeghi, R., & Kokini, J. (2018). Human exposure to nanoparticles through trophic transfer and the biosafety concerns that nanoparticle-contaminated foods pose to consumers. *Trends in Food Science & Technology, 75*, 129–145. https://doi.org/10.1016/j.tifs.2018.03.012

Khan, I., Saeed, K., & Khan, I. (2019). Nanoparticles: Properties, applications and toxicities. *Arabian Journal of Chemistry, 12*(7), 908–931. https://doi.org/10.1016/j.arabjc.2017.05.011

Khan, M. U., Ullah, H., Honey, S., Talib, Z., Abbas, M., Umar, A., Ahmad, T., Sohail, J., Sohail, A., Makgopa, K., Ahmad, J., & Asim, J. (2023). Metal nanoparticles: Synthesis approach, types and applications–a mini review. *Nano-Horizons: Journal of Nanosciences and Nanotechnologies, 2*, 1–21.

Khan, S. K., Dutta, J., Rather, M. A., Ahmad, I., Nazir, J., Khan, I. A., & Wani, G. B. (2024a). Silver nanoparticle toxicity in rainbow trout: Insights into physiological and molecular responses. *Toxicology Mechanisms and Methods*, 1–16.

Khan, S. K., Dutta, J., Ahmad, I., & Rather, M. A. (2024b). Nanotechnology in aquaculture: Transforming the future of food security. *Food Chemistry, X*, 101974.

Khosravi-Katuli, K., Prato, E., Lofrano, G., Guida, M., Vale, G., & Libralato, G. (2017). Effects of nanoparticles in species of aquaculture interest. *Environmental Science and Pollution Research, 24*, 17326–17346.

Khursheed, S., Dutta, J., Ahmad, I., Rather, M. A., Badroo, I. A., Bhat, T. A., Ahmad, I., Amin, A., Shah, A., Qadri, T., & Habib, H. (2023). Biogenic silver nanoparticles: Synthesis, applications and challenges in food sector with special emphasis on aquaculture. *Food Chemistry: X, 20*, 101051. https://doi.org/10.1016/j.fochx.2023.101051

Kim, S. T., Saha, K., Kim, C., & Rotello, V. M. (2013). The role of surface functionality in determining nanoparticle cytotoxicity. *Accounts of Chemical Research, 46*(3), 681–691. https://doi.org/10.1021/ar3000647

Kim, J., Kang, S. H., Choi, Y., Lee, W., Kim, N., Tanaka, M., Kang, S. H., & Choi, J. (2023). Antibacterial and biofilm-inhibiting cotton fabrics decorated with copper nanoparticles grown on graphene nanosheets. *Scientific Reports, 13*(1), 11947. https://doi.org/10.1038/s41598-023-38723-4

Kumar, R. (2019). Lipid-based nanoparticles for drug-delivery systems. In *Nanocarriers for Drug Delivery* (pp. 249–284). Elsevier.

Li, W., & Szoka, F. C. (2007). Lipid-based nanoparticles for nucleic acid delivery. *Pharmaceutical Research, 24*, 438–449.

Lim, S. B., Banerjee, A., & Onyuksel, H. (2012). Improvement of drug safety by the use of lipid-based nanocarriers. *Journal of Controlled Release, 163*(1), 34–45. https://doi.org/10.1016/j.jconrel.2012.06.002

Lim, E. K., Kim, T., Paik, S., Haam, S., Huh, Y. M., & Lee, K. (2015). Nanomaterials for theranostics: Recent advances and future challenges. *Chemical Reviews, 115*(1), 327–394. https://doi.org/10.1021/cr300213b

Liu, Y., Hardie, J., Zhang, X., & Rotello, V. M. (2017). Effects of engineered nanoparticles on the innate immune system. In *Seminars in Immunology* (Vol. 34, pp. 25–32). Academic. https://doi.org/10.1016/j.smim.2017.09.011

Liu, J., Liu, Z., Pang, Y., & Zhou, H. (2022). The interaction between nanoparticles and immune system: Application in the treatment of inflammatory diseases. *Journal of Nanobiotechnology, 20*(1), 1–25. https://doi.org/10.1186/s12951-022-01343-7

Loiseau, A., Asila, V., Boitel-Aullen, G., Lam, M., Salmain, M., & Boujday, S. (2019). Silver-based plasmonic nanoparticles for and their use in biosensing. *Biosensors, 9*(2), 78–117. https://doi.org/10.3390/bios9020078

Look, M., Bandyopadhyay, A., Blum, J. S., & Fahmy, T. M. (2010). Application of nanotechnologies for improved immune response against infectious diseases in the developing world. *Advanced Drug Delivery Reviews, 62*(4–5), 378–393. https://doi.org/10.1016/j.addr.2009.11.011

Losada-Barreiro, S., Sezgin-Bayindir, Z., Paiva-Martins, F., & Bravo-Díaz, C. (2022). Biochemistry of antioxidants: Mechanisms and pharmaceutical applications. *Biomedicine, 10*(12), 3051–3098. https://doi.org/10.3390/biomedicines10123051

Ma, S., & Lin, D. (2013). The biophysicochemical interactions at the interfaces between nanoparticles and aquatic organisms: Adsorption and internalization. *Environmental Science: Processes & Impacts, 15*(1), 145–160.

Mahmoodi, S., Elmi, A., & Hallaj-Nezhadi, S. (2018). Copper nanoparticles as antibacterial agents. *Journal of Molecular Pharmaceutics & Organic Process Research, 6*(1), 1–7. https://doi.org/10.4172/2329-9053.1000140

Makabenta, J. M. V., Nabawy, A., Li, C. H., Schmidt-Malan, S., Patel, R., & Rotello, V. M. (2021). Nanomaterial-based therapeutics for antibiotic-resistant bacterial infections. *Nature Reviews Microbiology, 19*(1), 23–36. https://doi.org/10.1038/s41579-020-0420-1

Marquez, J. C. M., Partida, A. H., del Carmen, M., Dosta, M., Mejia, J. C., & Martinez, J. A. B. (2018). Silver nanoparticles applications (AgNPs) in aquaculture. *International Journal of Fisheries and Aquatic Studies, 6*(2), 5–11.

McCracken, C., Dutta, P. K., & Waldman, W. J. (2016). Critical assessment of toxicological effects of ingested nanoparticles. *Environmental Science: Nano, 3*(2), 256–282.

Meghani, N., Dave, S., & Kumar, A. (2020). Introduction to nanofood. *Nano-food Engineering: Volume One, 1-23*. https://doi.org/10.1007/978-3-030-44552-2_1

Mitarotonda, R., Giorgi, E., Desimone, M. F., & De Marzi, M. C. (2019). Nanoparticles and immune cells. *Current Pharmaceutical Design, 25*(37), 3960–3982. https://doi.org/10.2174/1381612825666190926161209

Mitragotri, S., & Stayton, P. (2014). Organic nanoparticles for drug delivery and imaging. *MRS Bulletin, 39*(3), 219–223. https://doi.org/10.1557/mrs.2014.11

Moges, F. D., Patel, P., Parashar, S. K. S., & Das, B. (2020). Mechanistic insights into diverse nano-based strategies for aquaculture enhancement: A holistic review. *Aquaculture, 519*, 734770. https://doi.org/10.1016/j.aquaculture.2019.734770

Mu, Q., Jiang, G., Chen, L., Zhou, H., Fourches, D., Tropsha, A., & Yan, B. (2014). Chemical basis of interactions between engineered nanoparticles and biological systems. *Chemical Reviews, 114*(15), 7740–7781. https://doi.org/10.1021/cr400295a

Mulder, W. J., Strijkers, G. J., van Tilborg, G. A., Griffioen, A. W., & Nicolay, K. (2006). Lipid-based nanoparticles for contrast-enhanced MRI and molecular imaging. *NMR in Biomedicine: An International Journal Devoted to the Development and Application of Magnetic Resonance Invivo, 19*(1), 142–164. https://doi.org/10.1002/nbm.1011

Mulvaney, P. (2015). Nanoscience vs nanotechnology – Defining the field. *ACS Nano, 9*(3), 2215–2217. https://doi.org/10.1021/acsnano.5b01418

Murphy, P. (2010). Nanotechnology: Public engagement with health, environment and social issues. In *DCU Online Research Access Service* (pp. 1–69).

Nasr-Eldahan, S., Nabil-Adam, A., Shreadah, M. A., Maher, A. M., & El-Sayed Ali, T. (2021). A review article on nanotechnology in aquaculture sustainability as a novel tool in fish disease control. *Aquaculture International, 29*, 1459–1480.

Ng, K. K., & Zheng, G. (2015). Molecular interactions in organic nanoparticles for phototheranostic applications. *Chemical Reviews, 115*(19), 11012–11042. https://doi.org/10.1021/acs.chemrev.5b00140

Nickols-Richardson, S. M., & Piehowski, K. E. (2008). Nanotechnology in nutritional sciences. *Minerva Biotecnologica, 20*(3), 117–126.

Niesman, M. R., Khoobehi, B., Magin, R. L., & Webb, A. G. (1994). Liposomes and diagnostic imaging: The potential to visualize both structure and function. *Journal of Liposome Research, 4*(2), 741–768. https://doi.org/10.3109/08982109409018597

Ogunfowora, L. A., Iwuozor, K. O., Ighalo, J. O., & Igwegbe, C. A. (2021). Trends in the treatment of aquaculture effluents using nanotechnology. *Cleaner Materials, 2*, 100024. https://doi.org/10.1016/j.clema.2021.100024

Ogunkalu, O. A. (2019). Utilization of nanotechnology in aquaculture and seafood sectors. *Eurasian Journal of Food Science and Technology, 3*(1), 26–33.

Okeke, E. S., Chukwudozie, K. I., Nyaruaba, R., Ita, R. E., Oladipo, A., Ejeromedoghene, O., Atakpa, E. O., Agu, C. V., & Okoye, C. O. (2022). Antibiotic resistance in aquaculture and aquatic organisms: A review of current nanotechnology applications for sustainable management. *Environmental Science and Pollution Research, 29*(46), 69241–69274. https://doi.org/10.1007/s11356-022-22319-y

Oksel, C., Subramanian, V., Semenzin, E., Ma, C. Y., Hristozov, D., Wang, X. Z., Hunt, N., Costa, A., Fransman, W., Marcomini, A., & Wilkins, T. (2016). Evaluation of existing control measures in reducing health and safety risks of engineered nanomaterials. *Environmental Science: Nano, 3*(4), 869–882.

Parsai, T., & Kumar, A. (2020). Tradeoff between risks through ingestion of nanoparticle contaminated water or fish: Human health perspective. *Science of the Total Environment, 740*, 140140. https://doi.org/10.1016/j.scitotenv.2020.140140

Pateiro, M., Gomez, B., Munekata, P. E., Barba, F. J., Putnik, P., Kovacevic, D. B., & Lorenzo, J. M. (2021). Nanoencapsulation of promising bioactive compounds to improve their absorption, stability, functionality and the appearance of the final food products. *Molecules, 26*(6), 1547. https://doi.org/10.3390/molecules26061547

Patra, A., Chandaluri, C. G., & Radhakrishnan, T. P. (2012). Optical materials based on molecular nanoparticles. *Nanoscale, 4*(2), 343–359. https://doi.org/10.1039/C1NR11313E

Pelgrift, R. Y., & Friedman, A. J. (2013). Nanotechnology as a therapeutic tool to combat microbial resistance. *Advanced Drug Delivery Reviews, 65*(13–14), 1803–1815. https://doi.org/10.1016/j.addr.2013.07.011

Popoola, O. M., Behera, B. K., & Kumar, V. (2023). Dietary silver nanoparticles as immunostimulant on rohu (*Labeo rohita*): Effects on the growth, cellular ultrastructure, immune-gene expression, and survival against *Aeromonas hydrophila*. *Fish and Shellfish Immunology Reports, 4*, 100080. https://doi.org/10.1016/j.fsirep.2022.100080

Pradhan, D., Biswasroy, P., Goyal, A., Ghosh, G., & Rath, G. (2021). Recent advancement in nanotechnology-based drug delivery system against viral infections. *AAPS PharmSciTech, 22*, 47–65. https://doi.org/10.1208/s12249-020-01908-5

Priya, S., Desai, V. M., & Singhvi, G. (2022). Surface modification of lipid-based nanocarriers: A potential approach to enhance targeted drug delivery. *ACS Omega, 8*(1), 74–86. https://doi.org/10.1021/acsomega.2c05976

Pulit-Prociak, J., & Banach, M. (2016). Silver nanoparticles–a material of the future…? *Open Chemistry, 14*(1), 76–91. https://doi.org/10.1515/chem-2016-0005

Qie, Y., Yuan, H., Von Roemeling, C. A., Chen, Y., Liu, X., Shih, K. D., Knight, J. A., Tun, H. W., Wharen, R. E., Jiang, W., & Kim, B. Y. S. (2016). Surface modification of nanoparticles enables selective evasion of phagocytic clearance by distinct macrophage phenotypes. *Scientific Reports, 6*(1), 26269.

Ramesh, A. S., Tripathy, A., Meghana, N. V. P., Karthik, L., Kiran, G., & Arivarasan, V. K. (2023). Introduction of nanotechnology intervention in aquaculture. In *Nanotechnological Approaches to the Advancement of Innovations in Aquaculture* (pp. 99–114). Springer International Publishing.

Ramsden, J. J. (2018). Nanotechnology in a modern economy. In J. J. Ramsden (Ed.), *Applied nanotechnology. The conversion of research results to products* (pp. 61–80). William Andrew Applied Science Publishers.

Rassaei, L., Marken, F., Sillanpaa, M., Amiri, M., Cirtiu, C. M., & Sillanpaa, M. (2011). Nanoparticles in electrochemical sensors for environmental monitoring. *TrAC Trends in Analytical Chemistry, 30*(11), 1704–1715. https://doi.org/10.1016/j.trac.2011.05.009

Rather, M. A., Bhat, I. A., Sharma, N., & Sharma, R. (2018). Molecular and cellular toxicology of nanomaterials with related to aquatic organisms. *Cellular and Molecular Toxicology of Nanoparticles, 1048*, 263–284.

Rather, M. A., Ahmad, I., Shah, A., Hajam, Y. A., Amin, A., Khursheed, S.,Ahmad, I., & Rasool, S. (2024). Exploring opportunities of artificial intelligence in aquaculture to meet increasing food demand. *Food Chemistry, X*, 101309. https://doi.org/10.1016/j.fochx.2024.101309

Reilly, R. M. (2007). Carbon nanotubes: Potential benefits and risks of nanotechnology in nuclear medicine. *Journal of Nuclear Medicine, 48*(7), 1039–1042. https://jnm.snmjournals.org/authors.

Rennick, J. J., Johnston, A. P., & Parton, R. G. (2021). Key principles and methods for studying the endocytosis of biological and nanoparticle therapeutics. *Nature Nanotechnology, 16*(3), 266–276.

Roszczenko, P., Szewczyk, O. K., Czarnomysy, R., Bielawski, K., & Bielawska, A. (2022). Biosynthesized gold, silver, palladium, platinum, copper, and other transition metal nanoparticles. *Pharmaceutics, 14*(11), 2286–2304. https://doi.org/10.3390/pharmaceutics14112286

Roy, S., Kumar, V., Barman, D., & Mandal, S. (2012). Nanotechnology and its application in fisheries and aquaculture. *AQUA International, 20*(6), 2–3.

Sabourian, P., Yazdani, G., Ashraf, S. S., Frounchi, M., Mashayekhan, S., Kiani, S., & Kakkar, A. (2020). Effect of physico-chemical properties of nanoparticles on their intracellular uptake. *International Journal of Molecular Sciences, 21*(21), 8019–8038. https://doi.org/10.3390/ijms21218019

Sajid, M., Ilyas, M., Basheer, C., Tariq, M., Daud, M., Baig, N., & Shehzad, F. (2015). Impact of nanoparticles on human and environment: Review of toxicity factors, exposures, control strategies, and future prospects. *Environmental Science and Pollution Research, 22*, 4122–4143. https://doi.org/10.1007/s11356-014-3994-1

Samrot, A. V., Ram Singh, S. P., Deenadhayalan, R., Rajesh, V. V., Padmanaban, S., & Radhakrishnan, K. (2022). Nanoparticles, a double-edged sword with oxidant as well as antioxidant properties – A review. *Oxygen, 2*(4), 591–604. https://doi.org/10.3390/oxygen2040039

Sarkar, B., Mahanty, A., Gupta, S. K., Choudhury, A. R., Daware, A., & Bhattacharjee, S. (2022). Nanotechnology: A next-generation tool for sustainable aquaculture. *Aquaculture, 546*, 737330. https://doi.org/10.1016/j.aquaculture.2021.737330

Scioli Montoto, S., Muraca, G., & Ruiz, M. E. (2020). Solid lipid nanoparticles for drug delivery: Pharmacological and biopharmaceutical aspects. *Frontiers in Molecular Biosciences, 7*, 319–342. https://doi.org/10.3389/fmolb.2020.587997

Seaton, A., Tran, L., Aitken, R., & Donaldson, K. (2010). Nanoparticles, human health hazard and regulation. *Journal of the Royal Society Interface, 7*(1), S119–S129. https://doi.org/10.1098/rsif.2009.0252.focus

Shah, B. R., & Mraz, J. (2020). Advances in nanotechnology for sustainable aquaculture and fisheries. *Reviews in Aquaculture, 12*(2), 925–942. https://doi.org/10.1111/raq.12356

Sharma, V. K., Sayes, C. M., Guo, B., Pillai, S., Parsons, J. G., Wang, C., Yan, B., & Ma, X. (2019). Interactions between silver nanoparticles and other metal nanoparticles under environmentally relevant conditions: A review. *Science of the Total Environment, 653*, 1042–1051. https://doi.org/10.1016/j.scitotenv.2018.10.411

Shenashen, M. A., El-Safty, S. A., & Elshehy, E. A. (2014). Synthesis, morphological control, and properties of silver nanoparticles in potential applications. *Particle & Particle Systems Characterization, 31*(3), 293–316. https://doi.org/10.1002/ppsc.201300181

Shirodkar, R. K., Kumar, L., Mutalik, S., & Lewis, S. (2019). Solid lipid nanoparticles and nanostructured lipid carriers: Emerging lipid based drug delivery systems. *Pharmaceutical Chemistry Journal, 53*, 440–453.

Sirelkhatim, A., Mahmud, S., Seeni, A., Kaus, N. H. M., Ann, L. C., Bakhori, S. K. M., Hasan, H., & Mohamad, D. (2015). Review on zinc oxide nanoparticles: Antibacterial activity and toxicity mechanism. *Nano-Micro Letters, 7*, 219–242.

Smith, I. O., Liu, X. H., Smith, L. A., & Ma, P. X. (2009). Nanostructured polymer scaffolds for tissue engineering and regenerative medicine. *Wiley Interdisciplinary Reviews: Nanomedicine and Nanobiotechnology, 1*(2), 226–236. https://doi.org/10.1002/wnan.26

Stone, V., Nowack, B., Baun, A., van den Brink, N., von der Kammer, F., Dusinska, M., Handy, R., Hankin, S., Hassellov, M., Joner, E., & Fernandes, T. F. (2010). Nanomaterials for environmental studies: Classification, reference material issues, and strategies for physico-chemical characterisation. *The Science of the Total Environment, 408*(7), 1745–1754. https://doi.org/10.1016/j.scitotenv.2009.10.035

Tayel, A. A., Elsaied, B. E., & Diab, A. M. (2019). Nanotechnology for aquaculture. In *Nanoscience for Sustainable Agriculture* (pp. 479–544).

Technavio. (2019). *Global food nanotechnology market 2019–2023: Growing applications in nutraceuticals to boost the market*. Technavio.

Thiruvengadam, M., Rajakumar, G., & Chung, I. M. (2018). Nanotechnology: Current uses and future applications in the food industry. *3. Biotech, 8*(1), 74. https://doi.org/10.1007/s13205-018-1104-7

Torchilin, V. P. (2006). *Nanoparticulates as drug carriers*. Imperial College Press.

Vinay, T. N., Bhat, S., Gon Choudhury, T., Paria, A., Jung, M. H., Shivani Kallappa, G., & Jung, S. J. (2018). Recent advances in application of nanoparticles in fish vaccine delivery. *Reviews in Fisheries Science & Aquaculture, 26*(1), 29–41. https://doi.org/10.1080/23308249.2017.1334625

Vogel, C., Susanne, C. M., Roger, E. K., & Geoffrey, D. D. (2012). Linking vulnerability, adaptation, and resilience science to practice: Pathways, players and partnerships1. In *Integrating Science and Policy* (pp. 97–127). Routledge.

Vrignaud, S., Benoit, J. P., & Saulnier, P. (2011). Strategies for the nanoencapsulation of hydrophilic molecules in polymer-based nanoparticles. *Biomaterials, 32*(33), 8593–8604. https://doi.org/10.1016/j.biomaterials.2011.07.057

Wang, J., Chen, G., Jiang, H., Li, Z., & Wang, X. (2013). Advances in nano-scaled biosensors for biomedical applications. *Analyst, 138*(16), 4427–4435. https://doi.org/10.1039/C3AN00438D

Wang, L., Hu, C., & Shao, L. (2017). The antimicrobial activity of nanoparticles: Present situation and prospects for the future. *International Journal of Nanomedicine, 12*, 1227–1249. https://doi.org/10.2147/IJN.S121956

Wang, Y., Yang, Y., Shi, Y., Song, H., & Yu, C. (2020). Antibiotic-free antibacterial strategies enabled by nanomaterials: Progress and perspectives. *Advanced Materials, 32*(18), 1904106. https://doi.org/10.1002/adma.201904106

World Health Organization. (2010). *FAO/WHO expert meeting on the application of nanotechnologies in the food and agriculture sectors: Potential food safety implications: Meeting report* (p. 134). World Health Organization.

World Health Organization. (2013). *State of the art on the initiatives and activities relevant to risk assessment and risk management of nanotechnologies in the food and agriculture sectors: FAO/WHO technical paper* (p. 60). World Health Organization. https://www.fao.org/3/i3281e/i3281e.pdf

Wu, L. P., Ficker, M., Christensen, J. B., Trohopoulos, P. N., & Moghimi, S. M. (2015). Dendrimers in medicine: Therapeutic concepts and pharmaceutical challenges. *Bioconjugate Chemistry, 26*, 1198–1211. https://doi.org/10.1021/acs.bioconjchem.5b00031

Yaqoob, A. A., Parveen, T., Umar, K., & Mohamad Ibrahim, M. N. (2020). Role of nanomaterials in the treatment of wastewater: A review. *Water, 12*(2), 495–524. https://doi.org/10.3390/w12020495

Yung, M. M., Kwok, K. W., Djurisic, A. B., Giesy, J. P., & Leung, K. M. (2017). Influences of temperature and salinity on physicochemical properties and toxicity of zinc oxide nanoparticles to the marine diatom *Thalassiosira pseudonana*. *Scientific Reports, 7*(1), 3662–3670. https://doi.org/10.1038/s41598-017-03889-1

Yusuf, A., Almotairy, A. R. Z., Henidi, H., Alshehri, O. Y., & Aldughaim, M. S. (2023). Nanoparticles as drug delivery systems: A review of the implication of nanoparticles' physicochemical properties on responses in biological systems. *Polymers, 15*(7), 1596–1621. https://doi.org/10.3390/polym15071596

Zahra, Z., Habib, Z., Chung, S., & Badshah, M. A. (2020). Exposure route of TiO2 NPs from industrial applications to wastewater treatment and their impacts on the agro-environment. *Nanomaterials, 10*(8), 1469. https://doi.org/10.3390/nano10081469

Zehri, A. H., Nylander, A., Ye, L., & Liu, J. (2019). Graphene-coated copper nanoparticles for thermal conductivity enhancement in water-based nanofluid. In *2019 22nd European Microelectronics and Packaging Conference & Exhibition (EMPC)* (pp. 1–7). IEEE.

Zhang, L., Pornpattananangkul, D., Hu, C. M., & Huang, C. M. (2010). Development of nanoparticles for antimicrobial drug delivery. *Current Medicinal Chemistry, 17*(6), 585–594. https://doi.org/10.2174/092986710790416290

Zharov, V. P., & Lapotko, D. O. (2005). Photothermal imaging of nanoparticles and cells. *IEEE Journal of Selected Topics in Quantum Electronics, 11*(4), 733–751. https://doi.org/10.1109/JSTQE.2005.857382

Zhou, X., Wang, Y., Gu, Q., & Li, W. (2009). Effects of different dietary selenium sources (selenium nanoparticle and selenomethionine) on growth performance, muscle composition and glutathione peroxidase enzyme activity of crucian carp (*Carassius auratus gibelio*). *Aquaculture, 291*(1–2), 78–81. https://doi.org/10.1016/j.aquaculture.2009.03.007

Zielinska, A., Carreiro, F., Oliveira, A. M., Neves, A., Pires, B., Venkatesh, D. N., Durazzo, A., Lucarini, M., Eder, P., Silva, A. M., Santini, A., & Souto, E. B. (2020). Polymeric nanoparticles: Production, characterization, toxicology and ecotoxicology. *Molecules, 25*(16), 3731. https://doi.org/10.3390/molecules25163731

Zoroddu, M. A., Medici, S., Ledda, A., Nurchi, V. M., Lachowicz, J. I., & Peana, M. (2014). Toxicity of nanoparticles. *Current Medicinal Chemistry, 21*(33), 3837–3853.

Chapter 23
Genomic Innovations in Aquafeed: Unraveling the Nutrigenomics of Sustainable Aquaculture

Noorin Zafar, Sana Naz, and Mukhtar A. Khan

1 Introduction

Aquaculture is a fast-growing food production system that plays an important role in achieving food and nutritional security. It is a complex production system in which feed is a major factor for the growth of the fish (Mohanty et al., 2020). Understanding the specific nutritional requirements of different fish species is essential to formulate optimized feeds that promote healthy growth, improve feed conversion efficiency, and reduce waste, ultimately contributing to more sustainable aquaculture practices. Nutrient requirements of aquatic organisms are species as well as stage specific.

The interaction of nutrients, environment, and genetics plays a crucial role in determining the overall physiological function and health of an organism. Genetics and genomics are valuable tools for studying fish biology, offering insights into their biological makeup. From its inception, genomics has been a multidisciplinary field, integrating principles and methods from biochemistry, molecular biology, genetics, statistics, informatics, and more (Lu & Luo, 2020). In aquaculture, genetic selection focuses on enhancing desirable traits such as improved growth rates, disease resistance, feed conversion efficiency, and product quality, all of which contribute to more sustainable and productive farming practices. Genomics allows for broader genome-wide studies, better detection capabilities, and cost-effectiveness. Nutrigenomics, an emerging field in nutritional research, is the study of the interaction of the nutrients with genes. It is a broad field of science that focuses on the interaction between genes and the activity of common dietary nutrients, as well as their function in achieving a balance between disease and health by altering the

N. Zafar · S. Naz · M. A. Khan (✉)
Fish Nutrition Research Laboratory, Department of Zoology, Aligarh Muslim University, Aligarh, Uttar Pradesh, India

I. Ahmed, I. Ahmad (eds.), *Aquaculture: Enhancing Food Security and Nutrition*, https://doi.org/10.1007/978-3-031-92858-1_23

structure or expression of an organism's genetic composition (Hakim et al., 2018). From a nutrigenomics point of view, nutrients are considered as feeding signals detected by the cellular system that influence the expression of genes, translated into proteins, and the production of metabolites (Okani-Onyejiaka et al., 2024; Dragsted, 2010). Expression of related genes can be modified with different dietary nutrients. Understanding the nutritional requirements and need-based dietary manipulation are perhaps the most effective and sustainable way of dealing with most of the challenges faced in aquaculture. Gene expression is influenced by food and environment, and it could be helpful in maintaining health and preventing illness. It is also necessary to understand the various aspects of fish physiology and health influenced by nutrition and various environmental factors like habitat, stress, and environmental contaminants.

2 Genomics

In order to achieve the desirable goals of sustainable aquaculture, understanding of genomic structure and organization, phenomic variations, and genetic basis of characters of aquaculture species is crucial. In addition, it is also important to understand the mechanisms of regulation and evolutionary conservation at the levels of genome, transcriptome, proteome, epigenome, and systems biology. Significant efforts have been taken in aquaculture genomics for dozens of fish and shellfish species, which include the development of genetic linkage maps, physical maps, microarrays, single-nucleotide polymorphism (SNP) arrays, transcriptome databases, and various stages of genome reference sequences (Abdelrahman et al., 2017). Genomic technologies such as massive parallel sequencing, high-throughput genotyping, genome selection, and gene editing, combined with highly efficient computational methods, can accelerate the rate of genetic progress in animal breeding. Thus, such technologies can help us meet the needs for protein sources for human consumption in the upcoming years (Yáñez et al., 2022). Genomic information provides powerful tools to enhance physiological research, the results of which may be used for optimization of husbandry practices, feeding and feed formulations, breeding technologies, or nongenetic selection or screening such as epigenetics, proteomics, and metabolomics. Whole genome sequences, in various states of assembly, are now available for many aquaculture species, enabling the identification of genomic variations such as insertions or deletions, single-nucleotide polymorphisms (SNPs), copy number variations, and differentially methylated regions. The degree to which genomic technologies can be applied for the achievement of desirable traits varies across the different species.

Genome editing technologies can be used to introduce an immediate improvement in a phenotype in a single generation; hence, these technologies hold great promise for improving aquaculture.

3 Nutrigenomics

Nutrients such as carbohydrates, lipids, proteins, vitamins, and minerals undergo complex metabolic transformations in the body, providing energy for growth, maintenance, and physiological functions (Monfoulet et al., 2021; Kyriakaki et al., 2022). Nutrigenomics, one of the fastest-growing fields of research, investigates how these nutrients impact genome functioning through epigenetic modifications and genetic expression patterns. The benefits of nutrigenomics in fish nutrition are substantial, including improved fish yield, feed efficiency, reproductive performance, and immunity (Alhoshy et al., 2022). Nutrients serve as signals that are sent to specific cells in the body. When these cells interpret the information, they transmit biochemical signals from genes or DNA (genomics) to RNA (transcriptomics), which are then translated into proteins (proteomics), ultimately influencing metabolic pathways (metabolomics) (Fig. 23.1). All biological functions depend on the flow of information through this sequence, where genetic makeup drives these pathways. However, external factors such as environmental toxins, disease challenges, and diet can significantly impact the regulation of these processes.

Transcriptomics is the study of the complete set of RNA transcripts produced by the genome at any given time, providing insight into gene expression and how genes are regulated in response to various conditions, such as environmental factors or nutritional inputs (Lowe et al., 2017). Microarray or RNASeq is used in transcriptomics, which is usually revalidated for any erroneous data for accurate interpretation of any changes in genetic expression by doing real-time quantitative polymerase chain reaction (qPCR). Proteomics is the large-scale study of proteins, particularly their functions, structures, and interactions within a cell or organism, aiming to understand how proteins contribute to biological processes, health, and disease (Veenstra, 2021). It is used in aquaculture operations for knowing key proteins related to growth and production, identifying biomarkers for enhancing fish productivity, and also developing novel therapeutic agents for the treatment of various fish diseases. Metabolomics is the comprehensive study of metabolites within a biological system, aiming to understand how these molecules change in response to genetic, environmental, or nutritional factors and how they influence health and disease (Singh et al., 2023). It utilizes analytical platforms like gas chromatography/mass spectrometry (GC/MS), liquid chromatography with tandem mass spectrometry (LC/MS/ MS), and nuclear magnetic resonance (NMR) to assess hundreds of metabolites present in a sample. Another field called epigenomics is the study of modifications to DNA that do not change the genetic code but can still influence gene activity. These omics tools offer valuable insights into how a genome absorbs nutritional factors and responds through unique metabolic processes, thus playing a vital role in the broader field of nutrigenomics (Bordoni & Gabbianelli, 2019).

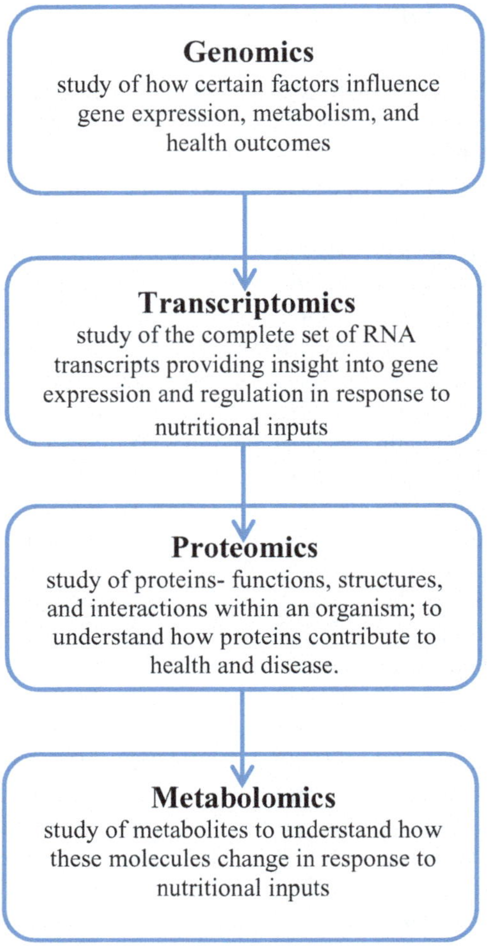

Fig. 23.1 Omics tools used in nutrigenomics

4 Tools in Nutrigenomics

Although fish farming has been practiced for 4000 years and aquaculture research dates back to about 1870, the traditional methods of biotechnology were introduced only recently in nutritional research (Hakim et al., 2018). Contemporary genomic approaches, which are often adapted from human or biomedical research, such as cDNA cloning and sequencing, cDNA microarrays, and functional genomics, have opened up new possibilities for aquaculture biotechnologists to improve fish growth rates, increase resistance to pathogens and stressors, improve the quality of the brood stocks, and create the opportunity to make new or different products by altering their genetic makeup. In aquaculture, most of the nutrigenomics studies at present involve the transcriptomic approach in which the transcription products, such as mRNA, are investigated, which provides information on the effect of nutrients on the gene. The transcriptomic studies are based on the reverse transcription PCR process. The total mRNA is extracted and is reverse transcribed to the cDNA. This cDNA can then be used for a number of PCR types. Semi-quantitative PCR and real-time PCR are the two most commonly used techniques for measuring the differential transcriptomic abundance in the different treatments following nutritional interference. The DNA microarray is an upcoming technique in fish nutrigenomics, which gives the overall genomic rather than a single gene response of the organism to the nutrient or diet. However, its use in fish nutrition is limited because it requires genomic sequences. The candidate gene approach involves selecting the possible risk gene based on the knowledge of biological function. Although this approach has limited applications, it can be applied in fish nutrition studies. This is because the candidate gene approach involves in vitro cell culture techniques to establish the positive and negative correlation between the expression of the gene and the effect it has on nutrient metabolism. Genome-wide association study (GWAS) involves the comparison of two groups, which show the varied responses to a nutrient. GWAS typically focuses on the association between the SNPs and the disease occurrence (Melamed et al., 2002; Adams & Thompson, 2006; Overturf, 2010).

5 Applications of Nutrigenomics in Nutritional Studies

Nutrigenomics provides insights into how dietary nutrients influence gene expression in aquatic animals, which is crucial for understanding the regulation of biological processes such as growth, immunity, and metabolism. By studying this, researchers can identify which genes are modulated by specific nutrients (Lv et al., 2025; Tseng et al., 2025). This knowledge can be used to tailor diets that influence gene expression in ways that promote desirable traits like enhanced growth rate, improved disease resistance, or better feed conversion efficiency. Ahmed et al. (2024) illustrated that dietary tryptophan at a level of 4.0 g/kg positively regulated gene expression related to muscle growth, enhancing skeletal muscle development in rainbow trout (*Oncorhynchus mykiss*) fingerlings. This was evidenced by the upregulation of TOR and 4E-BP1 compared to the control group. This information can guide the development of more effective feed formulations designed to enhance growth rate in farmed fish. In another experiment on European catfish (*Silurus glanis L.*), dietary inclusion of insect oil upregulated the expression of genes encoding lipoprotein lipase (*lpl*), carnitine palmitoyl transferase (*cpt1a*), and hepatocyte nuclear factor 4 alpha (*hnf4α)* involved in lipid metabolism in the livers of fish fed the 100% black soldier fly larvae oil in the diet (Egessa et al., 2024).

In aquaculture, environmental factors like water temperature, salinity, and oxygen levels can significantly affect nutrient utilization and metabolism in fish. Nutrigenomics helps to explore how climate change may influence the expression of genes involved in nutrient absorption and metabolism, which could help form climate-resilient feeds for aquaculture operations (Ge et al., 2024; Mai et al., 2025; Quirós-Pozo et al., 2025). For example, when *Pangasianodon hypophthalmus* fish were exposed to different combinations of high temperature, ammonia, and arsenic, genes associated with biotic and abiotic stress, such as HSP 70, cytochrome P450 (CYP 450), metallothionein (MT), DNA damage-inducible protein (DDIP), caspase

(CAS), tumor necrosis factor (TNFα), toll-like receptor (TLR), interleukin (IL), inducible nitric oxide synthase (iNOS), catalase (CAT), superoxide dismutase (SOD), and glutathione peroxidase (GPx), were highly upregulated. Supplementing manganese at 8 mg/kg diet significantly downregulated these genes (Kumar et al., 2024).

Disease outbreaks are a significant concern in aquaculture, and genetic factors influencing immune responses play a crucial role in disease resistance. Nutrigenomics helps to identify genetic variations that affect how fish respond to specific nutrients, including those that boost immune function or reduce susceptibility to infections. This knowledge can inform strategies for improving disease resistance through nutritional interventions (Dawood et al., 2024; Jastaniah et al., 2024). In a recent study, the effects of dietary probiotics on the expression of immune-related genes in *Oncorhynchus tshawytscha* were investigated. Results showed that the dietary supplementation with probiotics decreased mortality when challenged with *Vibrio anguillarum* and decreased expression of *cldn1* and *ocln*, which suggested that probiotic treatment confers better protection and intestinal integrity, resulting in improved survival (Soto-Dávila et al., 2024). Additionally, nutrigenomics could help develop targeted feed additives that activate specific immune-related genes in response to disease threats, reducing the need for antibiotics and promoting healthier fish populations.

Farmed fish often experience various forms of stress, such as handling, overcrowding, or toxins. Stress negatively impacts their immune function, growth, and overall health. Nutrigenomics can help researchers understand how different nutrients affect gene expression related to the fish's stress response (Chekani et al., 2021; Hamed & Abdel-Tawwab, 2021). For example, dietary supplementation with astaxanthin induced the expression levels of functional genes, such as SOD, Cu/ZnSOD, ferritin, ACP, and CYC in the hepatopancreas and increased the survival rate of *Babylonia areolata* under ammonia stress (Li et al., 2024).

The efficiency of feed conversion is critical in aquaculture. The ability of fish to convert the nutrients in their diet into growth and energy can be influenced by genetic factors. Nutrigenomics can help identify genetic markers associated with better nutrient utilization and more efficient feed conversion. This knowledge can be used to optimize feed formulations for improved nutrient utilization, reducing waste and the environmental impact of aquaculture (Jin et al., 2015; Lee et al., 2019). By understanding the genetic underpinnings of feed conversion efficiency, researchers could develop diets that are more precisely matched to the metabolic needs of the fish, improving feed efficiency and minimizing resource use in aquaculture operations. An experiment conducted on Chinese mitten crab (*Eriocheir sinensis*) to replace fish meal with cottonseed protein concentrate (CPC) found that the expression levels of genes related to protein metabolism, such as *pept1* gene involved with peptide transport and *slc7a5* involved in amino acid transport, and lipid metabolism, such as *srebp1, cpt1, Δ9fad, caat,* were downregulated on increasing dietary CPC level, resulting in insufficient uptake of amino acids and insignificant changes in crude lipid of whole-crab, respectively (Liu et al., 2024).

The integration of nutrigenomics into traditional nutritional research has opened up new avenues in aquaculture nutrition, representing a significant shift toward more personalized and targeted approaches to improving growth rates, health outcomes, and environmental sustainability in aquaculture operations.

6 Challenges with Implementing Nutrigenomics

While genomics research has progressed rapidly in humans and agricultural species like crops and livestock, aquaculture species have not received similar attention. This lack of functional genomics research limits our understanding of the genetic factors that influence key traits like growth, disease resistance, feed conversion efficiency, and reproductive performance in aquaculture species (Allal & Nguyen, 2022). Although fish genomes can be found in the genome database at the US National Center for Biotechnology Information (NCBI), which has published the genomic data of 265 fish species as of late 2019, the biological complexity of aquatic organisms, the high diversity of species, and the lack of a well-established genetic framework for many species in aquaculture make research challenging (Houston et al., 2020; Lu & Luo, 2020). Expanding genomics research in aquaculture species could lead to efficient nutrient use and better disease management, which could increase the sustainability and profitability of the aquaculture industry.

Different research teams often use varied methods for sample collection, data analysis, and result reporting. This lack of standardization makes it difficult to compare studies across laboratories and can impede the broader adoption of findings. Without standardized protocols, the results of different studies can be inconsistent, and the reproducibility of research findings becomes questionable. This is particularly problematic when applying genomic data to real-world aquaculture settings where consistency is crucial. Establishing standardized protocols and guidelines for genomics research in aquaculture would ensure that results are more reproducible, leading to stronger and more reliable conclusions that can be applied across the industry (You et al., 2020).

Omic technologies, such as high-throughput sequencing, metabolomics, and proteomics, are expensive, requiring specialized equipment and expertise. These technologies are often out of reach for smaller research institutions or aquaculture operations, particularly in developing countries (Matthews et al., 2016). Efforts to reduce the costs of these technologies through technological advancements, such as improved sequencing platforms and the development of open-source bioinformatics tools, could democratize access to omics data, benefiting more researchers and stakeholders in the aquaculture industry.

Integrating the findings of omic data with traditional knowledge of animal nutrition offers novel insights into the genetic factors influencing nutrition, which is necessary for practical applications. Nutritional science in aquaculture is based on established principles, and incorporating genomics requires expertise from both disciplines. Combining these two areas of knowledge can be difficult due to differences in methodologies, data types, and expertise. It also requires ongoing communication and collaboration between geneticists, nutritionists, and other specialists in the aquaculture field. A multidisciplinary approach, where genomic and traditional nutritional knowledge complement each other, could lead to the development of precise nutritional requirement data-based feeds and efficient feeding strategies, reducing costs and environmental impacts in aquaculture production (Luna et al., 2022).

Although nutrigenomics primarily focuses on genetic factors, aquaculture is a highly dynamic environment where water quality, temperature, salinity, and other external factors play a significant role in the nutritional and overall health of the fish. Focusing only on genetic factors might overlook how environmental factors interact with genetics to influence nutritional responses. This can limit the accuracy and effectiveness of nutrigenomics applications in real-world aquaculture environments (Rise et al., 2019). Taking a more holistic approach that integrates genetic, nutritional, and environmental data would provide a more complete understanding of how to optimize nutrition and health in aquaculture. This could involve systems biology approaches, where multiple layers of biological data are integrated to improve the overall sustainability of aquaculture.

7 Conclusion

While nutrigenomics offers exciting potential to revolutionize aquaculture, addressing the above-mentioned challenges is crucial for successful implementation. Solutions to these problems could lead to a more sustainable, efficient, and consumer-accepted aquaculture industry. These challenges highlight the need for continued research investment, interdisciplinary collaboration, and careful consideration of practical applications in aquaculture settings.

References

Abdelrahman, H., El-Hady, M., Alcivar-Warren, A., Allen, S., et al. (2017). Aquaculture genomics, genetics and breeding in the United States: Current status, challenges, and priorities for future research. *BMC Genomics, 18*, 191.

Adams, A., & Thompson, K. D. (2006). Biotechnology offers revolution to fish health management. *Trends in Biotechnology, 24*(5), 201–205.

Ahmed, I., Ahmad, I., & Malla, B. A. (2024). Effects of dietary tryptophan levels on growth performance, plasma profile, intestinal antioxidant capacity and growth related genes in rainbow trout (*Oncorhynchus mykiss*) fingerlings. *Aquaculture, 585*, 740710.

Alhoshy, M., Shehata, A. I., Habib, Y. J., Abdel-Latif, H. M., Wang, Y., & Zhang, Z. (2022). Nutrigenomics in crustaceans: Current status and future prospects. *Fish & Shellfish Immunology, 129*, 1–12.

Allal, F., & Nguyen, N. H. (2022). Genomic selection in aquaculture species. In *Genomic prediction of complex traits: Methods and protocols* (pp. 469–491). New York, NY, Springer.

Bordoni, L., & Gabbianelli, R. (2019). Primers on nutrigenetics and nutri (epi) genomics: Origins and development of precision nutrition. *Biochimie, 160*, 156–171.

Chekani, R., Akrami, R., Ghiasvand, Z., Chitsaz, H., & Jorjani, S. (2021). Effect of dietary dehydrated lemon peel (*Citrus limon*) supplementation on growth, hemato-immunolological and antioxidant status of rainbow trout (*Oncorhynchus mykiss*) under exposure to crowding stress. *Aquaculture, 539*, 736597.

Dawood, M. A., Abdo, S. E., El-Kassas, S., El-Naggar, K., et al. (2024). Chicken egg lysozyme enhanced the growth performance, feed utilization, upregulated immune-related genes, and mitigated the impacts of *Aeromonas hydrophila* infection in Nile tilapia (*Oreochromis niloticus*). *Fish & Shellfish Immunology, 146*, 109377.

Dragsted, L. O. (2010). Biomarkers of meat intake and the applications of nutrigenomics. *Meat Science, 84*, 301–307.

Egessa, R., Szűcs, A., Ardó, L., Biró, J., Lengyel-Kónya, É., Banjac, V., et al. (2024). Dietary inclusion of insect oil: Impact on growth, nutrient utilisation, lipid metabolism, antioxidant and immune-related responses in European catfish (*Silurus glanis* L). *Aquaculture, 592*, 741213.

Ge, K., Fan, Z., Huang, T., Gu, W., Wang, G., Liu, E., Pan, R., Li, D., Sun, Y., Yao, Z., Wang, L., Zhao, C., & Xu, G. (2024). Influence of increasing acclimation temperature on growth, digestion, antioxidant capacity, liver transcriptome and intestinal microflora of Ussruri whitefish *Coregonus ussuriensis* berg. *Fish & Shellfish Immunology, 151*, 109667.

Hakim, M. M., Ganai, N. A., Ahmad, S. M., Asmi, O., Akram, T., Hussain, S., Shabir, N., Shah, R., & Gora, A. H. (2018). Nutrigenomics: Omics approach in aquaculture research to mitigate the deficits in conventional nutritional practices. *Journal of Entomology and Zoology Studies, 6*(4), 582–587.

Hamed, H. S., & Abdel-Tawwab, M. (2021). Dietary pomegranate (*Punica granatum*) peel mitigated the adverse effects of silver nanoparticles on the performance, haemato-biochemical, antioxidant, and immune responses of Nile tilapia fingerlings. *Aquaculture, 540*, 736742.

Houston, R. D., Bean, T. P., Macqueen, D. J., Gundappa, M. K., Jin, Y. H., Jenkins, T. L., et al. (2020). Harnessing genomics to fast-track genetic improvement in aquaculture. *Nature Reviews Genetics, 21*(7), 389–409.

Jastaniah, S. D., Mansour, A. A., Al-Tarawni, A. H., El-Haroun, E., Munir, M. B., Saghir, S. A. M., Abdul Kari, Z., Téllez-Isaías, G., Bottje, W. G., et al. (2024). The effects of nano-curcumin on growth performance, feed utilization, blood biochemistry, disease resistance, and gene expression in European seabass (*Dicentrarchus labrax*) fingerlings. *Aquaculture Reports, 36*, 102034.

Jin, Y., Tian, L., Xie, S., Guo, D., Yang, H., Liang, G., & Liu, Y. (2015). Interactions between dietary protein levels, growth performance, feed utilization, gene expression and metabolic products in juvenile grass carp (*Ctenopharyngodon idella*). *Aquaculture, 437*, 75–83.

Kumar, N., Thorat, S. T., Kochewad, S. A., & Reddy, K. S. (2024). Manganese nutrient mitigates ammonia, arsenic toxicity and high temperature stress using gene regulation via NFkB mechanism in fish. *Scientific Reports, 14*(1), 1273.

Kyriakaki, P., Zisis, F., Pappas, A. C., Mavrommatis, A., & Tsiplakou, E. (2022). Effects of PUFA-rich dietary strategies on ruminants' mammary gland gene network: A nutrigenomics review. *Metabolites, 13*(1), 44.

Lee, S., Masagounder, K., Hardy, R. W., & Small, B. C. (2019). Effects of lowering dietary fishmeal and crude protein levels on growth performance, body composition, muscle metabolic gene expression, and chronic stress response of rainbow trout (*Oncorhynchus mykiss*). *Aquaculture, 513*, 734435.

Li, T., Zheng, P. H., Zhang, X. X., Zhang, Z. L., Li, J. T., Li, J. J., et al. (2024). Effects of dietary astaxanthin on growth performance, muscle composition, non-specific immunity, gene expression, and ammonia resistance of juvenile ivory shell (*Babylonia areolate*). *Fish & Shellfish Immunology, 145*, 109363.

Liu, X., Zou, D., Wang, Y., Zhuang, Y., Liu, Y., Li, Y., et al. (2024). Replacement of fish meal with cottonseed protein concentrate in Chinese mitten crab (*Eriocheir sinensis*): Nutrient digestibility, growth performance, free amino acid profile, and expression of genes related to nutrient metabolism. *Animal Nutrition, 17*, 447–462.

Lowe, R., Shirley, N., Bleackley, M., Dolan, S., & Shafee, T. (2017). Transcriptomics technologies. *PLoS Computational Biology, 13*(5), e1005457.
Lu, G., & Luo, M. (2020). Genomes of major fishes in world fisheries and aquaculture: Status, application and perspective. *Aquaculture and Fisheries, 5*(4), 163–173.
Luna, M., Llorente, I., & Cobo, A. (2022). Determination of feeding strategies in aquaculture farms using a multiple-criteria approach and genetic algorithms. *Annals of Operations Research, 314*(2), 551–576.
Lv, W., Wu, F., Sorensen, M., Tan, X., Wei, X., Yu, A., Zhong, C., Song, Y., & Luo, Z. (2025). Interaction between dietary zinc (Zn) and manganese (Mn) on growth performance, intestinal health and lipid metabolism in yellow catfish. *Aquaculture, 596*, 741809.
Mai, H., Li, Y., Song, Z., Zeng, Y., Lin, P., Sun, Z., Mai, K., Tan, B., & Ye, C. (2025). Effect of taurine on growth and immune response of Pacific white shrimp (*Litopenaeus vannamei*) cultured at different temperatures. *Aquaculture, 594*, 741393.
Matthews, H., Hanison, J., & Nirmalan, N. (2016). "Omics"-informed drug and biomarker discovery: Opportunities, challenges and future perspectives. *Proteomes, 4*(3), 28.
Melamed, P., Gong, Z., Fletcher, G., & Hew, C. L. (2002). The potential impact of modern biotechnology on fish aquaculture. *Aquaculture, 204*(3–4), 255–269.
Mohanty, B. P., Ganguly, S., Mahanty, A., Mitra, T., & Mohanty, S. (2020). Nutrigenomics and fish. *CAB Reviews, 2020*, 15.
Monfoulet, L. E., Ruskovska, T., Ajdžanović, V., Havlik, J., Vauzour, D., Bayram, B., Krga, I., Corral-Jara, K. F., Kistanova, E., Abadjieva, D., & Massaro, M. (2021). Molecular determinants of the cardiometabolic improvements of dietary flavanols identified by an integrative analysis of nutrigenomic data from a systematic review of animal studies. *Molecular Nutrition & Food Research, 65*(16), 2100227.
Okani-Onyejiaka, M. C., Etuk, E. B., & Ogundu, U. E. (2024). Nutrigenomics in livestock research and production: Principles, applications and challenges. In *Proceedings of the 49th annual conference of the Nigerian society for animal production* (pp. 24–27). University of Ibadan.
Overturf, K. (2010). Convergence of aquaculture and molecular biology. In *Molecular research in aquaculture* (pp. 1–5). Wiley-Blackwell.
Quirós-Pozo, R., Roo, J., Izquierdo, M., Koven, W., Ramírez-Bolaños, S., Ventura-Castellano, A., Serradell, A., & Robaina, L. (2025). Effective complete replacement of fish oil by linseed oil in diets for thick-lipped grey mullet (*Chelon labrosus*) juveniles reared at three environmental salinities. *Aquaculture, 594*, 741425.
Rise, M. L., Martyniuk, C. J., & Chen, M. (2019). Comparative physiology and aquaculture: Toward omics-enabled improvement of aquatic animal health and sustainable production. *Comparative Biochemistry and Physiology Part D: Genomics and Proteomics, 31*, 100603.
Singh, S., Sarma, D. K., Verma, V., Nagpal, R., & Kumar, M. (2023). Unveiling the future of metabolic medicine: Omics technologies driving personalized solutions for precision treatment of metabolic disorders. *Biochemical and Biophysical Research Communications, 682*, 1.
Soto-Dávila, M., Webb, R. A., Rodríguez-Ramos, T., McDonald, G., Dang, X., Heath, J. W., et al. (2024). Effect of dietary supplementation of probiotic on growth, survival, and immune-related biomarkers in Chinook Salmon (*Oncorhynchus tshawytscha*) challenged with *Vibrio anguillarum*. *Aquaculture, 583*, 740582.
Tseng, Y., Dominguez, D., Sivagurunathan, U., Zamorano, M. J., Philip, A. J. P., & Izquierdo, M. (2025). Effects of the dietary supplementation of manganese on growth, skeletal development, oxidative stress and bone metabolism related genes in gilthead seabream (*Sparus aurata*) larvae. *Aquaculture, 594*, 741439.
Veenstra, T. D. (2021). Omics in systems biology: Current progress and future outlook. *Proteomics, 21*(3–4), 2000235.
Yáñez, J. M., Xu, P., Carvalheiro, R., & Hayes, B. (2022). Genomics applied to livestock and aquaculture breeding. *Evolutionary Applications, 15*, 517–522.
You, X., Shan, X., & Shi, Q. (2020). Research advances in the genomics and applications for molecular breeding of aquaculture animals. *Aquaculture, 526*, 735357.

Index

I. Ahmed, I. Ahmad (eds.), *Aquaculture: Enhancing Food Security and Nutrition*, https://doi.org/10.1007/978-3-031-92858-1

The manufacturer's authorised representative in the EU is Springer Nature Customer Service Centre GmbH, Europaplatz 3, 69115 Heidelberg, Germany. If you have any concerns regarding our products, please contact ProductSafety@springernature.com

Printed and bound by CPI Group (UK) Ltd, Croydon, CR0 4YY
30/09/2026
02225562-0016